BIOMEMBRANES

A Multi-Volume Treatise

Volume 6 • 1996

TRANSMEMBRANE RECEPTORS AND CHANNELS

BIOMEMBRANES

A Multi-Volume Treatise

TRANSMEMBRANE RECEPTORS AND CHANNELS

Editor: A. G. LEE
Department of Biochemistry
University of Southampton
Southampton, England

VOLUME 6 • 1996

JAI PRESS INC.

Greenwich, Connecticut *London, England*

55 Old Post Road No. 2
Greenwich, Connecticut 06836

JAI PRESS LTD.
38 Tavistock Street
Covent Garden
London WC2E
England

ISBN: 1-55938-663-0

Transferred to digital printing 2006

CONTENTS

LIST OF CONTRIBUTORS

Ralph A. Bradshaw	Department of Biological Chemistry College of Medicine University of California, Irvine
Neil J. Cook	ELIAS Freiburg, Germany
Andrew H. Dittman	Department of Pharmacology University of Washington School of Medicine Seattle
Steven K. Dower	Section of Molecular Medicine The University of Sheffield
Declan A. Doyle	Department of Crystallography Birkbeck College University of London
Roy J. Duhe	National Cancer Institute Frederick, Maryland
M. Edery	INSERM Faculte de Medecine Necker Paris
J. Finidori	INSERM Faculte de Medecine Necker Paris
Michael J. Fry	Institute of Cancer Research Haddow Laboratories Surrey, England

Teiichi Furuichi — Department of Molecular Neurobiology
Institute of Medical Science
University of Tokyo

Ira D. Goldfine — Mount Zion Medical Center
University of California, San Francisco

Hubert Hondermarck — Laboratoire de Biologie du Developpement
Universite de Lille
Cedex, France

P.A. Kelly — INSERM
Facuite de Medecine Necker
Paris

Kathleen W. Kinnally — Wadsworth Center
New York State Department of Health
Albany, New York

A.G. Lee — Department of Biochemistry
University of Southampton
Southampton, England

C.A. Mannella — Wadsworth Center
New York State Department of Health
Albany, New York

John L. Mershon — Department of Pharmacology and Cell Biophysics
University of Cincinnati College of Medicine

Gabor Mikala — Institute of Molecular Pharmacology and Cell Biophysics
University of Cincinnati College of Medicine

Katsuhiko Mikoshiba — Department of Molecular Neurobiology
Institute of Medical Science
University of Tokyo

Atsushi Miyawaki
Department of Molecular Neurobiology
Institute of Medical Science
University of Tokyo

Olaf Pongs
Zentrum fur Molekulare Neurobiologie
Institut fur Neurale Signalverarbeitung
Hamburg, Germany

M.C. Postel-Vinay
INSERM
Faculte de Medecine Necker
Paris

Aiala Reizer
Department of Biology
University of California, San Diego

Jonathan Reizer
Department of Biology
University of California, San Diego

Milton H. Saier, Jr.
Department of Biology
University of California, San Diego

Arnold Schwartz
Institute of Molecular Pharmacology and Cell Biophysics
University of Cincinnati College of Medicine

John E. Sims
Department of Biochemistry
Immunex Research and Development Corporation
Seattle

Daniel R. Storm
Department of Pharmacology
University of Washington School of Medicine
Seattle

Chin K. Sung
Mount Zion Medical Center
University of California, San Francisco

Matthew L. Thomas
Department of Pathology
Washington University School of Medicine
St. Louis, Missouri

B.A. Wallace	Department of Crystallography Birkbeck College University of London
Paulus W. Wohlfart	HMR TA Research Cardiovascular Agents Hoechst Marion Roussel Frankfurt, Germany
Edward C.C. Wong	Department of Pathology Washington University School of Medicine St. Louis, Missouri
Terry A. Woodford-Thomas	Department of Pathology Washington University School of Medicine St. Louis, Missouri
Zhiliang Wu	Department of Pharmacology University of Washington School of Medicine Seattle

PREFACE

The quantity of information available about membrane proteins is now too large for any one person to be familiar with anything but a very small part of the primary literature. A series of volumes concentrating on molecular aspects of biological membranes therefore seems timely. The hope is that, when complete, these volumes will provide a convenient introduction to the study of a wide range of membrane functions.

Volume 6 of *Biomembranes* covers transmembrane receptors and channels. A particularly important role for the membrane is that of passing messages between a cell and its environment. Part I of this volume covers receptors for hormones and growth factors. Here, as in so many other areas of cell biology, the application of the methods of molecular biology have led to the recognition of a number of families of receptors. Typically, such receptors contain an extracellular ligand binding domain, a transmembrane domain, and an intracellular catalytic domain whose activation, as a result of ligand binding, leads to generation of second messengers within the cell and stimulation of a range of cytosolic enzymes. An alternative signaling strategy, exploited in particular in the nervous system, is to use ion channels to allow controlled movement of monovalent (Na^+, K^+) or divalent (Ca^{2+}) cations in or out of the cell, resulting in changes in membrane potential or alterations in the intracellular concentration of Ca^{2+}. Part II of this volume is concerned with these ion channels and with other, often simpler, ion channel systems whose study can throw light on channel mechanism.

As editor, I wish to thank all the contributors for their efforts and the staff of JAI Press for their professionalism in seeing everything through to final publication.

A. G. Lee
Editor

PART I

RECEPTORS FOR HORMONES AND GROWTH FACTORS

INSULIN RECEPTOR SIGNALING

Chin K. Sung and Ira D. Goldfine

I. INSULIN RECEPTOR (IR)

Insulin regulates the general metabolism of most differentiated cells (Goldfine, 1981; Jacobs and Cuatrecasas, 1981; Kahn, 1985; Reaven, 1988). In the major target cells—myocytes, hepatocytes, and adipocytes—insulin has specific effects on the metabolism of carbohydrates, lipids, and proteins. In other cells, insulin is a general anabolic hormone (Goldfine, 1981). The initial interaction of insulin is with the insulin receptor (IR) protein that is located on the plasma membrane. After insulin binds, the IR initiates biological responses. Accordingly, the nature of the

Biomembranes
Volume 6, pages 3–15.

ISBN: 1-55938-663-0.

IR has been intensively studied, and significant progress has been made in understanding this protein (Goldfine, 1987; Ebina et al., 1985; Ullrich et al., 1985; Seino et al., 1989).

The IR gene is located on the short arm of chromosome 19 (Goldfine, 1987; Seino et al., 1989). The IR gene is greater than 120 kilobases in length and is comprised of 22 exons ranging from 36 to >2,500 base pairs (Seino et al., 1989). The mature receptor on the plasma membrane is an $\alpha_2\ \beta_2$ tetramer composed of two extracellular α-subunits that bind the hormone, and two transmembrane β-subunits that have intracellular tyrosine kinase activity (Goldfine, 1987). One α-subunit (130 kDa) and one β-subunit (95 kDa) are derived from a common precursor of 1382 amino acids. After translation and N-glycosylation, the receptor precursor is transferred to the Golgi apparatus where it is split into separate subunits, the sugar residues modified, and the $\alpha_2\ \beta_2$ tetramer formed; the mature receptor is then transported to the cell surface where it initiates the actions of insulin (Goldfine, 1987).

II. ADULT ONSET DIABETES MELLITUS AND IR SIGNALING

Diabetes mellitus exists in two major forms (Harris et al., 1987). One form is insulin dependent diabetes mellitus (IDDM; also known as juvenile onset diabetes mellitus). This disease has a prevalence of 0.3% to 0.5% in the general population, and is due to an autoimmune destruction of pancreatic β cells. The other form is non-insulin dependent diabetes mellitus (NIDDM; also known as adult onset diabetes mellitus). This disease is 10-fold more common than IDDM, and has a prevalence in the general population of 3% to 5%. It is estimated that 10 to 15 million individuals in the United States have NIDDM (Harris et al., 1987). The prevalence rate of NIDDM is higher in certain populations such as Hispanic Americans and Native Americans (Zimmet, 1992). One group, Pima Indians, has a prevalence of NIDDM approaching 50% (Bogardus and Lillioja, 1992; Howard, 1993). In NIDDM patients, there is a decreased insulin secretory response of β cells to blood glucose (Halter and Porte, 1981). In addition, there is resistance to insulin in key target tissues including muscle (Olefsky, 1980; Reaven, 1988, Bogardus and Lillioja, 1992; Myers and White, 1993). Studies have suggested that the insulin resistance is genetically determined, and in most instances this resistance precedes the abnormalities in insulin secretion (Bogardus and Lillioja, 1992). In many NIDDM patients, the IR is normal and the defect in insulin action is at the post-receptor level. However, patients with defects in IR expression and function have also been described, and approximately up to 5% of NIDDM patients have been estimated to have defects in IR function and/or expression (Taylor, 1992).

III. TYROSINE KINASE FAMILY

The IR is a member of the tyrosine kinase family (Yarden and Ullrich, 1988). A number of growth factors stimulate cellular mitogenesis by interacting with a family of cell-surface receptors that possess an intrinsic ligand-sensitive protein tyrosine kinase activity. Tyrosine phosphorylation of key cellular proteins initiates changes in cell growth. Tyrosine kinase receptors are typically composed of an extracellular ligand binding domain that is linked to a cytoplasmic catalytic domain, which not only transduces the growth factor or hormonal signal, but also generates mitogenic second messengers. There are four subclasses of receptor tyrosine kinases (I, II, III, IV); and non-receptor tyrosine kinases (V) (Figure 1; Yarden and Ullrich, 1988).

Subclass I includes the epidermal growth factor-receptor (EGF-R), which is activated by the ligands EGF, TGF-α, and the closely related HER-2/*neu* receptor. Subclass II receptors include the IR and the closely related insulin like growth factor-1-receptor (IGF-1-R), which are activated by insulin and IGF-1, respectively; and the insulin receptor-related receptor (IRR), whose ligand is unknown. Subclass III receptors include the platelet-derived growth factor-receptor (PDGF-R), the colony stimulating factor-1-receptor (CSF-1-R), and the protooncogene, c-kit. Subclass IV receptors include the fibroblast growth factor-receptor (FGF-R) and its

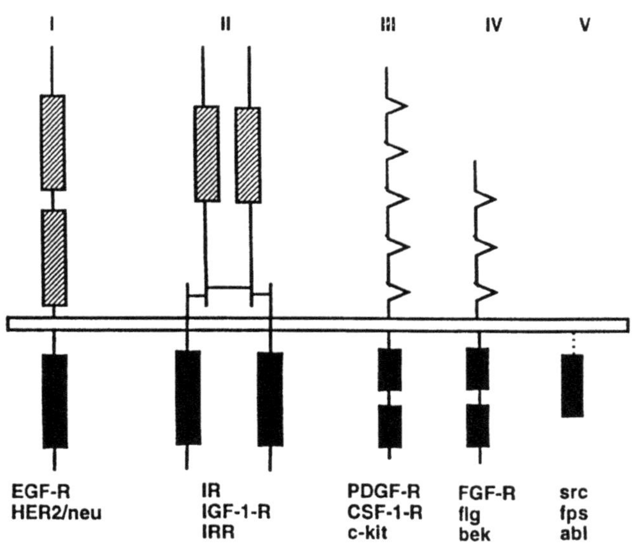

Figure 1. Schematic diagram of receptor and non-receptor tyrosine kinases. Receptor tyrosine kinases are classified into 4 subclasses (I, II, III, IV). Non-receptor tyrosine kinase subclass (V) includes viral oncogene tyrosine kinases. Hatched areas = cysteine rich regions; solid areas = tyrosine kinase domain; jagged lines = amino acid sequence repeats.

relatives, including flg and bek (Lee et al., 1989). In addition to these protooncogene encoded tyrosine kinase receptors, there are also receptor-derived viral oncogene products. v-*erb* B is derived from the EGF-R, and v-*fms* from the CSF-1-R. In general, these viral oncogene products differ from their normal receptor counterparts in that they have either amino acid deletions or substitutions that enable them to possess ligand-independent (and thus constitutively activated) tyrosine kinase activity. Subclass V includes non-receptor tyrosine kinases such as src, fps and abl.

IV. RECEPTOR TYROSINE KINASES AND SH2 PROTEINS

Several major clues in IR signaling have come from studies of other related receptor tyrosine kinases, such as the EGF-receptor (Skolnik et al., 1991) and the PDGF-receptor (Kaplan et al., 1990). In the case of these receptors, it has been demonstrated that intracellular adaptor and effector molecules attach to specific phosphorylated tyrosines of the receptors via Src homology 2 (SH2) domains (Table 1; Moran et al., 1990; Songyang et al., 1993). The SH2 domain is a 100 amino acid consensus sequence that was originally described in the oncogene product src and has the ability to bind to phosphotyrosines in specific motifs (Koch et al., 1991). At least five molecules containing SH2 domains bind to either the EGF or the PDGF receptors via phosphotyrosines, and become activated. These molecules include: phospholipase C, an enzyme that hydrolyzes phosphatidylinositol-4,5-bisphosphate to generate inositol-1,4,5-trisphosphate and 1,2-diacylglycerol; the p85 regulatory subunit of phosphatidylinositol-3-kinase (PIK) whose p110 catalytic subunit phosphorylates inositol at the 3 position; Ras GTPase activating protein (GAP) that

Table 1. SH2 Containing Proteins

I. Effectors	*II. Adaptors*
src, abl, syk[a]	p85 of PIK
PTPIC	c-crk
PLCγ	shc
	nck
GAP	Sem-5/Grb2
vav	

Note: [a]src,abl, syk are non-receptor tyrosine kinases. PTPIC is a phosphotyrosine phosphatase. PLCγ is a phospholipase. GAP accelerates Ras GTPase activity. vav has possible guanine nucleotide exchanger activity. p85 is a regulatory subunit of PIK. Sem-5/Grb 2 is an adaptor molecule involved in the Ras pathway. c-crk, shrc, and nck are adaptor molecules involved in cellular signaling.

promotes hydrolysis of Ras-GTP to Ras-GDP; phosphotyrosine phosphatase; and growth factor receptor bound protein 2 (Grb2), an intracellular adaptor molecule (Ullrich and Schlessinger, 1990, Cantley et al., 1991; Skolnik et al., 1991; Feng et al., 1993).

Although association of these molecules with the IR has been observed *in vitro* (Pronk et al., 1992; Yonezawa et al., 1992), none of the intracellular molecules mentioned previously have been documented to bind to phosphotyrosines of the IR in intact cells (Hadari et al., 1992; Songyang et al., 1993). However, insulin stimulation of the IR has been demonstrated to regulate PIK and Ras activity (*vida infra*), raising the possibility that the IR may indirectly interact with one or more of these regulatory proteins containing SH2 domains.

V. IRS-1 AND PHOSPHATIDYLINOSITOL-3-KINASE (PIK)

A major breakthrough in the studies of IR signaling has been the cloning and sequencing of insulin receptor substrate-1 (IRS-1; Sun et al., 1991). IRS-1 is a cytoplasmic protein with MW 160–190 kD and is a major cellular substrate for both the IR and the IGF-1 receptors (Myers and White, 1993). IRS-1 contains 20 tyrosine phosphorylation consensus sequences, six of which appear in YMXM (Tyr–Met–Xaa–Met) motifs (Sun et al., 1991). These motifs interact with SH2 domains of the p85 regulatory subunit of PIK (Myers and White, 1993; Songyang et al., 1993). PIK has a cytoplasmic location and phosphorylates inositol at the 3 position to yield potential intracellular signaling compounds such as PI-3,4-bisphosphate and PI-3,4,5-trisphosphate (Cantley et al., 1991). PIK activity is stimulated by oncogenes such as v-src and is believed to play a role in mitogenesis (Cantley et al., 1991). PIK consists of two subunits: a p110 catalytic subunit (Hiles et al., 1992) and a p85 regulatory subunit that contains two SH2 domains and one SH3 domain (Escobedo et al., 1991; Otsu et al., 1991; Skolnik et al., 1991). When tyrosine phosphorylated IRS-1 interacts with p85 of PIK, p110 is activated. This activation can be demonstrated *in vitro* (Myers and White, 1993; Carpenter et al., 1993). While the SH2 domain of p85 links PIK to tyrosine phosphorylated IRS-1, the role of the SH3 domain of the p85 remains to be defined.

VI. THE RAS SIGNALING PATHWAY AND RECEPTOR TYROSINE KINASES

Recent studies have strongly suggest that Ras, a monomeric membrane-bound GTP-binding protein, may be an intermediate in the IR signaling pathway and regulate certain biological functions of insulin. Evidence for this concept includes: (1) microinjection of anti-Ras antibodies inhibits insulin-induced maturation of *Xenopus* oocytes (Korn et al., 1987); (2) over-expression of a dominant inhibitory Ras mutant blocks insulin action on both gene expression and differentiation of 3T3-L1 cells to adipocytes (McCormick, 1993; Skolnik et al., 1991); (3) insulin

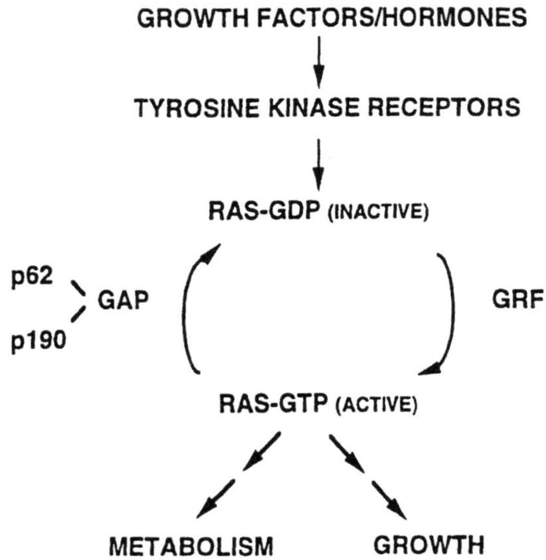

Figure 2. Regulation of Ras activity. In response to growth factors and hormones, receptor tyrosine kinases are activated. Activation of these kinases lead to activation of Ras by converting inactive Ras-GDP to active Ras-GTP. Two regulatory proteins, GRF and GAP, are identified. The activated Ras results in stimulation of metabolism and growth. GRF = guanine nucleotide releasing factors; GAP = GTPase activating protein; p62 and p190 = GAP-associated proteins.

treatment of intact cells increases the steady-state level of active endogenous Ras–GTP (McCormick, 1993; Porras et al., 1992); (4) activation of Ras also stimulates insulin-induced glucose transport via Glut-4 translocation (Kozma et al., 1993). Recently, it has been reported that a new member of the Ras-related family, Rad, is over-expressed in muscles of patients with NIDDM and may play a role in the insulin resistance of NIDDM (Kahn, 1993).

The content of active Ras-GTP is regulated by two classes of proteins (Wigler, 1990; Marx, 1992; Feig, 1993; Figure 2). One class, guanine nucleotide releasing factors (GRF), such as *Drosophila* Sos, exchanges GDP for GTP to activate Ras. Conversely, another class of proteins, such as GAP, promotes the hydrolysis of Ras-GTP to Ras-GDP to inactivate Ras. Two proteins, p62 and p190, are associated with GAP (Ellis et al., 1990; Settleman et al., 1992; Wong et al., 1992). It has been proposed that p62 and p190 regulate GAP activity. It should be also noted that GAP itself has been proposed as an effector molecule (Moran et al., 1991; Medema et al., 1992). Once Ras is activated by either activation of GRFs or inhibition of GAPs (or both), the activated Ras can trigger cascade of serine/threonine protein kinases (Figure 3). These protein kinases include: raf-1 kinase, which binds to and is activated by Ras-GTP; microtubule associated protein (MAP) kinase kinase; MAP kinase;

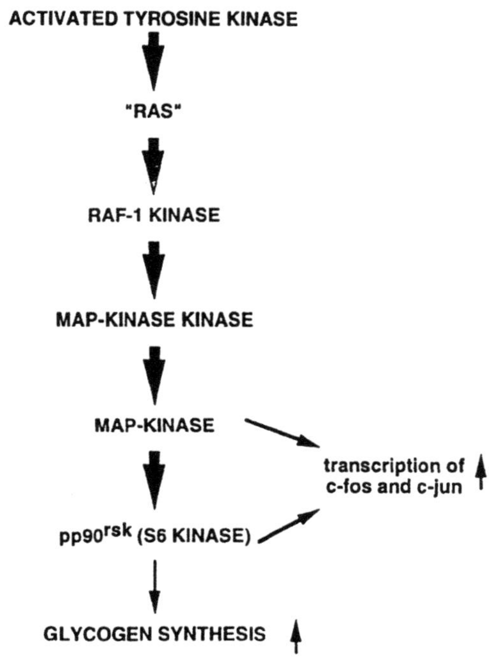

Figure 3. Tyrosine kinase induced serine/threonine kinase cascade via Ras activation. Ras activated by tyrosine kinases (either receptor or non-receptor bound) will active a series of serine/threonine kinases. Some of these kinases are reported to play a role in regulating transcription of nuclear genes and protein phosphatases leading to glycogen synthesis. MAP = microtubule associated protein.

and p90 ribosomal S6 kinase (Roberts, 1992). Activation of these protein kinases regulates various cellular functions including: c-fos and c-jun transcriptional activity, glycogen synthetase activity and cell proliferation (Roberts, 1992).

The major question arises, therefore, as to what is the link between activation of the receptor tyrosine kinases and activation of the Ras signaling pathway. In the case of the PDGF-receptor, GAP directly binds to a phosphotyrosine of the receptor via an SH2 domain, providing a direct link between the receptor and Ras (Kaplan et al., 1990). In the case of the EGF-receptor, it has been reported that the intracellular adaptor molecule, Grb2, binds to a phosphotyrosine of the EGF-receptor via its SH2 domain (Skolnik et al., 1991). Grb2 is a homolog of Sem-5 in *Caenorhabditis elegans* and drk in *Drosophila* (McCormick, 1993). Grb2 then interacts with mSos1 via its SH3 domain (a 50 amino acid consensus sequence found in src) and a proline-rich region of mSos1. mSos1 is the mammalian homolog of *Drosophila* Sos. Interaction of Grb2 with mSos1 activates Ras leading to receptor signaling (Egan et al., 1993; Li et al., 1993; Rozakis-Adcock et al., 1993). Other intracellular molecules containing SH2/3 domains may also interact with tyrosine kinase receptors and regulate Ras activity. These molecules include shc, crk, and nck (Table 1; Feig, 1993).

VII. THE RAS SIGNALING PATHWAY AND IR SIGNALING

The identification of adaptor proteins, such as Grb2, provides new insights into the signaling mechanisms of receptor tyrosine kinases. It has recently been reported

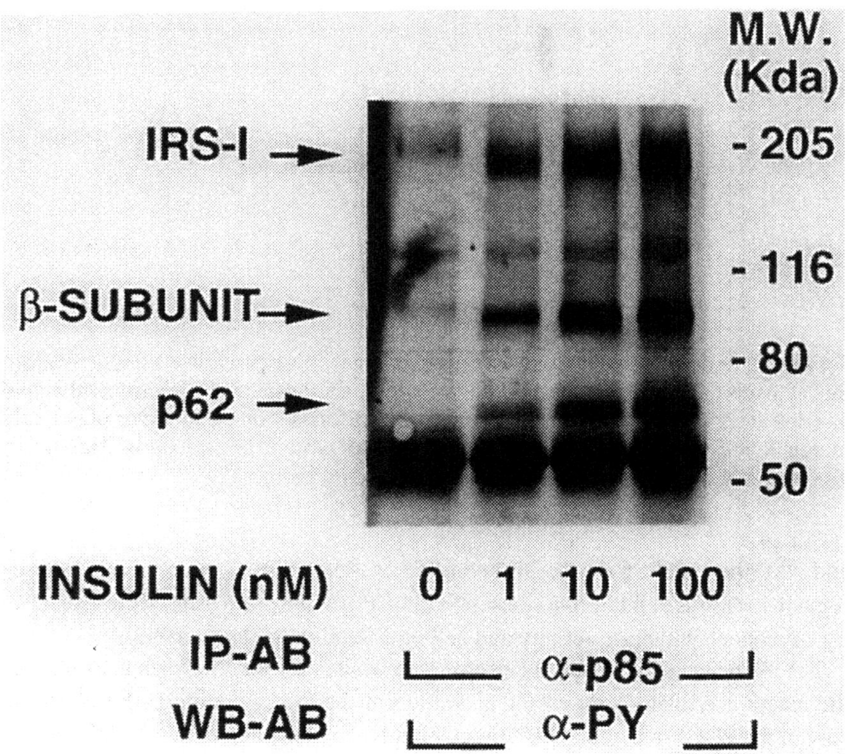

Figure 4. Tyrosine-phosphorylated proteins associated with α-p85 immunoprecipitates. Rat HTC hepatoma cells expressing human insulin receptors were treated for 5 min with increasing concentrations of insulin (Sung and Goldfine, 1992). The soluble cellular lysates were immunoprecipitated with an antibody to the p85 subunit of PIK (α-p85). Tyrosine phosphorylated proteins were then identified by western blotting analysis with an anti-phosphotyrosine antibody, α-PY. IRS-1 = insulin receptor substrate 1; β-subunit = insulin receptor β-subunit; IP-AB = immunoprecipitating antibody; WB-AB = western blotting antibody.

that Grb2 can bind to IRS-1 (Sun et al., 1991) and as a result, GRF activity (hSos1, dSos) and subsequently Ras activity are increased (Baltensperger, 1993; Skolnik, 1993). Our preliminary studies suggest that, in addition to Grb2 and GRF, GAP may also be involved in IR signaling. We have observed that when cells are stimulated with insulin and immunoprecipitated with antibodies to the p85 subunit of PIK, signaling complexes form that contain the IR, IRS-1, PIK, p62 GAP-associated protein, and GAP (Sung and Goldfine, 1992). p62 is tyrosine phosphorylated in a dose-dependent manner. Effects are seen with insulin concentrations as low as 1 nM. In response to insulin, IRS-1, IR β-subunit, and p62 are tyrosine phosphorylated (Figure 4).

Others have also reported that a similar 60–70 kD phosphoprotein is found in α-p85 immunoprecipitates following insulin stimulation (Kelly and Ruderman, 1993; Lavan and Lienhard, 1993). These studies suggest that p85 regulatory subunit

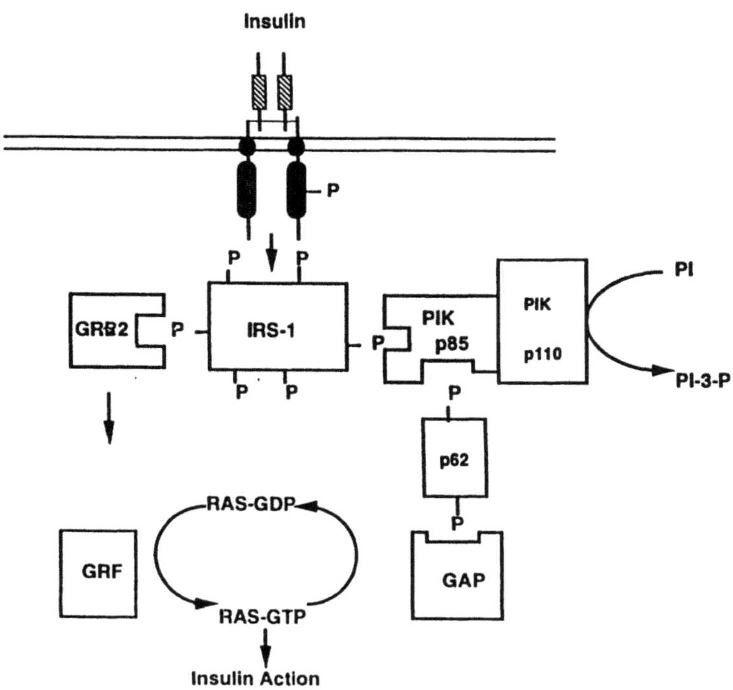

Figure 5. Proposed model of insulin receptor signaling. In response to insulin, IR is activated and phosphorylates IRS-1. Phosphorylated IRS-1 interacts with the p85 regulatory subunit of PIK, and then forms a complex with p62 GAP-associated protein and GAP. This complex may then negatively regulate GAP activity to stimulate the Ras pathway. In addition, phosphorylated IRS-1 interacts with Grb2 that directly links to GRF (Baltensperger, 1993; Skolnik, 1993).

of PIK may serve as an adaptor molecule, linking the activated IR with GAP via p62. In concert with our observations, PIK activity was reported to be associated with p21 Ras protein in a Ras-transformed rat liver epithelial cell line treated with insulin (Sjolander et al., 1991). These observations raise a possibility that, in certain cells, the IR may regulate Ras activity via IRS-1 by two mechanisms: (1) Grb2 and GRF, and (2) p85, p62, and GAP (Figure 5).

VIII. SUMMARY

In conclusion, major advances in the mechanisms of IR signaling have occurred during the past year. The observation that the IR functions, at least in part, by activating members of the Ras family allows new strategies for elucidating the specific effects of the IR on protein, lipid, and carbohydrate metabolism. Hopefully, these studies will lead to a greater understanding of the defects in IR signaling that occur in NIDDM.

REFERENCES

Baltensperger, K., Kozma, L. M., Cherniack, A. D., Klarlund, J. K., Chawla, A., Banerjee, U., Czech, M. P. (1993). Binding of the Ras Activator Son of Sevenless to Insulin Receptor Substrate-1 Signaling Complexes. Science 260, 1950–1952.

Bogardus, C., & Lillioja, S. (1992). Pima Indians and the genetics of NIDDM. J. Cell. Biol. 48, 337–343.

Cantley, L., Auger, K., Carpenter, C., Duckworth, B., Graziani, A., Kapeller, R., & Soltoff, S. (1991). Oncogenes and signal transduction. Cell 64, 281–302.

Carpenter, C. L., Auger, K. R., Chanudhuri, M., Yoakim, M., Schaffhausen, B., Shoelson, S., & Cantley, L. C. (1993). Phosphoinositide 3-kinase activated by phosphopeptides that bind to the SH2 domains of the 85-kDa subunit. J. Cell. Biol. 268, 9478–9783.

Ebina, Y., Ellis, L., Jarnagin, K., Edery, M., Graf, L, Clauser, E., Ou, J. H., Maslarz, F., Kan, Y. W., Goldfine, I. D., Roth, R. A., Z. Rutter, W. J. (1985). The human insulin receptor cDNA: The structural basis for hormone-activated transmembrane signaling. Cell 40, 747–758.

Egan, S. E., Giddings, B. W., Brooks, M. W., Buday, L., Sizeland, A. M., & Weinberg, R. A. (1993). Association of Sos Ras exchange protein with Grb2 is implicated in tyrosine kinase signal transduction and transformation. Nature 363, 45–51.

Ellis, C., Moran, M., McCormick, F., & Pawson, T. (1990). Phosphorylation of GAP and GAP-associated proteins by transforming and mitogenic tyrosine kinases. Nature 343, 377–381.

Escobedo, J., Navankasattusas, S., Kavanaugh, W., Milfay, D., Fried, V., & Williams, L. (1991). cDNA cloning of a novel 85 kd protein that has SH2 domains and regulates binding of PI3-kinase to the PDGF β-receptor. Cell 65, 75–82.

Feig, L. A. (1993). The many roads that lead to Ras. Science 260, 767–768.

Feng, G.-S., Hui, C.-C., & Pawson, T. (1993). SH2-containing phosphotyrosine phosphatase as a target of protein-tyrosine kinases. Science 259, 1607–1611.

Goldfine, I. D. (1981). Effects of insulin on intracellular functions. In: Biochemical Actions of Hormones (Litwack, G. ed.). Vol. B, pp. 273–305. Academic Press, New York.

Goldfine, I. D. (1987). The insulin receptor: Molecular biology and transmembrane signaling. Endocrine Reviews 8, 235–255.

Hadari, Y., Tzahar, E., Nadiv, O., Rothenberg, P., Roberts, C., LeRoith, D., Yarden, Y., & Zick, Y. (1992). Insulin and insulin nomimetiz agents induce activation of phosphatidylinositol 3-kinase upon its association with pp185 (IRS-1) in intact rat livers. J. Biol. Chem. 267, 17483–17486.

Halter, J. B., & Porte, D., Jr. (1981). Current concepts of insulin secretion in diabetes mellitus. In: Diabetes Mellitus: Diagnosis and Treatment (Rifkin, H., & Raskin, P., eds.). American Diabetes Association, New York.

Harris, M., Hadden, W., Knowles, W., & Bennett, P. (1987). Prevalence of diabetes and impaired glucose tolerance and plasma glucose levels in U.S. population aged 20–74. Diabetes 36, 523–534.

Hiles, I., Otsu, M., Volinia, S., Fry, M., Gout, I., Dhand, R., Panayotou, G., Ruiz-Larrea, F., Thompson, A., Totty, N., Hsuan, J., Courtneidge, S., Parker, P., & Waterfield, M. (1992). Phosphatidylinositol 3-kinase: Structure and expression of the 100 kd catalytic subunit. Cell 70, 419–429.

Howard, B. (1993). Diabetes and plasma lipoproteins in Native Americans: Studies of the Pima Indians. Diabetes Care 16, 284–291.

Jacobs, S., & Cuatrecasas, P. (1981). Insulin receptor: Structure and function. Endocrine Reviews 2, 251–263.

Kahn, C. R. (1985). The molecular mechanisms of insulin action. Annual Reviews in Medicine 36, 429–451.

Kahn, C. R. (1993). Insulin action, diabetogenes, and the cause of Type II diabetes. In: Banting Lecture at the 53rd Annual American Diabetes Association Meeting, June 12–15, Las Vegas, NV.

Kaplan, D. R., Morrison, D. K., Wong, G., McCormick, F., & Williams, L. T. (1990). PDGF β-receptor stimulates tyrosine phosphorylation of GAP and association of GAP with a signaling complex. Cell 61, 125–133.

Kelly, K. L., & Ruderman, N. B. (1993). Insulin-stimulated phosphatidylinositol 3-kinase. J. Biol. Chem. 268, 4391–4398.

Koch, C., Anderson, D., Moran, M., Ellis, C., & Pawson, T. (1991). SH2 and SH3 domains: Elements that control interactions of cytoplasmic signaling proteins. Science 252, 668–674.

Korn, L., Siebel, C., McCormick, F., & Roth, R. (1987). Ras p21 as a potential mediator of insulin action in *Xenopus* oocytes. Science 236, 840–843.

Kozma, L. M., Baltensperger, K., Klarlund, J. K., Porras, A., Santos, E., & Czech, M. P. (1993). The Ras signaling pathway mimics insulin action on glucose transporter translocation. Proc. Natl. Acad. Sci. USA 90, 4460–4464.

Lavan, B. E., & Lienhard, G. E. (1993). The insulin-elicited 60-kDa phosphotyrosine protein in rat adipocytes is associated with phosphatidyl inositol 3-kinase. J. Biol. Chem., 268, 5921–5928.

Lee, P., Johnson, D., Cousens, L., Fried, V., & Williams, L. (1989). Purification and complementary DNA cloning of a receptor for basic fibroblast growth factor. Science 245, 57–60.

Li, N., Batzer, A., Daly, R., Yajnik, V., Skolnik, E., Chardin, P., Bar-Sagi, D., Margolis, B., & Schlessinger, J. (1993). Guanine-nucleotide-releasing factor hSos1 binds to Grb2 and links receptor tyrosine kinase to Ras signaling. Nature 363, 85–88.

Marx, J. (1992). New genes may shed light on cell growth control. Science 257, 484–485.

McCormick, F. (1993). How receptors turn Ras on. Nature 363, 15–16.

Medema, R. H., De Laat, W. L., Martin, G. A., McCormick, F., & Bos, J. L. (1992). GTPase-activating protein SH2–SH3 domains induce gene expression in a Ras-dependent fashion. Molec. Cell. Biol. 12, 3425–3430.

Moran, M. F., Koch, C. A., Anderson, D., Ellis, C., England, L., Martin, G. S., & Pawson, T. (1990). Src homology region 2 domains direct protein–protein interactions in signal transduction. Proc. Natl. Acad. Sci. USA 87, 8622–8626.

Moran, M. F., Polakis, P., McCormick, F., Pawson, T., & Ellis, C. (1991). Protein-tyrosine kinases regulate the phosphorylation, protein interactions, subcellular distribution, and activity of p21 ras GTOase-activating protein. Mol. Cell. Biol. 11, 1804–1812.

Myers, M. G., Jr., & White, M. F. (1993). Insulin receptor substrate-1 and proteins with SH2 domains. Diabetes 42, 643–650.

Olefsky, J. M. (1980). Insulin resistance and insulin action. An *in vitro* and *in vivo* perspective. Diabetes 30, 148–162.

Otsu, M., Hiles, I., Gout, I., Fry, M. J., Ruiz-Larrea, F., Panayotou, G., Thompson, A., Dhand, R., Hsuan, J., Totty, N., Smith, A. D., Morgan, S. J., Courtneidge, S. A., Parker, P. J., & Waterfield, M. D. (1991). Characterization of two 85 kd proteins that associate with receptor tyrosine kinases, middle-T/pp60c-src complexes, and PI3-kinase. Cell 65, 91–104.

Porras, A., Nebreda, A. R., Benito, M., & Santos, E. (1992). Activation of Ras by insulin in 3T3 L1 cells does not involve GTPase-activating protein phosphorylation. J. Biol. Chem. 267, 21124–21131.

Pronk, G. J., Medema, R. H., Burgering, B. M. T., Clark, R., McCormick, F., & Bos, J. L. (1992). Interaction between the p21ras GTPase activating protein and the insulin receptor. J. Biol. Chem. 267, 24058–24063.

Reaven, G. M. (1988). Role of insulin resistance in human disease. Diabetes 37, 1595–1607.

Roberts, T. M. (1992). A signal chain of events. Nature 360, 534–535.

Rozakis-Adcock, M., Fernley, R., Wade, J., Pawson, T., & Bowtell, D. (1993). The SH2 and SH3 domains of mammalian Grb2 couple the EGF receptor to the Ras activator mSos1. Nature 363, 8385.

Seino, S., Seino, M., Nishi, S., & Bell, G. I. (1989). Structure of the human insulin receptor gene and characterization of it promoter. Proc. Natl. Acad. Sci. USA 86, 114–118.

Settleman, J., Narasimhan, V., Foster, L. C., & Weinberg, R. A. (1992). Molecular cloning of cDNAs encoding the GAP-associated protein p190: Implications for a signaling pathway from Ras to the nucleus. Cell 69, 539–549.

Sjolander, A., Yamamoto, K., Huber, B., & Lapetina, E. (1991). Association of $p21^{ras}$ with phosphatidylinositol 3-kinase. Proc. Natl. Acad. Sci. USA 88, 7908–7912.

Skolnik, E. Y., Batzer, A., Li, N., Lee, C.-H., Lowenstein, E., Mohammadi, M., Margolis, B., Schlessinger, J. (1993). The function of GRB2 in linking the insulin receptor to Ras signaling pathways. Science 260, 1953–1955.

Skolnik, E. Y., Margolis, B., Mohammadi, M., Lowenstein, E., Fischer, R., Drepps, A., Ullrich, A., & Schlessinger, J. (1991). Cloning of PI3 kinase-associated 085 utilizing a novel method for expression/cloning of target proteins for receptor tyrosine kinases. Cell 65, 83–90.

Songyang, Z., Shoelson, S. E., Chaudhuri, M., Gish, G., Pawson, T., Haser, W. G., King, F., Roberts, T., Ratnofsky, S., Lechleider, R. J., Neel, B. G., Birge, R. B., Fajardo, J. E., Chou, M. M., Hanafusa, H., Schaffhausen, B., & Cantley, L. C. (1993). SH2 domains recognize specific phosphopeptide sequences. Cell 72, 767–778.

Sun, X.-J., Rothenberg, P., Kahn, C., Backer, J., Araki, E., Wilden, P., Cahill, D., Goldstein, B., & White, M. (1991). Structure of the insulin receptor substrate IRS-1 defines a unique signal transduction protein. Nature 352, 73–77.

Sung, C. K., & Goldfine, I. D. (1992). Phosphatidylinositol-3-kinase is a non-tyrosine phosphorylated member of the insulin receptor signaling complex. Biochem. Biophys. Res. Comm. 189, 1024–1030.

Sung, C. K. & Goldfine, I. D. (1993). Insulin receptor signaling complex. Proceedings of the Endocrine Society. Abstract #662, June 1993.

Taylor, S. (1992). Lessons from patients with mutations in the insulin-receptor gene. Diabetes 41, 1473–1490.

Ullrich, A., & Schlessinger, J. (1990). Signal transduction by receptors with tyrosine kinase activity. Cell 61, 203–212.

Ullrich A., Bell, J. R., Chen, E. Y., Herrera, R., Petruzzelli, L. M., Dull, T. J., Gray, A., Coussens, L., Liao, Y-C., Tsubokawa, M., Mason, A., Seeburg, P. H., Grunfeld, C., Rosen, O. M., Ramachandran, J. (1985). Human insulin receptor and its relationship to the tyrosine kinase family of oncogenes. Nature 313, 756–761.

Wigler, M. H. (1990). GAPs in understanding Ras. Nature 346, 696–697.

Wong, G., Muller, O., Clark, R., Conroy, L., Moran, M. F., Polakis, P., & McCormick, F. (1992). Molecular cloning and nucleic acid binding properties of the GAP-associated tyrosine phosphoprotein p62. Cell 69, 551–558.

Yarden, Y., & Ullrich, A. (1988). Growth factor receptor tyrosine kinase. Annual Reviews of Biochemistry 57, 443–478.

Yonezawa, K., Yokono, K., Shii, K., Ogawa, W., Ando, A., Hara, K., Baba, S., Kaburagi, Y., Yamamoto-Honda, R., Momomura, K., Kadowaki, T., & Kasuga, M. (1992). *In vitro* association of phosphatidylinositol 3-kinase activity with the activated insulin receptor tyrosine kinase. J. Biol. Chem. 267, 440–446.

Zimmet, P. Z. (1992). Kelly West Lecture 1991: Challenges in diabetes epidemiology—from west to the rest. Diabetes Care 15, 232–252.

GROWTH FACTOR RECEPTOR TYROSINE KINASES

Michael J. Fry

Biomembranes
Volume 6, pages 17–76.

ISBN: 1-55938-663-0.

I. INTRODUCTION

Since the first description of tyrosine phosphorylation as a minor inducible modification of cellular proteins, a plethora of protein-tyrosine kinases (PTK) have been identified. This number is still on the increase, boosted by recent polymerase chain reaction (PCR) strategies designed for cloning of related genes (Wilks, 1989; Lai and Lemke, 1991). The PTKs isolated to date have been divided into two broad classes: the receptor PTKs possess extracellular ligand-binding domains, span the plasma membrane once, and have a large intracellular kinase domain. The non-receptor PTKs are cytoplasmic, although they may be found in association with membranes and/or transmembrane proteins, but do not span cell membranes themselves. Results obtained over the last few years suggest that the current classification needs to be re-addressed due to the identification of a number of receptor PTKs (mainly by PCR cloning) which do not readily fall into one of the known subclasses or families.

Receptor PTKs respond to a diverse group of peptide factors and mediate a host of cellular responses including proliferation, differentiation, migration, and cell survival (Rozengurt, 1986). This large family of receptors show, great diversity of structure based on a relatively small number of structural units which have been utilized in a number of combinations to produce discrete receptors, each possessing a narrow range of ligand-binding specificities. What has become clear with the rapid expansion of signaling pathways via molecular, biochemical, and genetic studies, is that the underlying themes by which this array of cell-surface receptors transmit intracellular signals appear surprisingly uniform. A common set of downstream signaling molecules such as the small GTP-binding protein, Ras, phospholipase Cγ (PLCγ), phosphatidylinositol (PI) 3-kinase, the GTPase-activating protein (GAP) for Ras, *src*-family cytosolic PTKs, and Raf and MAP serine/threonine protein kinases are frequently activated (reviewed in Fry et al., 1993). Interestingly, despite these similarities, the apparent activation of common signaling pathways results in changes in gene expression and biological responses which are highly specific for a particular ligand/receptor combination. Identification of the biochemical pathways responsible for the transduction of receptor signals is an essential first step towards understanding how they are able to regulate diverse cellular processes. In this chapter, the current status of our understanding of PTK receptor (PTKR) structure and function will be examined with special emphasis on the role of the PTKs as integral components of membrane signaling complexes. Due to the breadth of the subject matter, it is impossible to include everything or to cite all of the relevant literature. Those wishing further information in specific areas should consult some of the many excellent reviews on different aspects of receptor PTKs and the intracellular signaling field quoted herein.

II. PROTEIN–TYROSINE KINASE RECEPTOR STRUCTURE

Primary sequence data provided by the cDNA cloning of a large number of receptor PTKs has led to the description of a common structure and membrane topology for this family of growth factor and hormone receptors (see Figure 1). The basic structural motifs of the major families of receptor PTKs will be described briefly here (for an in-depth discussion of their structural characteristics, see Yarden and Ullrich, 1988; Ullrich and Schlessinger, 1990) and will be examined in more detail in the following sections describing the specific features of the individual receptor families.

A. Extracellular Ligand Binding Domain

Targeting of growth factors is achieved by the cell-specific expression of high-affinity receptors in responsive cells. All receptor PTKs identified to date possess an N-terminal extracellular domain, which binds to its cognate ligand (see Figure 1). Potential ligands might be secreted peptide growth factors, hormones, membrane-associated tethered ligands, extracellular matrix proteins, or cell-surface adhesion molecules. The extracellular domain is often extensively glycosylated. It is this extracellular domain which provides the specificity of response to external signals. The distinct structural motifs found within this domain have been used as the major basis for grouping of receptors into the families described below. Common re-occurring structural motifs found within extracellular domains include clusters of cysteine residues with distinct spacing patterns, immunoglobulin (Ig)-like domains, epidermal growth factor (EGF)-like repeats, and fibronectin III-like repeats. The functional significance of many of these domains is still unclear. In the case of the EGF- and insulin-receptors, some progress has been made towards identifying the key regions of the extracellular domain which are critical for ligand binding (reviewed in Yarden and Ullrich, 1988; Ullrich and Schlessinger, 1990). Such studies are important as ligand binding to the extracellular domain is essential for activation of the intracellular PTK domain and the subsequent transmission of

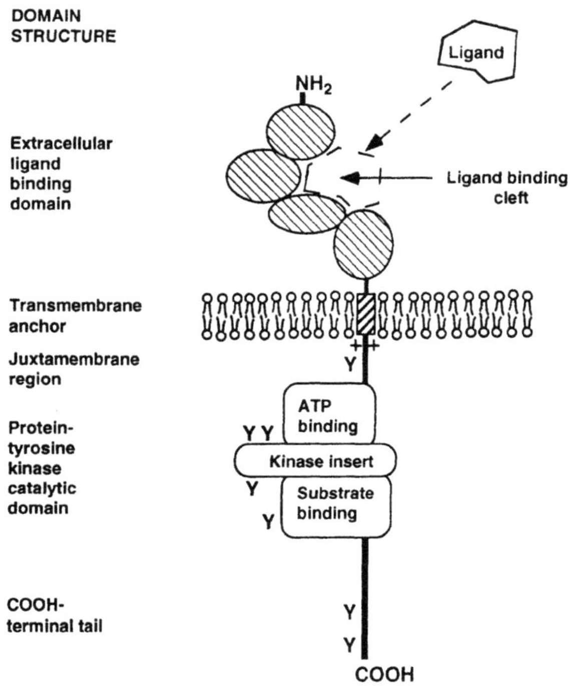

Figure 1. Structure/function schematic of a generic receptor protein-tyrosine kinase. All PTK receptors are orientated with their N-termini extracellular and their C-termini intracellular. The extracellular domain of PTK receptors have a multidomain structure, often composed of a number of distinct structural motifs. These are arranged in such a manner to form a binding cleft for the cognate ligand. Some or all of the extracellular structural motifs may be involved in the formation of this ligand-binding region. Ligand is shown both unbound (bold) and bound (dashes outline). A single hydrophobic amino acid stretch (dark stripes) spans the plasma membrane. A number of basic residues (+) form a stop-transfer sequence on the cytoplasmic side of the membrane. The intracellular portion of the receptor can be divided into three distinct domains. The juxtamembrane and C-terminal tail bracket the PTK domain and are involved in both receptor and signal regulation. These domains possess a number of important phosphorylation sites both for receptor autophosphorylation and for phosphorylation by other regulatory kinases. The PTK domain can also be divided into several distinct regions including an ATP-binding domain and substrate binding and catalytic motifs. Additionally some receptor PTKs (e.g., members of the PDGFR family) possess noncatalytic sequences which break up the PTK domain homology and are nonessential for catalysis. These inserts have been termed KI regions and are involved in receptor-specific substrate interactions. Potential areas for autophosphorylation are

extracellular domain may also exert negative control over the PTK domain. This theory has arisen from the study of viral oncogenes derived from receptor PTKs, e.g., v-*erb*B and v-*kit*, where deletion of the extracellular domain (and some C-terminal sequences) results in a constitutively activated PTKR (Downward et al., 1984; Qui et al., 1988; reviewed in Fry et al., 1992).

B. Transmembrane Domain

The extracellular ligand-binding domain is connected via a single short hydrophobic membrane-spanning segment of 23–26 amino acids to the cytoplasmic signaling domain. This domain is predicted to be predominantly α-helical in character. The transmembrane domain anchors the receptor in the plasma membrane and provides a link between the extracellular environment and the intracellular compartment of the cell. Additionally, the transmembrane domain may also play a critical role in transmitting the signal generated upon ligand binding into the interior of the cell with the resultant activation of the PTK domain and signaling functions. This proposal arises from studies of an activated form of c-*erb*B-2/*neu*, which has a single point mutation (Val664Glu) within its transmembrane domain that is responsible for converting this protein into an active oncogene (Bargmann and Weinberg, 1988). This mutation is thought to have a stabilizing effect on receptor dimer formation, resulting in the constitutive activation of the receptor's PTK domain. Escobedo et al. (1988) have examined the role of the platelet-derived growth factor (PDGF) β-receptor (PDGFβR) transmembrane domain by replacing it with the equivalent transmembrane domains of other proteins, including the low density lipoprotein receptor and both the normal and oncogenic form of *neu*. The resultant mutant receptors were down-regulated upon ligand binding, but were found to be unable to stimulate the activity of their PTK domains or any other intracellular events. These results suggest that the transmembrane domain may play functional roles beyond merely anchoring the receptor within the plasma membrane.

C. Juxtamembrane Domain

While the role of the extracellular domain is to provide specificity of the response to external signals, the intracellular portion of the receptor is responsible for generating receptor-specific signals. This part of the receptor can be subdivided into several recognizable domains of variable length depending upon the receptor class. The juxtamembrane domain lies between the plasma membrane and the start of the PTK domain. It contains the membrane stop–transfer sequence consisting of several charged amino acids and in some receptors, phosphorylation sites both for autophosphorylation and for regulatory phosphorylation by other kinases. This domain is highly divergent between the receptor subclasses, but is generally conserved within a receptor family. There is substantial evidence for a role for this

is termed receptor transmodulation. Activation of protein kinase C (PKC) by a number of growth factors (e.g., PDGF or bombesin) results in phosphorylation of the EGF receptor at sites within this domain, including Thr654. This phosphorylation results in a decreased affinity of the receptor for EGF and a reduced kinase activity, although it is likely that these two responses are brought about by distinct phosphorylation events (Lin et al., 1986). Mutations in the juxtamembrane sequence of the insulin receptor are also observed to have dramatic affects on signaling responses independent of PTK activity. Mutation of Tyr960 of the insulin receptor abolishes the ability of the receptor to phosphorylate the major substrate, IRS-1, without affecting the receptor's intrinsic PTK activity (White et al., 1988). Clearly this is a distinct effect from that involving the EGF receptor, but this serves to illustrate the range of potential roles for the juxtamembrane domain in regulation of receptor function.

D. Protein–Tyrosine Kinase Catalytic Domain

The PTK domain is 250–300 amino acids in length, giving a minimal domain size of approximately 30 kD as defined by mutagenesis studies. This domain is responsible for the intrinsic catalytic activity and for initiation of receptor-specific intracellular signals. The PTK domain exhibits the most extensive sequence homology among the members of the PTK family and shows weaker homology to the protein-serine/threonine kinases and phosphoinositide kinases (Hanks, 1991; Hanks et al., 1988; Hiles et al., 1992). While containing structural motifs which define the family of PTKs, this domain also contains several family-specific features which have aided the classification of the known PTKs (Hanks et al., 1988). The PTK domain can be subdivided into highly conserved subdomains numbered I to XI with gaps or regions of lower sequence identity lying in between. Starting from the N-terminus of the domain, the characteristic sequences include Gly–x–Gly–x–x–Gly (subdomain I) followed by a Lys 14–23 residues downstream which together form part of the ATP-binding site (subdomain II). These sequences are followed, 80–180 residues C-terminal to these motifs, by a sixty amino acid region containing several short highly conserved motifs, Arg–Asp–Leu (subdomain VI), Asp–Phe–Gly (subdomain VII), and Ala–Pro–Glu (subdomain VIII) (Hanks, 1991; Hanks et al., 1988), which make up the central core of the kinase domain. Several other highly conserved amino acids are scattered throughout the domain. Subdomains VI and VIII contain PTK-and protein-serine/threonine kinase-specific sequences characteristic of these two subfamilies of protein kinases. Presumably, some of these residues are involved in recognition of the different hydroxyl amino acid substrates.

E. Kinase Insert Region

In some receptors, the region of homology which defines the PTK domain is

length (ranging from 5–100 amino acids) which is neither conserved nor required for PTK catalytic activity (Taylor et al., 1989). This has been termed the kinase insert (KI) region and is diagnostic for several receptor classes, in particular members of Class III, the PDGFR family. It seems likely that the KI region forms a distinct surface feature of the kinase domain. The KI region often contains autophosphorylation sites and as described in the following section, in some receptors this domain plays an important role in interactions with substrates. Identification of these substrates is currently the major focus of PTKR research.

F. C-Terminal Regulatory Tail

The C-terminal tail varies considerably in length between species from just a few to more than 100 amino acids. It is also the domain which exhibits the greatest degree of sequence variation even within a given PTKR class. This domain has been suggested to be involved in generating intracellular signals, in the regulation of PTK activity, and in the clearing of activated receptor from the cell surface (Chen et al., 1989). Tyrosine autophosphorylation sites have been mapped to this domain in a number of different receptors and again vary from receptor to receptor even within a receptor family. These autophosphorylation sites are extremely important to PTKR function. Some are involved in positive aspects of signaling, forming sites of association for downstream substrates (see following), while other sites may exert negative control over the receptor PTK domain (Ullrich and Schlessinger, 1990; Roussel et al., 1987; Haley et al., 1989; Helin et al., 1991).

III. PROTEIN–TYROSINE KINASE RECEPTOR FUNCTION

Before examining the structural and functional characteristics of the individual PTKR families, let us first consider some of the basic features of receptor-mediated signaling which have been defined over the last few years. What follows is a brief description of a generalized activation cycle for a ligand-stimulated PTKR together with an examination of how receptor activation leads to the recruitment and stimulation of target signaling molecules. This is shown schematically in Figure 2. How this scheme varies for the different receptor classes will then be discussed. What will become apparent, however, is the underlying similarity to signaling strategies employed by the whole spectrum of receptor PTKs.

A. Ligand Binding and Receptor Dimerization

The first step in receptor activation involves the specific interaction and binding of a ligand. Following ligand binding, it has been suggested that the receptor undergoes a conformational change resulting in receptor dimerization (Schlessinger, 1988; Canals, 1992). Evidence for the existence of such dimers has been presented for several receptor classes to support this view and has been reviewed

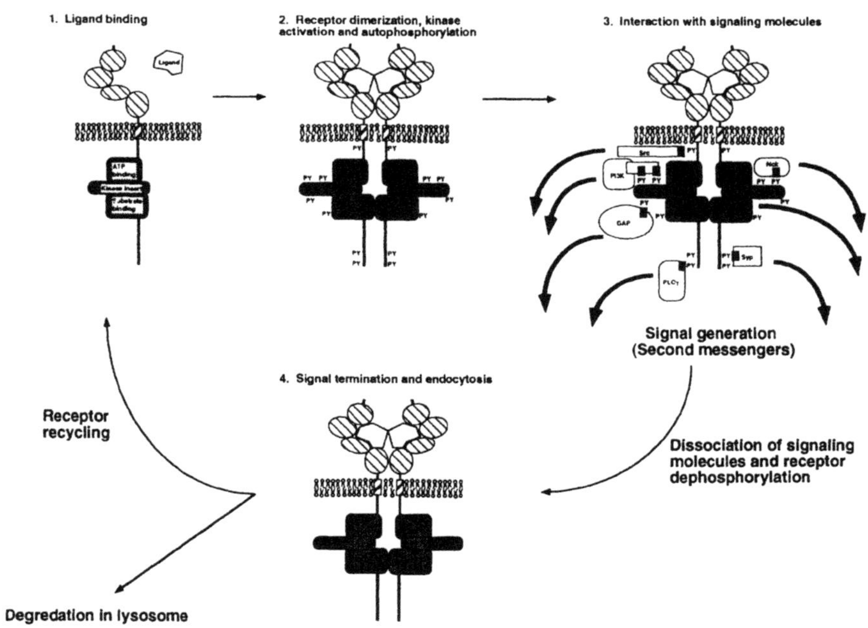

Figure 2. Model for signal transduction via receptor protein-tyrosine kinases. The events following activation of a generalized receptor PTK by its cognate ligand are shown schematically for a generalized receptor in this figure. (**1**) Ligand binding. An inactive monomeric cell surface PTKR is shown prior to interaction with ligand. (**2**) Receptor activation. Ligand binding drives receptor dimerization which presumably results in a conformational change. This conformational change is indicated by the exposure of the KI domain of the receptor and leads to kinase domain activation and trans-autophosphorylation of the cytoplasmic domain of the receptor at multiple sites. (**3**) Substrate binding. The phosphorylated tyrosine residues (PY) provide target binding sequences for specific signaling molecules. A number of signaling molecules (Src, PI3K, GAP, etc) are shown interacting with the activated receptor via SH2 domain (black box)–phosphotyrosine (PY) interactions. The thick curved arrows indicate the generation of specific intracellular signals by both the receptor's kinase domain and by the associated signaling molecules. The second messengers generated by these signaling molecules stimulate cytoplasmic enzymes such as protein kinases, for example, PKC, and GTP-binding proteins, for example, Ras, leading to receptor-specific and cell type-specific events. (**4**) Signal termination. The signaling molecules dissociate from the activated receptor possibly driven by their tyrosine phosphorylation. Signals from the receptor are terminated by its dephosphorylation and redistribution within the cell. In some cases, receptor recycling to the cell surface may be observed, but most of the activated receptor pool is targeted to the lysosome for

the PDGF, it is possible to readily conceive a scheme whereby receptor dimerization is driven by interaction with the bivalent ligand. However in other cases, such as with monomeric EGF, the mechanism is not so clearly apparent. Presumably ligand binding induces a conformational change in the receptor which exposes surface determinants that make receptor dimerization a favorable event. This may involve both the extracellular and transmembrane domains (see preceding). A further complication is that in addition to the formation of receptor homodimers, receptors heterodimers have also been noted. This phenomenon has been observed with the EGFR and PDGFR families and greatly extends the range of potential signaling responses which may be elicited (Hart et al., 1988; Connelly and Stern, 1990).

B. Autophosphorylation

Irrespective of the precise mechanism, ligand binding and dimerization would seem to result in the activation of the cytoplasmic PTK domain relative to that of the monomeric receptor. The formation of dimers facilitates the ability of the receptors to phosphorylate themselves on tyrosine residues (autophosphorylation). Where examined, this has been found to occur by an intermolecular trans-phosphorylation reaction (Heldin et al., 1989; Honegger et al., 1989, 1990; Bellot et al., 1991). The importance of this event is reflected by the apparent inability of heterodimers consisting of wild-type and kinase-inactive mutants to generate any downstream signals (Honegger et al., 1990; Treadway et al., 1991). This observation has resulted in the identification of a useful strategy for the construction of so called dominant–negative receptor mutants which may prove useful for examining receptor function and signaling *in vivo* (Redemann et al., 1992; Ueno et al., 1993; Werner et al., 1993).

C. Signal Generation

Considerable effort has been made to identify substrates and the intracellular signaling pathways which enable receptor PTKs to induce changes in gene expression, cell division, and differentiation. Early efforts focused on identifying protein substrates which were tyrosine-phosphorylated in either growth factor-stimulated cells, or in cells transformed by oncogenic forms of PTKs. This approach initially met with minimal success and many of the proteins identified by these studies appear not to be key substrates (Cooper and Hunter, 1983). However, a number of physiological substrates have been identified. These are beyond the scope of this review, but were recently examined by Glenney (1992).

The emphasis of PTK research has now shifted to examining which of the known second messenger generating signaling pathways are coupled to, and/or activated by, specific PTKRs. These studies have paved the way for our current understanding of how cell-surface PTKRs are linked to these intracellular signaling pathways. Among the early key observations were that PI 3-kinase and PLCγ could be found appropriate growth factors (Kaplan et al., 1987; Wahl et al., 1988). These discoveries were rapidly followed by reports that other signaling molecules were recruited into receptor complexes upon growth factor stimulation. RasGAP (Kazlauskas et al., 1990) and members of the *src*-family of non-receptor PTKs (Kypta et al., 1990) were subsequently implicated. These observations were made at about the same time as a report that the viral oncogene product, v-Crk, was directly associated with a wide-range of phosphotyrosine-containing proteins (Matsuda et al., 1990). A common feature of these proteins (PI 3-kinase, RasGAP, PLCγ, $pp60^{c\text{-}src}$, and v-Crk) is that they possess two short regions of sequence similarity known as the Src-homology 2 (SH2) and SH3 domains (reviewed in Pawson and Schlessinger, 1993; Koch et al., 1991; Musacchio et al., 1992). The structure and function of these domains have been the subject of intense investigation and the concepts that have emerged have completely altered our understanding of the ways in which PTKs function to bring about coordinated, and specific activation of the signaling pathways which lie downstream (reviewed in Fry et al., 1993).

The SH2 Domain: A Phosphotyrosine Recognition Module

SH2 domains have been identified in many proteins with seemingly diverse functions (reviewed in Pawson and Schlessinger, 1993; Koch et al., 1991), and can be classified into two broad subfamilies: 1) proteins which possess a known intrinsic enzymatic or regulatory activity (e.g., $pp60^{c\text{-}src}$ and PLCγ); 2) proteins which lack intrinsic enzymatic activity, but possess modules which mediate protein–protein interactions, and therefore, presumably play an adaptor role (e.g., the 85 kD subunit of the PI 3-kinase, Crk, Grb2, Nck, and Shc). Many of the proteins from both groups also contain SH3 domains, but these two protein modules can exist independently and serve functionally distinct roles. Mutational studies performed on the v-Crk oncogene and the Abl PTK, both of which possess SH2 and SH3 domains, have shown that their SH2 domains were of primary importance for their association with phosphotyrosine-containing proteins (Matsuda et al.,1991). These observations led to the suggestion that the function of the SH2 domain was to recognize phosphotyrosine residues when displayed within a particular peptide sequence context. This novel role for protein phosphorylation in the regulation of intracellular events has since been shown to function in diverse systems (Anderson et al., 1990; Moran et al., 1990; Escobedo et al., 1991) and has revolutionized our thinking about regulatory and interactive functions of phosphorylation (reviewed in Pawson and Schlessinger, 1993; Koch et al., 1991).

Formation of PTK Receptor Signaling Complexes

Following receptor autophosphorylation, specific SH2 domain-containing proteins are recruited from the cytosol into complexes with activated PTKRs (see Figure 2). This recruitment is mediated by phosphotyrosine-SH2 domain interac-

(reviewed in Fry et al., 1993). The critical requirement for phosphotyrosine has been demonstrated by the mutation of autophosphorylation sites on PTKRs to phenylalanine residues. Such changes have no effect on the overall PTK activity, but abolish both receptor autophosphorylation and the recruitment of the SH2 domain-containing protein which would normally bind to the phosphorylation sites in question (Kashishian et al., 1992; Valius and Kazlauskas, 1993). The binding of SH2 domain-containing proteins to phosphotyrosine-containing sequences is a precise interaction between specific SH2 domains and a limited-range of phosphorylated target sequences. Short tyrosine-phosphorylated peptides based on autophosphorylation sites of the PDGFβR have been shown to competitively inhibit binding of specific SH2 domain-containing proteins to phosphorylated PDGFβRs (Escobedo et al., 1991; Fantl et al., 1992). Specific binding sites have been identified for a number of SH2 domain-containing signaling molecules (see Figure 3). The binding sequences which mediate interactions with the SH2 domains of the PI 3-kinase are well defined. A Tyr–x–x–Met (where x can be a wide range of possible residues) motif appears to be critical for PI 3-kinase binding (Cantley et al., 1991).

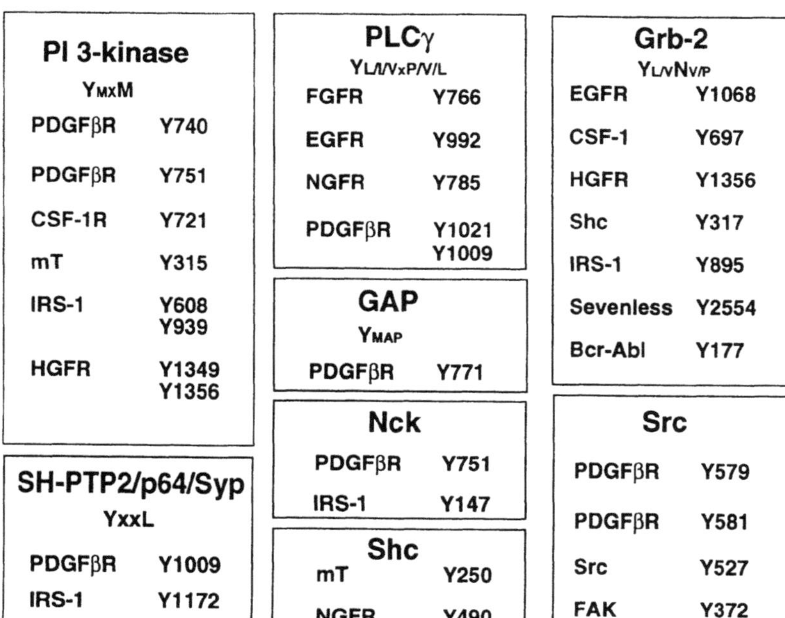

Figure 3. SH2 domain binding sites. The SH2 domain binding sites which have been identified on PTKRs, and in a few cases on other signaling proteins, are shown. Only those sites which had been rigorously confirmed by peptide mapping and sequence mutation as of January 1994 have been included. Additional more speculative SH2

Most SH2 domain binding motifs have yet to be defined. Songyang et al. (1993) have attempted to address this by binding short random phosphotyrosyl-peptide mixtures to recombinant SH2 domains. They then identified those SH2 domain-bound peptide species by protein sequence analysis. This approach identified the major PI 3-kinase binding site as the Tyr–x–x–Met motif and a consensus binding site for Grb2 of Tyr–x–Asn–x was suggested, which would seem to hold true in a number of systems (see Figure 3). This study indicated that amino acids at positions +1 to +3 relative to the phosphotyrosine may influence the binding of different SH2 domains. Additional SH2 domain binding sites will be examined in greater detail with the individual receptor families.

SH2 Domain-Mediated Signaling

Once an SH2 domain-containing protein has been recruited into a signaling complex, the activity of the associated protein can be modulated by a number of distinct mechanisms. The basic modes of regulation have been reviewed by Panayotou and Waterfield (1993). Regulatory mechanisms include translocation, phosphorylation, and induced conformational change following binding to phosphorylated sequences.

Complex formation between SH2 domain-containing proteins and membrane-bound receptors results in their movement from the cytosol to a juxtamembrane location. In many cases, this brings them into close proximity with their known substrates, or with other proteins with which they are known to interact, for example, membrane phosphoinositides for PLCγ and PI 3-kinase, and Ras in the case of GAP. While translocation alone might be sufficient to cause activation of a signaling pathway, it is likely that secondary events are required for more specific regulation and this has, indeed, been shown to be the case with PLCγ. Phosphorylation is a classical mechanism for modulating enzyme activity. Ample evidence supports a role for tyrosine phosphorylation operating in concert with SH2 domain binding in the regulation of PLCγ. PLCγ has also been shown to become phosphorylated at residues Tyr783 and Tyr1254 following binding to activated PTKRs. This phosphorylation is essential for achieving maximal activation of PLCγ (reviewed in Rhee, 1991). A number of reports demonstrate activation of PI 3-kinase following the binding of short phosphotyrosine-containing peptides to its SH2 domains (Backer et al.,1992; Carpenter et al., 1993). Similar activation can be achieved by the addition of tyrosine phosphorylated IRS-1 to preparations of PI 3-kinase (Backer et al., 1992). Presumably, following binding of the phosphopeptide to one or both of the SH2 domains of the p85 subunit of PI 3-kinase, a conformational change takes place which is transmitted to the associated p110 catalytic subunit resulting in an increase in PI 3-kinase activity.

The available evidence supports the concept that recruitment of signaling proteins to receptors mediated by specific phosphotyrosine-SH2 domain interactions pro-

required for full activation of the signaling molecule. These mechanisms are clearly not mutually exclusive and regulation of the function and binding of many SH2 domain-containing proteins probably occurs by a combination of effects acting in concert.

Coordinate Regulation of Ras Function Through SH2 and SH3 Domains

A second protein domain which has been found in many of the signaling proteins is the SH3 domain (reviewed in Koch et al., 1991; Musacchio et al., 1992). This domain functions in a similar fashion to the SH2 domain in mediating protein–protein interactions, but in this case, the recognition site has been shown to be composed of proline-rich regions of proteins of the type x–Pro–x–x–Pro–Pro–Pro–x–x–Pro (reviewed in Pawson and Schlessinger, 1993; Fry et al., 1993). As SH2 and SH3 domains are often found within the same protein, an important question is: do these domains function together in the regulation of intracellular signals? A description of concerted SH2 and SH3 domains function within cellular signaling pathways has come from studies on activation of the small GTP-binding protein, Ras.

The *let-60* Ras-related gene product functions in a developmental pathway leading to the induction of vulval cell fate (Horvitz and Sternberg, 1991). A cell-signaling gene in *Caenorhabditis elegans, sem-5*, which encodes a protein composed of a central SH2 domain flanked by two SH3 domains was identified by Clark et al. (1992). Genetic analysis suggested that this gene product functioned downstream of the *let-23* PTKR and upstream of the *let-60* product (Clark et al., 1992). Independent mutations in the SH2 and SH3 domains impaired *sem-5* activity, suggesting a role for both domains. Subsequently, a mammalian homolog of *sem-5*, Grb2, was isolated by an expression library screening technique designed to identify SH2 domain-containing proteins (Lowenstein et al., 1992; Margolis et al., 1992).

Grb2 binds to PTKRs in a phosphotyrosine/SH2 domain-dependent manner and can stimulate DNA synthesis, but only when introduced into cells together with a functional Ras protein (Lowenstein et al., 1992). If the SH2 domain of Grb2 was involved in its recruitment by receptor PTKs, then it was assumed that the SH3 domains would provide the link to Ras. The proteins involved in this interaction were identified as members of the Sos (Son of sevenless) protein family, which bind via proline-rich regions of sequence in their C-termini to the SH3 domains of Grb2 (Buday and Downward, 1993; Egan et al.,1993; Gale et al., 1993; Li et al., 1993; Olivier et al., 1993; Rozakis-Adcock et al., 1993; Simon et al., 1993). This was the first clear description of a physiological ligand for an SH3 domain. In addition to proline-rich motifs, Sos contains sequences related to known Ras guanine-nucleotide exchange proteins. Grb2 and Sos proteins appear to be constitutively associated, but only function to activate Ras upon receptor stimulation. Recruitment of the Grb2/Sos complex to the membrane results in phosphorylation of Sos, but not of Grb2 (Rozakis-Adcock et al., 1993). Sos converts the inactive form of Ras to an active GTP-bound form by nucleotide exchange. The details of the mechanism whereby recruitment of the Grb2/Sos complex to the membrane results in Ras activation remains to be elucidated. However, Grb2 and Sos would seem to be the missing components between receptor PTKs and Ras.

Further complexity is evident, since Grb2 forms complexes with proteins other than autophosphorylated receptors, including tyrosine phosphorylated forms of Shc (Pelicci et al., 1992; Rozakis-Adcock et al., 1992; Egan et al., 1993; Skolnik et al., 1993). Shc proteins have a C-terminal SH2 domain, an N-terminal glycine/proline-rich region and have been shown to be transforming when overexpressed in fibroblasts (Pelicci et al.,1992). Shc becomes tyrosine phosphorylated in response to growth factor stimulation (Pelicci et al., 1992; Ruff-Jamison et al., 1993) and in oncogenic PTK-transformed cells (McGlade et al., 1992; Egan et al., 1993). Shc/Grb2/Sos complexes have been identified in *src*-transformed Rat1 cells and it may be that this provides a mechanism for stimulating Ras in the absence of an activated PTKR to which Grb2 can bind directly (Egan et al.,1993). This would appear to be the case in cells stimulated by insulin where both Grb2 and Shc have been found in complex with IRS-1 (Skolnik et al., 1993). Such multiple pathways for Ras activation by a single receptor have been predicted for some time. Interestingly, recent studies on the PDGFβR have suggested that this receptor can mediate activation of Ras through distinct pathways in different cell types examined (Satoh et al.,1993) and that several of the known PDGFR activated signaling pathways may function upstream of Ras within a single cell type (Valius and Kazlauskas, 1993; Li et al., 1994).

Most of the remaining gaps in the PTKR-Ras pathway between the cell surface and the nucleus also appear to have been filled, as complexes containing activated (Gly12Val) mutants or GTP-bound forms of Ras together with the downstream protein kinase, Raf-1 and MAP kinase kinase have been isolated (Moodie et al., 1993). The specificity of these interactions is supported by the fact that effector mutants of Ras (e.g., Ile36Ala) fail to interact with these protein kinases. This would directly link Ras into a protein kinase cascade which ultimately leads to the phosphorylation of known nuclear proto-oncogenes, such as c-Fos, SRF, and c-Jun (reviewed in Roberts, 1992; Fry et al., 1993).

D. Signal Termination

Termination of PTKR signals is brought about by both dephosphorylation of the receptor by specific protein-tyrosine phosphatases and also by the rapid clearance of the ligand-bound receptor complexes from the cell surface (see Figure 2). While some of the receptors which enter this pathway are recycled, the majority are targeted to the lysosome where degradation occurs. It is not clear whether the receptor-bound signaling molecules also meet this fate or whether they are recycled. PI 3-kinase certainly enters the early stages of the internalization pathway with activated PDGFRs (Kapeller et al., 1993). Some data has been presented suggesting that internalization is an essential part of the signal transduction process. Sequences required for targeting the receptors into internalization pathways are located in their cytoplasmic domains (Riedel et al., 1989). PTK activity toward exogenous substrates is essential for targeting to lysosomes, but is not required for internalization (Chen et al., 1987; Honegger et al., 1987; Russel et al., 1987). Indeed, kinase-negative mutants are internalized at the same rate as wild-type receptors upon ligand binding, but are rapidly recycled back to the cell surface (Honegger et al., 1987). Thus, we complete a signaling cycle. With this information in mind, let us now look at the structural and functional aspects of signaling as mediated by the different PTKR families.

IV. PROTEIN–TYROSINE KINASE RECEPTOR FAMILIES

By comparison of known PTKR sequences, their general structural features have been used to group them into a number of families or classes (reviewed in Yarden and Ullrich, 1988; Ullrich and Schlessinger, 1990). However, with the increasingly rapid identification of novel PTKR genes over the last few years as retroviral (e.g., *ros*) and cellular oncogenes have been isolated by DNA transfection (e.g., *trk* A and B, *ret*, and *met*), by PCR (Wilks, 1989; Lai and Lemke, 1991), and by low stringency hybridization (Shibuya et al., 1990), it is clear that this original classification scheme is no longer sufficient to encompass all the known and putative receptors. It is, however, still a valid aid in the comparison of structure–function relationships between receptors. The original classification can now be expanded to several further-related receptor families. An extended scheme was recently presented by Fantl et al. (1993). The current range of distinct PTKR structural families are illustrated in Figure 4 and are described in the following sections.

A. The Epidermal Growth Factor Receptor Family

Originally termed Class I, this family is represented by the EGF receptor (EGFR, HER, c-*erb*B; Downward et al., 1984; Ullrich et al., 1984; reviewed in Carpenter and Cohen, 1990), its known mammalian relatives *neu*/c-*erb*B-2 (Bargmann et al., 1986a), c-*erb*B-3/HER3 (Kraus et al., 1989; Plowman et al., 1990), and c-*erb*B-4/HER4 (Plowman et al., 1993a), the *Drosophila* EGF-related gene, DER (Livneh et al., 1985), the product of the *let*-23 gene of *C. elegans* (Aroian et al., 1990), and *Xmrt*, the product of the *Xiphorphorus Tu* locus (Wittbrodt et al., 1989). The EGFR is the prototype for this class and its structure is shown schematically in Figure 4. All four mammalian receptors display strikingly similar features, including overall size, a large glycosylated extracellular ligand-binding domain with two signature cysteine-rich clusters, and a highly related, uninterrupted PTK domain followed by a long C-terminal extension which contains the major autophosphorylation sites. The three major mechanisms for activation of the transforming potential of a gene,

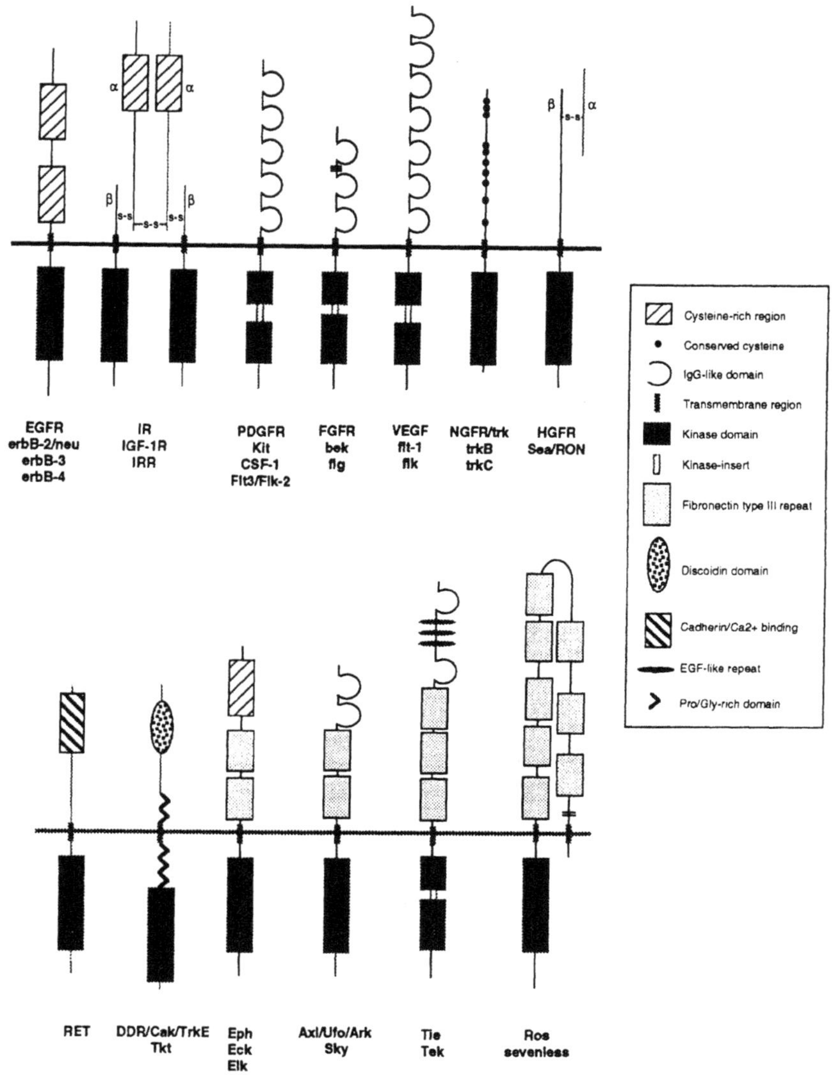

Figure 4. Receptor protein-tyrosine kinase structure and classification. This figure shows the diversity of distinct subfamilies of receptor PTKs based on the classification of Ullrich and Schlessinger (1990) and subsequently extended by Fantl et al. (1993). This classification is based mainly on the relatedness of their extracellular domain structures. The many structural features identified in these receptors are indicated in the key. The upper tier are receptors for which ligands are known and well characterized, while the lower tier are all orphan receptors for which ligands are currently lacking. The exception to this is the *Drosophila* Sevenless receptor which has a novel seven transmembrane-spanning ligand. No such ligand has as yet been described for the mammalian homolog, Ros.

addition/deletion of sequences, point mutation, and over-expression of the receptor protein, are all found to affect various members of this receptor family and have been discussed in Fry et al. (1992).

EGF Receptor (EGFR/HER/erbB)

The EGFR is able to bind to a number of related ligands including EGF, transforming growth factor-α, amphiregulin, heparin-binding EGF, and vaccinia virus growth factor (Carpenter and Cohen, 1990). The mature receptor is a single polypeptide chain of 170–180 kD, with approximately 40 kD of this being contributed by glycosylation of the extracellular domain at 10–11 sites (Ullrich et al., 1984). The two extracellular cysteine-rich domains flank the ligand-binding domain as defined by a receptor chimera (Ullrich and Schlessinger, 1990). The transmembrane and intracellular portions of this receptor are closely related to the avian oncogene, v-*erb*B, which was derived by truncation of the avian gene for this receptor (Downward et al., 1984). In addition to sequences comprising the PTK domain, there are two other distinct regions to the cytoplasmic portion of the receptor. First the juxtamembrane domain which contains Thr654, a regulatory site phosphorylated by PKC which is involved in transmodulation of the receptor in response to heterologous signals (reviewed in Ullrich and Schlessinger, 1990). Second, the long C-terminal tail, another characteristic of this family of receptors, contains a number of autophosphorylation sites which may be involved as both an autoinhibitory sequence prior to activation and also as a positive regulator of biological signals in the activated receptor (Helin et al., 1991).

The EGFR has been shown to interact directly with or to phosphorylate a number of known downstream signaling molecules including PLCγ (Wahl et al., 1988), GAP (Ellis et al., 1990), PI 3-kinase (although an order of magnitude less efficiently than observed with the PDGFR—see below) (Bjorge et al., 1990), Grb2 (Rozakis-Adcock et al., 1993), Shc (Gotoh et al., 1994), and the transcription factor p91 which mediates c-*fos* gene promoter activation (Fu and Zhang, 1993). The phosphotyrosine containing sequences responsible for binding these SH2 domain-containing proteins are indicated in Figure 3. For the EGFR family of receptors, many of these binding sites lie in the C-terminal tail. The importance of the C-terminus in defining cell-type specificity and function has been demonstrated genetically using the EGFR-like receptor, *let-23*, of *C. elegans*. Mutations within the C-terminus were shown to define at least three functional domains that contribute to receptor function in different cell types (Aroian et al., 1994). In light of the significant weight attached to the phosphotyrosine-SH2 domain interactions in PTKR mediated signaling, a recent report that a point-mutated EGFR lacking five C-terminal autophosphorylation sites is capable of eliciting a mitogenic response should also be noted (Decker, 1993). These mutant receptors were unable to activate, or to phosphorylate PI 3-kinase or PLCγ. The preceding observations suggest that a significant proportion of EGFR signaling can take place in the absence of interactions with SH2 domain-containing proteins. Furthermore, these mutant receptors were capable of phosphorylating Shc proteins and activating Ras and MAP kinase, thus demonstrating that a direct physical binding interaction with SH2 domain-containing proteins is not always essential to ensure their activation (Gotoh et al., 1994). These studies raise the question as to what fraction of the signaling pathways activated by receptor PTKs occurs through protein–protein interactions involving SH2 domains and tyrosine phosphorylated sequences.

C-erbB-2/HER2/neu

C-*erb*B-2 has been the subject of intense scrutiny due to a possible link with some forms of human breast and ovarian cancer. Approximately 25% of primary breast and ovarian tumors over-express c-*erb*B-2 and this over-expression is correlated with poor prognosis (Slamon et al., 1987). Initially identified as a rat oncogene called *neu*, it became clear upon cloning that the product of this gene was related to the EGFR (Bargmann et al., 1986a; Yamamoto et al., 1986). C-*erb*B-2/*neu* is predicted to produce a glycosylated protein of 185 kD. Sequence similarity is observed throughout the length of EGFR and c-*erb*B-2 proteins with the highest level of identity in the PTK domain and lowest in the C-terminal tail sequences, although the major sites of autophosphorylation within this latter domain are conserved. The rat *neu* oncogene was found to differ from the normal nontransforming cellular form by a single amino acid change, a Val664Glu conversion within the transmembrane domain (Bargmann et al., 1986b). This mutation results in constitutive activation of the PTK domain (Bargmann and Weinberg, 1988) and in the stabilization of receptor dimers (Cao et al., 1992; Lofts et al., 1993). The transmembrane domain of this receptor is, therefore, proposed to play an important role in receptor dimerization and mechanisms leading to PTK domain activation via specific protein–protein interactions (Cao et al., 1992; Lofts et al., 1993). Insertion of the activated *neu* transmembrane sequence into some (e.g., the insulin receptor), but not all receptors, resulted in activation of the PTK domain (Escobedo et al., 1988; Cheatham et al., 1993).

Despite the high degree of structural similarity, c-*erb*B-2 does not bind any of the known EGFR ligands. It has, however, been shown to become phosphorylated in response to EGF via the formation of heterodimers with the EGFR (Stern and Kamps, 1988; King et al., 1988). This occurs by transphosphorylation by the activated EGFR and it has been suggested that this indirect phosphorylation and activation of c-*erb*B-2 by EGF may represent a receptor PTK phosphorylation cascade (Connelly and Stern, 1990). This is not the only mechanism of activation of this receptor as EGFR-independent functions of c-*erb*B-2 have also been described (Chazin et al., 1992). A number of putative ligands for c-*erb*B-2 have been identified, but the precise nature of these factors remains a point of intense study. Recently, a family of related ligands has been described based on their ability to induce tyrosine phosphorylation of c-*erb*B-2 (reviewed in Peles and Yarden,

1993). These ligands have variously been called heregulins (Holmes et al., 1992), Neu differentiation factor (Wen et al., 1992), glial growth factors (Marchionni et al., 1993), and ARIA (Falls et al., 1993). However, discrepancies in their ability to activate and stimulate the c-*erb*B-2 protein in all expressing cell types has resulted in the discovery that these factors in fact bind to the closely related c-*erb*B-4 and not to c-*erb*B-2 (Culouscou et al.,1993; Peles et al., 1993; Plowman et al., 1993b; see below), with the activation and phosphorylation of c-*erb*B-2 presumably being due to the formation of c-*erb*B-2/c-*erb*B-4 heterodimers as described previously for the EGFR and c-*erb*B-2 (Stern and Kamps,1988). True peptide ligands for c-*erb*B-2 remain to be identified. Interestingly, in light of its close link with breast and ovarian cancers, 17β-estradiol has been shown to function as a potential ligand for c-*erb*B-2, stimulating tyrosine phosphorylation in MCF7 cells (Matsuda et al., 1993).

In the absence of defined peptide ligands, some preliminary studies have been carried out to address the distinct signaling properties of the individual members of the EGFR family using receptor chimeras. EGFR/c-*erb*B-2 chimeras have been constructed and expressed in various cell types. The kinase activity, mitogenic and transforming signals of these c-*erb*B-2 chimera can be regulated in an EGF-dependent manner (Lee et al., 1989a; Lehväslaiho et al., 1989). These chimeras have been used to study ligand-dependent interactions with substrate proteins (Segatto et al., 1992). Both common and unique substrates have been identified for the distinct EGFR family members. Noncovalent interactions involving SH2 domains have been detected with PLCγ, GAP, PI 3-kinase, and Shc proteins (Fazioli et al., 1991; Peles et al., 1991, 1992; Segatto et al., 1993). Furthermore, differences between these chimeras have also been noted in terms of their respective rates of internalization. This event is mediated via the C-terminal tail of c-*erb*B-2 and not by its PTK domain. These results suggest that in addition to specific signals generated following the binding of SH2 domain-containing proteins, the duration of these signals may also play a significant role in defining the cells response to different receptors (Sorkin et al., 1993). Further differences in terms of receptor-specific signals appear to be determined by the juxtamembrane regions of EGFR and c-*erb*B-2 (Segatto et al., 1991; Di Fiore et al., 1992). This is based on the ability to identify receptor-specific signaling properties when these two receptors are expressed in different targets cells (Segatto et al., 1991). The molecular basis for the role of the juxtamembrane regions in receptor-specific signaling remains to be elucidated.

C-erbB-3/HER3

The c-*erb*B-3 gene was isolated by reduced stringency hybridization of human genomic DNA using a v-*erb*B probe (Kraus et al., 1989; Plowman et al., 1990). The full-length cDNA possesses all the structural features of the EGFR family and encodes a polypeptide of 148 kD. Analysis of expressing cells with antibodies have revealed glycosylated proteins in the range of 160–180 kD (Prigent et al., 1992;

Kraus et al.,1993). The PTK domain is the most highly conserved, exhibiting >60% amino acid identity with other family members. However, there are a number of single nonconservative amino acid substitutions within the PTK domain at residues which are invarient among other PTKs (Hanks, 1991). One of these substitutions, a Asp to Asn in subdomain VI, has been found to give rise to the W^{42} kinase-inactive mutation in the c-Kit receptor (Tan et al., 1990). The functional significance of these changes are unclear as c-*erb*B-3 appears to encode a functional PTK kinase (Kraus et al., 1993). The most divergent region between c-*erb*B-3 and the other EGFR family members is in the C-terminal tail which is long at 364 residues in length (Kraus et al., 1993). The tail contains putative autophosphorylation sites which may be critical for interactions with substrates. C-*erb*B-3 is expressed in most tissues, but is absent in hematopoietic tissues. The highest levels were detected in epithelial tissues and brain (Kraus et al., 1989; Prigent et al.,1992). While this is similar to that described for EGFR and c-*erb*B-2, when examined in detail, the expression patterns for each of the three receptors was found to be distinct (Prigent et al., 1992). Like c-*erb*B-2, it has also been found to be over-expressed and constitutively activated in a high proportion of mammary tumor cell lines (Kraus et al.,1989, 1993). c-*erb*B-3 does not bind any of the known EGF-related ligands. Preliminary signaling studies have therefore utilized receptor chimeras. An EGFR–c-*erb*B-3 chimera has been shown to possess EGF-dependent PTK activity, to transmit a EGF-dependent mitogenic signal in NIH 3T3 cells, and to promote EGF-dependent growth in soft agar (Kraus et al., 1993; Fedi et al., 1994). This EGFR–c-*erb*B-3 chimera couples efficiently to PI 3-kinase, Shc, and Grb2, but not to PLCγ or GAP, thus distinguishing c-*erb*B-3 signaling from that generated by the other EGFR family members (Fedi et al., 1994). Indeed, the EGFR–c-*erb*B-3 chimera induces tyrosine phosphorylation of a distinct set of intracellular substrates from the EGFR.

C-erbB-4/HER4

This is the fourth member of the EGFR family to be isolated (Plowman et al., 1993a). A glycosylated protein of 180 kD, its extracellular domain is most closely related to that of c-*erb*B-3, while the kinase domain shows 79% amino acid identity with the EGFR. Again, this receptor shows similar structural features to those described for the other family members. A notable difference is the absence of a residue homologous to Thr654 in the juxtamembrane region, the major site of PKC induced phosphorylation and transmodulation (Ullrich and Schlessinger, 1990). The C-terminal tail is 282 amino acids in length and contains 18 tyrosine residues. This domain shows only limited homology to the other family members, although the sequences surrounding the major tyrosine autophosphorylation sites of the EGFR are conserved. c-*erb*B-4 is expressed in skeletal muscle, heart, pituitary, and brain. Like c-*erb*B-2, c-*erb*B-4 is predominantly expressed in a number of breast tumor cell lines (Plowman et al., 1993a). Little is currently known about the signaling properties of this receptor, but the recent identification of a putative c-*erb*B-2 ligand as the ligand for c-*erb*B-4 suggests that results will soon be forthcoming (Culouscou et al., 1993; Plowman et al., 1993b; see preceding). With respect to its function, preliminary studies suggest that activation of c-*erb*B-4 may be able to promote phenotypic differentiation of human mammary cell lines (Culouscou et al., 1993; Plowman et al., 1993a,b).

B. The Insulin Receptor Family

Class II PTKRs are defined by the insulin receptor (IR) (Ebina et al., 1985; Ullrich et al., 1985). This receptor is virtually ubiquitously expressed, although the actual receptor number in different cell populations can vary from less than 100 per cell in erythrocytes to greater than 200,000 in adipocytes and hepatocytes (reviewed in Tavaré and Siddle, 1993). Two other closely related family members exist, the insulin-like growth factor-1 (IGF-1) receptor (Ullrich et al., 1986) and the insulin-related receptor (IRR; Shier and Watt,1989; Zhang and Roth, 1992). The IR and IGF-1R exhibit the highest degree of similarity among the ectodomains of PTKs and have been found to share partial ligand specificity. Insulin binds to the IR with high affinity and to the IGF-1R with a lower affinity, while IGF-1 binds its own receptor with high affinity and the IR with very low affinity. These interactions may be further complicated by the demonstration of the formation of IR–IGF-1R hybrids from half receptors of each type (Soos et al., 1990). The ligand for IRR is currently unknown and this receptor binds neither insulin nor IGF-1. Unlike Class I and III receptors, this family has a heterotetrameric structure (see Figure 2; reviewed in Tavaré and Siddle, 1993). The holoreceptor molecule is composed of two 135 kD extracellular α-chains, which contribute the ligand-binding domain. Each α-chain contains a single cysteine-rich domain. These α-chains are disulphide bonded to two 95 kD β-chains,which span the plasma membrane and possess intrinsic PTK activity, resulting in the formation of an $\alpha 2\beta 2$ heterotetramer. An alternative splice site in exon 11 of the IR results in the formation of two receptor isoforms differing at 12 amino acids near the C-terminus of the α-subunit. The C-terminal portion of the α-subunit and the N-terminal extracellular region of the β-subunit contain two regions resembling fibronectin-type III repeats. The β subunit undergoes autophosphorylation on several tyrosine residues upon ligand binding. In particular, autophosphorylation of all three tyrosine residues in the catalytic domain within the motif Tyr–x–x–x–Tyr–Tyr stimulates the PTK activity 10–20-fold. Lesser degrees of phosphorylation within this motif can result in reduced activity. Both chains are encoded by a single gene and the mature product is produced by proteolytic cleavage at a Arg–Arg–Lys–Arg sequence. All three receptors are closely related to one another throughout their entire length, except in their extreme C-termini. The C-terminal tail of IRR (at 57 or 60 residues) is considerably shorter than that of either the IR (98 residues) or the IGF-1R (107 residues). Despite containing a number of autophosphorylation sites, the tail appears dispensable for signaling. In light of the results described above for the

EGFR family, this may be important for the generation of receptor-specific signals within this family. The mutational studies carried out on this family of receptors has recently been extensively reviewed by Tavaré and Siddle (1993).

The Class II receptors use a somewhat different approach to the other receptor families in interacting with many downstream signaling molecules. Unlike many of the other PTKRs, members of the IR family do not appear to interact directly with many of their substrates. The key discovery in elucidating a key signaling mechanism by this family of receptors was the identification and cDNA cloning of the Insulin Receptor Substrate, IRS-1 (reviewed in White and Kahn, 1994). This is a cytosolic protein of relative molecular weight 165–185 kD which rapidly becomes tyrosine phosphorylated upon stimulation of responsive cells with insulin. IRS-1 is also an *in vivo* substrate for the IGF-1R and the IRR (Zhang and Roth, 1992; Myers et al., 1993). The importance of IRS-1 to insulin-dependent signaling was suggested by the observation that a Tyr960Phe point mutation within a Asn–Tyr–Glu–Tyr^{960} motif of the juxtamembrane domain of the IR abolished tyrosine phosphorylation of IRS-1 without affecting the PTK activity of the receptor (White et al., 1988). These mutant receptors are unable to stimulate either DNA synthesis or glycogen synthesis. Tyr960 is not a site of autophosphorylation on the IR, but facilitates interaction with intracellular substrates of the IR-PTK (White and Kahn, 1994; White et al., 1988). This tyrosine residue and flanking sequences are conserved in the IGF-1R. It is possible that this region forms a substrate recognition domain although the basis for interaction is not clear. Additionally, a study utilizing antisense IRS-1 RNA has also pointed to the essential nature of the IRS-1 molecule for insulin-stimulated mitogenic signaling, suggesting its likely central importance in intracellular signaling pathways activated by this class of receptors (Waters et al., 1993).

IRS-1 contains 21 potential tyrosine phosphorylation sites, nine of which lie in Try–x–x–Met motifs, the consensus site for PI 3-kinase association (Sun et al., 1991). Stimulation of cells with insulin was shown to lead to IRS-1 tyrosine phosphorylation and its subsequent association with PI 3-kinase (Sun et al., 1991; Myers et al., 1993). This interaction results in activation of the PI 3-kinase without its direct tyrosine phosphorylation (Backer et al., 1992). These observations suggest that the activated IR does not interact directly with the PI 3-kinase, but transmits an activating signal via IRS-1. IRS-1, therefore, acts as an insulin-responsive "docking protein" for SH2 domain-containing proteins such as PI 3-kinase. IRS-1 has also been shown to act as a docking protein for other signaling molecules. Binding sites have been mapped for the phosphotyrosyl protein phosphatase Syp (Kuhne et al., 1993), Grb2 (Skolnik et al., 1993), and Nck, an SH2/SH3 domain adaptor protein (Lee et al., 1993; see Figure 3). All of these proteins recognize and bind to, distinct motifs on IRS-1. It remains to be determined whether activation of IGF-1R and IRR results in the recruitment of a distinct set of signaling molecules to tyrosine phosphorylated IRS-1.

Insulin acts primarily as a regulator of the nutritional state of the organism, stimulating glucose uptake and metabolism in muscle and adipocytes and inhibiting gluconeogenesis in the liver, while IGF-1 acts as a mitogen. Both, however, appear to stimulate IRS-1 phosphorylation and the activation of a very similar set of signaling pathways. It is not clear how insulin and IGF-1 elicit these different cellular responses. Their intracellular domains are highly homologous, but small differences are observed in the juxtamembrane, the short (9 residues in the IGF-1R) KI region and in the extreme of the C-terminal tail. Presumably one or all of these regions will be demonstrated to interact with receptor-specific substrates and signaling pathways.

C. The Platelet-derived Growth Factor Receptor Family

Class III is defined by the PDGFRs (α- and β-subtypes), the CSF-1 receptor (CSF-1R, c-Fms), c-Kit, the receptor for the steel ligand, and Flt3/Flk-2. This family is quite distinct from Class I and II receptors, with the PDGFβR as the prototype. Instead of cysteine-rich repeat motifs in the extracellular domain, this subclass has a different signature which includes regions related to Ig domains. Another distinct feature of this family of receptors are the large KI regions of 60–100 residues (see Figure 4). This domain has been suggested to play a role in catalysis, modulation of the kinase activity, regulation of intracellular transport of the receptor, or recognition of cellular substrate proteins. Of these possibilities, the available evidence supports the last suggestion in that it has been shown that PI 3-kinase and RasGAP bind to phosphorylated tyrosine residues within this domain of both the PDGFR and CSF-1R (Kazlauskas et al., 1990; Shurtleff et al., 1990).

The PDGF α- and β-Receptors

PDGF has been found to be a mitogen for a wide-range of mesenchymal cell types. In addition to mitogenic effects, PDGFs have also been shown to elicit a motility response involving actin reorganization, membrane ruffling, and chemotaxis. PDGF exists in three forms as the disulphide-linked dimer products of two genes which encode a PDGF A-chain, and a PDGF B-chain, giving rise to PDGF-AA, PDGF-BB, and PDGF-AB (reviewed in Heldin, 1992). There are also two distinct receptors for these factors termed the PDGFαR (Heldin et al., 1988; Claesson-Welsh et al., 1989; Matsui et al., 1989) and the PDGFβR (Yarden et al., 1986; Claesson-Welsh et al., 1988; Gronwald et al., 1988). These receptors are also the products of distinct genes. The two receptors are highly homologous throughout their length with the major known differences described to date being their ligand-binding properties and cell type expression. Data from binding studies suggest that the PDGF A-chain can only bind to the PDGFαR, while the PDGF B-chain can bind either receptor subtype (Gronwald et al., 1988; Hart et al., 1988). Therefore, the nature of ligand-driven activated receptor dimer formed will depend upon both the relative concentrations of the two PDGFR types within the target cell and also on the form of PDGF dimer present. PDGFRs are found on mesenchymal cells, such as fibroblasts and glial cells and are developmentally regulated in many cell types.

Signaling by the PDGFRs has been recently reviewed (Claesson-Welsh, 1994). While the two PDGFRs are very similar, some differences in function have been noted suggesting independent signaling roles. Both the PDGFαR and PDGFβR are capable of mediating PDGF-stimulated mitogenesis, but only the PDGFβR is able to mediate cytoskeletal changes and a chemotactic response to PDGF (Eriksson et al., 1992). These differences are reflected by small differences in the patterns of tyrosine phosphorylation induced by the two receptor subtypes (Eriksson et al., 1992) and in their relative affinities for SH2 domain-containing substrates such as RasGAP (Heidaran et al., 1993).

Much of the early evidence for SH2 domain-mediated signaling complexes came from studies on the PDGFβR. The PDGFβR has been shown to bind directly to PI 3-kinase (Kaplan et al., 1987), GAP (Kazlauskas et al., 1990), PLCγ (Kumjian et al., 1989), *src*-family kinase (Kypta et al., 1990), Nck (Nishimura et al., 1993), and Syp (Kazlauskas et al., 1993; Lechleider et al., 1993). In the cases of PI 3-kinase and Syp, binding to the phosphorylated receptor sequence alone appears to be sufficient for their activation (Carpenter et al., 1993; Lechleider et al., 1993), while in the case of PLCγ, subsequent tyrosine phosphorylation is also essential (Rhee, 1991). Several SH2 domain binding sites required for these interactions have been identified on the human PDGFβR. PI 3-kinase has been shown to associate with the human PDGFβR via Tyr740 and Tyr751 which lie in the KI region of this receptor (Fantl et al., 1992; Kashishian et al., 1992). Other mapped sites include Tyr771 in the KI region for interaction with RasGAP (Kazlauskas et al., 1992), Tyr1021 in the C-terminal tail for interaction with PLCγ (Rönnstrand et al.,1992; Valius et al., 1993), Tyr1009—the binding site for the phosphotyrosine phosphatase, Syp (Feng et al., 1993; Valius et al., 1993), and Tyr579 and Tyr581 in the juxtamembrane segment N-terminal to the catalytic domain, as the binding site for *src* family PTKs (Mori et al., 1993). Grb2 has also been shown to associate with the PDGFβR, but by a novel method. In contrast to the situation found with the EGFR, Grb2 does not appear to bind directly to the PDGFR, but binds to the receptor complex indirectly via the SH2-containing phosphatase, Syp, which itself binds directly to the receptor (Li et al., 1994). This result sheds light on the surprising results of Valius and Kazlauskas (1993) who found that Tyr1009 was only one of several Tyr residues which were linked, presumably via different SH2 domain-containing proteins, to Ras activation. This suggests that Ras activation by the PDGFR may occur by more than one independent signaling route. Interestingly, the adaptor molecule, Nck, has recently been shown to bind via Tyr751, a site earlier identified as being involved in PI 3-kinase binding (Fantl et al., 1992). Similarly, Tyr1009 has been reported as a binding site for both PLCγ and Syp (Kashiashian and Cooper, 1993; Kazlauskas et al., 1993). The observation that more than one signaling molecule may share a binding site indicates that data from mutagenesis of these sites will have to be interpreted carefully. Fantl et al. (1992) have reported that peptides which block PI 3-kinase binding to the PDGFβR, such as a phosphopeptide spanning Tyr751, are able to block a PDGF-induced mitogenic response. This was originally interpreted as demonstrating a critical role for PI 3-kinase in mitogenic signaling, but recent data may also suggest an involvement of Nck, whose function is currently unknown. The observation that a single receptor can use the same mechanism to interact with such a diverse range of signaling molecules (and this list is probably not yet complete) demonstrates the flexibility and functional diversity of SH2 domain-phosphotyrosine interactions. With the mapping of these sites to the receptor, the task of defining the contribution of various signaling molecules to distinct effects of PDGF can begin. In this respect Wennström et al. (1994) have recently demonstrated that the binding sites for PI 3-kinase in the KI region are required for the membrane ruffling and chemotaxis induced by the PDGFβR. Kundra et al. (1994) have extended these studies and demonstrated that the chemotactic response to PDGF mediated by the PDGFβR is brought about by a delicate balance of recruitment and activation of PI 3-kinase, PLCγ, and RasGAP. The former two molecules would appear to play a role in promoting migration while GAP may serve a migration suppressing function.

In addition to the ligand binding domain, the lowest degree of homology between the two PDGFRs lie in the domains which interact with downstream signaling molecules, the KI region (35% identity) and the C-terminal tail (27% identity). However, the PDGFαR has been shown to bind to a similar spectrum of substrates including PI 3-kinase, PLCγ, Src-family kinases, and Syp (Yu et al., 1991; Bazenet and Kazlauskas, 1993; Claesson-Welsh, 1994) with similar affinities to the PDGFβR. However, it has a fivefold lower binding affinity for GAP with respect to PDGFβR (Bazenet and Kazlauskas, 1993; Heidaran et al., 1993). The PDGFαR has conserved the Tyr residue in the KI region, which was shown to be responsible for binding to GAP in the PDGFβR, but the surrounding sequence has significantly diverged. This result highlights the slowly emerging differences between the two PDGFRs and the signaling pathways with which they interact. Further studies of this kind will be required to elucidate the distinct targets and signaling roles of these two receptors.

The CSF-1 Receptor/c-Fms

The v-*fms* oncogene was identified as the transforming gene of two feline retroviruses, where it was probably transduced from feline cellular DNA by recombination events with a feline leukemia virus (Donner et al., 1982; Besmer et al., 1986a). Hampe et al. (1984) sequenced the v-*fms* gene and demonstrated its homology to the family of PTKs. Sherr et al. (1985) have since shown that this oncogene is related to the colony-stimulating factor-1 (CSF-1, also called macrophage colony stimulating factor) receptor (CSF-1R). CSF-1 regulates the survival, growth, and differentiation of mononuclear phagocytic cells. The structure, function, and transforming ability of this receptor has been extensively reviewed (Sherr, 1990; Fry et al., 1992). The human CSF-1R is a 150 kD integral transmembrane glycoprotein. It possesses a 512 amino acid extracellular ligand-binding domain, composed of Ig-like repeats, which is linked by a 25 amino acid membrane-spanning segment to a 435 amino acid intracellular domain containing a PTK domain, which is split by a KI region (see Figure 4). The contribution of the 68 amino acid KI region to the enzymatic and transforming activities of both c- and v-*fms* have been investigated by Taylor et al. (1989). Three potential autophosphorylation sites are located within the KI domain. Partial or complete deletion of this domain had no effect on the kinase activity or autocrine transforming ability of c-*fms* in NIH 3T3 cells, nor on the transforming ability of v-*fms* in the same cell type (Taylor et al., 1989). However, Reedijk et al. (1990) found that deletion of the KI region from the murine CSF-1R decreased its ability to induce both morphological changes and mitogenesis in Rat-2 fibroblasts. This latter result has been born out by point mutational studies of specific tyrosine residues within the KI region, suggesting an important role in signaling as described above for the PDGFRs (van der Geer and Hunter, 1993). It seems likely that this domain will also play a specific role in cells where the CSF-1R is normally expressed, for example, in monocytes or trophoblasts. The CSF-1R is highly related to other members of the PDGFR family and its head to tail localization with the PDGFβR on chromosome 5 suggests that the two genes arose by gene duplication followed by sequence divergence (Roberts et al., 1988).

The CSF-1R shows several differences with the PDGFβR in terms of the signaling molecules with which it is capable of interacting. The binding site for PI 3-kinase lies within the KI region (Tyr721), as it does in both the PDGFαR and PDGFβR (Shurtleff et al., 1990; Reedijk et al., 1992). Also in common with the PDGFR, members of the *src* family of PTKs are activated by CSF-1 (Courtneidge et al., 1993). Grb2 binds to the CSF-1R via Tyr697, which together with the PI 3-kinase binding site, are essential for mitogenesis mediated by this receptor in fibroblasts (van der Geer and Hunter, 1993). In contrast to the PDGFRs, neither GAP nor PLCγ bind to the activated CSF-1R (Downing et al., 1989; Reedijk et al., 1990). The failure of the latter signaling molecule to bind is in accordance with data which shows that CSF-1 stimulation of its receptor does not induce a rapid hydrolysis of phosphoinositides (Downing et al., 1989). In contrast to the PDGFR, the C-terminal tail of the CSF-1R contains only a single tyrosine residue (Tyr969). This is not known to be a site of autophosphorylation.

The Steel Receptor/c-Kit

The c-*kit* proto-oncogene encodes a transmembrane PTKR whose ligand has been identified as a 30 kD product of the *steel* locus (reviewed in Witte, 1990; Galli et al., 1994). This factor, also called stem cell factor (SCF), Kit ligand (KL), and mast cell growth factor (MGF), stimulates hematopoiesis, the proliferation and survival of myeloid and lymphoid cells, and is a potent synergistic factor with other cytokines. The c-Kit receptor was first identified as the transforming gene of the Hardy–Zuckerman 4 feline sarcoma virus, where it was expressed as a truncated $p80^{gag-kit}$ fusion protein (Besmer et al., 1986b). The extensive nature of the deletions involved in generating $p80^{gag-kit}$ result in a minimal PTK domain being fused to 340 amino acids of FeLV *gag*. That the v-*kit* gene was derived from a novel PTKR was demonstrated with the cloning of the human (Yarden et al., 1987) and murine (Qui et al., 1988) c-*kit* genes. The whole of the extracellular and transmembrane domain, 17 amino acids of the juxtamembrane domain and 49 amino acids from the C-terminal tail are deleted in v-*kit*. These alterations in structure probably account for the oncogenic activation of v-*kit* (Qiu et al., 1988). The *gag–kit* protein is able to associate with cellular membranes, as are other *gag–onc* fusion proteins, probably involving *gag*-linked myristylation (Besmer et al., 1986b; Majumder et al., 1990). Comparison of the resulting juxtamembrane domains of v-*kit* and c-*kit* suggest that there may be a qualitatively different interaction between the PTKs of c-Kit and v-Kit with cellular membranes, which might contribute to the transforming potential of v-*kit*. v-*kit* is, therefore, distinct from most other known oncogenic variants of PTKR genes where the transmembrane domain is not deleted and so the subcellular localization is not necessarily affected.

C-Kit encodes a 145 kD transmembrane glycoprotein which is most closely related to the CSF-1R and also to a lesser degree, the PDGFRs (Yarden et al., 1987). It is comprised of a signal peptide followed by a 500 amino acid extracellular domain composed of five Ig-like domains characteristic of this receptor class. The cytoplasmic portion of the receptor is 433 amino acids in length with the PTK domain possessing the other characteristic feature of this receptor class, a 77 amino acid hydrophilic KI region. C-Kit is expressed in cell lineages derived from the neural crest (melanocytes and brain), the bone marrow, and germ cell lineages. c-*kit* is allelic with the mouse W locus and its product is presumably required for the development of stem cells of the melanocyte, hematopoietic, and germ cell lineages (Chabot et al., 1988; Geissler et al., 1988). Severe W alleles are the result of large deletions within the gene while more subtle phenotypes are brought about by point mutations within the PTK domain (Tan et al., 1990).

Some steps have been taken to investigate the signaling pathways with which the c-Kit receptor interacts. Both PI 3-kinase and PLCγ have been shown to associate with activated EGFR-c-Kit chimera (Rottapel et al., 1991; Lev et al., 1992). No association is observed with either GAP or Vav, although the latter is a substrate for c-Kit-induced tyrosine phosphorylation (Rottapel et al., 1991; Alai et al., 1992). This is distinct from the set of proteins which associate with either the PDGFRs or with the CSF-1R (see above). Within a single cell type, ligand stimulated c-Kit, but not CSF-1R, has been shown to bind to the SH2 domain-containing hematopoietic cell phosphatase, PTP1C (Yi and Ihle, 1993). Both factors stimulate weak tyrosine phosphorylation of this phosphatase. This difference may be the first step in determining the differences in signaling via these two receptors of the hematopoietic system.

Flt3/Flk-2

Flk-2/Flt3 are the most recently isolated members of the PDGFR family (Matthews et al., 1991; Rosnet et al., 1991). These two proteins are nearly identical over most of their length except at the C-terminus where they differ completely over a stretch of approximately 31 residues. It is likely that these two receptors are the products of a single gene. The product of the *flt3* gene has been shown to be a 155 kD glycoprotein possessing intrinsic PTK activity (Lyman et al., 1993a; Maroc et al., 1993). This receptor is expressed in the hematopoietic system and also in neural, gonadal, hepatic, and placental tissues of the mouse (Maroc et al., 1993). A membrane-spanning ligand for the Flk-2/Flt3 receptor has been cloned using a soluble form of the receptor to identify a suitable source. The Flk-2/Flt3 ligand is able to function as a proliferative factor for primitive hematopoietic cells (Lyman et al., 1993b). The above noted sequence differences in the extreme C-terminal tails of the two receptor proteins may have interesting implications for intracellular signaling. The Flk-2 sequence lacks C-terminal autophosphorylation sites, while Flt3 possesses two such sites which fall into consensus binding sites of PI 3-kinase and Grb2 (Matthews et al., 1991; Rosnet et al., 1991). A preliminary study of the signaling characteristics of Flk-2 has been carried out using a chimeric receptor molecule possessing the extracellular domain of the human CSF-1R fused to the transmembrane and intracellular domains of murine Flk-2 (Dosil et al., 1993). This chimeric receptor displayed both receptor-specific (compared to PDGFR and CSF-1R) and cell type-specific tyrosine phosphorylation of intracellular substrates. PLCγ, GAP, PI 3-kinase, Shc, Grb2, Vav, Fyn, and pp60$^{c\text{-}src}$ were all demonstrated to be potential signaling components lying on the Flk-2-regulated pathways. Of these proteins, PLCγ, PI 3-kinase, Shc, Grb2, and the *src* family kinases were able to physically associate with the cytoplasmic domain of the activated CSF-1R–Flk-2 chimera. The availability of a ligand for the Flk-2/Flt3 receptor will allow further studies of signaling by the endogenous receptor(s) to be carried out.

D. The Fibroblast Growth Factor Receptor Family

At least seven polypeptides, sharing significant sequence homology, comprise the fibroblast growth factor (FGF) family of ligands. These seven polypeptides are: acidic FGF (aFGF), basic FGF (bFGF), Int-2 (FGF3), Hst (FGF4), FGF5, FGF6, and keratinocyte growth factor. These factors exhibit angiogenic, neurotrophic, cell survival, and mitogenic effects on a range of cell types derived from the mesoderm and neuroectoderm (reviewed in Jaye et al., 1992). Heparin sulphate proteoglycans constitute low-affinity binding sites for FGFs both on the plasma membrane and in the extracellular matrix. FGFs interact with heparin-like glycoaminoglycans which increase the affinity of the peptide ligand for the receptor. This review will focus only on the high-affinity receptors for these factors which possess intrinsic PTK activity and form the Class IV receptor family which is composed of at least five distinct genes: FGFR1 (*flg*, Cek1), FGFR2 (*bek*, Cek3), FGFR3 (Cek2), FGFR4, and *flg*-2. These FGFRs are differentially expressed during development. The FGFRs are observed as glycosylated proteins in the range of 110–150 kD and are closely related to the Class III receptors described previously, having Ig-like repeats within their extracellular domains and a shorter KI region within their PTK domains (see Figure 4; Pasquale, 1990; Pasquale and Singer, 1989; Kornbluth et al., 1988; Lee et al., 1989b; Ruta et al., 1989; Dionne et al., 1990; Safran et al., 1990; Avivi et al., 1991; Keegan et al.; 1991; Partanen et al., 1991). This class of PTKRs has only two or three Ig-like repeats, depending upon differential splicing of the receptor mRNA (rather than 5 Ig repeats as in Class III receptors). The effect of the absence of the third N-terminal Ig-like domain on ligand binding and receptor function is not clear as the two-domain variant is still functional. A domain referred to as the acidic box is located between the first and second Ig-like domain and is typical of FGFRs (Lee et al., 1989b). The second Ig-like domain contains sequences required for binding heparin sulphate proteoglycans and this domain has been shown to also be essential for binding of FGFs (Kan et al., 1993). A 50 amino acid variable region lying within the C-terminal half of the third Ig-like domain determines the ligand-binding specificities of the different receptor family members (Yayon et al., 1992). This region is also observed to vary by alternative splicing, presenting a novel method for generating receptor diversity. The PTK domain is split by a short, highly variable, 14 amino acid KI region and then followed by a relatively short C-terminal tail of 55–69 amino acids. Multiple forms of FGFR1 (at least 12 distinct forms can be encoded by this gene) and FGFR2 have been described which are produced by alternative splicing and cleavage of the extracellular ligand binding domain, of the PTK domain, and also of the intracellular interaction regions which bracket the PTK domain (Champion-Arnaud et al., 1991; reviewed in Jaye et al., 1992). The FGFRs respond to different members of the FGF family. Thus, while aFGF binds with high affinity to FGFRs 1–4, bFGF binds with high affinity to FGFR1 and FGFR2, with low affinity to FGFR3, and not at all to FGFR4. The ligand–binding characteristics of Flg-2 are currently unknown. The FGFR family are able to form both homodimers and heterodimers following ligand stimulation, thus possibly creating further diversity of signaling potential (Bellot et al., 1991).

Most studies so far have highlighted the similarities in the signaling pathways activated by the different FGFR family members. Signaling molecules such as PI 3-kinase and GAP, which become associated with many activated PTKRs, do not seem to associate with FGFRs (Molley et al., 1989; Mohammadi et al., 1992). FGFs are known to stimulate PI turnover and elevate intracellular calcium levels via activation of PLCγ. Upon activation of the FGFR1, PLCγ becomes autophosphorylated at Tyr766. This acts as a binding site for the SH2 domains of PLCγ. There is some data which supports specificity of FGFR function. FGFR1 and FGFR4 have been shown to phosphorylate distinct sets of polypeptides within a single cell type (the BaF3 cell line), but little progress has been made towards identifying these substrates (Vainikka et al., 1992). Tyrosine phosphorylation of 80–90 kD proteins in particular are characteristic of FGFR1, but not FGFR4 activation. FGFR4 has also been shown to phosphorylate PLCγ on tyrosine, but to a lesser extent than FGFR1 (Vainikka et al., 1992; Wang et al., 1994). The role of PLCγ-induced PI breakdown in FGF-treated cells remains to be determined. Phospholipid turnover would seem to be nonessential for the mitogenic properties of the FGFRs (Mohammadi et al., 1992; Peters et al., 1992) and it has been suggested that FGF-induced activation of PLCγ may, therefore, be involved in one of the nonmitogenic responses to FGFs such as changes in morphology or the regulation of cell differentiation. Activation of FGFR1, but not FGFR4, also resulted in the phosphorylation of Shc proteins and MAP kinase, further highlighting the differences in signaling mediated by these two members of the FGFR family (Wang et al., 1994). These differences are reflected by the ability of only FGFR1 activation to promote cell survival and growth (Wang et al., 1994). In light of the role for tyrosine autophosphorylation sites on receptors in regulating signaling interactions, it should be noted that FGFR4 lacks three such sites conserved between FGFR1, FGFR2, and FGFR3. The PLCγ binding site is conserved and is common to all five receptors. However, it is deleted from many of the known C-terminal splice variants. It is possible that such variation of the intracellular domain may play a role in the regulation of signaling interactions mediated by these receptors under different circumstances.

The complexity which can be envisaged from five receptors which are able to cross-interact with seven growth factors is unique to the FGF/FGFR family. The diversity and the regulation of signaling pathways and receptor function may be further extended by the presence of the extensive repertoire of splice variants of the FGFRs and their ability to form heterodimers (Shi et al., 1993; reviewed in Jaye et al., 1992). Clearly, extensive study will be required before a complete understanding of the regulation and functioning of FGF/FGFR interactions is attained.

E. The Vascular Endothelial Cell Growth Factor Receptor Family

Vascular endothelial growth factor (VEGF) is structurally related to the PDGF family of growth factors and is a specific mitogen for vascular endothelial cells, while inducing migration and chemotaxis in some other cell types (reviewed in Neufeld et al., 1994). Two closely related receptors for VEGF have recently been cloned (*flt* and *flk*1) and these have many features in common with the Class III PDGFR family including an extracellular domain composed of Ig-like repeats and a PTK domain which is split by a large (67 amino acid) insert sequence (see Figure 4; Shibuya et al., 1990; De Vries et al., 1992; Millauer et al., 1993; Oelrichs et al., 1993). Flt4 appears to be a third orphan member of this family, which exhibits isoforms with distinct C-terminal tails having different potential autophosphorylation sites (Finnerty et al., 1993; Galland et al., 1993; Pajusola et al.,1993). These receptors are distinguished from the PDGFR family by the presence of seven, rather than five, Ig-like domains making up their extracellular ligand-binding domain. Flt, Flk1, and Flt4 are all expressed in endothelial cells, Flt exclusively so. Flt and Flk-1 have been shown to function as a major regulator of vasculogenesis and angiogenesis (Millauer et al., 1993; Millauer et al., 1994). Little is known about VEGF receptor-mediated signaling pathways, but the data currently available on VEGFs and their receptors was recently reviewed by Neufeld et al. (1994).

F. The Hepatocyte Growth Factor/Scatter Factor Receptor Family

Two independently isolated peptide factors, hepatocyte growth factor (HGF) and scatter factor (SF), have recently been shown to be identical ligands for the c-Met receptor (Class V) PTK (Bottaro et al., 1991; Naldini et al., 1991a; reviewed in Cooper, 1992; Rubin et al., 1993). The effects of HGF/SF on different cell types are interesting in that while promoting proliferation of some epithelial cells and melanocytes, it also acts as a motility agent increasing the migration of other epithelial and endothelial cells. Additionally HGF/SF may have morphogenic properties on particular cell populations in an appropriate context (Rubin et al., 1993; Tsarfaty et al., 1994).

Met was originally isolated as an activated oncogene from a N-methyl-N′-nitro-N-nitrosoguanidine-treated human osteogenic sarcoma cell line (Cooper et al., 1984). Preliminary characterization suggested that the *met* oncogene product was related to PTKRs, most closely to the IR and to *abl* within the PTK domain, but showing little or no homology to other proteins in its putative ligand-binding domain (Dean et al., 1985). Further analysis revealed that the *met* oncogene had arisen as a result of rearrangement of DNA sequences at unlinked loci, fusing a translocated promoter and N-terminus of the *tpr* (translocated promoter region) locus to the *met* PTK domain (Park et al., 1986). This resulted in the expression of a 65 kD Trp–Met fusion protein with PTK activity (Gonzatti-Haces et al., 1988). Isolation of the c-*met* proto-oncogene cDNA confirmed early suspicions of its relationship with PTKRs (Park et al., 1987).

The c-*Met* gene encodes a protein which possesses a 24 amino acid signal sequence followed by a large putative external domain (926 amino acids) with a characteristic array of cysteine residues distinct from both the Class I and Class III PTKRs (Figure 4). A single small cysteine-rich cluster is observed with the majority of the cysteine residues dispersed throughout the extracellular domain. The mature c-Met protein is a glycosylated αβ heterodimer of 190 kD, consisting of a 145 kD β-subunit, possessing autophosphorylating activity, disulphide linked to a 50 kD α-subunit which is exposed at the cell surface (Giordano et al., 1989a). Biosynthesis studies indicate that both the 145- and 50-kD chains are derived from a single precursor glycoprotein which is cleaved to generate two subunits during maturation (Giordano et al., 1989b). This cleavage is essential for correct function. Absence of cleavage results in a constitutively activated receptor (Mondino et al., 1991). In light of this, it is of interest to note that alternatively splicing of approximately 10%

of c-*met* transcripts results in a receptor with 18 additional amino acids in its extracellular domain. This form of the receptor is not cleaved and is found as at the cell surface as a 170 kD monomer. It is not known whether this alternate form has distinct ligand-binding properties. These structural observations suggest that c-Met, rather than being closely related to the IR family as proposed in the evolutionary tree of Hanks et al. (1988), is probably the prototype for a new class of αβ dimeric PTKRs (see Figure 2). Two other c-Met-related PTKs which may be members of this proposed family have been identified; v-*Sea*, the transforming gene of the S13 avian erythroblastosis virus (Hayman et al., 1985; Smith et al., 1989) and RON, a recently described human orphan receptor-like PTK (Ronsin et al., 1993). Amino acid sequence comparison of RON with c-Met reveals homology throughout the length of the two products (72% in the PTK domain) and, although the overall homology in the extracellular domain falls to 25%, all the characteristics of the c-Met extracellular domain are observed in RON including the cysteine-rich cluster and a putative cleavage site to generate a αβ heterodimer (Ronsin et al., 1993). It would appear that RON is the human homolog of avian c-Sea (Huff et al., 1993; Ronsin et al., 1993). c-Sea/RON shows a highly restricted tissue distribution being expressed only in an unidentified population of white blood cells (Huff et al., 1993).

Analysis of signal transduction via c-Met has recently begun and interactions with a number of signaling molecules has been reported. Ligand stimulation of c-Met leads to autophosphorylation at a number of sites (Gonzatti-Haces et al.,1988; Naldini et al., 1991b). Within the kinase domain, autophosphorylation of Tyr1234 and Tyr1235 appears to be essential for activation of the PTK domain in response to ligand binding (Longati et al., 1994). A site of negative regulation or transmodulation is provided by a serine residue in the juxtamembrane domain, which is a site of action for PKC (Gandino et al., 1990). This results in the C-terminal truncation of the receptor removing the PTK domain. Ligand-activated c-Met has been found to associate with PI 3-kinase (Graziani et al., 1991; Ponzetto et al., 1993), RasGAP (Bardelli et al., 1992), PLCγ (Bardelli et al., 1992), *src*-family kinases (Bardelli et al., 1992), and an unidentified protein-tyrosine phosphatase (Villa-Moruzzi et al., 1993). HGF/SF is also known to activate the Ras pathway (Graziani et al., 1993) and the related small GTP-binding protein, Rho, has been implicated in HGF/SF-induced cell motility (Takaishi et al., 1994).

HGF/SF can elicit either mitogenesis and/or motogenesis via c-Met, depending upon the target cell. Both effects are mediated through the cytoplasmic and PTK domains of c-Met (Komada and Kitamura, 1993). It may also be involved in mesenchymal to epithelial cell conversion during normal development (Tsarfaty et al., 1994). The different biological effects must ultimately be due to the activation or modification of distinct intracellular signal cascades. It is not clear, however, whether these differences are brought about by the interaction of different SH2 containing proteins/substrates for tyrosine phosphorylation or by distinct mediators downstream of the initially receptor-mediated events which are specific for certain cell types.

The only substrates of Sea identified to date are the Shc proteins (Crowe et al., 1994). These were found to be tyrosine phosphorylated by, but not associated with, the *env-sea* oncoprotein, suggesting a possible role for Shc proteins in signaling via this oncoprotein. This does not however necessarily imply that the Sea receptor will also interact with this particular substrate in the course of normal signaling. Additionally, the C-terminal tail of the *Sea* product contains putative PI 3-kinase (Tyr–Ile–Asn–Met) and Grb2 (Tyr–Ile–Gln–Met/Tyr–Val–Asn–Leu) consensus binding sequences. Binding of Grb2 to v-Sea protein has recently been detected by Crowe et al. (1994).

G. The Nerve Growth Factor/Neurotrophin Receptor Family

The *trk* oncogene (originally named *onc*D) was isolated as a gene from a human colon carcinoma which was found to transform transfected NIH 3T3 cells (Pulciani et al., 1982). Molecular cloning of the transforming gene demonstrated that *trk* was the product of a fusion between 221 amino acids of a truncated nonmuscle tropomyosin gene with the transmembrane domain and a PTK domain of a novel PTKR (Martin-Zanca et al., 1986). The product of the tropomyosin–*trk* fusion was identified as a 70 kD cytosolic protein which is able to phosphorylate proteins on tyrosine residues (Mitra et al., 1987). Coulier et al. (1990) have shown that a single point mutation of a conserved cysteine residue (Cys345Ser) in the extracellular domain results in the generation of a novel *trk* oncogene, illustrating that small mutations within this locus can also lead to its activation. This is reminiscent of a mutation of the v-*fms* gene described previously. These results suggest that alteration of the normal extracellular domain, probably resulting in a constitutively active, ligand-independent PTK, is more important than specific structural features of the N-terminal sequences involved in the generation of a transforming *trk* species.

Cloning of the normal allele confirmed that the *trk* proto-oncogene has a molecular structure similar to that of other known PTKR genes (Martin-Zanca et al., 1989). The gene codes for a 790 amino acid protein composed of a 32 amino acid signal peptide, a 375 amino acid external domain, which contains highly conserved cysteine residues, but no cysteine-rich regions. This is linked, via a transmembrane domain, to a PTK domain, with a short KI sequence (14 residues) and a characteristic very short C-terminal tail (15 residues). *Trk* shows no particular homology with any of the other PTKR families. The best alignment is with members of the IR family and *ros*. The mature Trk glycoprotein has a molecular weight of 140 kD and possesses *in vitro* PTK activity. The extracellular ligand-binding domain shows no homology to known proteins. Subsequently, two other *trk*-related genes were cloned, *trk*B (Klein et al., 1989), and *trk*C (Lamballe et al., 1991), which are closely related to *trk* throughout their length. All three encode PTKRs with molecular weights in the range 130–145 kD.

Studies on the products of these three genes have shown them to be the receptors for a family of neurotrophic factors including nerve growth factor (NGF), brain-derived neurotrophic factor (BDNF), neurotrophin-3 (NT-3), and neurotrophin-4 (NT-4; reviewed in Barbacid et al., 1993). These factors play important roles in the growth, differentiation, and survival of neurons. Trk was first described to become tyrosine phosphorylated in response to NGF (Kaplan et al., 1991a), and was then subsequently demonstrated to be the high-affinity receptor for this factor (Kaplan et al., 1991b; Klein et al., 1991; Loeb et al.,1991). TrkB is a signaling receptor for BDNF and NT-4 (Soppet et al., 1991), while TrkC is the receptor for NT-3 (Lamballe et al., 1991). Whether these Trk proteins alone are sufficient for all the effects mediated by these neurotrophic factors, or whether a second low-affinity receptor for all of these factors, termed p75, is also required, remains a point of debate (Hempstead et al., 1991; Klein et al., 1991). The arguments for and against p75 involvement were the focus of a recent review by Barbacid (1993).

Unlike Trk, whose expression is confined to a subpopulation of sensory neurons in the peripheral nervous system which originate from the neural crest (Martin-Zanca et al.,1990), TrkB is widely expressed in both neuronal and non-neuronal cells which make up the central and peripheral nervous system (Klein et al., 1989, 1990a). Targeted disruption of *trk*B has been shown to result in nervous system disorders including defects of both the central and peripheral nervous system. These animals grew to term, but did not display feeding activity and died soon after birth (Klein et al., 1993). Another distinction between Trk and TrkB are that two classes of TrkB receptors have been described, the full length 145 kD glycoprotein and a shorter species which lacks the catalytic domain (Klein et al., 1990b; Middlemas et al., 1991). The truncated form, composed of the extracellular domain, the transmembrane domain, and a 21 or 23 residue cytoplasmic domain lacking PTK domain motifs, is expressed at high levels in adult brain (Klein et al., 1990b). The significance of this is unclear. TrkC is expressed in the central, peripheral, and enteric nervous system and preferentially within distinct structures of the brain, with a pattern complementary to TrkB (Lamballe et al., 1991). As observed with TrkB, the *trk*C locus encodes multiple forms of this receptor. However, in this case, the forms differ in the PTK domain where inserts of 14 and 25 amino acids are found between kinase subdomains VII and VIII (Lamballe et al., 1993). Transcripts for these alternate forms are found in various structures of the brain. These receptors have been shown to display distinct biological properties and substrate specificities. Only TrkC, which lacks these inserts, is mitogenic when expressed in murine fibroblasts. Indeed, the forms of this receptor with inserts in the PTK domain fail to interact with the substrate signaling molecules, PLCγ and PI 3-kinase (Lamballe et al., 1993). It has been suggested, therefore, that these receptor isoforms may interact with distinct signaling pathways.

Trk has been shown to associate in a ligand-dependent manner with PLCγ, GAP, and PI 3-kinase (Ohmichi et al., 1992, 1993a; Soltoff et al., 1992). The major site of Trk substrate interactions has been mapped to Tyr785 which lies in the short C-terminal tail of Trk (Obermeier et al., 1993a). Mutation of this tyrosine residue to phenylalanine resulted in reduced tyrosine phosphorylation of GAP and PI

3-kinase, and complete loss of association with PLCγ. Notably, the affinity of the Trk-PLCγ interaction is estimated to be 100-fold higher than that observed with the EGFR. Obermeier et al. (1993b) have mapped a potential PI 3-kinase binding site to Tyr751 in the C-terminal tail using the p85 subunit of this enzyme. This is, however, a relatively low-affinity binding site. Despite the close proximity of the apparent binding sites, both PI 3-kinase and PLCγ were shown to be able to bind the receptor simultaneously. However, other groups have been unable to detect a direct interaction between Trk and PI 3-kinase and have suggested an indirect coupling of the type observed with the insulin family of receptors involving an IRS-1-like molecule (Ohmichi et al., 1992). The extent to which IRS-1 and other related "SH2 domain-docking proteins" function in the regulation of receptor-mediated signal transduction pathways remains to be determined (Wang et al., 1993). NGF is known to activate $p21^{ras}$ dependent pathways by a mechanism distinct from EGF in PC12 cells (Qiu and Green, 1991; Thomas et al., 1992; Wood et al., 1992). Trk has recently been shown to form complexes with the potential Ras regulator, Shc (Obermeier et al., 1993b). The Shc binding site has been mapped *in vitro* to Tyr490 in the juxtamembrane domain and is of similar affinity to the Shc binding site on the EGFR. The modular adaptor molecule, Grb2, does not appear to bind to activated Trk (Suen et al., 1993). These observed differences between the EGF and NGF receptors will be discussed later. Few details of the signaling specificities of the other Trk family members have yet to be reported. Like Trk, TrkC has been shown to interact with and phosphorylate PLCγ and PI 3-kinase (Lamballe et al., 1993). The high level of homology observed between the cytoplasmic domains of these receptors suggests that they will interact with a similar set of target molecules.

H. The Eph Receptor Family

This is currently the largest known family of PTKRs with at least eleven unique members in vertebrate species including Eph (Hirai et al., 1987), Elk (Letwin and Pawson, 1988; Lhotak et al., 1991), Eck (Lindberg and Hunter, 1990), Eek (Chan and Watt, 1991), Erk (Chan and Watt, 1991), Cek5 (Pasquale, 1991), HEK/Cek4/MEK4 (Sajjadi and Pasquale, 1993; Wicks et al., 1992), mSek(Tyro 1) (Lai and Lemke, 1991; Gilardi-Hebenstreit et al.,1992), Ehk-1 and Ehk-2 (Maisonpierre et al., 1993b), and HEK2 (Böhme et al., 1993). The members of this family show a number of different tissue localizations with Eph expressed mainly in the lung, liver, kidney, and testes (Maru et al., 1988), Elk in the brain (Letwin and Pawson, 1988; Lhotak et al., 1991), and Eck in cells and tissues of epithelial origin (Lindberg and Hunter, 1990). The majority of the other members are expressed primarily in the developing and adult nervous system, but little is known as to their cellular functions. Ligands have yet to be described for any of the family members. All *Eph* family genes encode glycosylated proteins in the range of 130–135 kD. The extracellular portion of these receptors are distinguished by the presence of a conserved cysteine-rich region followed by two fibronectin-type III domains. Such extracellular domains have been reported in cell adhesion molecules and in receptor tyrosine phosphatases (Cunningham et al., 1987; Streuli et al., 1989). As with other proteins containing fibronectin type III domains, it has been speculated that these receptors may play a role in cell adhesion and in cell–cell communication. The juxtamembrane region contains two highly conserved tyrosine residues which are found in all family members except Eph. Their PTK domains are more closely related to the cytoplasmic PTKs than to the other families of transmembrane PTKRs. The prototype family member, Eph, was originally identified using a cDNA probe from the kinase domain of the cytosolic PTK c-*fps* at low stringency (Hirai et al., 1987). The PTK domains lack inserts of the type found typified by the PDGFR family. The PTK domain homology region is followed by C-terminal tails of 90–100 amino acids in length. Little is currently known as to the signal transduction pathways with which members of this family might interact. A single study using a chimeric receptor composed of the EGFR extracellular domain coupled to the Elk intracellular domain has been performed (Lhotak and Pawson., 1993). This EGFR–Elk chimera was shown to undergo autophosphorylation and to tyrosine phosphorylate other intracellular substrates in response to EGF. EGF also induced mitogenesis in fibroblasts expressing this chimeric receptor suggesting that, at least in fibroblast cell types, the intracellular portion of Elk is able to interact with signaling pathways sufficient for mitogenesis.

I. The Axl/Ufo/ARK Receptor Family

The Axl/Ufo/Ark PTKR has been independently isolated by three groups (Janssen et al., 1991; O'Bryan et al., 1991; Rescigno et al., 1991). The human Sky PTKR forms a second family member which is expressed predominantly in the brain (Ohashi et al., 1994). On the basis of sequence similarity and the described expression pattern, the PTKR, Brt, is probably the murine homolog of Sky (Fujimoto and Yamamoto, 1994). Further potential family members (Tyro 3, Tyro 7, and Tyro 12) have been isolated as a PCR-derived PTK domain fragments by Lai and Lemke (1991). The extracellular domain of this family is defined by the presence of two Ig-like repeats amino-terminal of two fibronectin-type III repeats. A single transmembrane domain connects the extracellular domain to the putative cytoplasmic signaling domain. The PTK domain is closely related to that of the Eph/Elk family and to members of the IR family. Short KI regions, 8–9 amino acids in length, are present. Putative PI 3-kinase consensus binding sites are present in the C-terminal tails. The functions or signaling interactions performed by this new class of orphan PTKRs are currently unknown.

J. The Tie/Tek Receptor Family

Tie-1 (JTK14) and Tek (variously termed *hyk*, HPK-6, Tie-2) would appear to define another new PTKR subfamily about which little is currently known (Dumont et al., 1992; Partanen et al., 1992; Dumont et al., 1993; Maisonpierre et al., 1993a;

Sato et al., 1993; Ziegler et al., 1993). Both members have been found to be expressed in vascular endothelial cells and in their presumptive precursors. Based on its expression pattern, a role for Tie has been suggested during embryonic angiogenesis (Korhonen et al., 1994). This family of receptors is defined by the novel structure of their extracellular domains which contain two Ig-like domains which flank a cluster of three cysteine-rich EGF-like repeats. The EGF-like domain is widely found at the cell surface of extracellular proteins involved in protein–protein interactions (Davis, 1990). These motifs are separated from the membrane by three fibronectin-type III repeats (Partanen et al., 1992; Dumont et al.,1993). All three motifs have been shown to function as cell adhesion-associated domains in other proteins. A single membrane spanning region connects to a short juxtamembrane region. The PTK domains contain a short KI sequence (14 amino acids in Tie, 21 residues in Tek) which lacks any putative autophosphorylation sites. The C-terminal tails are also short at 31 amino acids in length. Nothing is currently known as to the ligands or signaling molecules with which these novel receptors interact.

K. The Ret Receptor

Ret was originally isolated as an oncogene encoding a fusion protein which probably resulted from recombination between two unlinked human genes during transfection of NIH 3T3 cells with DNA from a human T cell lymphoma (Takahashi et al., 1985). Cloning of the c-*ret* gene from the THP-1 human monocyte leukemia cell line indicated that it encodes a protein structurally related to the PTKR family (Takahashi et al., 1988). The extracellular domain has a novel sequence similarity with cadherins in a region known to be important for Ca^{2+}-dependent homophilic binding of cadherins (Schneider, 1992). Clear adhesive properties for this domain in c-Ret have yet to be demonstrated (Takahashi et al., 1993).

Comparison of c-*ret* with the transforming *ret* gene revealed that in addition to the N-terminal deletion, the last 51 C-terminal amino acids have been replaced by nine unrelated amino acids. Focus formation assays indicated that *ret* cDNAs containing both the N- and C-terminal deletions scored higher in the transformation assay than *ret* cDNAs containing the N-terminal deletion alone (Takahashi et al., 1988). This suggests a role for both N-and C-terminal changes in the activation of the full transforming potential of *ret*, similar to that noted for other oncogenic forms of PTKR genes (reviewed in Fry et al., 1992).

Several other activated *ret* genes have been isolated by DNA transfection and cloned including RetII (Ishizaka et al., 1988), retTPC (Ishizaka et al., 1990), and PTC (Papillary Thyroid Carcinoma gene; Grieco et al., 1990). The human c-*ret* locus has been mapped to chromosome 10q11–q12 (Donghi et al., 1989; Grieco et al., 1990), a region previously identified as the possible locus responsible for multiple endocrine neoplasia type 2A (MEN2A). Recently, germ-line mutations of the *ret* proto-oncogene have been identified in 20 of 23 cases of MEN2A (Mulligan et al., 1993). Mutations in Ret have also been linked to familial medullary thyroid carcinoma (FMTC), multiple endocrine neoplasia type 2B (MEN 2B), and Hirschsprung's disease (Edery et al., 1994; Hofstra et al., 1994; Romeo et al., 1994).

Little is known about the normal function of the c-Ret protein. A recent study has found that while mice lacking c-Ret develop to term, they exhibit severe defects of the kidney and enteric nervous system and die shortly after birth (Schuchardt et al., 1994). The identification of several oncogenic forms of Ret suggests that it may be involved in the regulation of cell proliferation. Expression studies in human tumor cell lines have shown *ret* to be restricted to neuroblastoma cell lines (11/11 neuroblastoma cell lines tested), suggesting a role for *ret* in cellular functions specific to this cell type (Ikeda et al., 1990; Tahira et al., 1991). Expression of c-*ret* in normal tissue appears to be limited to cells of neuroendocrine origin (Ikeda et al., 1990; Tahira et al., 1991). In cell lines derived from such cell types, *ret* expression is observed to be elevated upon simulation along the differentiation pathway, suggesting a role in pathways involved in differentiative signals (Tahira et al., 1991). In the absence of a physiological ligand for c-Ret, recent attempts to address its function have utilized chimeras possessing the extracellular ligand-binding domain of the EGFR coupled to the cytoplasmic sequences of c-Ret. This EGFR–Ret chimera has been demonstrated to be capable of transducing EGF-dependent mitogenic and transforming signals (Santoro et al., 1994). The EGFR–Ret chimera interacted with a distinct spectrum of signaling proteins from those observed for the wild-type EGFR (Santoro et al., 1994).

L. The Ros/Sevenless Receptor

The acutely transforming chicken retrovirus, UR2, encodes a fusion protein, p68$^{gag-ros}$, which has intrinsic PTK activity (Feldman et al., 1982), although UR2-transformed cells show no significant increase in phosphotyrosine content (Cooper and Hunter, 1983). Nucleotide sequence analysis revealed that the v-*ros* oncogene of UR2 encodes a PTK domain (Neckameyer and Wang, 1985). This was later shown to be most related to the amino acid sequence of the PTK domain of the IR (Ebina et al., 1985; Ullrich et al., 1985) and to the *Drosophila* PTKR, sevenless (Hafen et al., 1987; Simon et al., 1989).

Both the chicken and human c-*ros* genes have been isolated and shown to encode a transmembrane PTK with an extremely large putative ligand-binding domain similar to other cell-surface receptors for growth or differentiation factors (Birchmeier et al., 1986; Matsushime et al., 1986; Neckameyer et al., 1986). The strongest homology is seen with the *Drosophila sevenless* receptor throughout the length of c-Ros (Hafen et al., 1987; Simon et al., 1989; Chen et al., 1991). Both receptors are extremely large, in the range of 280–290 kD. Most of this is extracellular, with the intracellular domain being of average size. Eight fibronectin type-III-like repeats are found in the ectodomain of both proteins in identical relative positions. Currently, there is no known ligand for c-Ros. The ligand for Sevenless however has been identified as the product of the *Boss* gene (Hart et al., 1993). This is a unique ligand for PTKRs described to date, possessing a large extracellular domain, seven transmembrane domains, and a C-terminal cytoplasmic tail. Signaling via Sevenless is, therefore, triggered by direct cell–cell interaction. It will be interesting to find out whether this aspect of Sevenless signaling has been retained by the c-Ros signaling pathway. Although structurally highly homologous, the physiological roles of these two receptors appear to have diverged extensively. The Sevenless receptor is involved in morphogenesis and differentiation of the R7 photoreceptor cell in the *Drosophila* eye, while the best estimate of c-Ros functioning suggests a role in morphogenesis and differentiation of epithelia during development (reviewed in Birchmeier et al., 1993).

Very little is known about signaling via c-Ros. A chimera consisting of the ligand-binding domain of the IR fused to the transmembrane and catalytic domains of v-*ros* has been constructed. Although this IR–Ros chimera bound insulin and activated the Ros PTK domain, no glucose metabolism or mitogenesis was observed, suggesting that Ros and the IR may interact with distinct intracellular pathways (Ellis et al., 1987). More is known about the signaling pathways downstream of Sevenless. Ras1 activation has been shown to mimic Sevenless signaling (Fortini et al., 1992) and Sevenless has recently been shown to activate the Ras pathway via a direct interaction with *Drosophila* homolog of the SH2 domain-containing protein Grb2, Drk (Olivier et al., 1993; Simon et al., 1993).

M. The DDR/Cak/TrkE Receptor Family

A representative of this receptor family was originally isolated as a PCR-derived PTK domain fragment (Tyro 10) by Lai and Lemke (1991). Subsequently, at least two distinct family members have been cloned by a number of groups. These receptors have variously been named the Discoidin Domain Receptor (DDR) (Johnson et al., 1993), Cell adhesion kinase (Cak; Perez et al., 1994), Tyrosine kinase-related to Trk (TKT; Karn et al., 1993), Tyro10 (Lai and Lemke, 1994), TrkE (DiMarco et al., 1993), and NEP (Zerlin et al., 1993). DDR/Cak and TrkE, although exhibiting minor sequence differences, would all appear to be independent isolates of the same PTKR. Protein sequence comparison suggests that NEP may be the murine homolog of this receptor. Human TKT, and its murine homolog Tyro 10, is a second member of this receptor family. Further support comes from chromosomal mapping of DDR/Cak to human chromosome 6, while TKT has been located on chromosome 1 (Karn et al., 1993). DDR/Cak/TrkE/NEP is expressed in most adult tissues with highest levels in the brain and lung. Expression was also observed in some epidermal and breast carcinoma cell lines (Johnson et al., 1993). TKT/Tyro 10 expression is highest in the heart and lung with lower levels in brain, placenta, liver, skeletal muscle, pancreas, and kidneys (Karn et al., 1993; Lai and Lemke, 1994).

This family of putative receptors is distinguished by a novel ectodomain lacking motifs found in the extracellular domains of other PTKRs (DiMarco et al., 1993; Johnson et al., 1993; Karn et al., 1993; Lai and Lemke, 1994; Perez et al., 1994). Within the 400 amino acids of this ectodomain lies an N-terminal region which is related to a putative phospholipid-binding domain found in proteins such as the *Xenopus* neuronal A5 antigen, a milk fat globule membrane protein, the *Dictyostelium* cell adhesion molecule, discoidin I, and in coagulation factors V and VIII (Perez et al., 1994). On the basis of this, it has been suggested that members of this receptor family may play a role in cell–cell interaction and recognition events and hence the name Cak—Cell adhesion kinase. A putative protease cleavage site for furin is also present between this domain and the transmembrane domain. The juxtamembrane region, at 176 amino acids in length, is the largest described to date and is rich in proline and glycine residues. This proline/glycine-rich sequence is quite distinct from those found in other PTKRs and may, therefore, play a novel role in the interactions of this particular PTKR with specific components of the signal transduction pathways. In light of the recent description of SH3 domains binding to proline-rich regions, this is an attractive possibility which deserves investigation. Examination of the PTK domains of these receptors are closely related to Trk and to the IR. However, despite the naming of this gene *Trk*E by DiMarco et al. (1993), this receptor family shows no homology to the Trk outside of the PTK domain and are, therefore, not a true member of the neurotrophin receptor family. The cytoplasmic tails are very short and lack potential autophosphorylation sites ruling out a role for this domain in the interaction with SH2 domain-containing proteins. Due to the novel nature of this family of PTKRs, a description of their signaling properties will be awaited with great interest.

N. Cytoplasmic Protein–Tyrosine Kinases as Receptor Subunits

Although, there is not space in this review to do justice to this area, I feel that it should be briefly mentioned and the reader pointed in the direction of several recent papers and reviews. As well as acting as downstream signaling targets for several transmembrane PTKRs such as described above for *src*-family members and the PDGFβR and CSF-1R, members of the cytoplasmic PTK families have recently been shown to function as intracellular subunits of receptor complexes in association with membrane-spanning subunits which lack any intracellular kinase or other catalytic activity themselves. *Src* family kinases have been found in association with CD4/CD8 (Lck) (Rudd et al., 1993), the IL-2 receptor (Rudd et al., 1993), IgE receptor (Eiseman and Bolen, 1992), and with the T cell receptor (Weiss, 1993). Another family of cytoplasmic PTKs, the JAK kinases, have also been recently described as forming intracellular subunits of the erythropoietin (Witthuhn et al., 1993), growth hormone receptors (Argetsinger et al., 1993), and interferon α/β signaling system (Valazquez et al., 1992). It is possible that this arrangement of separate extracellular ligand binding and transmembrane subunit coupled by noncovalent association with an intracellular PTK signaling domain may reflect a more primitive, ancestral form of PTKR. The extent to which this type of interaction occurs is not yet apparent, but is probably widespread and blurs the previous distinction of PTKs into receptor and nonreceptor classes.

V. CONCLUDING REMARKS

Significant advances have been made in our understanding of the role that this large family of PTKRs plays in the regulation of intracellular signal transduction pathways. The concept of PTKR-mediated signaling complexes involving SH2 and SH3 domain-containing proteins has now become well accepted as a major mechanism underlying receptor function. However, as can be seen from the preceding sections, the similarities between the pathways stimulated by different receptors currently would appear to outweigh the differences. Many of the PTKRs studied activate the Ras pathway and identical SH2 domain-containing signaling proteins, while often eliciting distinct cellular responses. The most important question which now needs to be addressed is where is the signaling specificity encoded? The first possible control of specificity lies in ligand–receptor interactions. At this level, signaling is regulated by the presentation of a particular receptor on a specific cell type at an appropriate time and on the availability of the cognate ligand to this receptor. This would be sufficient to give rise to specific cellular responses to the activation of a given PTKR if the different receptors within a particular cell type interacted with distinct intracellular signaling pathways. To date, where examined, this does not seem to be the case. An often quoted example of this dilemma is the situation observed with EGF and NGF stimulation of PC12 cells. Both of these factors can stimulate an almost identical set of early interactions including substrate phosphorylation, recruitment of SH2 domain-containing molecules, and stimulation of Ras and second messenger activation, but they have very distinct effects on these cells. EGF is mitogenic for PC12 cells, while NGF stimulates neurite outgrowth and differentiation (Chao, 1992). However, closer investigation reveals underlying differences in the signaling despite the similarities. While both EGF and NGF activate PLCγ, the extent of the interaction varies for the two receptors with a 100-fold higher affinity of PLCγ for the EGFR over the NGFR being observed (Obermeier et al., 1993a). Similarly, both factors activate Ras, but only the EGFR interacts directly with Grb2, a putative mediator in Ras activation (Qui and Green, 1991; Suen et al., 1993). The pattern of tyrosine phosphorylation induced by the two peptide factors are related, but distinct for each factor. The nature of the different substrates, however, remains to be determined. Related signaling differences have been noted between the PDGFαR and the PDGFβR. In this case, the two receptors show very similar binding characteristics for PI 3-kinase, PLCγ, and Syp, and both can activate $p21^{ras}$, but there is a five-fold difference in their abilities to associate with RasGAP (Bazenet and Kazlauskas, 1993; Heidaran et al., 1993). Therefore, qualitative differences of the extent to which different signaling pathways are activated and the duration of the signals may be interpreted differently by the cell and result in different responses. Dramatically different cell type-specific responses to a single ligand receptor combination have been observed, arguing in favor of the receptors in question interacting with distinct signaling molecules/pathways in different cell types in addition to the ubiquitous substrates. Differences clearly do exist and perhaps the methodologies utilized to isolated signaling molecules have thus far only revealed the more common steps in the signaling pathways, with more cell type-specific factors remaining to be discovered. The library screening protocol described to clone Grb2 may prove useful in identifying the putative cell-type specific signaling proteins for receptors (Lowenstein et al., 1992; Margolis et al., 1992). What is obvious, however, is that despite the considerable advances that have been made, there is still much left to be unravelled before we can claim to have any clear understanding of the true complexity of the interconnecting network of pathways which regulate intracellular signaling.

ACKNOWLEDGMENTS

I would like to thank Ritu Dhand, Liza Ho, and Rushika Sumathipala for critical reading of the manuscript and for providing helpful insights, and Dr. George Panayotou for assistance in preparation of Figures 3 and 4. The author is supported by the Leopold Muller Trust.

REFERENCES

Alai, M., Mui, A. L., Cutler, R. L., Bustelo, X. R., Barbacid, M., & Krystal, G. (1992). Steel factor stimulates the tyrosine phosphorylation of the proto-oncogene product, p95^{vav}, in human hemopoietic cells. J. Biol. Chem. 267, 18021–18025.

Anderson, D., Koch, C. A., Grey, L., Ellis, C., Moran, M. F., & Pawson, T. (1990). Binding of SH2 domains of phospholipase Cγ1, GAP and Src to activated growth factor receptors. Science 250, 979–982.

Argetsinger, L. S., Campbell, G. S., Yang, X., Witthuln, B. A., Silvennoinen, O., Ihle, J. N., & Carter-Su, C. (1993). Identification of JAK2 as a growth hormone receptor-associated tyrosine kinase. Cell 74, 237–244.

Aroian, R. V., Koga, M., Mendel, J. E., Ohshima, Y., & Sternberg, P. W. (1990). The *let-23* gene necessary for *Caenorhabditis elegans* vulval induction encodes a tyrosine kinase of the EGF receptor subfamily. Nature 348, 693–698.

Aroian, R. V., Lesa, G. M., & Sternberg, P. W. (1994). Mutations in the *Caenorhabditis elegans let-23* EGFR-like gene define elements important for cell type specificity and function. EMBO J. 13, 360–366.

Avivi, A., Zimmer, Y., Yayon, A., Yarden, Y., & Givol, D. (1991). Flg-2, a new member of the family of fibroblast growth factor receptors. Oncogene 6, 1089–1092.

Backer, J. M., Myers, M. G. Jr., Shoelson, S. E., Chin, D. J., Sun, X. J., Margolis, B., Skolnik, E. Y., Schlessinger, J., & White, M. F. (1992). Phosphatidylinositol 3′-kinase is activated by association with IRS-1 during insulin stimulation. EMBO J. 11, 3469–3479.

Barbacid, M. (1993). Nerve growth factor: A tale of two receptors. Oncogene 8, 2033–2042.

Bardelli, A., Maina, F., Gout, I., Fry, M. J., Waterfield, M. D., Comoglio, P. M., & Ponzetto, C. (1992). Autophosphorylation promotes complex formation of recombinant hepatocyte growth factor receptor with cytoplasmic effectors containing SH2 domains. Oncogene 7, 1973–1978.

Bargmann, C. I., & Weinberg, R. A. (1988). Increased tyrosine kinase activity associated with the activated *neu* oncogene encoded protein. Proc. Natl. Acad. Sci. USA 85, 5394–5398.

Bargmann, C. I., Hung, M. C., & Weinberg, R. A. (1986a). The *neu* oncogene encodes an epidermal growth factor receptor-related protein. Nature 319, 226–230.

Bargmann, C. I., Hung, M.-C., & Weinberg, R. A. (1986b). Multiple independent activations of the *neu* oncogene by point mutation altering the transmembrane domain of p185. Cell 45, 649–657.

Bazenet, C. E., & Kazlauskas, A. (1993). The PDGF receptor α subunit activates p21ras and triggers DNA synthesis without interacting with rasGAP. Oncogene 9, 517–525.

Bellot, F., Crumley, G., Kaplow, J. M., Schlessinger, J., Jaye, M., & Dionne, C. A. (1991). Ligand induced transphosphorylation between different FGF receptors. EMBO J. 10, 2849–2854.

Besmer, P., Lader, E., George, P. C., Bergold, P. J., Qui, F.-J., Zuckerman, E. E., & Hardy, W. D. (1986a). A new acutely transforming feline retrovirus with *fms* homology specifies a C-terminally truncated version of the c-*fms* protein that is different from SM-feline sarcoma virus. J. Virol. 60, 194–203.

Besmer, P., Murphy, J. E., George, P. C., Qui, F., Bergold, P. J., Lederman, L., Snyder, Jr., H. W., Brodeur, D., Zuckerman, E. E., & Hardy, W. D. (1986b). A new acute feline retrovirus and relationship of its oncogene v-*kit* with the protein kinase gene family. Nature 320, 415–421.

Birchmeier, C., Birnbaum, D., Waitches, G., Fasano, O., & Wigler, M. (1986). Characterization of an activated human *ros* gene. Mol. Cell. Biol. 6, 3109–3116.

Birchmeier, C., Sonnenberg, E., Weidner, K. M., & Walter, B. (1993). Tyrosine kinase receptors in the control of epithelial growth and morphogenesis during development. Bioessays 15, 185–190.

Bjorge, J. D., Chan, T.-O., Antczak, M., Kung, H.-J., & Fujita, D. J. (1990). Activated type I phosphatidylinositol kinase is associated with the epidermal growth factor (EGF) receptor following EGF stimulation. Proc. Natl. Acad. Sci. USA 87, 3816–3820.

Böhme, B., Holtrich, U., Wolf, G., Luzius, H., Grzeschik, K.-H., Strebhardt, K., & Rübsamen-Waigmann, H. (1993). PCR mediated detection of a new human receptor-tyrosine kinase, HEK2. Oncogene 8, 2857–2862.

Bottaro, D. P., Rubin, J. S., Faletto, D. L., Chan, A. M.-L., Kmiecik, T. E., Vande Woude, G. F., & Aaronson, S. A. (1991). Identification of the hepatocyte growth factor receptor as the c-*met* proto-oncogene product. Science 251, 802–804.

Buday, L., & Downward, J. (1993). Epidermal growth factor regulates p21ras through the formation of a complex of receptor, Grb2 adaptor protein, and Sos nucleotide exchange factor. Cell 73, 611–620.

Canals, F. (1992). Signal transduction by epidermal growth factor receptor: Coincidence of activation and dimerization. Biochemistry 31, 4493–4501.

Cantley, L. C., Auger, K. R., Carpenter, C., Duckworth, B., Graziani, A., Kappeller, R., & Soltoff, S. (1991). Oncogenes and signal transduction. Cell 64, 281–302.

Cao, H., Bangalore, L., Bormann, B. J., & Stern, D. F. (1992). A subdomain in the transmembrane domain is necessary for $p185^{neu*}$ activation. EMBO J. 11, 923–932.

Carpenter, G., & Cohen, S. (1990). Epidermal growth factor. J. Biol. Chem. 265, 7709–7712.

Carpenter, C. L., Auger, K. R., Chanudhuri, M., Yoakim, M., Schaffhausen, B., Shoelson, S., & Cantley, L. C. (1993). Phosphoinositide 3-kinase is activated by phosphopeptides that bind to the SH2 domains of the 85 kD subunit. J. Biol. Chem. 268, 9478–9483.

Chabot, B., Stephenson, D. A., Chapman, V. M., Besmer, P., & Bernstein, A. (1988). The proto-oncogene c-kit encoding a transmembrane tyrosine kinase receptor maps to the mouse W locus. Nature 335, 684–687.

Champion-Arnaud, P., Ronsin, C., Gilbert, E., Gesnel, M. C., Houssaint, E., & Breathnach, R. (1991). Multiple mRNAs code for proteins related to the BEK fibroblast growth factor receptor. Oncogene 6, 979–987.

Chan, J., & Watt, V. M. (1991). *eek* and *erk*, new members of the *eph* subclass of receptor protein-tyrosine kinases. Oncogene 6, 1057–1061.

Chao, M. V. (1992). Growth factor signaling: Where is the specificity? Cell 68, 995–997.

Chazin, V. R., Kaleko, M., Miller, A. D., & Slamon, D. J. (1992). Transformation mediated by the human *HER-2* gene independent of the epidermal growth factor receptor. Oncogene 7, 1859–1866.

Cheatham, B., Shoelson, S. E., Yamada, K., Goncalves, E., & Kahn, C.R. (1993). Substitution of the erbB-2 oncoprotein transmembrane domain activates the insulin receptor and modulates the action of insulin and insulin-receptor substrate 1. Proc. Natl. Acad. Sci. USA 90, 7336–7340.

Chen, W. S., Lazar, C. S., Poenie, M., Tsien, R. Y., Gill, G. N., & Rosenfeld, M. G. (1987). Requirement for intrinsic protein tyrosine kinase in the intermediate and late actions of the EGF receptor. Nature 328, 820–823.

Chen, W. S., Lazar, C. S., Lund, K. A., Welsh, J. B., Chang, C.-P., Walton, G. M., Der, C. J., Wiley, H. S., Gill, G. N., & Rosenfeld, M. G. (1989). Functional independence of the epidermal growth factor receptor from a domain required for ligand-induced internalization and calcium regulation. Cell 59, 33–43.

Chen, J., Heller, D., Poon, B., Kang, L., & Wang, L.-H. (1991). The proto-oncogene c-ros codes for a transmembrane tyrosine protein kinase sharing sequence and structural homology with *sevenless* protein of *Drosophila melanogaster*. Oncogene 6, 257–264.

Claesson-Welsh, L., Eriksson, A., Moren, A., Severinsson, L., Ek, B. Östman, A., Betsholtz, C., & Heldin, C.-H. (1988). cDNA cloning and expression of a human platelet-derived growth factor (PDGF) receptor specific for the B-chain-containing PDGF molecule. Mol. Cell. Biol. 8, 3476–3486.

Claesson-Welsh, L., Eriksson, A., Westermark, B., & Heldin, C.-H. (1989). cDNA cloning and expression of the human A-type platelet-derived growth factor (PDGF) receptor establishes structural similarity to the B-type PDGF receptor. Proc. Natl. Acad. Sci. USA 86, 4917–4921.

Claesson-Welsh, L. (1994). Signal transduction by the PDGF receptors. Progress in Growth Factor Research 5, 37–54.

Clark, S. G., Stern, M. J., & Horvitz, H. R. (1992). *C. elegans* signalling gene *sem5* encodes a protein with SH2 and SH3 domains. Nature 356, 340–344.

Connelly, P. A., & Stern, D. F. (1990). The epidermal growth factor receptor and the product of the *neu* protooncogene are members of a receptor tyrosine phosphorylation cascade. Proc. Natl. Acad. Sci. USA 87, 6054–6057.

Cooper, C. S. (1992). The *met* oncogene: From detection by transfection to transmembrane receptor for hepatocyte growth factor. Oncogene 7, 3–7.

Cooper, J. A., & Hunter, T. (1983). Regulation of cell growth and transformation by tyrosine specific protein kinases: the search for important cellular substrate proteins. Current Topics in Microbiology and Immunology 107, 125–161.

Cooper, C. S., Park, M., Blair, D. G., Tainsky, M. A., Huebner, K., Croce, C. M., & Van de Woude, G. F. (1984). Molecular cloning of a new transforming gene from chemically transformed human cell line. Nature 311, 29–33.

Coulier, F., Kumar, R., Ernst, M., Klein, R., Martin-Zanca, D., & Barbacid, M. (1990). Human trk oncogenes activated by point mutations, in-frame deletion and duplication of the tyrosine kinase domain. Mol. Cell. Biol. 10, 4202–4210.

Courtneidge, S. A., Dhand, R., Pilat, D., Twamley, G. M., Waterfield, M. D., & Roussel, M. F. (1993). Activation of Src family kinases by colony stimulating factor-1, and their association with its receptor. EMBO J. 12, 943–950.

Crowe, A. J., McGlade, J., Pawson, T., & Hayman, M. J. (1994). Phosphorylation of the SHC proteins on tyrosine correlates with transformation of fibroblasts and erythroblasts by the v-sea tyrosine kinase. Oncogene 9, 537–544.

Culouscou, J.-M., Plowman, G. D., Carlton, G. W., Green, J. M., & Shoyab, M. (1993). Characterization of a breast cancer cell differentiation factor that specifically activates the HER4/p180^{erbB4} receptor. J. Biol. Chem. 268, 18407–18410.

Cunningham, B. A., Hemperly, J. J., Murray, B. A., Prediger, E. A., Brackenbury, R., & Edelmann, G. M. (1987). Neural cell adhesion molecule: Structure, immunoglobulin-like domains, cell surface modulation, and alternative RNA splicing. Science 236, 799–806.

Davis, C. G. (1990). The many faces of epidermal growth factor repeats. New Biol. 2, 410–419.

Dean, M., Park, M., Le Beau, M. M., Robins, T. S., Diaz, M. O., Rowley, J. D., Blair, D. G., & Van de Woude, G. F. (1985). The human *met* oncogene is related to tyrosine kinase oncogenes. Nature 318, 385–388.

Decker, S. J. (1993). Transmembrane signaling by epidermal growth factor receptors lacking autophosphorylation sites. J. Biol. Chem. 268, 9176–9179.

De Vries, C., Escobedo, J. A., Ueno, H., Houck, K., Ferrara, N., & Williams, L. T. (1992). The *fms*-like tyrosine kinase, a receptor for vascular endothelial growth factor. Science 255, 989–991.

Di Fiore, P. P., Helin, K., Kraus, M. H., Pierce, J. H., Artrip, J., Segatto, O., & Pottaro, D. P. (1992). A single amino acid substitution is sufficient to modify the mitogenic properties of the epidermal growth factor receptor to resemble that of gp185$^{erbB\text{-}2}$. EMBO J. 11, 3927–3933.

Di Marco, E., Cutuli, N., Guerra, L., Cancedda, R., & De Luca, M. (1993). Molecular cloning of *trk*E, a novel *trk*-related putative tyrosine kinase receptor isolated from normal human keratinocytes and widely expressed by normal human tissues. J. Biol. Chem. 268, 24290–24295.

Dionne, C., Crumley, G., Bellot, F., Kaplow, J. M., Searfoss, G., Ruta, M., Burgess, W. H., Jaye, M., & Schlessinger, J. (1990). Cloning and expression of distinct high affinity receptors cross reacting with acidic and basic fibroblast growth factors. EMBO J. 9, 2685–2692.

Donghi, R., Sozzi, G., Pierotti, M. A., Biunno, I., Miozzo, M., Fusco, A., Grieco, M., Santoro, M., Vecchio, G., Spurr, N. K., & Della Porta, G. (1989). The oncogene associated with human papillary thyroid carcinoma (PTC) is assigned to chromosome 10q11–12 in the same region as multiple endocrine neoplasia type 2A (MEN2A). Oncogene 4, 521–523.

Donner, L., Fedele, L. A., Garon, C. F., Anderson, S. J., & Sherr, C. J. (1982). McDonough feline sarcoma virus: Characterization of the molecularly cloned provirus and its feline oncogene (v-*fms*). J. Virol. 41, 489–500.

Dosil, M., Wang, S., & Lemischka, I. R. (1993). Mitogenic signalling and substrate specificity of the Flk2/Flt3 receptor tyrosine kinase in fibroblasts and interleukin 3-dependent hemopoietic cells. Mol. Cell. Biol. 13, 6572–6585.

Downing, J. R., Margolis, B. L., Zilberstein, A., Ashman, R. A., Ullrich, A., Sherr, C. J., & Schlessinger, J. (1989). Phospholipase Cγ, a substrate for PDGF receptor kinase, is not phosphorylated on tyrosine during a mitogenic response to CSF-1. EMBO J. 8, 3345–3350.

Downward, J., Yarden, Y., Mayes, E., Scrace, G., Totty, N., Stockwell, P., Ullrich, A., Schlessinger, J., & Waterfield, M. D. (1984). Close similarity of epidermal growth factor receptor and v-*erbB* oncogene protein sequences. Nature 307, 521–527.

Dumont, D. J., Yamaguchi, T. P., Conlon, R. A., Rossant, J., & Breitman, M. L. (1992). tek, a novel tyrosine kinase gene located on mouse chromosome 4, is expressed in endothelial cells and their presumptive precursors. Oncogene 7, 1471–1480.

Dumont, D. J., Gradwohl, G. J., Fong, G.-H., Auerbach, R., & Breitman, M. L. (1993). The endothelial-specific receptor tyrosine kinase, *tek*, is a member of a new subfamily of receptors. Oncogene 8, 1293–1301.

Ebina, Y., Ellis, L., Jarnagin, K., Edery, M., Graf, L., Clauser, E., Ou, J., Masiarz, F., Kan, Y. W., Goldfine, I. D., Roth, R. A., & Rutter, W. J. (1985). The human insulin receptor cDNA: the structural basis for hormone activated transmembrane signalling. Cell 40, 747–758.

Edery, P., Lyonnet, S., Mulligan, L. M., Pelet, A., Dow, E., Abel, L., Holder, S., Nihoul-Fékété, C., Ponder, B. A. J., & Munnich, A. (1994). Mutations of the *RET* proto-oncogene in Hirschsprung's disease. Nature 367, 378–380.

Egan, S. E., Giddings, B. W., Brooks, M. W., Buday, L., Sizeland, A. M., & Weinberg, R. A. (1993). Association of Sos Ras exchange protein with Grb2 is implicated in tyrosine kinase signal transduction and transformation. Nature 363, 45–51.

Eiseman, E., & Bolen, J. B. (1992). Engagement of the high-affinity IgE receptor activates src protein-related tyrosine kinase. Nature 355, 78–80.

Ellis, L., Morgan, D. O., Jong, S.-M., Wang, L.-H., Roth, R. A., & Rutter, W. J. (1987). Heterologous transmembrane signalling by a human insulin receptor v-*ros* hybrid in Chinese hamster ovary cells. Proc. Natl. Acad. Sci. USA 84, 5101–5105.

Ellis, C., Moran, M., McCormick, F., & Pawson, T. (1990). Phosphorylation of GAP and GAP-associated proteins by transforming and mitogenic tyrosine kinases. Nature 343, 377–381.

Eriksson, A., Siegbahn, A., Westermark, B., Heldin, C.-H., & Claesson-Welsh, L. (1992). PDGF α- and β-receptors activate unique and common signal transduction pathways. EMBO J. 11, 543–550.

Escobedo, J. A., Barr, P. J., & Williams, L. T. (1988). Role of the tyrosine kinase and membrane-spanning domains in signal transduction by the platelet-derived growth factor receptor. Mol. Cell. Biol. 8, 5126–5131.

Escobedo, J. A., Kaplan, D. R., Kavanaugh, W. M., Turck, C. W., & Williams, L. T. (1991). A phosphatidylinositol 3-kinase binds to platelet-derived growth factor receptors through a specific receptor sequence containing phosphotyrosine. Mol. Cell. Biol. 11, 1125–1132.

Falls, D. L., Rosen, K. M., Corfas, G., Lane, W. S., & Fischbach, G. D. (1993). ARIA, a protein that stimulates acetylcholine receptor synthesis, is a member of the Neu ligand family. Cell 72, 801–815.

Fantl, W. J., Escobedo, J. A., Martin, G. A., Turck, C. W., del Rosario, M., McCormick, F., & Williams, L. T. (1992). Distinct phosphotyrosines on growth factor receptor bind to specific molecules that mediate different signalling pathways. Cell 69, 413–423.

Fantl, W. J., Johnson, D. E., & Williams, L. T. (1993). Signalling by receptor tyrosine kinases. Annu. Rev. Biochem. 62, 453–481.

Fazioli, F., Kim, U.-H., Rhee, S. G., Molley, C. J., Segatto, O., & Di Fiore, P. P. (1991). The *erb*B-2 mitogenic signaling pathway: Tyrosine phosphorylation of phospholipase C-γ and GTPase-activating protein does not correlate with *erb*B-2 mitogenic potency. Mol. Cell. Biol. 11, 2040–2048.

Fedi, P., Pierce, J. H., Di Fiore, P. P., & Kraus, M. H. (1994). Efficient coupling with phosphatidylinositol 3-kinase, but not phospholipase Cγ or GTPase-activating protein, distinguishes ErbB-3 signaling from that of other ErbB/EGFR family members. Mol. Cell. Biol. 14, 492–500.

Feldman, R. A., Wang, L.-H., Hanafusa, H., & Balbuzzi, P. C. (1982). Avian sarcoma virus UR2 encodes a transforming protein which is associated with a unique protein kinase activity. J. Virol. 42, 228–236.

Feng, G.-S., Hui, C.-C., & Pawson, T. (1993). SH2-containing phosphotyrosine phosphatase as a target of protein-tyrosine kinases. Science 259, 1607–1611.

Finnerty, H., Kelleher, K., Morris, G. E., Bean, K., Merberg, D. M., Kriz, R., Morris, J. C., Sookdeo, H., Turner, K. J., & Wood, C. R. (1993). Molecular cloning of murine FLT and FLT4. Oncogene 8, 2293–2298.

Fortini, M. E., Simon, M. A., & Rubin, G. M. (1992). Signalling by the *sevenless* protein tyrosine kinase is mimicked by Ras1 activation. Nature 355, 559–561.

Fry, M. J., Panayotou, G., & Waterfield, M. D. (1992). Oncogenes and their relation to growth regulating receptors. In: Receptor Subunits and Complexes (Barnard, E. A., Burgen, A. S. V. & Roberts, G. C. K., eds.) pp. 419–446. Cambridge University Press, Cambridge, UK.

Fry, M. J., Panayotou, G., Booker, G. W., & Waterfield, M. D. (1993). New insights into protein-tyrosine kinase receptor signaling complexes. Protein Science 2, 1785–1797.

Fu, X.-Y., & Zhang, J.-J. (1993). Transcription factor p91 interacts with the epidermal growth factor receptor and mediates activation of the c-*fos* gene promoter. Cell 74, 1135–1145.

Fujimoto, J., & Yamamoto, T. (1994). *brt*, a mouse gene encoding a novel receptor-type protein-tyrosine kinase, is preferentially expressed in the brain. Oncogene 9, 693–698.

Gale, N. W., Kaplan, S., Lowenstein, E. J., Schlessinger, J., & Bar-Sagi, D. (1993). Grb2 mediates EGF-dependent activation of guanine nucleotide exchange on Ras. Nature 363, 88–92.

Galland, F., Karamysheva, A., Pebusque, M.-J., Borg, J.-P., Rottapel, R., Dubreuil, P., Rosnet, O., & Birnbaum, D. (1993). The *FLT4* gene encodes a transmembrane tyrosine kinase related to the vascular endothelial growth factor receptor. Oncogene 8, 1233–1240.

Galli, S. J., Zsebo, K. M., & Geissler, E. M. (1994) The Kit ligand, stem cell factor. In: Advances in Immunology (Dixon, F. J., ed.) Vol 55, pp. 1–96. Academic Press, Inc., London.

Gandino, L., Di Renzo, M. F., Giordano, S., Bussolino, F., & Comoglio, P. M. (1990). Protein kinase-c activation inhibits tyrosine phosphorylation of the c-*met* protein. Oncogene 5, 721–725.

Geissler, E. N., Ryan, M. A., & Housman, D. E. (1988). The dominant-white spotting (W) locus of the mouse encodes the c-kit proto-oncogene. Cell 55, 185–192.

Gilardi-Hebenstreit, P., Nieto, M. A., Frain, M., Mattei, M.-G., Chestier, A., Wilkinsojn, D. G., & Charney, P. (1992). An Eph-related receptor protein-tyrosine kinase gene segmentally expressed in the developing mouse hind brain. Oncogene 7, 2499–2506.

Giordano, S., Di Renzo, M. F., Narsimhan, R. P., Cooper, C. S., Rosa, C., & Comoglio, P. M. (1989a). Biosynthesis of the protein encoded by the c-*met* proto-oncogene. Oncogene 4, 1383–1388.

Giordano, S., Ponzetto, C., Di Renzo, M. F., Cooper, C. S., & Comoglio, P. M. (1989b). Tyrosine kinase receptor indistinguishable from the c-*met* protein. Nature 339, 155–156.

Glenney, J. R. Jr. (1992). Tyrosine-phosphorylated proteins: mediators of signal transduction from the tyrosine kinases. Biochim. Biophys. Acta. 1134, 113–127.

Gonzatti-Haces, M., Seth, A., Park, M., Copeland, T., Oroszlan, S., & Van de Woude, G. F. (1988). Characterization of the TPR-MET oncogene p65 and the Met protooncogene p140 protein-tyrosine kinase. Proc. Natl. Acad. Sci. USA 85, 21–25.

Gotoh, N., Tojo, A., Muroya, K., Hashimoto, Y., Hattori, S., Nakamura, S., Takenawa, T., Yazaki, Y., & Shibuya, M. (1994). Epidermal growth factor receptor mutant lacking the autophosphorylation sites induces phosphorylation of Shc protein and Shc-Grb-2/ASH association and retains mitogenic activity. Proc. Natl. Acad. Sci. USA 91, 167–171.

Graziani, A., Gramaglia, D., Cantley, L. C., & Comoglio, P. M. (1991). The tyrosine–phosphorylated hepatocyte growth factor/scatter factor receptor associates with phosphatidylinositol 3-kinase. J. Biol. Chem. 266, 22087–22090.

Graziani, A., Gramaglia, D., Dalla Zonca, P., & Comoglio, P. M. (1993). Hepatocyte growth factor/scatter factor stimulates the ras-guanine nucleotide exchanger. J. Biol. Chem. 268, 9165–9168.

Grieco, M., Santoro, M., Berlingieri, M. T., Melillo, R. M., Donghi, R., Bongarzone, I., Pierotti, M. A., Della Porta, G., Fusco, A., & Vecchio, G. (1990). PTC is a novel rearranged form of the ret proto-oncogene and is frequently detected *in vivo* in human thyroid papillary carcinomas. Cell 60, 557–563.

Gronwold, R. G. K., Grant, F. J., Haldeman, B. A., Hart, C. E., O'Hara, P. J., Hagen, F. S., Ross, R., Bowen-Pope, D. F., & Murrey, M. J. (1988). Cloning and expression of a cDNA coding for the human platelet-derived growth factor receptor: evidence for more than one receptor class. Proc. Natl. Acad. Sci. USA 85, 3435–3439.

Hafen, E., Basler, K., Edstroem, J.-E., & Rubin, G. M. (1987). *sevenless*, a cell-specific homeotic gene of *Drosophila*, encodes a putative transmembrane receptor with a tyrosine kinase domain. Science 236, 55–63.

Haley, J. D., Hsuan, J. J., & Waterfield, M. D. (1989). Analysis of mammalian fibroblast transformation by normal and mutated EGF receptors. Oncogene 4, 273–283.

Hampe, A., Gobot, M., Sherr, C. J., & Galibert, F. (1984). The nucleotide sequence of the feline retroviral oncogene v-fms shows unexpected homology with oncogenes encoding tyrosine-specific protein kinases. Proc. Natl. Acad. Sci. USA 81, 85–89.

Hanks, S. K. (1991). Eukaryotic protein kinases. Curr. Op. Struct. Biol. 1, 369–383.

Hanks, S. K., Quinn, A. M., & Hunter, T. (1988). The protein kinase family: Conserved features and deduced phylogeny of the catalytic domains. Science 241, 42–52.

Hart, A. C., Krämer, H., & Zipursky, S. L. (1993). Extracellular domain of the boss transmembrane ligand acts as an antagonist of the sev receptor. Nature 361, 732–736.

Hart, C. E., Forsteom, J. W., Kelly, J. D., Seifert, R. A., Smith, R. A., Ross, R., Murray, M., & Bowen-Pope, D. F. (1988). Two classes of PDGF receptor recognize different isoforms of PDGF. Science 240, 1529–1531.

Hayman, M. J., Kitchener, G., Vogt, P. K., & Beug, H. (1985). The putative transforming protein of S13 avian erythroblastosis virus is a transmembrane glycoprotein with an associated protein kinase activity. Proc. Natl. Acad. Sci. USA 82, 8237–8241

Heidaran, M. A., Beeler, J. F., Yu, J.-C., Ishibashi, T., LaRochelle, W. J., Pierce, J., & Aaronson, S. A. (1993). Differences in substrate specificities of α and β platelet-derived growth factor (PDGF) receptors. J. Biol. Chem. 268, 9287–9295.

Heldin, C.-H. (1992). Structural and functional studies on platelet-derived growth factor. EMBO J. 11, 4251–4259.

Heldin, C.-H., Bäckstrom, G., Östman, A., Hammacher, A., Rönnstrand, L., Rubin, K., Nister, M., & Westermark, B. (1988). Binding of different dimeric forms of PDGF to human fibroblasts: evidence for two separate receptor types. EMBO J. 7, 1389–1393.

Heldin, C-H., Ernlund, A., Rorsmen, C., & Rönnstrand, L. (1989). Dimerization of the B-type platelet-derived growth factor receptors occurs after ligand binding and is closely associated with receptor kinase activation. J. Biol. Chem. 264, 8905–8912.

Helin, K., Velu, T., Martin, P., Vass, W. C., Allevato, G., Lowy, D. R., & Beguinot, L. (1991). The biological activity of the human epidermal growth factor receptor is positively regulated by its C-terminal tyrosines. Oncogene 6, 825–832.

Hempstead, B. L., Martin-Zanca, D., Kaplan, D. R., Parada, L. F., & Chao, M. V. (1991). High-affinity NGF binding requires coexpression of the *trk* proto-oncogene and the low-affinity NGF receptor. Nature 350, 678–683.

Hiles, I. D., Otsu, M., Volinia, S., Fry, M. J., Gout, I., Dhand, R., Panayotou, G., Ruiz-Larrea, F., Thompson, A., Totty, N. F., Hsuan, J. J., Courtneidge, S. A., Parker, P. J., & Waterfield, M. D. (1992). Phosphatidylinositol 3-kinase: Structure and expression of the 110 kd catalytic subunit. Cell 70, 419–429.

Hirai, H., Maru, Y., Hagiwara, K., Nishida, J., & Takaku, F. (1987). A novel putative tyrosine kinase receptor encoded by the *eph* gene. Science 238, 1717–1720.

Hofstra, R. M. W., Landsvater, R. M., Ceccherini, I., Stulp, R. P., Stelwagen, T., Luo, Y., Pasini, B., Höpener, J. W. M., van Amstel, H. K. P., Romeo, G., Lips, C. J. M., & Buys, C. H. C. M. (1994). A mutation in the *RET* proto-oncogene associated with multiple endocrine neoplasia type 2B and sporadic medullary thyroid carcinoma. Nature 367, 375–376.

Holmes, W. W., Sliwkowski, M. X., Akita, R. W., Henzel, W. J., Lee, J., Park, J. W., Yansura, D., Abadi, N., Raab, H., Lewis, G. D., Shepard, H. M., Kuang, W.-J., Wood, W. I., Goeddel, D. V., & Vandlen, R. L. (1992). Identification of heregulin, a specific activator of $p185^{erbB2}$. Science 256, 1205–1210.

Honegger, A. M., Dull, T., Felder, S., Van Obberghen, E., Bellot, F., Szapary, D., Schmidt, A., Ullrich, A., & Schlessinger, J. (1987). Point mutation at the ATP binding site of EGF receptor abolishes protein tyrosine kinase activity and alters cellular routing. Cell 51, 199–209.

Honegger, A. M., Kris, R. M., Ullrich, A., & Schlessinger, J. (1989). Evidence that autophosphorylation of solubilized receptors for epidermal growth factor is mediated by intermolecular cross-phosphorylation. Proc. Natl. Acad. Sci. USA 86, 925–929.

Honegger, A. M., Schmidt, A., Ullrich, A., & Schlessinger, J. (1990). Evidence for epidermal growth factor (EGF)-induced intermolecular autophosphorylation of the EGF receptors in living cells. Mol. Cell. Biol. 10, 4035–4044.

Horvitz, H. R., & Sternberg, P. W. (1991). Multiple intercellular signalling systems control the development of the *Caenorhabditis elegans* vulva. Nature 351, 535–541.

Huff, J. L., Jelinek, M. A., Borgman, C. A., Lansing, T. J., & Parsons, J. T. (1993). The protooncogene c-*sea* encodes a transmembrane protein-tyrosine kinase related to the Met/hepatocyte growth factor/scatter factor receptor. Proc. Natl. Acad. Sci. USA 90, 6140–6144.

Ikeda, I., Ishizaka, Y., Tahira, T., Sugimura, T., & Nagao, M. (1990). Specific expression of the ret proto-oncogene in human neuroblastoma cell lines. Oncogene 6, 1291–1296.

Ishizaka, Y., Tahira, T., Ochiai, M., Ikeda, I., Sugimura, T., & Nagao, M. (1988). Molecular cloning and characterization of human retII oncogene. Oncogene Res. 3, 193–197.

Ishizaka, Y., Ushijima, T., Sugimura, T., & Nagao, M. (1990). cDNA cloning and characterization of *ret* activated in a human papillary carcinoma cell line. Biochim. Biophys. Res. Commun. 168, 402–408.

Janssen, J. W. G., Schultz, A. S., Steenvoorden, A. C. M., Schmidberger, M., Strehl, S., Ambros, P. F., & Bartram, C. R. (1991). A novel putative tyrosine kinase receptor with oncogenic potential. Oncogene 6, 2113–2120.

Jaye, M., Schlessinger, J., & Dionne, C. A. (1992). Fibroblast growth factor receptor tyrosine kinases: molecular analysis and signal transduction. Biochem. Biophys. Acta 1135, 185–199.

Johnson, J. D., Edman, J. C., & Rutter, W. J. (1993). A receptor tyrosine kinase found in breast carcinoma cells has an extracellular discoidin I-like domain. Proc. Natl. Acad. Sci. USA 90, 5677–5681.

Kan, M., Wang, F., Xu, J., Crabb, J. W., Hou, J., & McKeehan, W. L. (1993). An essential heparin-binding domain in the fibroblast growth factor receptor kinase. Science 259, 1918–1921.

Kapeller, R., Chakrabarti, R., Cantley, L., Fay, F., & Corvera, S. (1993). Internalization of activated platelet-derived growth factor receptor-phosphatidylinositol-3′ kinase complexes: Potential interactions with the microtubule cytoskeleton. Mol. Cell. Biol. 13, 6052–6063.

Kaplan, D. R., Whitman, M., Schaffhausen, B., Pallas, D. C., White, M., Cantley, L., & Roberts, T. M. (1987). Common elements in growth factor stimulation and oncogenic transformation: 85 kd phosphoprotein and phosphatidylinositol kinase activity. Cell 50, 1021–1029.

Kaplan, D. R., Martin-Zanca, D., & Parada, L. F. (1991a). Tyrosine phosphorylation and tyrosine kinase activity of the *trk* proto-oncogene product induced by NGF. Nature 350, 158–160.

Kaplan, D. R., Hempstead, B. L., Martin-Zanca, D., Chao, M. V., & Parada, L. F. (1991b). The *trk* proto-oncogene product: A signal transducing receptor for nerve growth factor. Science 252, 554–558.

Karn, T., Holtrich, U., Bräuninger, A., Böhme, B., Wolf, G., Rübesamen-Waigmann, H., & Strebhardt, K. (1993). Structure and chromosomal mapping of TKT from man and mouse: A new subclass of receptor tyrosine kinases with a factor VIII-like domain. Oncogene 8, 3433–3440.

Kashishian, A., & Cooper, J. A. (1993). Phosphorylation sites at the C-terminus of the platelet-derived growth factor receptor bind phospholipase Cγ 1. Mol. Biol. Cell 4, 49–57.

Kashishian, A., Kazlauskas, A., & Cooper, J. A. (1992). Phosphorylation sites in the PDGF receptor with different specificities for binding GAP and PI 3-kinase *in vivo*. EMBO J. 11, 1373–1382.

Kazlauskas, A., Ellis, C., Pawson, T., & Cooper, J. A. (1990). Binding of GAP to activated PDGF receptors. Science 247, 1578–1581.

Kazlauskas, A., Kashishian, A., Cooper, J. A., & Valius, M. (1992). GTPase-activating protein and phosphatidylinositol 3-kinase bind to distinct regions of the platelet-derived growth factor receptor beta-subunit. Mol. Cell. Biol. 12, 2534–2544.

Kazlauskas, A., Feng, G.-S., Pawson, T., & Valius, M. (1993). The 64-kD protein that associates with the platelet-derived growth factor receptor β subunit via Tyr-1009 is the SH2-containing phosphotyrosine phosphatase Syp. Proc. Natl. Acad. Sci. USA 90, 6939–6942.

Keegan, K., Johnson, D. E., Williams, L. T., & Hayman, M. J. (1991). Isolation of an additional member of the fibroblast growth factor receptor family, FGFR-3. Proc. Natl. Acad. Sci. USA 88, 1095–1099.

King, C. R., Borrello, I., Bellot, F., Comoglio, P., & Schlessinger, J. (1988). EGF binding to its receptor triggers a rapid tyrosine phosphorylation of the *erb*B-2 protein in the mammary tumor cell line SK-BR-3. EMBO J. 7, 1647–1651.

Klein, R., Parada, L., Coulier, F., & Barbacid, M. (1989). *trk*B, a novel tyrosine protein kinase receptor expressed during mouse neural development. EMBO J. 8, 3701–3709.

Klein, R., Martin-Zanca, D., Barbacid, M., & Parada, L. F. (1990a). Expression of the tyrosine kinase receptor gene *trk*B is confined to the murine embryonic and adult nervous system. Development 109, 845–850.

Klein, R., Conway, D., Parada, L. F., & Barbacid, M. (1990b). The *trk*B tyrosine kinase gene codes for a second neurogenic receptor that lacks the catalytic kinase domain. Cell 61, 647–656.

Klein, R., Jing, S., Nanduri, V., O'Rourke, E., & Barbacid, M. (1991). The *trk* proto-oncogene encodes a receptor for nerve growth factor. Cell 65, 189–197.

Klein, R., Smeyne, R. J., Wurst, W., Long, L. K., Auerbach, B. A., Joyner, A. L., & Barbacid, M. (1993). Targeted disruption of the *trk*B neurotrophin receptor gene results in nervous system lesions and neonatal death. Cell 75, 113–122.

Koch, C. A., Anderson, D., Moran, M. F., Ellis, C., & Pawson, T. (1991). SH2 and SH3 domains: Elements that control interactions of cytoplasmic signalling proteins. Science 252, 668–674.

Komada, M., & Kitamura, N. (1993). The cell dissociation and motility triggered by scatter factor/hepatocyte growth factor are mediated through the cytoplasmic domain of the c-Met receptor. Oncogene 8, 2381–2390.

Korhonen, J., Polvi, A., Partanen, J., & Alitalo, K. (1994). The mouse *tie* receptor tyrosine kinase gene: Expression during embryonic angiogenesis. Oncogene 9, 395–403.

Kornbluth, S., Paulson, K. E., & Hanafusa, H. (1988). Novel tyrosine kinase identified by phosphotyrosine antibody screening of cDNA libraries. Mol. Cell. Biol. 8, 5541–5544.

Kraus, M. H., Issing, W., Miki, T., Popescu, N.C., & Aaronson, S. A. (1989). Isolation and characterization of *ERBB3*, a third member of the *ERBB*/epidermal growth factor receptor family: Evidence for overexpression in a subset of human mammary tumors. Proc. Natl. Acad. Sci. USA 86, 9193–9197.

Kraus, M. H., Fedi, P., Starks, V., Muraro, R., & Aaronson, S. A. (1993). Demonstration of ligand-dependent signaling by the *erbB-3* tyrosine kinase and its constitutive activation in human breast tumor cells. Proc. Natl. Acad. Sci. USA 90, 2900–2904.

Kuhne, M. R., Pawson, T., Lienhard, G. E., & Feng, G.-S. (1993). The insulin receptor substrate 1 associates with the SH2-containing phosphotyrosine phosphatase Syp. J. Biol. Chem. 268, 11479–11481.

Kumjian, D. A., Wahl, M. I., Rhee, S. G., & Daniel, T. O. (1989). Platelet-derived growth factor (PDGF) binding promotes physical association of PDGF receptor with phospholipase C. Proc. Natl. Acad. Sci. USA 86, 8232–8236.

Kundra, V., Escobedo, J. A., Kazlauskas, A., Kim, H. K., Rhee, S. G., Williams, L. T., & Zetter, B. R. (1994). Regulation of chemotaxis by the platelet-derived growth factor receptor-β. Nature 367, 474–476.

Kypta, R. M., Goldberg, Y., Ulug, E. T., & Courtneidge, S. A. (1990). Association between the PDGF receptor and members of the *src* family of tyrosine kinases. Cell 62, 481–492.

Lai, C., & Lemke, G. (1991). An extended family of protein-tyrosine kinase genes differentially expressed in the vertebrate nervous system. Neuron 6, 691–704.

Lai, C., & Lemke, G. (1994). Structure and expression of the Tyro 10 receptor tyrosine kinase. Oncogene 9, 877–883.

Lamballe, F., Klein, R., & Barbacid, M. (1991). *trk*C, a new member of the *trk* family of tyrosine protein kinases, is a receptor for neurotrophin-3. Cell 66, 967–979.

Lamballe, F., Tapley, P., & Barbacid, M. (1993). *trk*C encodes multiple neurotrophin-3 receptors with distinct biological properties and substrate specificities. EMBO J. 12, 3083–3094.

Lechleider, R. J., Sugimoto, S., Bennett, A. M., Kashishian, A. S., Cooper, J. A., Shoelson, S. E., Walsh, C. T., & Neel, B. G. (1993). Activation of the SH2-containing phosphotyrosine phosphatase SH-PTP2 by its binding site, phosphotyrosine 1009, on the human platelet-derived growth factor receptor b. J. Biol. Chem. 268, 21478–21481.

Lee, J., Dull, T. J., Lax, I., Schlessinger, J., & Ullrich, A. (1989a). HER2 cytoplasmic domain generates normal mitogenic and transforming signals in a chimeric receptor. EMBO J. 8, 167–173.

Lee, P. L., Johnson, D. E., Cousens, L. S., Fried, V. A., & Williams, L. T. (1989b). Purification and complementary DNA cloning of a receptor for basic fibroblast growth factor. Science 245, 57–60.

Lee, C.-H., Li, W., Nishimura, R., Zhou, M., Batzer, A. G., Myers, M. G. Jr., White, M. F., Schlessinger, J., & Skolnik, E. Y. (1993). Nck associates with the SH2 domain-docking protein IRS-1 in insulin-treated cells. Proc. Natl. Acad. Sci. USA 90, 11713–11717.

Lehväslaiho, H., Lehtola, L., Sistonen, L., & Alitalo, K. (1989). A chimeric EGFR-*neu* proto-oncogene allows EGF to regulate *neu* tyrosine kinase and cell transformation. EMBO J. 8, 159–166.

Letwin, K., Y, S.-P., & Pawson, T. (1988). Novel protein-tyrosine kinase cDNAs related to *fps/fes* and *eph* cloned using anti-phosphotyrosine antibody. Oncogene 3, 621–627.

Lev, S., Givol, D., & Yarden, Y. (1992). Interkinase domain of kit contains the binding site for phosphatidylinositol 3′ kinase. Proc. Natl. Acad. Sci. USA 89, 678–682.

Lhotak, V., & Pawson, T. (1993). Biological and biochemical activities of a chimeric epidermal growth factor-Elk receptor tyrosine kinase. Mol. Cell. Biol. 13, 7071–7079.

Lhotak, V., Greer, P., Letwin, K., & Pawson, T. (1991). Characterization of Elk, a brain-specific receptor tyrosine kinase. Mol. Cell. Biol. 11, 2496–2502.

Li, N., Batzer, A., Daly, R., Yajnik, V., Skolnik, E., Chardin, P., Bar-Sagi, D., Margolis, B., & Schlessinger, J. (1993). Guanine-nucleotide releasing factor, hSos1 binds to Grb2 and links receptor tyrosine kinases to Ras signalling. Nature 363, 85–88.

Li, W., Nishimura, R., Kashishian, A., Batzer, A. G., Kim, W. J. H., Cooper, J. A., & Schlessinger, J. (1994). A new function for a phosphotyrosine phosphatase: Linking GRB2-Sos to a receptor tyrosine kinase. Mol. Cell. Biol. 14, 509–517.

Lin, C. R., Chen, W. S., Lazar, C. W., Carpenter, C. D., Gill, G. N., Evans, R. M., & Rosenfeld, M. G. (1986). Protein kinase C phosphorylation at Thr 654 of the unoccupied EGF receptor and EGF binding regulate functional receptor loss by independent mechanisms. Cell 44, 839–848.

Lindberg, R. A., & Hunter, T. (1990). cDNA cloning and characterization of *eck*, an epithelial cell receptor protein-tyrosine kinase in the *eph/elk* family of protein kinases. Mol. Cell. Biol. 10, 6316–6324.

Livneh, E., Glaser, L., Segal, D., Schlessinger, J., & Shilo, B.-Z. (1985). The *Drosophila* EGF receptor gene homolog: Conservation of both hormone binding and kinase domains. Cell 40, 599–607.

Loeb, D. M., Maragos, J., Martin-Zanca, D., Chao, M. V., Parada, L. F., & Greene, L. A. (1991). The trk proto-oncogene rescues NGF responsiveness in mutant NGF-nonresponsive PC12 cell lines. Cell 66, 961–966.

Lofts, F. J., Hurst, H. C., Sternberg, M. J. E., & Gullick, W. J. (1993). Specific short transmembrane sequences can inhibit transformation by the mutant *neu* growth factor receptor *in vitro* and *in vivo*. Oncogene 8, 2813–2820.

Longati, P., Bardelli, A., Ponzetto, C., Naldinin, L., & Comoglio, P. M. (1994). Tyrosines$^{1234-1235}$ are critical for activation of the tyrosine kinase encoded by the *MET* proto-oncogene (HGF receptor). Oncogene 9, 49–57.

Lowenstein, E. J., Daly, R., Batzer, A. G., Li, W., Margolis, B., Lammers, R., Ullrich, A., Skolnik, E. Y., Bar-Sagi, D., & Schlessinger, J. (1992). The SH2 and SH3 domain-containing protein GRB2 links receptor tyrosine kinases to ras signalling. Cell 70, 431–442.

Lyman, S. D., James, L., Zappone, J., Sleath, P. R., Beckmann, M. P., & Bird, T. (1993a). Characterization of the protein encoded by the flt3 (flk2) receptor-like tyrosine kinase gene. Oncogene 8, 815–822.

Lyman, S. D., James, L., Vanden Bos, T., de Vries, P., Brasel, K., Gliniak, B., Hollingsworth, L. T., Picha, K. S., McKenna, H. J., Splett, R. R., Fletcher, F. A., Maraskovsky, E., Farrah, T., Foxworthe, D., Williams, D. E., & Beckmann, M. P. (1993b). Molecular cloning of a ligand for the flt3/flk-2 tyrosine kinase receptor: A proliferative factor for primitive hematopoietic cells. Cell 75, 1157–1167.

Maisonpierre, P. C., Goldfarb, M., Yancopoulos, G. D., & Gao, G. (1993a). Distinct rat genes with related profiles of expression define a TIE receptor tyrosine kinase family. Oncogene 8, 1631–1637.

Maisonpierre, P. C. Barrezueta, N. X., & Yancopoulos, G. D. (1993b). Ehk-1 and Ehk-2: two novel members of the Eph receptor-like tyrosine kinase family with distinctive structures and neuronal expression. Oncogene 8, 3277–3288.

Majumder, S., Ray, P., & Besmer, P. (1990). Tyrosine protein kinase activity of the HZ4-feline sarcoma virus $p80^{gag-kit}$ transforming protein. Oncogene Research 5, 329–335.

Marchionni, M. A., Goodearl, A. D. J., Chen, M. S., Bermingham-McDonogh, O., Kirk, C., Hendricks, M., Danehy, F., Misumi, D., Sudhalter, J., Kobayashi, K., Wroblewski, D., Lynch, C., Baldassare, M., Hiles, I., Davis, J. B., Hsuan, J. J., Totty, N. F., Otsu, M., McBurney, R. N., Waterfield, M. D., Stroobant, P., & Gwynne, D. (1993). Glial growth factors are alternatively spliced erbB2 ligands expressed in the nervous system. Nature 362, 312–318.

Margolis, B., Silvennoinen, O., Comoglio, F., Roonprapunt, C., Skolnik, E., Ullrich, A., & Schlessinger, J. (1992). High efficiency expression cloning of epidermal growth factor receptor binding proteins with Src homology 2 domains. Proc. Natl. Acad. Sci. USA 89, 8894–8898.

Maroc, N., Rottapel, R., Rosnet, O., Marchetto, S., Lavezzi, C., Mannoni, P., Birnbaum, D., & Dubreuil, P. (1993). Biochemical characterization and analysis of the transforming potential of the FLT3/Flk2 receptor tyrosine kinase. Oncogene 8, 909–918.

Martin-Zanca, D., Hughes, S. H., & Barbacid, M. (1986). A human oncogene formed by fusion of truncated tropomyosin and protein kinase sequences. Nature 319, 743–748.

Martin-Zanca, D., Oskam, R., Mitra, G., Copeland, T., & Barbacid, M. (1989). Molecular and biochemical characterization of the human *trk* proto-oncogene. Mol. Cell. Biol. 9, 24–33.

Martin-Zanca, D., Barbacid, M., & Parada, L. F. (1990). Expression of the *trk* proto-oncogene is restricted to the sensory cranial and spinal ganglia of neural crest origin in mouse development. Genes Dev. 4, 683–694.

Maru, Y., Hirai, H., Yoshida, M., & Takaku, F. (1988). Evolution, expression and chromosomal location of a novel receptor tyrosine kinase, *eph*. Mol. Cell. Biol. 8, 3770–3776.

Matsuda, M., Mayer, B. J., Fukui, Y., & Hanafusa, H. (1990). Binding of transforming protein $p47^{gag-crk}$ to a broad range of phosphotyrosine containing proteins. Science 248, 1537–1539.

Matsuda, M., Mayer, B. J., & Hanafusa, H. (1991). Identification of domains of the v-crk oncogene product sufficient for association with phosphotyrosine-containing proteins. Mol. Cell. Biol. 11, 1607–1613.

Matsuda, S., Kadowaki, Y., Ichino, M., Akiyama, T., Toyoshima, K., & Yamamoto, T. (1993). 17b-Estradiol mimics ligand activity of the c-*erb*B2 protooncogene product. Proc. Natl. Acad. Sci. USA 90, 10803–10807.

Matsui, T., Heidaran, M., Miki, T., Popescu, N., La Rochelle, W., Kraus, M., Pierce, J., & Aaronson, S. (1989). Isolation of a novel receptor cDNA establishes the existence of two PDGF receptor genes. Science 243, 800–804.

Matsushime, J., Wang, L.-H., & Shibuya, M. (1986). Human c-*ros*-1 gene homologous to the v-*ros* sequence of UR2 sarcoma virus encodes for a transmembrane receptor-like molecule. Mol. Cell. Biol. 6, 3000–3004.

Matthews, W., Jordan, C. T., Wiegand, G. W., Pacol, C., & Lemischka, I. I. (1991). A receptor tyrosine kinase specific to hematopoietic stem and progenitor cell populations. Cell 65, 1143–1152.

McGlade, J., Cheng, A., Pelicci, G., Pelicci, P. G., & Pawson, T. (1992). Shc proteins are phosphorylated and regulated by the v-Src and v-Fps protein-tyrosine kinases. Proc. Natl. Acad. Sci. USA 89, 8869–8873.

Middlemas, D. S., Lindberg, R. A., & Hunter, T. (1991). *trk*B, a neural receptor protein-tyrosine kinase: Evidence for a full-length and two truncated receptors. Mol. Cell. Biol. 11, 143–153.

Millauer, B., Wizigmann-Voos, S., Schnurch, H., Martinez, R., Moller, N. P. H., Risau, W., & Ullrich, A. (1993). High affinity VEGF binding and developmental expression suggest Flk-1 as a major regulator of vasculogenesis and angiogenesis. Cell 72, 835–846.

Millauer, B., Shawver, L. K., Plate, K. H., Risau, W., & Ullrich, A. (1994). Glioblastoma growth inhibited *in vivo* by a dominant-negative Flk-1 mutant. Nature 367, 576–579.

Mitra, G., Martin-Zanca, D., & Barbacid, M. (1987). Identification and biochemical characterization of $p70^{trk}$, product of the human TRK oncogene. Proc. Natl. Acad. Sci. USA 84, 6707–6711.

Mohammadi, M., Dione, C. A., Li, W., Li, N., Spivak, T., Honegger, A. M., Jaye, M., & Schlessinger, J. (1992). Point mutation in FGF receptor eliminates phosphatidylinositol hydrolysis without affecting mitogenesis. Nature 358, 681–684.

Molley, C. J., Bottaro, D. P., Fleming, T. P., Marshall, M. S., Gibbs, J. B., & Aaronson, S. A. (1989). PDGF induction of tyrosine phosphorylation of GTPase activating protein. Nature 342, 711–714.

Mondino, A., Giordano, S., & Comoglio, P. M. (1991). Defective posttranslational processing activates the tyrosine kinase encoded by the *MET* proto-oncogene (hepatocyte growth factor). Mol. Cell. Biol. 11, 6084–6092.

Moodie, S. A., Willumsen, B. M., Weber, M. J., & Wolfman, A. (1993). Complexes of RasGTP with raf-1 and mitogen activated protein kinase. Science 260, 1658–1661.

Moran, M. F., Koch, C. A., Anderson, D., Ellis, C., England, L., Martin, G. S., & Pawson, T. (1990). Src homology region 2 domains direct protein–protein interactions in signal transduction. Proc. Natl. Acad. Sci. USA 87, 8622–8626.

Mori, S., Rönnstrand, L., Yokote, K., Engström, Å, Courtneidge, S. A., Claesson-Welsh, L., & Heldin, C.-H. (1993). Identification of two juxtamembrane autophosphorylation sites in the PDGF β-receptor; involvement in the interaction with Src family tyrosine kinases. EMBO J. 12, 2257–2264.

Mulligan, L. M., Kwok, J. B. J., Healey, C. S., Elsdon, M. J., Eng, C., Gardner, E., Love, D. R., Mole, S. E., Moore, J. K., Papi, L., Ponder, M. A., Telenius, H., Tunnacliffe, A., & Ponder, B. A. J. (1993). Germ-line mutations of the *RET* proto-oncogene in multiple endocrine neoplasia type 2A. Nature 363, 458–460.

Musacchio, A., Gibson, T., Lehto, V.-P., & Saraste, M. (1992). SH3—an abundant protein domain in search of a function. FEBS Lett. 307, 55–61.

Myers, M. G. Jr., Sun, X. J., Cheatham, B., Jachna, B. R., Glasheen, E. M., Backer, J. M., & White, M. F. (1993). IRS-1 is a common element in insulin and insulin-like growth factor-1 signaling to the phosphatidylinositol 3′-kinase. Endocrinology 132, 1421–1432.

Naldini, L., Weidner, K. M., Vigna, E., Gaudino, G., Bardelli, A., Ponzetto, C., Narsimhan, R. P., Hartmann, G., Zarnegar, R., Michalopoulos, G. M., Birchmeier, W., & Comoglio, P. M. (1991a). Scatter factor and hepatocyte growth factor are indistinguishable ligands for the *MET* receptor. EMBO J. 10, 2867–2878.

Naldini, L., Vigna, E., Narsimhan, R. P., Gaudino, G., Zarnegar, R., Michalopoulos, G. K., & Comoglio, P. M. (1991b). Hepatocyte growth factor (HGF) stimulates the tyrosine kinase activity of the receptor encoded by the proto-oncogene c-*Met*. Oncogene 6, 501–504.

Neckameyer, W. S., & Wang, L.-H. (1985). Nucleotide sequence of the avian sarcoma virus UR2 and comparison of its transforming gene with other members of the tyrosine protein kinase oncogene family. J. Virol. 53, 879–884.

Neckameyer, W. S., Shibuya, M., Hsu, M.-T., & Wang, L.-H. (1986). Proto-oncogene c-*ros* codes for a molecule with structural features common to those of growth factor receptors and displays tissue specific and developmentally regulated expression. Mol. Cell. Biol. 6, 1478–1486.

Neufeld, G., Tessler, S., Gitay-Goren, H., Cohen, T., & Levi, B.-Z. (1994). Vascular endothelial growth factor and its receptors. Progress in Growth Factor Research 5, 89–97.

Nishimura, R., Li, W., Kashishian, A., Mondino, A., Zhou, M., Cooper, J., & Schlessinger, J. (1993). Two signaling molecules share a phosphotyrosine-containing binding site in the platelet-derived growth factor receptor. Mol. Cell. Biol. 13, 6889–6896.

Obermeier, A., Halfter, H., Wiesmüller, K.-H., Jung, G., Schlessinger, J., & Ullrich, A. (1993a). Tyrosine 785 is a major determinant of Trk-substrate interaction. EMBO J. 12, 933–941.

Obermeier, A., Lammers, R., Wiesmüller, K.-H., Jung, G., Schlessinger, J., & Ullrich, A. (1993b). Identification of Trk binding sites for SHC and phosphatidylinositol 3′-kinase and formation of a multimeric signaling complex. J. Biol. Chem. 268, 22963–22966.

O'Bryan, J. P., Frye, R. A., Cogswell, P. C., Neubauer, A., Kitch, B., Prokop, C., Espinosa III, R., Le Beau, M.M., Earp, H.S., & Liu, E. T. (1991). *axl*, a transforming gene isolated from primary human myeloid leukaemia cells, encodes a novel receptor tyrosine kinase. Mol. Cell. Biol. 11, 5016–5031.

Oelrichs, R. B., Reid, H. H., Bernard, O., Ziemiecki, A., & Wilks, A. F. (1993). *NYK/FLK*-1: A putative receptor protein tyrosine kinase isolated from E10 embryonic neuroepithelium is expressed in endothelial cells of the developing embryo. Oncogene 8, 11–18.

Ohashi, K., Mizuno, K., Kuma, K.-I., Miyata, T., & Nakamura, T. (1994). Cloning of the cDNA for a novel receptor tyrosine kinase, Sky, predominantly expressed in brain. Oncogene 9, 699–705.

Ohmichi, M., Decker, S. J., & Saltiel, A. R. (1992). Activation of phosphatidylinositol-3 kinase by nerve growth factor involves indirect coupling of the *trk* proto-oncogene with src homology 2 domains. Neuron 9, 769–777.

Olivier, J. P., Raabe, T., Henkemeyer, M., Dickson, B., Mbamalu, G., Margolis, B., Schlessinger, J., Hafen, E., & Pawson, T. (1993). A *Drosophila* SH2–SH3 adaptor protein implicated in coupling the sevenless tyrosine kinase to an activator of Ras guanine nucleotide exchange, Sos. Cell 73, 179–191.

Pajusola, K., Aprelikova, O., Armstrong, E., Morris, S., & Alitalo, K. (1993). Two human FLT4 receptor tyrosine kinase isoforms with distinct carboxy terminal tails are produced by alternative processing of primary transcripts. Oncogene 8, 2931–2937.

Panayotou, G., & Waterfield, M. D. (1993). The assembly of signalling complexes by receptor tyrosine kinases. BioEssays 15, 171–177.

Park, M., Dean, M., Cooper, C. S., Schmidt, M., O'Brian, S. J., Blair, D. G., & Van de Woude, G. F. (1986). Mechanism of *met* oncogene activation. Cell 45, 895–904.

Park, M., Dean, M., Kaul, K., Braun, M. J., Gonda, M. A., & Van de Woude, G. F. (1987). Sequence of met proto-oncogene cDNA has features characteristic of the tyrosine kinase family of growth factor receptors. Proc. Natl. Acad. Sci. USA 84, 6379–6383.

Partanen, J., Mäkelä, T. P., Eerola, E., Korhonen, J., Hirvonen, H., Claesson-Welsh, L., & Alitalo, K. (1991). FGF-R4, a novel acidic fibroblast growth factor receptor with distinct expression pattern. EMBO J. 10, 1347–1454.

Partanen, J., Armstrong, E., Makela, T.P., Korhonen, J., Sandberg, M., Renkonen, R., Knuutila, S., Huebner, K., & Alitalo, K. (1992). A novel endothelial cell surface receptor tyrosine kinase with extracellular epidermal growth factor homology domains. Mol. Cell. Biol. 12, 1698–1701.

Pasquale, E. B. (1990). A distinctive family of embryonic protein-tyrosine kinase receptors. Proc. Natl. Acad. Sci. USA 87, 5812–5816.

Pasquale, E. B. (1991). Identification of chicken embryo kinase 5, a developmentally regulated receptor-type tyrosine kinase of the Eph family. Cell Regulation 2, 523–534.

Pasquale, E. B., & Singer, S. J. (1989). Identification of a developmentally regulated protein-tyrosine kinase by using anti-phosphotyrosine antibodies to screen a cDNA expression library. Proc. Natl. Acad. Sci. USA 86, 5449–5453.

Pawson, T., & Schlessinger, J. (1993). SH2 and SH3 domains. Current Biology 3, 434–442.

Peles, E., & Yarden, Y. (1993). *Neu* and its ligands: From an oncogene to neural factors. BioEssays 15, 815–824.

Peles, E., Lamprecht, R., Ben-Levy, R., Tzahar, E., & Yarden, Y. (1992). Regulated coupling of the *Neu* receptor to phosphatidylinositol 3′-kinase and its release by oncogenic activation. J. Biol. Chem. 267, 12266–12274.

Peles, E., Ben-Levy, R., Tzahar, E., Liu, N., Wen, D., & Yarden, Y. (1993). Cell-type specific interaction of Neu differentiation factor (NDF/heregulin) with Neu/HER-2 suggests complex ligand–receptor relationships. EMBO J. 12, 961–971.

Pelicci, G., Lanfrancone, L., Grignani, F., McGlade, J., Cavallo, F., Forni, G., Nicoletti, I., Pawson, T., & Pelicci, P. G. (1992). A novel transforming protein (Shc) with an SH2 domain is implicated in mitogenic signal transduction. Cell 70, 93–104.

Perez, J. L., Shen, X., Finkernagel, S., Sciorra, L., Jenkins, N. A., Gilbert, D. J., Copeland, N. G., & Wong, T. W. (1994). Identification and chromosomal mapping of a receptor tyrosine kinase with a putative phospholipid binding sequence in its ectodomain. Oncogene 9, 211–219.

Peters, K. G., Marie, J., Wilson, E., Ives, H. E., Escobedo, J., Del Rosario, M., Mirda, D., & Williams, L. T. (1992). Point mutation of an FGF receptor abolishes phosphatidylinositol turnover and calcium flux but not mitogenesis. Nature 358, 678–681.

Plowman, G. D., Whitney, G. S., Neubauer, M. G., Green, J. M., McDonald, V. L., Todaro, G. J., & Shoyab, M. (1990). Molecular cloning and expression of an additional epidermal growth factor receptor-related gene. Proc. Natl. Acad. Sci. USA 87, 4905–4909.

Plowman, G. D., Culouscou, J.-M., Whitney, G. S., Green, J. M., Carlton, G. W., Foy, L., Neubauer, M. G., & Shoyab, M. (1993a). Ligand-specific activation of HER4/p180^{erbB4}, a fourth member of the epidermal growth factor receptor family. Proc. Natl. Acad. Sci. USA 90, 1746–1750.

Plowman, G. D., Green, J. M., Culouscou, J.-M., Carlton, G. W., Rothwell, V. M., & Buckley, S. (1993b). Heregulin induces tyrosine phosphorylation of HER4/p180^{erbB4}. Nature 366, 473–475.

Ponzetto, C., Bardelli, A., Maina, F., Longati, P., Panayotou, G., Dhand, R., Waterfield, M. D., & Comoglio, P. M. (1993). A novel recognition motif for phosphatidylinositol 3-kinase binding mediates its association with the hepatocyte growth factor/scatter factor receptor. Mol. Cell. Biol. 13, 4600–4608.

Prigent, S. A., Lemoine, N. R., Hughes, C. M., Plowman, G. D., Selden, C., & Gullick, W. J. (1992). Expression of the c-*erb*B-3 protein in normal human adult and fetal tissues. Oncogene 7, 1273–1278.

Pulciani, S., Santos, E., Lauver, A. V., Long, L. K., Aaronson, S. A., & Barbacid, M. (1982). Oncogenes in solid human tumors. Nature 300, 539–542.

Qui, F., Ray, P., Brown, K., Barker, P. E., Jhawer, S., Ruddle, R. H., & Besmer, P. (1988). Primary structure of c-*kit*: relationship with the CSF-1/PDGF receptor kinase family: Oncogenic activation of v-*kit* involves deletion of extracellular domain and C-terminus. EMBO J. 7, 1003–1011.

Qiu, M.-S., & Green, S. H. (1991). NGF and EGF rapidly activate p21^{ras} in PC12 cells by distinct, convergent pathways involving tyrosine phosphorylation. Neuron 7, 937–946.

Redemann, N., Holzmann, B., von Rüden, T., Wagner, E. F., Schlessinger, J., & Ullrich, A. (1992). Anti-oncogenic activity of signalling-defective epidermal growth factor receptor mutants. Mol. Cell. Biol. 12, 491–498.

Reedijk, M., Liu, X., & Pawson, T. (1990). Interactions of phosphatidylinositol kinase, GTPase-activating protein (GAP), and GAP-associated proteins with the colony stimulating factor 1 receptor. Mol. Cell. Biol. 10, 5601–5608.

Reedijk, M., Liu, X., van der Geer, P., Letwin, K., Waterfield, M. D., Hunter, T., & Pawson, T. (1992). Tyr721 regulates specific binding of the CSF-1 receptor kinase insert to the PI 3'-kinase SH2 domains—a model for SH2-mediated receptor target interactions. EMBO J. 11, 1365–1372.

Rescigno, J., Manukhani, A., & Basilico, C. (1991). A putative receptor tyrosine kinase with unique structural topology. Oncogene 6, 1909–1913.

Rhee, S. G. (1991). Inositol phospholipid-specific phospholipase C: Interaction of the γ1 isoform, with tyrosine kinases. Trend Biochem. Sci. 16, 297–301.

Riedel, H., Dull, T., Honegger, A. M., Schlessinger, J., & Ullrich, A. (1989). Cytoplasmic domains determine signal specificity, cellular routing characteristics and influence ligand binding of epidermal growth factor and insulin-receptors. EMBO J. 8, 2943–2954.

Roberts, T. M. (1992). A signal chain of events. Nature 360, 534-535.

Roberts, W. M., Look, A. T., Roussel, M. F., & Sherr, C. J. (1988). Tandem linkage of human CSF-1 receptor (c-*fms*) and PDGF receptor genes. Cell 55, 655–661.

Romeo, G., Ronchetto, P., Luo, Y., Barone, V., Seri, M., Ceccherini, I., Pasini, B., Bocciadi, R., Lerone, M., Kääriäinen, H., & Martucciello, G. (1994). Point mutations affecting the tyrosine kinase domain of the *RET* proto-oncogene in Hirschsprung's disease. Nature 367, 377–378.

Rönnstrand, L., Mori, S., Arridsson, A.-K., Eriksson, A., Wernstedt, C., Hellman, U., Claesson-Welsh, L., & Heldin, C.-H. (1992). Identification of the two C-terminal autophosphorylation sites in the PDGF β-receptor: Involvement in the interaction with phospholipase C-γ. EMBO J. 11, 3911–3919.

Ronsin, C., Muscatelli, F., Mattei, M.-G., & Breathnach, R. (1993). A novel putative receptor protein-tyrosine kinase of the met family. Oncogene 8, 1195–1202.

Rosnet, O., Marchetto, S., deLapeyriere, O., & Birnbaum, D. (1991). Murine *Flt*3, a gene encoding a novel tyrosine kinase receptor of the PDGFR/CSF1R family. Oncogene 6, 1641–1650.

Rottapel, R., Reedijk, M., Williams, D. E., Lyman, S. D., Anderson, D. M., Pawson, T., & Bernstein, A. (1991). The *steel*/W transduction pathway: kit autophosphorylation and its association with a unique subset of cytoplasmic signaling proteins is induced by the Steel factor. Mol. Cell. Biol. 11, 3043–3051.

Roussel, M. F., Dull, T. J., Rettenmier, C. W., Ralph, P., Ullrich, A., & Sherr, C. J. (1987). Transforming potential of the c-*fms* proto-oncogene (CSF-1 receptor). Nature 325, 549–552.

Rozakis-Adcock, M., McGlade, J., Mbamalu, G., Pelicci, G., Daly, R., LI, W., Batzer, A., Thomas, S., Brugge, J., Pelicci, P. G., Schlessinger, J., & Pawson, T. (1992). Association of the Shc and Grb-2/Sem5 SH2-containing protein is implicated in activation of the Ras pathway by tyrosine kinases. Nature 360, 689–692.

Rozakis-Adcock, M., Fernley, R., Wade, J., Pawson, T., & Bowtell, D. (1993). The SH2 and SH3 domains of mammalian Grb2 couple the EGF receptor to the Ras activator mSos1. Nature 363, 83–85.

Rozengurt, E. (1986). Early signals in the mitogenic response. Science 234, 161–166.

Rubin, J. S., Bottaro, D. P., & Aaronson, S. A. (1993). Hepatocyte growth factor/scatter factor and its receptor, the c-*met* proto-oncogene product. Biochim. Biophys. Acta 1155, 357–371.

Rudd, C. E., Janssen, O., Prasad, K. V. S., Raab, M., da Silva, A., Telfer, J. C., & Yamamoto, M. (1993). *src*-related protein-tyrosine kinases and their surface receptors. Biochem. Biophys. Acta 1155, 239–266.

Ruff-Jamison, S., McGlade, J., Pawson, T., Chen, K., & Cohen, S. (1993). Epidermal growth factor stimulates the tyrosine phosphorylation of SHC in the mouse. J. Biol. Chem. 268, 7610–7612.

Russell, D. S., Gherzi, R., Johnson, E. L., Chou, C.-K., & Rosen, O. M. (1987). The protein-tyrosine kinase activity of the insulin receptor is necessary for insulin-mediated receptor down-regulation. J. Biol. Chem. 262, 11833–11840.

Ruta, M., Burgess, W., Givol, D., Epstein, J., Neiger, N., Kaplow, J., Crumley, G., Dionne, C., Jaye, M., & Schlessinger, J. (1989). Receptor for acidic fibroblast growth factor is related to the tyrosine kinase encoded by the *fms*-like gene (FLG). Proc. Natl. Acad. Sci. USA 86, 8722–8726.

Safran, A., Avivi, A., Orr-Urtereger, A., Neufeld, G., Lonai, P., Givol, D., & Yarden, Y. (1990). The murine *flg* gene encodes a receptor for fibroblast growth factor. Oncogene 5, 635–643.

Sajjadi, F. G., & Pasquale, E. B. (1993). Five novel avian Eph-related tyrosine kinases are differentially expressed. Oncogene 8, 1807–1813.

Santoro, M., Wong, W. T., Aroca, P., Santos, E., Matoskova, B., Grieco, M., Fusco, A., & Di Fiore, P. P. (1994). An epidermal growth factor receptor/*ret* chimera generates mitogenic and transforming signals: Evidence for *ret*-specific signaling pathways. Mol. Cell. Biol. 14, 663–675.

Sato, T. N., Qin, Y., Kozak, C. A., & Audus, K. L. (1993). *tie*-1 and *tie*-2 define another class of putative receptor tyrosine kinase genes expressed in early embryonic vascular system. Proc. Natl. Acad. Sci. USA 90, 9355–9358.

Satoh, T., Fantl, W. J., Escobedo, J. A., Williams, L. T., & Kaziro, Y. (1993). Platelet-derived growth factor mediates activation of ras through different signalling pathways in different cell types. Mol. Cell. Biol. 13, 3706–3713.

Schlessinger, J. (1988). Signal transduction by allosteric receptor oligomerization. TIBS 13, 443–447.

Schlessinger, J., & Ullrich, A. (1992). Growth factor signaling by receptor tyrosine kinases. Neuron, 9, 383–391.

Schneider, R. (1992). The human protooncogene *ret*: A communicative cadherin. TIBS 17, 468–469.

Schuchardt, A., D'Agati, V., Larsson-Blomberg, L., Constantini, F., & Pachnis, V. (1994). Defects in the kidney and enteric nervous system of mice lacking the tyrosine kinase receptor Ret. Nature 367, 380–383.

Segatto, O., Lonardo, F., Wexler, D., Fazioli, F., Pierce, J. H., Bottaro, D. P., White, M. F., & Di Fiore, P. P. (1991). The juxtamembrane regions of the epidermal growth factor receptor and gp185$^{erbB\text{-}2}$ determine the specificity of signal transduction. Mol. Cell. Biol. 11, 3191–3202.

Segatto, O., Londardo, F., Helin, K., Wexler, D., Fazioli, F., Rhee, S. G., & Di Fiore, P. P. (1992). *erb*B-2 autophosphorylation is required for mitogenic action and high-affinity substrate coupling. Oncogene 7, 1339–1346.

Segatto, O., Pelicci, G., Giuli, S., Digiesi, G., Di Fiore, P. P., McGlade, J., Pawson, T. &, Pelicci, P. G. (1993). Shc products are substrates of *erb*B-2 kinase. Oncogene 8, 2105–2112.

Sherr, C.J. (1990). Colony-stimulating factor-1 receptor. Blood 75, 1–12.

Sherr, C. J., Rettenmier, C. W., Sacca, R., Roussel, M. F., Look, A. T., & Stanley, E. R. (1985). The c-*fms* proto-oncogene product is related to the receptor for mononuclear phagocyte growth factor, CSF-1. Cell 41, 665–676.

Shi, E., Kan, M., Xu, J., Wang, F., Hou, J., & McKeehan, W. L. (1993). Control of fibroblast growth factor receptor kinase signal transduction by heterodimerization of combinatorial splice variants. Mol. Cell. Biol. 13, 3907–3918.

Shibuya, M., Yamaguchi, S., Yamane, A., Ikeda, T., Tojo, A., Matsushime, H., & Sato, M. (1990). Nucleotide sequence and expression of a novel human receptor type tyrosine kinase gene (*flt*) closely related to the *fms* family. Oncogene 5, 519–524.

Shier, P., & Watt, V. M. (1989). Primary structure of a putative receptor for a ligand of the insulin family. J. Biol. Chem. 264, 14605–14608.

Shurtleff, S. A., Downing, J. R., Rock, C. O., Hawkins, S. A., Roussel, M. F., & Sherr, C. J. (1990). Structural features of the colony stimulating factor-1 receptor that affects its association with phosphatidylinositol 3-kinase. EMBO J. 9, 2415–2421.

Simon, M. A., Bowtell, D. D. L., & Rubin, G. M. (1989). Structure and activity of the sevenless protein: a protein tyrosine kinase receptor required for photoreceptor development in *Drosophila*. Proc. Natl. Acad. Sci. USA 86, 8333–8337.

Simon, M. A., Dodson, G. S., & Rubin, G. M. (1993). An SH3–SH2–SH3 protein is required for p21ras1 activation and binds to Sevenless and Sos proteins *in vitro*. Cell 73, 169–177.

Skolnik, E. Y., Lee, C.-H., Batzer, A., Vicentini, L. M., Zhou, M., Daly, R., Myer, M. J., Jr., Backer, J. M., Ullrich, A., White, M. F., & Schlessinger, J. (1993). The SH2/SH3 domain-containing protein Grb-2 interacts with tyrosine phosphorylated IRS-1 and Shc: Implications for insulin control of Ras signalling. EMBO J. 12, 1929–1936.

Slamon, D. J., Clark, G. M., Wong, S. G., Levin, W. J., Ullrich, A., & McGuire, W. L. (1987). Human breast cancer: Correlation of relapse and survival with amplification of the HER-2/*neu* oncogene. Science 235, 177–182.

Smith, D. R., Vogt, P. T., & Hayman, M. J. (1989). The v-*sea* oncogene of avian erythroblastosis retrovirus S13: Another member of the protein-tyrosine kinase gene family. Proc. Natl. Acad. Sci. USA 86, 5291–5295.

Soltoff, S. P., Rabin, S. L., Cantley, L. C., Kaplan, D. R. (1992). Nerve growth factor promotes the activation of phosphatidylinositol 3-kinase and its association with the *trk* tyrosine kinase. J. Biol. Chem. 267, 17472–17477.

Songyang, Z., Shoelson, S. E., Chaudhuri, M., Gish, G., Pawson, T., Haser, W. G., King, F., Roberts, T., Rafnofsky, S., Lechleider, R. J., Neel, B. G., Birge, R. B., Fajardo, J. E., Chou, M. M., Hanafusa, H., Schaffhausen, B., & Cantley, L. C. (1993). SH2 domains recognise specific phosphopeptide sequences. Cell 72, 767–778.

Soos, M. A., Whittaker, J., Lammers, R., Ullrich, A., & Siddle, K. (1990). Receptors for insulin and insulin-like growth factor I can form hybrid dimers. Biochem. J. 720, 383–390.

Soppet, D., Escandon, E., Maragos, J., Middlemas, D. S., Reid, S. W., Blair, J., Burton, L. E., Stanton, B. R., Kaplan, D. R., Hunter, T., Nikolics, K., & Parada, L. F. (1991). The neurotrophic factors brain-derived neurotrophic factor and neurotrophin-3 are ligands for the *trk*B tyrosine kinase receptor. Cell 65, 895–903.

Sorkin, A., Di Fiore, P. P., & Carpenter, G. (1993). The carboxyl terminus of epidermal growth factor receptor/*erb*B-2 chimerae is internalization impaired. Oncogene 8, 3021–3028.

Stern, D. F., & Kamps, M. P. (1988). EGF-stimulated tyrosine phosphorylation of $p185^{neu}$: A potential model for receptor interactions. EMBO J. 7, 995–1001.

Streuli, M., Kreuger, N. X., Tsai, A. Y. M., & Saito, H. (1989). A family of receptor-linked protein tyrosine phosphatases in human and *Drosophila*. Proc. Natl. Acad. Sci. USA 86, 8698–8702.

Suen, K.-L., Bustelo, X. R., Pawson, T., & Barbacid, M. (1993). Molecular cloning of the mouse *grb*2 gene: Differential interaction of the Grb2 adaptor protein with epidermal growth factor and nerve growth factor receptors. Mol. Cell. Biol. 13, 5500–5512.

Sun, X.-J., Rothenberg, P., Kahn, C. R., Backer, J. M., Araki, E., Wilden, P. A., Cahill, D. A., Goldstein, B. J., & White, M. F. (1991). The structure of the insulin receptor substrate IRS-1 defines a unique signal transduction protein. Nature 352, 73–77.

Tahira, T., Ishizawa, Y., Itoh, F., Nakayasu, M., Sugimura, T., & Nagao, M. (1991). Expression of the ret proto-oncogene in human neuroblastoma cell lines and its increase during neuronal differentiation induced by retinoic acid. Oncogene 6, 2333–2338.

Takahashi, M., Ritz, J., & Cooper, G. M. (1985). Activation of a novel human transforming gene, ret, by DNA rearrangement. Cell 42, 581–588.

Takahashi, M., Buma, Y., Iwamoto, T., Inaguma, Y., Ikedo, H., & Hirai, H. (1988). Cloning and expression of the ret proto-oncogene encoding a tyrosine kinase with two potential transmembrane domains. Oncogene 3, 571–578.

Takahashi, M., Asai, N., Iwashita, T., Isomura, T., Miyazaki, K., & Matsuyama, M. (1993). Characterization of the *ret* proto-oncogene products expressed in mouse L cells. Oncogene 8, 2925–2929.

Takaishi, K., Sasaki, T., Kato, M., Yamochi, W., Kuroda, S., Nakamura, T., Takeichi, M., & Takai, Y. (1994). Involvement of *Rho* p21 small GTP-binding protein and its regulator in HGF-induced cell motility. Oncogene 9, 273–279.

Tan, J. C., Nocka, K., Ray, P., Traktman, P., & Besmer, P. (1990). The dominant W^{42} *spotting* phenotype results from a missense mutation in the c-*kit* receptor kinase. Science 247, 209–212.

Tavaré, J. M., & Siddle, K. (1993). Mutational analysis of insulin receptor function: Consensus and controversy. Biochim. Biophys. Acta 1178, 21–39.

Taylor, G. R., Reedijk, M., Rothwell, V., Rohrschneider, L., & Pawson, T. (1989). The unique insert of cellular and viral *fms* protein tyrosine kinase domains is dispensable for enzymatic and transforming activities. EMBO J. 8, 2029–2037.

Thomas, S. M., DeMarco, M., D'Arcangelo, G., Halegoua, S., & Brugge, J. S. (1992). Ras is essential for nerve growth factor- and phorbol ester-induced tyrosine phosphorylation of MAP kinases. Cell 68, 1031–1040.

Treadway, J. L., Morrison, B. D., Soos, M. A., Siddle, K., Olefsky, J., Ullrich, A., McClain, D. A., & Pessin, J. E. (1991). Transdominant inhibition of tyrosine kinase activity in mutant insulin/insulin growth factor I hybrid receptors. Proc. Natl. Acad. Sci. USA 88, 214–218.

Tsarfaty, I., Rong, S., Resau, J. H., Rulong, S., da Silva, P. P., & Vande Woude, G. F. (1994). The Met proto-oncogene mesenchymal to epithelial cell conversion. Science 263, 98–101.

Ueno, H., Escobedo, J. A., & Williams, L. T. (1993). Dominant–negative mutations of platelet-derived growth factor (PDGF) receptors. J. Biol. Chem. 268, 22814–22819.

Ullrich, A., & Schlessinger, J. (1990). Signal transduction by receptors with tyrosine kinase activity. Cell 61, 203–212.

Ullrich, A., Coussens, L., Hayflick, J. S., Dull, T. J., Gray, A., Tam, A. W., Lee, T. J., Yarden, Y., Liberman, T. A., Schlessinger, J., Downward, J., Mayes, E. L. V., Whittle, N., Waterfield, M. D., & Seeberg,

P. H. (1984). Human epidermal growth factor receptor cDNA sequence and aberrant expression of the amplified gene in A431 epidermoid carcinoma cells. Nature 309, 418–425.

Ullrich, A., Bell, J. R., Chen, E. Y., Herrera, R., Petruzzelli, L. M., Dull, T. J., Gray, A., Coussens, L., Liao, Y.-C., Tsubokawa, M., Mason, A., Seeberg, P. H., Grunfeld, C., Rosen, O. M., & Ramachandran, J. (1985). Human insulin receptor and its relationship to the tyrosine kinase family of oncogenes. Nature 313, 756–761.

Ullrich, A., Gray, A., Tam, A. W., Yeng-Feng, T., Tsubokawa, M., Collins, C., Henzel, W., Le Bon, T., Kathuria, S., Chen, E., Jacobs, S., Francke, U., Ramachandran, J., & Fujita-Yamaguchi, Y. (1986). Insulin-like growth factor I receptor primary structure: comparison with the insulin receptor suggests structural determinants that define functional specificity. EMBO J. 5, 2503–2512.

Vainikka, S., Partanen, J., Bellosta, P., Coulier, F., Basilico, C., Jaye, M., & Alitalo, K. (1992). Fibroblast growth factor receptor-4 shows novel features in genomic structure, ligand binding and signal transduction. EMBO J. 11, 4273–4280.

Valazquez, L., Fellous, M., Stark, G., & Pellegrini, S. (1992). A protein tyrosine kinase in the interferon α/β signaling pathway. Cell 70, 313–322.

Valius, M., & Kazlauskas, A. (1993). Phospholipase Cγ1 and phosphatidylinositol 3-kinase are downstream mediators of PDGF receptor's mitogenic signal. Cell 73, 321–334.

Valius, M., Bazenet, C., & Kazlauskas, A. (1993). Tyrosines 1021 and 1009 are phosphorylation sites in the carboxy terminus of the platelet-derived growth factor receptor β subunit and are required for binding phospholipase Cγ and a 64 kilodalton protein respectively. Mol. Cell. Biol. 13, 133–143.

van der Geer, P., & Hunter, T. (1993). Mutation of Tyr697, a GRB2-binding site, and Tyr721, a PI 3-kinase binding site, abrogates signal transduction by the murine CSF-1 receptor expressed in RAT-2 fibroblasts. EMBO J. 12, 5161–5172.

Villa-Moruzzi, E., Lapi, S., Prat, M., Gaudino, G., & Comoglio, P. M. (1993). A protein tyrosine phosphatase activity associated with the hepatocyte growth factor/scatter factor receptor. J. Biol. Chem. 268, 18176–18180.

Wahl, M. I., Daniel, T. O., & Carpenter, G. (1988). Anti-phosphotyrosine recovery of phospholipase C activity after EGF treatment of A431 cells. Science 241, 968–970.

Wang, L.-M., Myers, M. G. Jr., Sun, X.-J., Aaronson, S. A., White, M., & Pierce J. H. (1993). IRS-1: Essential for insulin- and IL-4-stimulated mitogenesis in hematopoietic cells. Science 261, 1591–1594.

Wang, J.-K., Gao, G., & Goldfarb, M. (1994). Fibroblast growth factor receptors have different signaling and mitogenic potentials. Mol. Cell. Biol. 14, 181–188.

Waters, S. B., Yamauchi, K., & Pessin, J. E. (1993). Functional expression of insulin receptor substrate-1 is required for insulin-stimulated mitogenic signaling. J. Biol. Chem. 268, 22231–22234.

Weiss, A. (1993). T cell antigen receptor signal transduction: A tail of tails and cytoplasmic protein-tyrosine kinases. Cell 73, 209–212.

Wen, D., Peles, E., Cupples, R., Suggs, S. V., Bacus, S. S., Luo, Y., Trail, G., Hu, S., Silbiger, S. M., Levy, R. B., Koski, R. A., Lu, H. S., & Yarden, Y. (1992). Neu differentiation factor: A transmembrane glycoprotein containing an EGF domain and an immunoglobulin homology unit. Cell 69, 559–572.

Wennström, S., Siegbahn, A., Yokote, K., Arvidsson, A.-K., Heldin, C.-H., Mori, S., & Claesson-Welsh, L. (1994). Membrane ruffling and chemotaxis transduced by the PDGF β-receptor require the binding site for phosphatidylinositol 3′ kinase. Oncogene 9, 651–660.

Werner, S., Weinberg, W., Liao, X., Peters, K. G., Blewssing, M., Yuspa, S. H., Weiner, R. L., & Williams, L. T. (1993). Targeted expression of a dominant–negative FGF receptor mutant in the epidermis of transgenic mice reveals a role of FGF in keratinocyte organization and differentiation. EMBO J. 12, 2635–2643.

White, M. F., Livingston, J. N., Backer, J. M., Lauris, V., Dull, T., Ullrich, A., & Kahn, C. R. (1988). Mutation of the insulin receptor at tyrosine 960 inhibits signal transmission but does not affect its tyrosine kinase activity. Cell 54, 641–649.

White, M. F., & Kahn, C. R. (1994). The insulin signaling system. J. Biol. Chem. 269, 1–4.

Wicks, I. P., Wilkinson, D., Salvaris, E., & Boyd, A. W. (1992). Molecular cloning of *HEK*, the gene encoding a receptor tyrosine kinase expressed by the human lymphoid tumor cell lines. Proc. Natl. Acad. Sci. USA 89, 1611–1615.

Wilks, A. F. (1989). Two putative protein-tyrosine kinases identified by application of the polymerase chain reaction. Proc. Natl. Acad. Sci. USA 86, 1603–1607.

Wittbrodt, J., Adam, D., Malitschek, B., Mäueler, B., Raulf, F., Telling, A., Robertson, S. M., & Schartl, M. (1989). Novel putative receptor tyrosine kinase encoded by the melanoma-inducing Tu locus in *Xiphophorus*. Nature 341, 415–421.

Witte, O. N. (1990). Steel locus defines new multipotent growth factor. Cell 63, 5–6.

Witthuhn, B. A., Quelle, F. W., Silvennoinen, O., Yi, T., Tang, B., Miyra, O., & Ihle, J. N. (1993). JAK2 associates with the erythropoietin receptor and is tyrosine phosphorylated and activated following stimulation with erythropoietin. Cell 74, 227–236.

Wood, K. W., Sarnecki, C., Roberts, T. M., & Blenis, J. (1992). ras mediates nerve growth factor receptor modulation of three signal transducing protein kinases: MAP kinase, Raf-1 and RSK. Cell 68, 1041–1050.

Yamamoto, T., Ikawa, S., Akiyama, T., Semba, K., Nomura, N., Mijajima, N., Saito, T., & Toyoshima, K. (1986). Similarity of protein encoded by the human c-erbB-2 gene to the epidermal growth factor receptor. Nature 319, 230–234.

Yarden, Y., & Ullrich, A. (1988). Growth factor receptor tyrosine kinases. Ann. Rev. Biochem. 57, 443–478.

Yarden, Y., Escobedo, J. A., Kuang, W.-J., Yang-Feng, T. L., Daniel, T. O., Tremble, P. M., Chen, E. Y., Ando, M. E., Harkins, R. N., Francke, Y., Fried, V. A., Ullrich, A., & Williams, A. (1986). Structure of the receptor for the platelet-derived growth factor receptor helps define a family of closely related growth factor receptors. Nature 323, 226–232.

Yarden, Y., Kuang, W.-J., Yang-Feng, T., Coussens, L., Munemitsu, S., Dull, T.J., Chen, E., Schlessinger, J., Francke, U., & Ullrich, A. (1987). Human proto-oncogene c-*kit*: A new cell surface receptor tyrosine kinase for an unidentified ligand. EMBO J. 6, 3341–3351.

Yayon, A., Zimmer, Y., Guo-Hong, S., Avivi, A., Yarden, Y., & Givol, D. (1992). A confined variable region confers ligand specificity on fibroblast growth factor receptors: Implications for the origin of the immunoglobulin fold. EMBO J. 11, 1885–1890.

Yi, T., & Ihle, J. N. (1993). Association of hematopoietic cell phosphatase with c-Kit after stimulation with c-Kit ligand. Mol. Cell. Biol. 13, 3350–3358.

Yu, J.-C., Heidaran, M. A., Pierce, J. H., Gutkind, J. S., Lombardi, D., Ruggiero, M., & Aaronson, S. A. (1991). Tyrosine mutations within the α platelet-derived growth factor receptor kinase insert domain abrogate receptor associated phosphatidylinositol-3 kinase activity without affecting mitogenic or chemotactic signal transduction. Mol. Cell. Biol. 11, 3780–3785.

Zerlin, M., Julius, M. A., & Goldfarb, M. (1993). NEP: A novel receptor-like tyrosine kinase expressed in proliferating neuroepithelia. Oncogene 8, 2731–2739.

Zhang, B., & Roth, R. A. (1992). The insulin receptor-related receptor. J. Biol. Chem. 267, 18320–18328.

Ziegler, S. F., Bird, T. A., Schneringer, J. A., Schooley, K. A., & Baum, P. R. (1993). Molecular cloning and characterization of a novel receptor protein tyrosine kinase from human placenta. Oncogene 8, 663–670.

TRANSMEMBRANE PROTEIN TYROSINE PHOSPHATASES

Edward C. C. Wong, Terry A. Woodford-Thomas, and Matthew L. Thomas

Biomembranes
Volume 6, pages 77–106.

ISBN: 1-55938-663-0.

I. INTRODUCTION

Cell surface receptors serve to monitor environmental signals and function to initiate events that result in changes in cellular physiology. One key regulatory mechanism effected by many cell surface receptors is reversible tyrosine phosphorylation. This important molecular switch controls many aspects of cellular activation, differentiation, and division. Tyrosine phosphorylation is controlled by the antagonistic action of protein tyrosine kinases (PTKs) and protein tyrosine phosphatases (PTPases). Transmembrane PTKs function as receptors for growth factors and regulate signaling by an increase in enzymatic activity upon ligand binding, thus effecting an increase in tyrosine phosphorylation on key cellular protein substrates. By analogy, it is possible that transmembrane PTPases could counter the effect of transmembrane PTKs by binding a ligand, increasing phosphatase activity, and thus, down-regulating the events initiated by growth factors (Charbonneau et al., 1988; Fischer et al., 1991). However, extracellular signals have not been shown to effect transmembrane PTPase activity and it is likely that the interactions and regulatory effects exerted by transmembrane PTPases are more intricate. Indeed, while ligand binding has been demonstrated for many transmembrane PTKs, and thus, they are known as receptor PTKs, the interactions of the extracellular domains for most transmembrane PTPases are not known. This review will discuss transmembrane phosphatases with regards to their potential roles in regulating cellular behavior.

II. DISCOVERY AND STRUCTURE

The importance of receptor PTKs in regulating cell growth led to the search for PTPases that may function to counter the effects of PTKs. This culminated in 1988 with the report by Fischer and colleagues of the isolation and sequence of a 32 kD soluble PTPase from human placenta, PTP1B (Charbonneau et al., 1988; Tonks et al., 1988b). Surprisingly, PTP1B did not resemble serine/threonine phosphatases, but demonstrated significant similarity to the previously reported sequence of the leukocyte cell surface glycoprotein, CD45 (Thomas et al., 1985). CD45 consists of a large, amino-terminal, heavily glycosylated extracellular domain, a single membrane spanning region, and a 700 amino acid cytoplasmic domain that contains two tandem subdomains, each with significant homology to PTP1B (Thomas, 1989). The implication from this observation was that there were both transmembrane and intracellular PTPases, and indeed, CD45 was quickly shown to have PTPase activity (Tonks et al., 1988a). Concurrent with this observation, a second transmembrane glycoprotein, LAR, was reported to have a cytoplasmic domain with similarity to CD45 (Streuli et al., 1988). However, LAR did not contain a heavily glycosylated extracellular domain, but had an extracellular domain reminiscent of the neural cell adhesion glycoprotein, N-CAM. Thus, these initial observations demonstrated that there were multiple diverse PTPases.

The conservation in sequence between PTPase domains resulted in the ability to design degenerate oligonucleotides for the cloning of additional PTPases cDNAs by reverse transcription and polymerase chain reaction (PCR). A large number of PTPase cDNAS have thus been isolated, and the PTPase family appears to be as large and as diverse as the PTKs. A comparison of PTPase sequences shows that a PTPase domain is approximately 260–300 amino acids in length and there are 'hallmark' residues conserved by all PTPases, some of which have been demonstrated by site-directed mutagenesis to be critical for activity (Figure 1). Catalytic activity proceeds through the formation of a thiol–phosphate intermediate with the thiol group being donated by a cysteine located in the highly conserved sequence, HCSAGXGRTG.

All but one transmembrane PTPase, PTPβ, contain two tandem PTPase domains. However, it appears that while some PTPase activity can be demonstrated for the second domain, the first domain is responsible for most of the activity and the function of the second domain is not known (Wang and Pallen, 1991, 1992; Streuli et al.,1990; Cho et al., 1991, 1992; Itoh et al., 1992; Johnson et al., 1992).

A large number of mammalian transmembrane PTPases have been identified over the past few years and they can be divided into 6 groups based on structural and sequence similarity (Figures 2 and 3). For the purpose of this review, the six groups will be classified by representative members and be discussed in the order of CD45; LRP family; LAR family, PTP μfamily; PTPγ family; and PTPβ. The transmembrane PTPases are evolutionarily conserved and comparisons between mammalian and invertebrate transmembrane PTPases will be discussed.

III. CD45

CD45 is a major lymphocyte glycoprotein that is ubiquitously expressed by all nucleated cells of hematopoietic origin and constitutes approximately 10% of the cell surface area (Thomas, 1989). Multiple isoforms exist due to the variable use of exon-encoding sequences near the amino-terminus. Three exons, 4, 5, and 6, are alternatively used to give rise to eight isoforms. In addition, a ninth isoform has

```
CD45  EFQSIP---R VFSKFPIKDA RKPHNQNKNR YVDILPYDYN RVELSEING- -DAGSTYINA SYID------ GFKEPRKYIA AQGPRDETVD DFWRMIWEQK
LRP   EFNALP---A CPIQATCEAA SKEENKEKNR YVNILPYDHS RVHLTPVEG- -VPDSDYINA SFIN------ GYQEKNKFIA AQGPKEETVN DFWRMIWEQN
PTPε  EFNSLP---S GHIQGTFELA NKEENREKNR YPNILPNDHS RVILSQLDG- -IPCSDYINA SYID------ GYKEKNKFIA AQGPKQETVN DFWRMVWEQK
LAR   EYESID---P GQ-QFTWENS NLEVNKPKNR YANVIAYDHS RVILTSIDG- -VPGSDYINA NYID------ GYRKQNAYIA TQGPLPETMG DFWRMVWEQR
PTPδ  EYESID---P GQ-QFTWEHS NLEVNKPKNR YANVIAYDHS RVILSAIEG- -IPGSDYVNA NYID------ GYRKQNAYIA TQGSLPETFG DFWRMIWEQR
PTPσ  EYESID---P GQ-QFTWEHS NLEANKPKNR YANVIAYDHS RVILQPLEG- -IMGSDYINA NYVD------ GYRRQNAYIA TQGPLPETFG DFWRMVWEQR
PTPμ  EYESFF---E GQ-SAPWDSA KKDENRMKNR YGNIIAYDHS RVRLQTIEG- -DTNSDYING NYID------ GYHRPNHYIA TQGPMQETIY DFWRMVWHEN
PTPκ  EYESFF---E GQ-SASWDVA KKDQNRAKNR YGNIIAYDHS RVILQPVED- -DPSSDYINA NYIDIWLYRD GYQRPSHYIA TQGPVHETVY DFWRMVWQEQ
PTPγ  EVQRCT---A DM-NITAEHS NHPENKHKNR YINILAYDHS RVKLRPLPGK DSKHSDYINA NYVD------ GYNKAKAYIA TQGPLKSTFE DFWRMIWEQN
PTPζ  EFYQEVQSCT VDLGITADSS NHPDNKHKNR YINIVAYDHS RVKLAQLAEK DGKLTDYINA NYVD------ GYNRPKAYIA AQGPLKSTAE DFWRMIWEHN
PTPβ  EYEELK---D VGRNQSCDIA LLPENRGKNR YNNILPYDAT RVKLSNVDD- -DPCSDYINA SYIP------ GNNFRREYIV TQGPLPGTKD DFWKMVWEQN
SAP-1 EYQQLS---L VGHSQSQMVA SASENNAKNR YRNVLPYDWS RVPLKPIHE- -EPGSDYINA SFMP------ GLWSPQEFIA TQGPLPQTVG DEWRLVWEQQ
Cons  EY.S..---. .....T...A ....N..KNR Y.NI.AYDHS RV.L....G- -.P.SDYINA NYID------ GY.....YIA TQGPL.ET.. DFWRMVWEQ.

CD45  EATVIVMVTR CEEGNRNKCA EYWPSMEEGT RAFKDIVVTI NDHKRCPDYI IQKL------ ----NVAHKK EKATGREVTH IQFTSWPDHG VPEDPHLLLK
LRP   EATIVMV-TN LKERKECKCA QYWP--DQGC WTYGNVRVSV EDVTVLVDYT VRKFCIQ--- ----QVGDVT NRKPQRLITQ FHFTSWPDFG VPFTPIGMLK
PTPε  EATIVML-TN LKERKEEKCH QYWP--DQGC WTYGNIRVCV EDCVVLVDYT IRKFCIQ--- ----PQLPDG CKAP-RLVSQ LHFTSWPDFG VPFTPIGMLK
LAR   EATVVMM-TR LEEKSRVKCD QYWP--ARGT ETCGLIQVTL LDTVELATYT VRTFALH--- ----K-SGSS EK---RELRQ FQFMAWPDHG VPEYPTPILA
PTPδ  EATVVMM-TK LEERSRVKCD QYWP--SRGT ETHGLVQVTL LDTVELATYC VRTFALY--- ----K-NGSS EK---REVRQ FQFTAWPDHG VPEHPTPFLA
PTPσ  EATVVMM-TR LEEKSRVKCD QYWP--NRGT ETYGFIQVTL LDTMELATFC VRTFSLH--- ----K-NGSS EK---REVRH FQFTAWPDHG VPEYPTPFLA
PTPμ  EASIIMV-TN LVEVGRVKCC KYWP---DDT EIYKDIKVTL IETELLAEYV IRTFAVE--- ----K-RGVH EI---REIRQ FHFTGWPDHG VPYHATGLLG
PTPκ  EACIVMV-TN LVEVGRVKCY KYWP---DDT EVYGDFKVTC VEHEPLAEYV VRTFTLE--- ----R-RGYN EI---REVKQ FHFTGWPDHG VPYHATGLLS
PTPγ  EGIIVMI-TN LVEKGRRKCD QYWPTENSE- -EYGNIIVTL KSTKIHACYT VRRFSIRNTK VKKGQKGNPK GRQNERVVIQ YHYTQWPDMG VPEYALPVLT
PTPζ  EVEVIVMITN LVEKGRRKCD QYWP--ADGS EEYGNFLVTQ KSVQVLAYYT VRNFTLR--- NTKIKKGSQK GRPSGRVVTQ YHYTQWPDMG VPEYSLPVLT
PTPβ  EHNIVMV-TQ CVEKGRVKCD HYWPADQDS- LYYGDLILQM LSESVLPEWT IREFKIC--- ----GEEQLD AH---RLIRH FHYTVWPDHG VPETTQSLIQ
SAP-1 EHTLVML-TN CMEAGRVKCE HYWP-LDSQP CTHGHLRVTL VGEEVMENWT VRELLLL--- ----QVEEQK TL----SVRQ FHYQAWPDHG VPSSPDTLLA
Cons  EAT.VM.-TN L.E..RVKC. QYWP--..G. ..YG...VT. .....LA.YT VR.F...--- ----...... ..---R.V.Q FHFT.WPDHG VPE.P...L.
```

CD45	ELRRRVNAF--	SNFFSGPIVV	HCSAGVGRTG	TYIGIDAMLE	GLEAEGKVDV	YGYVVKLRRQ	RCLMVQVEAQ	YILIHQALVE	YNQFGETQVN	LSELHSCLHN
LRP	EFLKKVKAC--	NPQYAGAIVV	HCSAGVGRTG	TFVVIDAMLD	MMHSERKVDV	YGFVSRIRAQ	RCQMVQTDMQ	YVFIYQALLE	HYLYGDTELE	VTSLETHLQK
PTPε	EFLKKVKTL--	NPVHAGPIVV	HCSAGVGRTG	TFIVIDAMMA	MMHAEQKVDV	FEFVSRIRNQ	RPQMVQTDMQ	YTFIYQALLE	YYLYGDTELD	VSSLEKHLQT
LAR	EFLRRVKAC--	NPLDAGPMVV	HCSAGVGRTG	CFIVIDAMLE	RMKHEKTVDI	YGHVTCMRSQ	RNYMVQTEDQ	YVFIHEALLE	AATCGHTEVP	ARNLYAHI--
PTPδ	EFLRRVKTC--	NPPDAGPMVV	HCSAGVGRTG	CFIVIDAMLE	RIKHEKTVDI	YGHVTLMRAQ	RNYMVQTEDQ	YIFIHDALLE	AVTCGNTEVP	ARNLYAYI--
PTPσ	EFLRRVKTC--	NPPDAGPVVV	HCSAGVGRTG	CFIVIDAMLE	RIRTEKTVDV	YGHVTLMRSQ	RNYMVQTEDQ	YSFIHEALLE	AVGCGNTEVP	ARSLYTYIQK
PTPμ	EFVRQVKSK--	SPPSAGPLVV	HCSAGAGRTG	CFIVIDIMLD	MAEREGVVDI	YNCVRELRSR	RVNMVQTEEQ	YVFIHDAILE	ACLCGDTSVP	ASQV-RSL--
PTPκ	EFIRRVKLS--	NPPSAGPIVV	HCSAGAGRTG	CYIVIDIMLD	MAEREGVVDI	YNCVKALRSR	RINMVQTEEQ	YIFIHDAILE	ACLCGETAIP	VCE-------
PTPγ	EFVRRSSAA--	RMPETGPVLV	HCSAGVGRTG	TYIVIDSMLQ	QIKDKSTVNV	LGFLKHIRTQ	RNYLVQTEEQ	YIFIHDALLE	AILGKETEVS	SNQLHSYVNS
PTPζ	EFVRKAAYA--	KRHAVGPVVV	HCSAGVGRTG	TYIVLDSMLQ	QIQHEGTVNI	FGFLKHIRSQ	RNYLVQTEEQ	YVFIHDTLVE	AILSKETEVL	DSHIHAYVNA
PTPβ	EFVRTVRDYIN	RSPGAGPTVV	HCSAGVGRTG	TFIALDRILQ	QLDSKDSVDI	YGAVHDLRLH	RVHMVQTECQ	YVYLHQCVRD	VLRARKLRSE	QENPLFPIYE
SAP-1	EFWRMLRQWLD	QTMEGGPPIV	HCSAGVGRTG	TLIALDVLLR	QLQSEGLLGP	FSFVRKMRES	RPLMVQTEAQ	YVFLHQCICG	SSNSQPRPQP	RRKS------
Cons	EF.R.VK..--	.P..AGP.VV	HCSAGVGRTG	TFIVID.ML.	....E..VD.	YG.V...R.Q	R..MVQTE.Q	Y.FIH.ALLE	A...G.TEV.	...L......

Figure 1. Sequence comparison of the first catalytic domain from mammalian transmembrane PTPases. Consensus residues are shown below and identical residues are shaded. Gaps for optimal alignment are shown by dashes. The single letter amino acid code is used. Sequences were derived from the following references: CD45 (Thomas et al., 1985), LRP (Matthews et al., 1990), PTPε (Krueger et al., 1990), LAR (Streuli et al., 1988), PTPδ (Krueger et al., 1990), PTPσ (Yan et al., 1993), PTPμ (Gebbink et al., 1991), PTPκ (Jiang et al., 1993), PTPγ (Krueger et al., 1990), PTPζ (Krueger et al., 1990), PTPβ (Krueger et al., 1990), and SAP-1 (Matozaki et al., 1994).

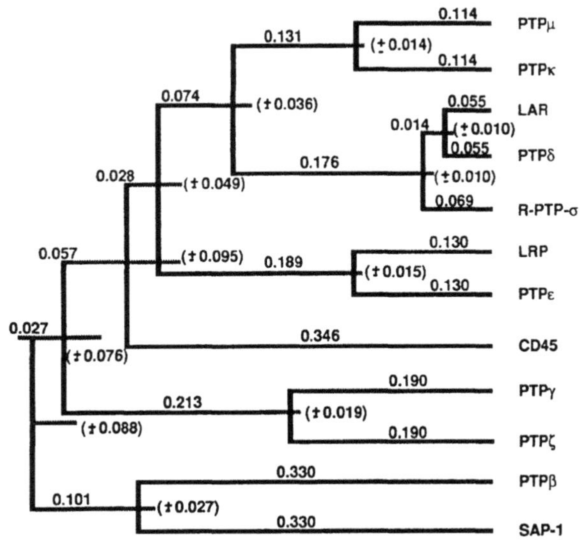

Figure 2. Tree alignment of the first catalytic domain from mammalian PTPases. Alignments were made using unweighted pair group method with arithmetic mean in the GeneWorks program (Intelligenetics). The length of the horizontal line is proportional to estimated genetic distance between the sequences. The shaded regions represent the standard error of the branch position.

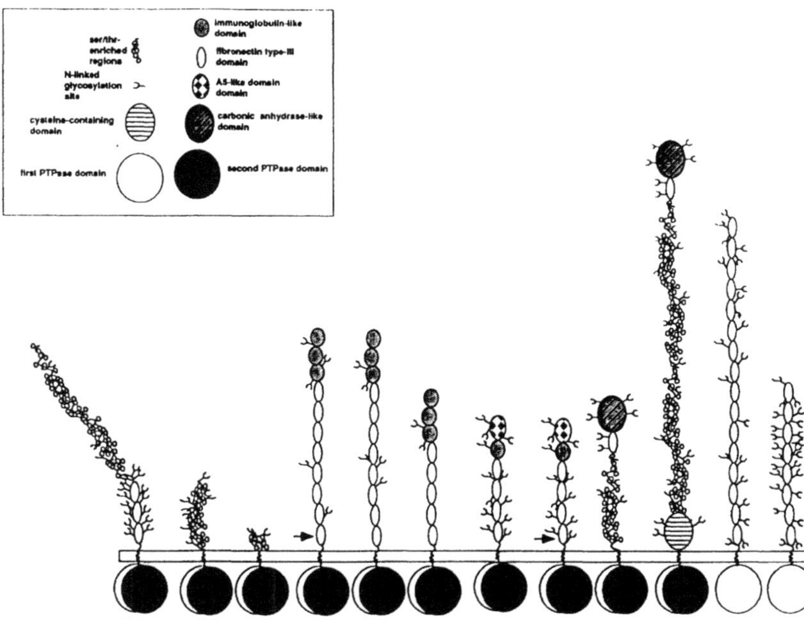

Figure 3. Cartoon representation of the mammalian transmembrane PTPases.

been described by PCR analysis that results from the splicing of exon 3 to 8 (Chang et al., 1991). The use of the variable exons is highly regulated during cellular differentiation and activation, implying that their use is important for function. This idea is supported by the observation that the pattern of variable exon usage is conserved in vertebrate evolution (Robb, Matthews, and Thomas, forthcoming).

A. The Requirement for CD45 in Lymphocyte Activation

CD45 function is the most well characterized of the transmembrane PTPases. Because CD45 is an abundant lymphocyte glycoprotein, it has been extensively studied for twenty years. However, it has only relatively recently been shown that CD45 is required for lymphocyte differentiation and activation. Studies using anti-CD45 monoclonal antibodies suggested that CD45 was involved in lymphocyte activation (Prickett and Hart, 1990; Kanner and Ledbetter, 1992; Schraven et al., 1989; King et al., 1990; Ledbetter et al., 1991; Nel et al., 1991; Goldman et al., 1992; Wagner et al., 1993). More compelling evidence was obtained by the use of CD45-deficient cell lines and mice (Pingel and Thomas, 1989; Koretzky et al., 1990, 1991; Justement et al., 1991; Peyron et al., 1991; Weaver et al., 1991; Deans et al., 1992; Shiroo et al., 1992; Volarevic et al., 1992; Kishihara et al., 1993; Bell et al., 1993). T cell clones deficient in CD45 expression fail to efficiently respond to antigen although they express the antigen receptor and other cell surface proteins required for activation (Pingel and Thomas, 1989; Weaver et al., 1991). Cross-linking of the antigen receptor, a normal mitogenic signal, also fails to elicit an efficient response indicating that the inability to signal through the antigen receptor was not due to the failure to recognize antigen. Similarly, in response to anti-antigen receptor antibodies, CD45-deficient T cell leukemic lines do not generate intracellular signals, including an increase in tyrosine phosphorylation, inositol (tris)phosphate, or intracellular calcium (Koretzky et al., 1990, 1991; Peyron et al., 1991; Deans et al., 1992; Volarevic et al., 1992). CD45-deficient B cells and natural killer cells are also unable to initiate intracellular signals in response to membrane immunoglobulin and Fc receptor cross-linking, respectively (Justement et al., 1991; Bell et al., 1993). Thus, the requirement of CD45 for activation appears to be a feature shared by all lymphocytes.

Activation can be rescued by treating either CD45-deficient T cell clones or T cell leukemic lines with agents that activate second messengers (such as phorbol esters in conjunction with calcium ionophores), suggesting that the defect in signal transduction is proximal to antigen recognition (Koretzky et al., 1991; Pingel et al., 1994). Indeed, since an increase in tyrosine phosphorylation is the earliest event upon antigen receptor activation, a failure to increase tyrosine phosphorylation suggests that a deficiency in the CD45 PTPase results in a failure to activate a PTK(s).

Direct evidence that CD45 regulates PTKs came from the analysis of CD45-deficient lymphoma cell lines which, unfortunately, could not be analyzed for defects in signal transduction (Mustelin et al., 1989; Ostergaard et al., 1989; Hurley et al., 1993). The Src-family member kinase, $p56^{lck}$, was increased in tyrosine phosphorylation in CD45-deficient lymphoma lines, suggesting that this kinase may be a direct substrate of CD45. Further analysis of these lines suggested that a second Src-family member, $p59^{fyn}$ was also increased in tyrosine phosphorylation (Hurley et al., 1993). Both $p56^{lck}$ and $p59^{fyn}$ are substrates for CD45 *in vitro*, and it is possible that CD45 directly dephosphorylates both kinases *in vivo* (Mustelin and Altman, 1990; Guttinger et al., 1992; Mustelin et al., 1992). In contrast, expression of $p60^{c\text{-}src}$ in a CD45-deficient lymphoma did not demonstrate an increase in tyrosine phosphorylation, indicating that not all Src-family members expressed by leukocytes are necessarily affected by CD45 (Hurley et al., 1993).

Src-family members contain two sites of tyrosine phosphorylation, a site within the kinase domain that potentiates kinase activity, and a site near the carboxy-terminus that functions as a negative-regulatory site (Hunter, 1987). In addition, all Src-family members contain an amino-terminal SH2 domain that serves to interact with tyrosine phosphorylated residues in a sequence-specific manner (Songyang et al., 1993). Recent data suggests that the carboxy-terminal tyrosine phosphorylation site of Src-family members interacts intramolecularly with the amino-terminal SH2 domain (reviewed in Mustelin and Burn, 1993), thus accounting for the ability of the carboxy-terminal phosphorylation site to inhibit activity.

Analysis of CD45-deficient T cell clones and leukemic lines defective in antigen receptor signaling demonstrates dysregulation of Src-family member kinases. $p56^{lck}$ and $p59^{fyn}$ from CD45-deficient T cell clones show increased tyrosine phosphorylation at the negative-regulatory site and accordingly, have decreased kinase activity (Cahir McFarland et al., 1993). Similarly, $p56^{lck}$ from the CD45-deficient leukemic line, Jurkat, does not bind a tyrosine phosphorylated peptide corresponding to the carboxy-terminal site, suggesting that the SH2 domain is blocked (Sieh et al., 1993). Furthermore, this line demonstrates an increase in tyrosine phosphorylation at the carboxy-terminal site. Interestingly, $p59^{fyn}$ is not dysregulated in these cells. An additional study shows that a CD45-deficient HPB.ALL cell line has decreased $p59^{fyn}$ kinase activity, but not $p56^{lck}$ kinase activity (Shiroo et al., 1992). Therefore, there is a discordance in regulation of Src-family by CD45 in various leukemic lines. Nevertheless, the data indicate that Src-family members are regulated by CD45 and correlate this regulation with the ability to transduce signals through the antigen receptor.

It is likely that CD45 directly dephosphorylates Src-family members *in vivo*; $p56^{lck}$ co-immunoprecipitates with CD45 and both $p56^{lck}$ and $p59^{fyn}$ will co-cap with CD45 upon cross-linking (Schraven et al., 1991, 1992; Guttinger et al., 1992; Mustelin et al., 1992; Koretzky et al., 1993). Furthermore, it is likely that the failure to signal through the antigen receptor in CD45-deficient cells is due to the inability to efficiently activate Src-family members since both $p56^{lck}$ and $p59^{fyn}$ have been implicated in T cell antigen receptor signaling (reviewed in Perlmutter et al., 1993).

To examine the role of CD45 in hematopoiesis, mice have been generated in which the CD45 gene is disrupted by replacement of exon 6 with the neomycin resistance gene (Kishihara et al., 1993). Because it was possible that CD45 gene ablation would result in a failure of hematopoiesis and embryonic lethality, exon 6, an alternatively spliced exon, was targeted to preserve expression of CD45 isoforms not using exon 6. However, exon 6-disrupted mice are virtually devoid of CD45 expression. While most hematopoietic lineages develop normally in these mice, thymocyte maturation is blocked, resulting in few peripheral T cells. The block is consistent with the failure to proceed with positive selection, although the mice do appear capable of negative selection. Interestingly, while other hematopoietic lineages develop normally, membrane immunoglobulin cross-linking of B cells does not result in an increase intracellular calcium, indicating that both T and B cells require CD45 for activation.

B. Regulation of CD45 Function

Certain anti-CD45 monoclonal antibodies can directly inhibit lymphocyte activation (Prickett and Hart, 1990; Kanner and Ledbetter, 1992; Schraven et al., 1989; King et al., 1990; Ledbetter et al., 1991; Nel et al., 1991; Goldman et al., 1992; Wagner et al., 1993). While the molecular basis for the inhibition is not known, it raises the possibility that ligand binding to the extracellular domain of CD45 may affect CD45 function. However, no ligands for CD45 are known and changes in CD45 function have not been demonstrated in response to physiological stimuli.

Indeed, the extracellular domain of CD45 does not appear to be required to initiate antigen-receptor signaling. CD45-deficient leukemic lines can be rescued for antigen-receptor-induced signaling by expressing, on the plasma membrane, chimeric proteins that contain the CD45 cytoplasmic domain, but not the extracellular domain (Mustelin et al., 1989; Desai et al., 1993; Hovis et al., 1993; Volarevic et al., 1993). These studies indicate that surface expression of the CD45 cytoplasmic domain is sufficient to generate the signals necessary for activation and that the extracellular domain is not required to activate phosphatase activity. This is supported by the observation that bacterial expressed CD45 cytoplasmic domain and CD45 isolated from splenocytes are 10–100 faster in catalytic activity than PTKs (Tonks et al., 1990; Cho et al., 1992). This observation suggests that unlike receptor PTKs, ligand binding may not be required to increase enzymatic activity.

It is possible that CD45 is biosynthesized in an active form and functions to constituitively dephosphorylate Src-family members. Comparison of $p56^{lck}$ kinase activity between extensively rested CD45-deficient and parental T cell clones shows that the kinase from the parental lines is 8–10 times more active (Pingel et al., 1994). In contrast, activation of lymphocytes by anti-Thy-1 (a stimulation that mimics T cell receptor-induced activation) results in a twofold increase in $p56^{lck}$ kinase activity. Likewise, anti-CD4 cross-linking results in a 4-fold increase in activity (Veillette et al., 1989). Therefore, the difference in $p56^{lck}$ kinase activity is greater between resting CD45-expressing and -deficient lines than that found upon activation. This is supported by the observation that CD45 function is continually required for signaling through the T cell antigen receptor (Desai et al., 1993). Expression of a chimeric protein consisting of the epidermal growth factor (EGF) receptor extracellular and transmembrane domains and the CD45 cytoplasmic domain will restore the ability of CD45-deficient HPB.ALL cells to increase intracellular calcium in response to antigen-receptor antibodies (Desai et al., 1993). Interestingly, addition of EGF at the time when the increase in intracellular calcium is maximal results in a dramatic and precipitous decrease in intracellular calcium. This appears due to EGF-induced dimerization of the receptor and suggests that CD45 function is required during antigen-induced signaling. Therefore, it is possible that CD45 functions to dephosphorylate Src-family members prior to and during signaling through the antigen receptor.

It is not known whether CD45 dimerizes or whether this is a potential means of regulation. However, studies utilizing fluorescence resonance energy transfer suggest that in resting lymphocytes CD45 exists as a monomer (Mittler et al., 1991). However, CD45 may associate with the cytoskeleton since CD45 is found in antibody-induced capping of cell surface molecules (Bourguignon et al., 1978). The cytoskeleton component, fodrin, will bind to CD45 and causes an increase in phosphatase activity *in vitro*. (Lokeshwar and Bourguignon, 1992). Thus, it is possible that CD45 function may be regulated by cytoskeletal interactions.

Treatment of thymocytes or T cell hybridomas with ionomycin results in a decrease in CD45 phosphatase activity and a corollary decrease in CD45 serine phosphorylation (Ostergaard and Trowbridge, 1991). Nonetheless, CD45 appears to be enzymatically active and functioning in resting cells, and regulation of activity has not been shown to coincide with any biological response.

C. Structural Implications of the Extracellular Domain

One of the pressing issues with regard to CD45 is how the extracellular domain affects function. CD45 variable exons encode regions enriched in O-linked glycosylation. Since O-linked regions formed rigid rods, the use of the variable exons greatly influences the shape of the molecules. Thus, the high- and low-molecular weight isoforms differ in the length of the extracellular region between 25 and 50 nm (McCall et al., 1992). Since CD45 is an abundant cell surface glycoprotein, changes in CD45 isoform expression may greatly affect the overall net surface charge and amount of carbohydrate present on the cell surface.

CD45 isoform expression is highly regulated in lymphocyte differentiation and activation and the patterns expressed are conserved throughout vertebrate evolution (reviewed in Thomas, 1989). This suggests that the isoforms expressed are important to function. The pattern of carbohydrates expressed by lymphocytes also changes during lymphocyte differentiation and activation and many of these changes affect CD45 glycosylation. Given the correlation between CD45 isoform expression and lymphocyte activation and differentiation, and given that corresponding changes in glycosylation during activation affect CD45, it is likely that carbohydrates are important in influencing the interactions of the extracellular domain. However, there are currently no data to demonstrate that carbohydrate interactions are important to CD45 function.

CD22 is a B cell transmembrane glycoprotein containing immunoglobulin-like domains that serves as a cell surface lectin and binds carbohydrate structures that bear in part α2,6-sialic acid (Powell et al., 1993). CD22 has been demonstrated to interact with CD45 and while the interactions are not specific to CD45 or any given isoform (Engel et al., 1993; Sgroi et al., 1993), it is perhaps this sort of interaction that may be important in directing cellular interactions of which CD45 is a part. While no given cell surface lectin may bind CD45 specifically, since CD45 is a major glycoprotein that is predicted to project an extended distance from the cell surface, interactions with the extracellular domain of CD45 may be important in mediating contact between cells. It is not clear, though, how this sort of interaction is important in regulating CD45 activity and function. Correlating CD45 extracellular domain interactions with CD45 function is one of the important phenomena that remains to be understood.

IV. LRP FAMILY

The LRP subgroup consists of two members, leukocyte common antigen-related phosphatase (LRP, also known as PTPα), and PTPε. Both are characterized by having short extracellular domains that contain numerous N- and O-linked glycosylation sites.

A. LRP

LRP is widely, if not ubiquitously expressed. Because of its widespread expression, the cDNA clone is frequently isolated in searches for novel PTPases (Jirik et al., 1990; Kaplan et al., 1990; Krueger et al., 1990; Matthews et al., 1990; Sap et al., 1990; Moriyama et al., 1992). The mRNA is 3.0 kb and the predicted human protein encodes a protein of 793 amino acids consisting of a 121 amino acid serine and threonine-rich extracellular domain, a 26 amino acid single transmembrane segment, and two tandem phosphatase domains (Jirik et al., 1990; Kaplan et al., 1990; Krueger et al., 1990). Human, mouse, and rat intracellular domains share nearly 99% amino acid identity. Generation of fusion proteins consisting of glutathione-S-transferase and portions of the human LRP intracellular region has shown that both intracellular phosphatase domains have phosphatase activity, although the carboxy-terminal domain has much lower activity toward artificial substrates (Wang and Pallen, 1991). A cDNA has been isolated that encodes an alternatively spliced exon which results in an insertion of 36 amino acids within the first phosphatase domain of LRP (Matthews et al., 1990). When the alternative spliced form of the LRP cytoplasmic domain is expressed as a glutathione-S-transferase fusion protein, no phosphatase activity toward para-nitrophenyl phosphate can be demonstrated (J.E. Mullersman and M.L. Thomas, unpublished observations), suggesting that use of the alternative exons could regulate phosphatase activity.

The extracellular domain of LRP contains a high percentage of serine and threonine residues, multiple N-linked glycosylation sites, and does not have significant protein sequence similarities to other known proteins. It is likely that the extracellular domain is highly glycosylated. The predicted unglycosylated molecular weight of LRP is 87.5 kD. However, immunoprecipitation experiments using antibodies to a carboxy-terminal peptide demonstrate a protein of approximately 130–140 kD, indicating extensive post-translational modification (Sap et al., 1990; Zheng et al., 1992). The expression of human LRP by Baculovirus in Sf9 cells defective in O-linked glycosylation renders a protein of only 114 kD suggesting that both N-linked and O-linked glycosylation occur under normal circumstances (Daum et al., 1991).

The human gene has been localized independently to similar regions on chromosome 20p12-pter (Jirik et al., 1992; Rao et al., 1992). The gene in mouse has been localized to chromosome 2 closely linked to the IL-1α and Bmp-2a loci (Sap et al., 1990). To date, there has yet to be an identified disease associated with mutations of the gene in either mouse or human.

The mouse LRP genomic structure consists of 22 exons spanning at least 95 kb (Wong et al., 1993). Comparison of LRP and CD45 gene structures (Hall et al., 1988; Saga et al., 1988; Johnson et al., 1989) shows extraordinary gene similarity within the phosphatase domains suggesting a recent evolutionary divergence. Interestingly, both phosphatase domains of CD45 and LRP require correct splicing to generate the conserved catalytic sequence, HCSAGXGRTG. Similar to many other widely expressed proteins, the region 5′ of the transcription initiation site is CG rich and contains no TATA or CCAAT box promoter elements.

While the function of LRP is unknown, transfection of human LRP in Rat 1 cells results in transformation of these cells (Zheng et al., 1992). The over-expressed human LRP dephosphorylates $p60^{c\text{-}src}$ at the site of negative regulation, Tyr 527, suggesting that LRP functions to regulate $p60^{c\text{-}src}$ activity analogous to CD45 regulation of $p56^{lck}$ and $p59^{fyn}$ (Zheng et al., 1992). Further support for this comes from the analysis of LRP expression in the stem-cell line, P19 EC (den Hertog et al., 1993). Transfection of LRP cDNA into P19 results in dephosphorylation and activation of $p60^{c\text{-}src}$. Interestingly, treatment of P19 cells expressing LRP with retinoic acid results in a preferential neuronal differentiation. Normally, LRP is transiently expressed during neuronal differentiation (den Hertog et al., 1993) and it is interesting to speculate that linkage of $p60^{c\text{-}src}$ regulation with LRP may be important in determining cell fate. Although LRP is expressed in virtually all cell types, the levels are usually very low. However, LRP is more highly expressed in neurons (G. Roy, W. Hickey, and M.L. Thomas, unpublished data) and the levels of expression may be important to cellular regulation. Further support for ideas that LRP expression may regulate cell fate comes from the observation that IL-1 and IL-6 down-regulate rat LRP mRNA levels suggesting a role in cytokine action or inflammation (Moriyama et al., 1992).

The extracellular domain of LRP does not possess any known adhesion protein motifs, or any immunoglobulin-like domains. Furthermore, only one exon encodes the mature extracellular domain, eliminating the possibility of extensive alternative splicing giving rise to functionally different isoforms. How the extracellular domain regulates LRP function is unknown but, similar to CD45, it may be that the carbohydrate groups are important in mediating cell–cell interactions.

B. PTPε

PTPε contains an extremely small extracellular domain, 27 amino acids, that is enriched in serines and threonines (Krueger et al., 1990). There are two predominant mRNA species seen by northern blot analysis, 2.4- and 5.2-kb and it is possible that an isoform with a larger extracellular domain may be found (Kume et al., 1994). PTPε and LRP phosphatase domains share 60–70% sequence identity. Expression of PTPε is much more restricted and has been demonstrated in testis (Yi et al., 1991), an erythro-leukemia line (Kume et al., 1994), and bone marrow cells (E.C.C.W. and M.L. Thomas, unpublished data). Differentiation of erythro-leukemia cells with dimethylsulfoxide results in an initial decrease in mRNA expression followed by a substantial increase in mRNA levels with an eventual down-regulation of the mRNA, suggesting that PTPε may play a role in erythrocyte differentiation (Kume et al., 1994).

V. LAR FAMILY

Members of the LAR family include leukocyte common antigen-related phosphatase LAR (Streuli et al., 1988), PTPδ (Krueger et al., 1990; Mizuno et al., 1993), and RPTP-σ (also known as PTP-P1 and PTP NE-3; Pan et al., 1993; Walton et al., 1993; Yan et al., 1993). The family is characterized by having an extracellular domain reminiscent of the neural adhesion molecule, N-CAM, and high similarity in the cytoplasmic domain sequence with approximately 85% identical residues.

A. LAR

LAR, the first member of this family to be characterized, was initially identified as a structural homolog of LCA (Streuli et al., 1988). Similarity between LAR and LCA exists even at the gene level with the partial known exon–intron organization of LAR being very similar to that of LCA (Streuli et al., 1988). LAR exons III, IV, VII, and VIII precisely correspond to LCA exons 23, 24, 29, and 30, respectively.

Although LAR is broadly expressed, expression levels are more pronounced in epithelial cells.

The extracellular region of LAR is composed of three immunoglobulin-like domains and eight fibronectin-type III repeats and variations of this structural theme are shared by other members of this subfamily. The extracellular domain of LAR has five potential N-linked glycosylation sites. These structural features are characteristic of the N-CAM family of cell adhesion molecules which includes N-CAM, Ng-CAM, contactin, and fasciclin II. This striking structural similarity of the extracellular region of LAR with cell adhesion molecules suggests that LAR may have the capacity to participate in extracellular protein interactions of either a homotypic or heterotypic nature.

Similar to CD45, residues important for catalytic activity have been well characterized for LAR. Like other PTPases, the catalytic activity of LAR towards phosphotyrosine-containing substrates requires the presence of thiol-reducing agents. Several sulfhydryl-directed inhibitors have been shown to inactivate LAR (Pot et al., 1991). Following the irreversible inactivation of rat LAR with [^{14}C]-iodoacetate, a stoichiometry of 0.8 mol of iodoacetate/mol purified LAR was demonstrated. Sequence analysis of a single [^{14}C]-labeled peptide generated following proteolytic digestion identified the highly reactive thiol to be Cys1522, located in the active-site sequence, HCSAGVGRS. Again, it is likely that catalysis proceeds through the formation of a covalent phosphoenzyme intermediate (Pot et al., 1991; Cho et al., 1992). A rapid denaturing technique followed by SDS-PAGE analysis was used to demonstrate a radioactively-labeled phosphocysteine intermediate of LAR (Pot et al., 1991). A radiolabeled LAR phosphoprotein was also shown by rapid chemical quench flow analysis and the phosphocysteine intermediate demonstrated by [^{31}P]-NMR analysis (Cho et al., 1992). Systematic mutagenesis of the conserved cysteine residues in the membrane-proximal PTPase domain I and the more distal PTPase domain II of LAR has shown that greater than 99% of the enzymatic activity resides in the first PTPase domain (Streuli et al., 1990; Pot et al., 1991). Additional point mutations within this active site region have revealed that amino acid substitutions at Ala1524, Gly1525, and Arg1528 also result in the loss of enzyme activity, whereas substitutions at Pro1517, Val1520, His1521, Ser1523, and Val1526 are tolerated to varying degrees depending upon the nature of the substitution (Streuli et al., 1990). The presence of domain II appears to affect the substrate preference of domain I, suggesting that domain II possesses a regulatory function.

Chemically-induced, temperature-sensitive mutants of LAR have been made in which amino acid changes have been mapped to domain I sequences 1329–1407 (Tsai et al., 1991). For example, mutation of Tyr 1379 to phenylalanine results in a temperature-sensitive PTPase phenotype. Interestingly, a second site mutation of Cys1446 to tyrosine could suppress several temperature-sensitive mutations and enhanced the folding of recombinant LAR (Tsai et al., 1991). The cytoplasmic region of LAR can be phosphorylated *in vitro* by both protein kinase C and multiple tyrosine kinases, although the implications of these phosphorylations *in vivo* are unknown (Pot et al., 1991).

The structure of the LAR extracellular domain suggests it may have a role in cell–cell interactions. How the extracellular domain affects LAR function is not known. Surprisingly, however, the extracellular domain undergoes proteolytic processing at a penta-arginine site in the extracellular domain and is expressed on the cell surface as a complex of two noncovalently linked subunits (Streuli et al., 1992). The larger of the two subunits, 150 kD, is derived from the N-CAM-like extracellular region. The 85 kD subunit encompasses the two PTPase domains, the transmembrane region, and a short external segment that interacts with the 150 kD subunit. The 150 kD subunit is apparently shed from the cell surface during growth in a density-dependent manner. It is not known how this effects cell–cell interactions, but implies, similar to CD45, that interactions of the extracellular domain may not be required to potentiate phosphatase activity.

B. PTPδ

PTPδ has been identified in both humans and mice (Krueger et al., 1990; Mizuno et al., 1993) and the cytoplasmic domains are 96% identical between the two species. Three different isoforms of mouse PTPδ have been reported, possibly generated by alternative splicing. Type A contains one Ig-like domain and four fibronectin-type III-like domains. Type B is composed of one Ig-like domain and eight fibronectin-III-like domains. Type C contains three Ig-like domains and eight fibronectin-III-like domains. The amino acid sequences of the extracellular region shows 90% identity to human PTPδ and 69% identity to LAR. The 5′ untranslated regions, as well as the leader peptide of types A and B, differ from those of type C. The three mRNA species (~ 7 kb) are all detected in brain, kidney, and heart. Antibody directed against the extracellular region of type A PTPδ detects a 210 kD protein in brain and kidney lysates. It is not known whether proteolytic processing of this protein occurs although a sequence similar to the basic residues found at the site of proteolytic processing in LAR are conserved in PTPδ. *In situ* hybridization studies have shown that PTPδ mRNA is present in specialized regions of the brain including the hippocampus, thalamic reticular nucleus, and piriform complex. PTPδ mRNA has not been detected in lymphoid tissues, but has been found in pre-B cell and certain B cell lines (Mizuno et al., 1993). The mouse PTPδ locus appears to be tightly linked to the *brown (b)* locus on mouse chromosome 4 (Mizuno et al., 1993). The syntenic region on human homologous chromosome 9q contains tumor suppressor genes for bladder carcinoma and endometrial carcinoma, although no direct link to PTPδ has been made.

C. RPTP-σ

The extracellular domain of RPTP-σ contains three Ig-like domains and four fibronectin type-III domains (Streuli et al., 1992; Pan et al., 1993; Walton et al.,

1993; Yan et al., 1993) and thus, has a smaller extracellular domain than LAR or PTPδ. However, there are multiple mRNAs found by northern blot analysis (Pan et al., 1993) and it is possible that other isoforms with larger extracellular domains may be found. RPTP-σ is unusual among the transmembrane PTPases in that a cDNA has been isolated that results from alternative exon splicing yielding a protein that contains just the first phosphatase domain (Pan et al., 1993). Both isoforms are enzymatically active and, therefore, it is not clear how each will differ in function.

Both isoforms are primarily expressed in the brain, however, the mRNA for the larger isoform is also detected in the heart and lung, and at lower levels in testis, kidney, and intestine (Streuli et al., 1992; Pan et al., 1993; Walton et al., 1993; Yan et al., 1993). Within the brain, RPTP-σ is expressed in the cortex, thalamus, cerebellum, olfactory bulb, and the hippocampal pyramidal cell layer and granular cell layer of the dentate gyrus.

The mouse gene has been mapped to the distal arm of chromosome 17 (Yan et al., 1993).

VI. PTPμ FAMILY

This family of transmembrane PTPases consists of two members, PTPμ (Gebbink et al., 1991) and R-PTP-κ (Jiang et al., 1993), and is similar to the LAR family in having extracellular domains that contain immunoglobulin-like domains and fibronectin type-III domains. In addition, at the amino-terminus is a domain that has similarities to *Xenopus* A5 protein, a protein thought to be involved in neuronal recognition. The cytoplasmic domains are approximately 80% identical.

A. PTPμ

Mouse PTPμ cDNA encodes for a protein of 1432 amino acids (excluding the signal peptide) with a predicted molecular weight of 162 kD (Gebbink et al., 1991). The human homolog has also been cloned and shows 99% amino acid identity to the mouse PTPase (Gebbink et al., 1991). The predicted structure consists of a 720 amino acid extracellular domain which contains 13 potential N-glycosylation sites, an amino-terminal extracellular region which contains a stretch of approximately 170 amino acids with similarity to a region of *Xenopus* A5 protein, a single Ig-like domain, and four fibronectin type-III- like repeats. PTPμ has a single transmembrane segment and a 688 amino acid intracellular domain containing two tandem catalytic domains. The intracellular domain is unusual in that the region between the transmembrane domain and the first catalytic domain is about twice the length of that in other receptor-like PTPases and is relatively rich in serine and threonine residues. The mRNA for PTPμ is approximately 5.7 kb and is abundantly expressed in lung tissue and at lower levels in brain and heart. Expression of mouse PTPμ in COS cells results in an immunoreactive protein of 195 kD (Gebbink et al., 1991). Interestingly, the possibility exists that the post-translational processing that occurs in LAR may also occur with PTPμ since a potential cleavage site exists in the corresponding position. The human PTPμ is localized on chromosome 18pter-q11 (Gebbink et al., 1991), a region which demonstrates cytogenetic aberrations frequently encountered in neoplastic conditions including colon carcinoma, synovial carcinoma, and malignant lymphoma.

PTPμ likely plays an important role in signaling cell–cell recognition. It has recently been shown that human PTPμ can mediate cell–cell adhesion (Brady-Kalnay et al., 1993; Gebbink et al., 1993). Wild-type PTPμ expressed in insect Sf9 cells using recombinant Baculovirus promoted cell–cell aggregation in a homophilic, Ca^{2+}-independent manner. Interestingly, the increased cell–cell interaction demonstrated a pH sensitivity that is uncommon among other cell adhesion molecules. Increased adhesion was not observed upon expression of a chimeric PTPμ containing the extracellular domain of the EGF receptor (Brady-Kalnay et al., 1993; Gebbink et al., 1993). In addition, cells expressing either a point-mutated catalytically inactive enzyme, a truncated form of the enzyme in which the entire catalytic region was deleted, or the extracellular segment alone showed increased adhesion properties similar to that observed with wild-type PTPμ. Thus, catalytic function was not essential for the PTPμ-mediated increase in cell adhesion and there does not appear to be an increase in catalytic activity upon interaction of the extracellular domain (Brady-Kalnay et al., 1993). PTPμ apparently mediates aggregation by a homophilic mechanism since uninfected Sf9 cells were not incorporated into aggregates formed by PTPμ-expressing cells (Gebbink et al., 1993). Homophilic aggregation has also been demonstrated using PTPμ-coated beads or in combination with MvLu cells expressing PTPμ, as well as bacterial-expressed extracellular domain adsorbed to a surface (Brady-Kalnay et al., 1993). A pivotal question which remains to be answered is how PTPμ homophilic interaction effects intracellular functions mediated by the phosphatase domains.

B. R-PTP-κ

The cDNA for R-PTP-κ encodes a 1,457 amino acid protein composed of a 752 amino acid extracellular region (including signal peptide) and a 683 amino acid intracellular region containing tandemly repeated PTPase domains. The extracellular domain is similar to PTPμ in that it contains a *Xenopus* A5 protein domain, an Ig-like domain, and four fibronectin type-III-like repeats (Jiang et al., 1993). The extracellular region contains 12 potential N-glycosylation sites, a histidine–alanine–valine sequence in the first fibronectin-type III domain, which has been implicated in cell–cell adhesion in members of the cadherin family of proteins, and serine–glycine motifs which may be candidate sites for chondroitin sulfate attachment. R-PTP-κ, like PTPμ, has an unusually large juxtamembrane region that is approximately 150 amino acids in length. The overall sequence identity at the amino acid level between R-PTP-κ and PTPμ is 77%.

R-PTP-κ is more widely expressed than PTPμ (Jiang et al., 1993). Two major transcripts of 5.3-and 7-kb are detectable at different levels in all adult mouse tissues examined with the exception of spleen and testis. The 5.3 kb transcript is particularly high in adult liver and kidney. *In situ* hybridization analysis indicates that R-PTP-κ expression is developmentally regulated in the central nervous system and shows interesting correlation with areas capable of neuronal plasticity. In the developing mouse embryo, R-PTP-κ mRNA levels are higher than in adult mouse brain. Significant R-PTP-κ mRNA expression is seen in the hippocampal formation, cerebellum, brain stem, and spinal cord, and also in embryonic liver, kidney, and intestine.

The R-PTP-κ cDNA, when transfected into mammalian cells, renders a protein of ~210 kD protein (Jiang et al., 1993). Like LAR, post-translational processing occurs which generates cleavage products of 110 kD, containing most of the extracellular domain, and 100 kD, consisting of the intracellular region, transmembrane region, and 113 amino acids of the extracellular domain. These proteins are found associated at the membrane. Evidence indicates that the likely cleavage site is a RTKR consensus sequence for the endopeptidase furin located in the fourth fibronectin-III repeat. It is possible that the amino-terminal 110 kD protein is shed from the membrane, rendering the membrane-bound 110 kD protein insensitive to modulation by protein binding in the extracellular environment.

Similar to PTPμ, R-PTP-κ mediates homophilic binding (Sap et al., 1994). Induced expression of R-PTP-κ protein in heterologous cells results in the formation of stable cellular aggregates strictly consisting of R-PTP-κ-expressing cells. The purified extracellular domain functions in adhesion as shown by its ability to induce aggregation of coated synthetic beads. Like PTPμ, R-PTP-κ-mediated intercellular adhesion is calcium-independent and does not require PTPase activity or post-translational proteolytic processing.

VII. PTPγ FAMILY

This family of transmembrane tyrosine phosphatases holds in common an extracellular domain containing an amino-terminal region of approximately 260 amino acids that is similar to carbonic anhydrase. In addition, there is a single fibronectin type-III domain and a large cysteine-free region. There are two members in this family, PTPγ (Kaplan et al., 1990; Krueger et al., 1990) and PTPζ (also known as R-PTP-β and PTP 18; Guan and Dixon, 1990; Kaplan et al., 1990; Krueger et al., 1990) which share approximately 75% identity in their phosphatase domains. Despite the fact that both PTPγ and PTPζ have two tandem phosphatase domains, only the amino-terminal domain has phosphatase activity. An aspartic acid residue replaces the active-site cysteine in the HCSAGAGRTG sequence, in the second domain of both PTPγ and PTPζ. Expression of glutathione-S-transferase fusion proteins of PTPζ demonstrates that mutation of the active-site cysteine in the first domain abolishes all enzymatic activity (Krueger and Saito, 1992).

A. PTPγ

The cDNAs for PTPγ have been cloned for both human and mouse and show 90% amino acid identity. The mRNA is widely expressed including brain, lung, kidney, heart, skeletal muscle, liver, spleen, and testis (Krueger et al., 1990; Barnea et al., 1993). PTPγ is highly expressed in the lung (Tsukamoto et al., 1992) and significant expression is found in newborn rat brain, predominantly in the hippocampal formation, in the sepal and midline thalamic nuclei, and in the cortex (Barnea et al., 1993). In the adult rat brain, expression appears to be localized to the hippocampal region (Barnea et al., 1993).

The PTPγ extracellular domain consists of approximately 730 amino acids and contains 6–8 potential sites of N-linked glycosylation. The carbonic anhydrase-like domain is an unusual feature for a transmembrane protein. It is, however, unlikely to catalyze the reaction, $CO_2 + H_2O \leftrightarrow HCO_3^- + H^+$, because of the loss of two of the three conserved histidine residues that are necessary for the zinc ion coordination that is, in turn, essential for enzymatic activity (Barnea et al., 1993). However, there is the possibility that ions which undergo octahedral coordination, such as manganese, may be bound (Barnea et al., 1993). Amino-terminal to the carbonic anhydrase-like domain in both PTPγ and PTPζ is a series of four basic amino acids, reminiscent to the proteolysis site in LAR and PTPκ, raising the possibility that this may also be a site of proteolytic cleavage. The possibility thus exists that proteolysis may be necessary for interactions with the carbonic anhydrase-like domain.

The fibronectin type-III domain is bounded by a conserved cysteine residue. This is followed by a large cysteine-free region that contains sequences enriched in serine and threonine and polar amino acids and is a potential region for O-linked glycosylation, similar to the CD45 and LRP extracellular domains.

The gene for human PTPγ has been localized to chromosome 3p14-21, a site of chromosomal deletion seen in renal cell and lung carcinomas, suggesting that PTPγ may act as a tumor suppressor (LaForgia et al., 1991). Interestingly, analysis of renal and lung carcinoma cell lines reveals that one PTPγ allele is lost in three of five, and five of ten lines examined, respectively. Two human lung cell lines contained abnormal size PTPγ mRNA. However, no mutations were discovered in the cytoplasmic portion of PTPγ in 31 human lung cancer lines, although mutations in other parts of the gene could not be ruled out (Tsukamoto et al., 1992). The original mouse L-cell line, derived from a methylcholanthrene-induced connective tissue tumor which produces sarcomas in syngeneic mice, contains the deletion of one PTPγ allele and has an intragenic deletion in the other allele (Wary et al., 1993). The combined data are consistent with the notion that PTPγ may function as a tumor suppressor.

B. PTPζ

PTPζ is very similar to PTPγ in structure (Krueger and Saito, 1992; Levy et al., 1993). The carbonic anhydrase-like domains have approximately 58% amino acid identity and the fibronectin type-III repeat shares 37% amino acid identity. Both extracellular domains contain a large cysteine-free domain and have approximately 75% amino acid identity in their cytoplasmic domain sequence. The gene for human PTPζ localizes to chromosome 7q31–33 and has not been implicated as a tumor suppressor gene (Levy et al., 1993).

In contrast to PTPγ, PTPζ has a restricted distribution. Northern analysis and reverse transcriptase PCR has shown that PTPζ is uniquely, if not exclusively, expressed in the brain (Krueger and Saito, 1992; Levy et al., 1993). In mouse embryos, PTPζ is expressed in the ventricular and subventricular zones and the spinal cord (Levy et al., 1993). There is lower expression in the adult mouse brain mainly localized in the Purkinje cell layer of the cerebellum, the dentate gyrus, and the sub-ependymal layer of the anterior horn of the lateral ventricle (Levy et al., 1993).

Three PTPζ transcripts of 6.4-, 7.5-, and 8.8-kb are detected by northern analysis. These transcripts encode at least two distinct isoforms. The 6.4 kb mRNA is derived from alternative splicing, resulting in a protein that deletes 860 amino acids from the cysteine-free region (Levy et al., 1993).

The extracellular domain of PTPζ is known to contain extensive post-translational modifications. PTPζ contains 21 potential sites of N-linked glycosylation and treatment with tunicamycin results in a decreased mobility by SDS-PAGE of 50 kD (Levy et al., 1993). Interestingly, the 7.5 kb mRNA appears to encode a secreted form of the extracellular domain. The soluble rat brain chondroitin sulfate proteoglycan appears to be the extracellular domain of rat PTPζ and is likely to be the product of the 7.5 kb mRNA (Barnea et al., 1994). The proteoglycan is known to interact with neural adhesion molecules, N-CAM and Ng-CAM. Thus, it possible that the extracellular domain functions as a proteoglycan in cell–cell adhesion. How this affects the phosphatase function is unknown.

VIII. PTPβ FAMILY

Currently, there are two complete structures known for the PTPβ group, PTPβ (Krueger et al., 1990) and SAP-1 (stomach associated PTP; Matozaki et al., 1994). This family is unusual in that members contain only one catalytic domain. It is likely that there is at least one additional family member. We have identified a mouse PCR-derived cDNA fragment, PTPβ2 (Thomas et al., 1993), that encodes a sequence highly similar to, but distinct from, the catalytic domain sequence of PTPβ and SAP-1 (Figure 4). A human PCR-derived cDNA fragment has been reported that is virtually identical to our mouse sequence, suggesting that the two cDNA fragments represent species homologs (Honda et al., 1993). In addition, another mouse PCR-derived cDNA sequence has been reported that is virtually identical to the human PTPβ sequence (Schepens et al., 1992), also supporting the idea that PTPβ2 is not the species homolog to PTPβ. PTPβ2, similar to PTPε, shows increased expression during differentiation of erythro-leukemia lines (Kume et al., 1994).

```
hPTPβ         QNEHNIVMVT QCVEKGRVKC DHYWPADQDS LYYGDLILQM LSES-VLPEW TIREFKICGE
mPTPβ           V                            P            V                 S
hSAP-1         Q  TL  L  N M A      E    L SQP CTH H RVTL VG E- MEN   V  LLLLQV
hF36-12         VYAII L  K   Q  T   EE   SK AQ D-   ITVA  T   I         D TVKNI
mPTPβ2        K VYAI  L  K   Q  T   EE   SK AQ D-   ITVA  T  V-         D VVKNM

hPTPβ         EQLDAHRLIR HFHYTVWPDH GVPETTQSLI QEFVRTVRDY INRSPGAGPT VV
mPTPβ
hSAP-1         EQKTL-SV  Q   QA        SSPDT L A  W ML QW LDQTMEG  P I
hF36-12       QTSES -PL  Q  F S        D  DL   N- RYL     MKQ  PESEI L
mPTPβ2        QNSES -PL  Q  F S        D  DL   N- RYL     MKQI PESP   V
```

Figure 4. Sequence comparison of potential PTPβ family members. Only nonidentical amino acids are shown. Human sequences are represented by "h" and mouse sequences by "m". Single letter amino acid code is used. Sequences were derived from the following references: hPTPβ (Krueger et al., 1990), mPTPβ (Schepens et al., 1992), hSAP-1 (Matozaki et al., 1994), hF36-12 (Honda et al., 1993), and mPTPβ2 (Thomas et al., 1993).

A. PTPβ

The extracellular domain of PTPβ consists of 16 fibronectin type-III domains, although an alternatively spliced isoform that encodes 15 fibronectin type-III domains has been identified as a cDNA (Krueger et al., 1990). Fibronectin type-III domains are potential protein–protein interaction motifs found in both extracellular matrix proteins and adhesion molecules, and it is possible that the PTPβ extracellular domain is involved in cell–cell binding. The extracellular domain is predicted to be heavily glycosylated with 27 potential sites for N-linked glycosylation.

The gene for PTPβ has been mapped to human chromosome 12q15–21. While cytogenetic abnormalities in this region have been shown for some benign tumors, no correlation between the PTPβ gene and genetic diseases are known.

Expression and isolation of the PTPβ catalytic domain demonstrates distinct substrate specificity when compared to the catalytic domains of CD45 and LAR and high catalytic efficiency (Wang and Pallen, 1992; Itoh et al., 1992; Cho et al., 1993). It is uncertain whether PTPβ functions in concert with intracellular PTPases to achieve the dual PTPase domain structure of other transmembrane PTPases or whether the high catalytic efficiency represents a requirement of PTPβ to maintain a low basal rate of phosphorylation of its respective substrates.

B. SAP-1

The cDNA for SAP-1 was isolated from a screen of PTPase cDNAs from gastric cancers (Matozaki et al., 1994). The mRNA is expressed in brain, liver, and lower levels in heart and stomach, and is highly expressed in pancreatic and colorectal cancer cell lines. The gene for SAP-1 maps to human chromosome 19q13.4, a region associated with carcinoma embryonic antigen, although it has not been determined whether they are the same molecule.

The extracellular domain of SAP-1 contains eight fibronectin type-III repeats and 24 potential sites of N-linked glycosylation (Matozaki et al., 1994). This structure is similar to PTPβ, although the sequence similarity between the two extracellular domains is low at 24%. The sequence identity between SAP-1 and PTPβ phosphatase domains is 50%, also somewhat lower than that seen when comparing members of other PTPase families.

IX. INVERTEBRATE TRANSMEMBRANE PTPases

Transmembrane PTPases are widely conserved in eukaryotic evolution and may be important for all multicellular organisms. Conservation of transmembrane PTPases was first noted in the isolation of a PTPases cDNA from the fruitfly, *Drosophila melanogaster* (Streuli et al., 1989). Four transmembrane PTPases from *Drosophila* have now been described (Streuli et al., 1989; Hariharan et al., 1991; Tian et al., 1991; Yang et al., 1991; Figure 5). A large number of cDNAs encoding fragments of PTPase domains have been isolated from the protochordate sea squirt, *Styela plicata* (Matthews et al., 1991). Many of the *Styela* PTPase sequences are highly similar to known mammalian transmembrane PTPase sequences suggesting that the transmembrane PTPase families, in general, are conserved in this higher invertebrate.

Interestingly, the four *Drosophila* transmembrane PTPases, DLAR, DPTP10D, DPTP99A, and DPTP69D are highly similar in sequence and/or structure to

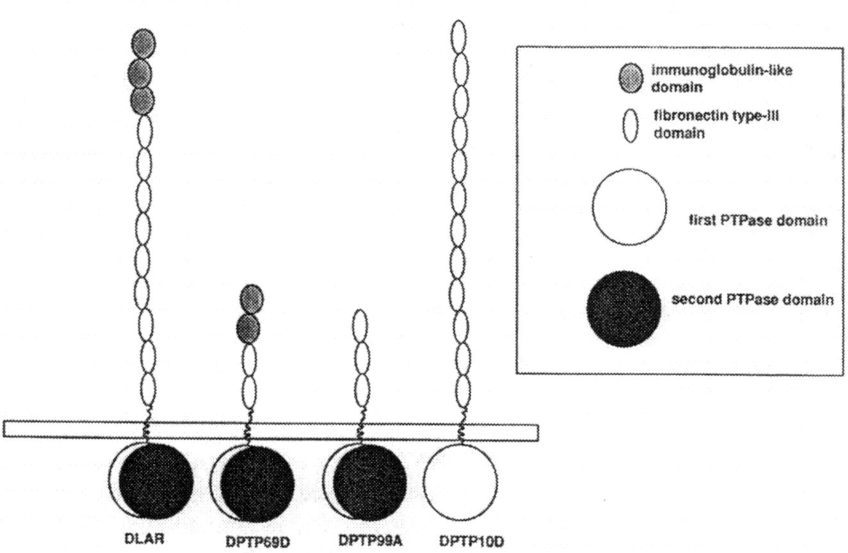

Figure 5. Cartoon representation of invertebrate transmembrane PTPases.

mammalian PTPases (Streuli et al., 1989; Hariharan et al., 1991; Tian et al., 1991; Yang et al., 1991). DLAR and LAR both contain N-CAM-like extracellular domains and are similar in sequence. DPTP10D and PTPβ have similar structures and sequences, and the phosphatase domains of DPTP99A show marked similarity to the phosphatase domains of PTPγ including an aspartic acid instead of a cysteine residue in the active-site sequence of the second domain.

All four *Drosophila* transmembrane PTPases are restricted in expression to the nervous system (Hariharan et al., 1991; Tian et al., 1991; Yang et al., 1991). So far, however, mutational analysis has not demonstrated any functional defects and the precise role in neurological function or development cannot be ascribed. Similar to the mammalian PTPases, the structure of the extracellular domain gives rise to the speculation that they may be involved in cell–cell interactions.

X. CONCLUSION

Transmembrane PTPases present a potentially novel and important mechanism for regulating extracellular signals. The interactions of the extracellular domains, for the most part, remain unknown and a key question in the field is determining how interactions of the extracellular domain affects PTPase function. Many of the extracellular domains are heavily glycosylated and many have characteristics of adhesion proteins or cell–cell interaction proteins. This leads to the speculation that they may function in directing or facilitating interactions between cells. This may be important in communicating and controlling tyrosine phosphorylation levels in the cell.

The best characterized transmembrane PTPase, CD45, functions to regulate Src-family member kinase activity and this function is essential for efficient signaling through lymphocyte antigen receptors. Importantly, the extracellular domain has not been shown to be essential for this function. Currently, it is thought that CD45 dephosphorylates and activates Src-family members prior to exposure to antigen, and thus the confinement of this interaction is important to activation. It will be interesting to determine how the CD45 extracellular domain affects Src-family members kinase activity.

Given the adhesion-like properties of the extracellular domain of LAR and PTPμ family members, they may function, in part, by providing a link between cell–cell contact and cellular signaling events that involve tyrosine phosphorylation. Although the homophilic binding capacities of PTPμ and R-PTP-κ have been demonstrated, the ability of the extracellular moieties of these PTPases to engage in heterophilic binding with other cell adhesion molecules remains unknown. Direct cell–cell contact between transmembrane PTPases on adjacent cells could modify activity through oligomerization, and thus alter tyrosine phosphorylation levels.

An important observation is that LAR and R-PTP-κ are proteolytically processed to yield two noncovalently linked chains; the chain derived from the amino-terminus contains virtually the entire extracellular domain and the other chain contains the transmembrane and cytoplasmic domain. It is likely that other members of these two families, PTPδ, RPTP-σ, and PTPμ, will be proteolytically processed in a similar way. The extracellular domains from these PTPases may be shed from the cell. Together with the observations that alternative exon usage results in secretion of the soluble extracellular domain of PTPζ, it is possible to speculate that the extracellular domains of many transmembrane PTPases may have important cell–cell interaction properties/functions that are distinct from the role of these molecules in regulating signal transduction. Determining the interactions and properties of transmembrane PTPases will be crucial in understanding how this family of enzymes function.

ACKNOWLEDGMENTS

Our research on transmembrane PTPases is supported by USPHS grant AI26263, The Council for Tobacco Research, and the Howard Hughes Medical Institute. Edward C.C. Wong was supported by NIH Training Grant 5T32 AI07163.

REFERENCES

Barnea, G., Silvennoinen, O., Shaanan, B., Honegger, A. M., Canoll, P. D., D'Eustachio, P., Morse, B., Levy, J. B., LaForgia, S., Huebner, K., Musacchio, J. M., Sap, J., & Schlessinger, J. (1993). Identification of a carbonic anhydrase-like domain in the extracellular region of RPTPγ defines a new subfamily of receptor tyrosine phosphatases. Mol. Cell. Biol. 13, 1497–1506.

Barnea, G., Grumet, M., Sap, J., Margolis, R. U., & Schlessinger, J. (1994). Close similarity between receptor-linked tyrosine phosphatase and rat brain proteoglycan. Cell 76, 205.

Bell, G. M., Dethloff, G. M., & Imboden, J. B. (1993). CD45-negative mutants of a rat natural killer cell line fail to lyse tumor target cells. J. Immunol. 151, 3646–3653.

Bourguignon, L. Y. W., Hyman, R., Trowbridge, I., & Singer, S. J. (1978). Participation of histocompatibility antigens in capping of molecularly independent cell surface components by their specific antibodies. Proc. Natl. Acad. Sci. USA 75, 2406–2410.

Brady-Kalnay, S. M., Flint, A. J., Tonks, N. K., Singer, A., & Westphal, H. (1993). Homophilic binding of PTPμ, a receptor-type protein tyrosine phosphatase, can mediate cell–cell aggregation. J. Cell Biol. 122, 961–972.

Cahir McFarland, E. D., Hurley, T. R., Pingel, J. T., Sefton, B. M., Shaw, A., & Thomas, M. L. (1993). Correlation between Src-family member regulation by the protein tyrosine phosphatase, CD45, and transmembrane signaling through the T-cell receptor. Proc. Natl. Acad. Sci. USA 90, 1402–1406.

Chang, H.-L., Lefrancois, L., Zaroukian, M. H., & Esselman, W. J. (1991). Developmental expression of CD45 alternative exons in murine T cells. J. Immunol. 147, 1687–1693.

Charbonneau, H., Tonks, N. K., Walsh, K. A., & Fischer, E. H. (1988). The leukocyte common antigen (CD45): A putative receptor-linked protein tyrosine phosphatase. Proc. Natl. Acad. Sci. USA 85, 7182–7186.

Cho, H., Ramer, S. E., Itoh, M., Winkler, D. G., Kitas, E., Bannwarth, W., Burn, P., Saito, H., & Walsh, C. T. (1991). Purification and characterization of a soluble catalytic fragment of the human leukocyte antigen related (LAR) protein tyrosine phosphatase. Biochemistry 30, 6210–6216.

Cho, H., Ramer, S.E., Itoh, M., Kitas, E., Bannwarth, W., Burn, P., Saito, H., & Walsh, C. T. (1992). Catalytic domains of the LAR and CD45 protein tyrosine phosphatases from *Escherichia coli* expression systems: Purification and characterization for specificity and mechanism. Biochemistry 31, 133–138.

Cho, H., Krishnaraj, R., Itoh, M., Kitas, E., Bannwarth, W., Saito, H., & Walsh, C. T. (1993). Substrate specificities of catalytic fragments of protein tyrosine phosphatases (HPTPβ, LAR, and CD45) towards phosphotyrosylpeptides substrates and thiophosphorylated peptides as inhibitors. Protein Science 2, 977–984.

Daum, G., Zander, N. F., Morse, B., Hurwitz, D., Schlessinger, J., & Fischer, E. H. (1991). Characterization of a human recombinant receptor-linked protein tyrosine phosphatase. J. Biol. Chem. 266, 12211–12215.

Deans, J. P., Kanner, S. B., Torres, R. M., & Ledbetter, J. A. (1992). Interaction of CD4:lck with the T cell receptor/CD3 complex induces early signaling events in the absence of CD45 tyrosine phosphatase. Eur. J. Immunol. 22, 661–668.

den Hertog, J., Pals, C. E. G. M., Peppelenbosch, M. P., Tertoolen, L. G. J., de Laat, S. W., & Kruijer, W. (1993). Receptor protein tyrosine phosphatase α activates $pp60^{c\text{-}src}$ and is involved in neuronal differentiation. EMBO J. 12, 3789–3798.

Desai, D. M., Sap, J., Schlessinger, J., & Weiss, A. (1993). Ligand-mediated negative regulation of a chimeric transmembrane receptor tyrosine phosphatase. Cell 73, 541–554.

Engel, P., Nojima, Y., Rothstein, D., Zhou, L.-J., Wilson, G. L., Kehr, J. H., & Tedder, T. F. (1993). The same epitope of CD22 of B lymphocytes mediates the adhesion of erythrocytes, T and B lymphocytes, neutrophils and monocytes. J. Immunol. 150, 4719–4732.

Fischer, E. H., Charbonneau, H., & Tonks, N. K. (1991). Protein tyrosine phosphatases: A diverse family of intracellular and transmembrane enzymes. Science 253, 401–406.

Gebbink, M. B. G., van Etten, I., Hateboer, G., Suijkerbuijk, R., Beijersbergen, R. L., van Kessel, A. G., & Moolenaar, W. H. (1991). Cloning, expression and chromosomal localization of a new putative receptor-like protein tyrosine phosphatase. FEBS Letters 290, 121–130.

Gebbink, M. F. B. G., Zondag, G. C. M., Wubbolts, R. W., Beijersbergen, R. L., van Etten, I., & Moolenaar, W. H. (1993). Cell–cell adhesion mediated by a receptor-like protein tyrosine phosphatase. J. Biol. Chem. 268, 16101–16104.

Goldman, S. J., Uniyal, S., Ferguson, L. M., Golan, D. E., Burakoff, S. J., & Kiener, P. A. (1992). Differential activation of phosphotyrosine protein phosphatase activity in a murine T cell hybridoma by monoclonal antibodies to CD45. J. Biol. Chem. 267, 6197–6204.

Guan, K., & Dixon, J. E. (1990). Protein tyrosine phosphatase activity of an essential virulence determinant in *Yersinia*. Science 249, 553–556.

Guttinger, M., Gassmann, M., Amrein, K. E., & Burn, P. (1992). CD45 phosphotyrosine phosphatase and $p56^{lck}$ protein tyrosine kinase: A functional complex crucial in T cell signal transduction. Int. Immunol. 4, 1325–1330.

Hall, L. R., Streuli, M., Schlossman, S. F., & Saito, H. (1988). Complete exon–intron organization of the human leukocyte common antigen (CD45) gene. J. Immunol. 141, 2781–2787.

Hariharan, I. K., Chuang, P.-T., & Rubin, G. M. (1991). Cloning and characterization of a receptor-class phosphotyrosine phosphatase gene expressed on central nervous system axons in *Drosophila melanogaster*. Proc. Natl. Acad. Sci. USA 88, 11266–11270.

Honda, H., Shibuya, M., Chiba, S., Yazaki, Y., & Hirai, H. (1993). Identification of novel protein- tyrosine phosphatases in a human leukemia cell line, F-36P. Leukemia 7, 742–746.

Hovis, R. R., Donovan, J. A., Musci, M. A., Motto, D. G., Goldman, F. D., Ross, S. E., & Koretzky, G. A. (1993). Rescue of signaling by a chimeric protein containing the cytoplasmic domain of CD45. Science 260, 544–546.

Hunter, T. (1987). A tail of two src's: Mutatis mutandis. Cell 49, 1–4.

Hurley, T. R., Hyman, R., & Sefton, B. M. (1993). Differential effects of expression of the CD45 tyrosine protein phosphatase on the tyrosine phosphorylation of the *lck, fyn,* and c-*src* tyrosine protein kinases. Mol. Cell. Biol. 13, 1651–1656.

Itoh, M., Streuli, M., Krueger, N. X., & Saito, H. (1992). Purification and characterization of the catalytic domains of the hyman receptor-linked protein tyrosine phosphatase HPTPβ, leukocyte common antigen (LCA), and leukocyte common antigen-related molecule (LAR). J. Biol. Chem. 267, 12356–12363.

Jiang, Y.-P., Wang, H., D'Eustachio, P., Musacchio, J. M., Schlessinger, J., & Sap, J. (1993). Cloning and characterization of R-PTP-κ, a new member of the receptor protein tyrosine phosphatase family with a proteolytically cleaved cellular adhesion molecule-like extracellular region. Mol. Cell. Biol. 13, 2942–2951.

Jirik, F. R., Janzen, N. M., Melhado, I. G., & Harder, K. W. (1990). Cloning and chromosomal assignment of a widely expressed human receptor-like protein-tyrosine phosphatase. FEBS Letters 273, 239–242.

Jirik, F. R., Anderson, L. L., & Duncan, A. M. V. (1992). The human protein-tyrosine phosphatase PTPα/LRP gene (PTPA) is assigned to chromosome 20p13. Cytogenet. Cell Genet. 60, 117–118.

Johnson, N. A., Meyer, C. M., Pingel, J. T., & Thomas, M. L. (1989). Sequence conservation in potential regulatory regions of the mouse and human leukocyte common antigen gene. J. Biol. Chem. 264, 6220–6229.

Johnson, P., Ostergaard, H. L., Wasden, C., & Trowbridge, I. S. (1992). Mutational analysis of CD45: A leukocyte-specific protein tyrosine phosphatase. J. Biol. Chem. 267, 8035–8041.

Justement, L. B., Campbell, K. S., Chien, N. C., & Cambier, J. C. (1991). Regulation of B cell antigen receptor signal transduction and phosphorylation by CD45. Science 252, 1839–1842.

Kanner, S. B., & Ledbetter, J. A. (1992). CD45 regulates TCR-induced signalling through tyrosine phosphorylation of phospholipase C gamma 1. Biochem. Soc. Trans. 20, 178–184.

Kaplan, R., Morse, B., Huebner, K., Croce, C., Howk, R., Ravera, M., Ricca, G., Jaye, M., & Schlessinger, J. (1990). Cloning of three human tyrosine phosphatases reveals a multigene family of receptor-linked protein-tyrosine-phosphatases expressed in brain. Proc. Natl. Acad. Sci. USA 87, 7000–7004.

King, P. D., Batchelor, A. H., Lawlor, P., & Katz, D. R. (1990). The role of CD44, CD45, CD45RO, CD46 and CD55 as potential anti-adhesion molecules involved in the binding of human tonsillar T cells to phorbol 12-myristate 13-acetate-differentiated U-937 cells. Eur. J. Immunol. 20, 363–368.

Kishihara, K., Penninger, J., Wallace, V. A., Kundig, T. M., Kawai, K., Wakeham, A., Timms, E., Pfeffer, K., Ohashi, P. S., Thomas, M. L., Furlonger, C., Paige, C. J., & Mak, T. W. (1993). Normal B lymphocyte development but impaired T cell maturation in CD45-exon6 protein tyrosine phosphatase-deficient mice. Cell 74, 143–156.

Koretzky, G. A., Picus, J., Thomas, M. L., & Weiss, A. (1990). Tyrosine phosphatase CD45 is essential for coupling T-cell antigen receptor to the phosphatidyl inositol pathway. Nature 346, 66–68.

Koretzky, G. A., Picus, J., Schultz, T., & Weiss, A. (1991). Tyrosine phosphatase CD45 is required for T-cell antigen receptor and CD2-mediated activation of a protein tyrosine kinase and interleukin 2 production. Proc. Natl. Acad. Sci. USA 88, 2037–2041.

Koretzky, G. A., Kohmetscher, M., & Ross, S. (1993). CD45-associated kinase activity requires *lck* but not T cell receptor expression in the Jurkat T cell line. J. Biol. Chem. 268, 8958–8964.

Krueger, N. X., & Saito, H. (1992). A human transmembrane protein-tyrosine-phosphatase, PTPζ, is expressed in brain and has an N-terminal receptor domain homologous to carbonic anhydrases. Proc. Natl. Acad. Sci. USA 89, 7417–7421.

Krueger, N. X., Streuli, M., & Saito, H. (1990). Structural diversity and evolution of human receptor-like protein tyrosine phosphatases. EMBO J. 9, 3241–3252.

Kume, T., Tsuneizumi, K., Watanabe, T., Thomas, M. L., & Oishi, M. (1994). Induction of specific protein tyrosine phosphatase transcripts during differentiation of mouse erythroleukemia cells. J. Biol. Chem. 269, 4709–4712.

LaForgia, S., Morse, B., Levy, J., Barnea, G., Cannizzaro, L. A., Li, F., Nowell, P. C., Boghosian-Sell, L., Glick, J., Weston, A., Harris, C. C., Drabkin, H., Patterson, D., Croce, C. M., Schlessinger, J.,

& Huebner, K. (1991). Receptor protein-tyrosine phosphatase γ is a candidate tumor suppressor gene at human chromosome region 3p21. Proc. Natl. Acad. Sci. USA 88, 5036–5040.

Ledbetter, J. A., Schieven, G. L., Uckun, F. M., & Imboden, J. B. (1991). CD45 cross-linking regulates phospholipase C activation and tyrosine phosphorylation of specific substrates in CD3/Ti-stimulated T cells. J. Immunol. 146, 1577–1583.

Levy, J. B., Canoll, P. D., Silvennoinen, O., Barnea, G., Morse, B., Honegger, A. M., Huang, J.-T., Cannizzaro, L. A., Park, S.-H., Druck, T., Huebner, K., Sap, J., Ehrlich, M., Musacchio, J. M., & Schlessinger, J. (1993). The cloning of a receptor-type protein tyrosine phosphatase expressed in the central nervous system. J. Biol. Chem. 268, 10573–10581.

Lokeshwar, V. B., & Bourguignon, L. Y. W. (1992). Tyrosine phosphatase activity of lymphoma CD45 (GP180) is regulated by a direct interaction with the cytoskeleton. J. Biol. Chem. 267, 21551–21557.

Matozaki, T., Suzuki, T., Uchida, T., Inazawa, J., Ariyama, T., Matsuda, K., Horita, K., Noguchi, H., Mizuno, H., Sakamoto, C., & Kasuga, M. (1994). Molecular cloning of a human transmembrane-type protein tyrosine phosphatase and its expression in gastrointestinal cancers. J. Biol. Chem. 269, 2075–2081.

Matthews, R. J., Cahir, E. D., & Thomas, M. L. (1990). Identification of an additional member of the protein-tyrosine-phosphatase family: Evidence for alternative slicing in the tyrosine phosphatase domain. Proc. Natl. Acad. Sci. USA 87, 4444–4448.

Matthews, R. J., Flores, E., & Thomas, M. L. (1991). Protein tyrosine phosphatase domains from the protochordate *Styela plicata*. Immunogenetics 33, 33–41.

McCall, M. N., Shotton, D. M., & Barclay, A. N. (1992). Expression of soluble isoforms of rat CD45. Analysis by electron microscopy and use in epitope of anti-CD45R monoclonal antibodies. Immunology 76, 310–317.

Mittler, R. S., Rankin, B. M., & Kiener, P. A. (1991). Physical associations between CD45 and CD4 or CD8 occur as late activation events in antigen receptor-stimulated human T cells. J. Immunol. 147, 3434–3440.

Mizuno, K., Hasegawa, K., Katagiri, T., Ogimoto, M., Ichikawa, T., & Yakura, H. (1993). MPTPδ, a putative murine homolog of HPTPδ, is expressed in specialized regions of the brain and in the B-cell lineage. Mol. Cell. Biol. 13, 5513–5523.

Moriyama, T., Fujiwara, Y., Imai, E., Takenaka, M., Kawanishi, S., Inoue, T., Noguchi, T., Tanaka, T., Kamada, T., & Ueda, N. (1992). cDNA cloning of rat LRP, a receptor like protein tyrosine phosphatase, and evidence for its gene regulation in cultured rat mesangial cells. Biochem. Biophys. Res. Comm. 188, 34–39.

Mustelin, T., & Altman, A. (1990). Dephosphorylation and activation of the T cell tyrosine kinase $p56^{lck}$ by the leukocyte common antigen (CD45). Oncogene 5, 809–813.

Mustelin, T., & Burn, P. (1993). Regulation of *src* family tyrosine kinases in lymphocytes. TIBS, 18, 215–220.

Mustelin, T., Coggeshall, K. M., & Altman, A. (1989). Rapid activation of the T-cell tyrosine protein kinase $pp56^{lck}$ by the CD45 phosphotyrosine phosphatase. Proc. Natl. Acad. Sci. USA 86, 6302–6306.

Mustelin, T., Pessa-Morikawa, T., Autero, M., Gassmann, M., Andersson, L., Gahmberg, C. G., & Burn, P. (1992). Regulation of the $p59^{fyn}$ protein tyrosine kinase by the CD45 phosphotyrosine phosphatase. Eur. J. Immunol. 22, 1173–1178.

Nel, A. E., Ledbetter, J. A., Williams, K., Ho, P., Akerley, B., Franklin, K., & Katz, R. (1991). Activation of MAP-2 kinase, activity by the CD2 receptor in Jurkat T cells can be reversed by CD45 phosphatase. Immunology 73, 129–133.

Ostergaard, H. L., & Trowbridge, I. S. (1991). Negative regulation of CD45 protein tyrosine phosphatase activity by ionomycin in T cells. Science 253, 1423–1425.

Ostergaard, H. L., Schackelford, D. A., Hurley, T. R., Johnson, P., Hyman, R., Sefton, B. M., & Trowbridge, I. S. (1989). Expression of CD45 alters phosphorylation of the *lck*-encoded tyrosine protein kinase in murine lymphoma T-cell lines. Proc. Natl. Acad. Sci. USA 86, 8959–8963.

Pan, M.-G., Rim, C., Lu, K. P., Florio, T., & Stork, P. J. S. (1993). Cloning and expression of two structurally distinct receptor-linked protein-tyrosine phosphatases generated by RNA processing from a single gene. J. Biol. Chem. 268, 19284–19291.

Perlmutter, R. M., Levin, S. D., Appleby, M. W., Anderson, S. J., & Alberola-Ila, J. (1993). Regulation of lymphocyte function by protein phosphorylation. Annu. Rev. Immunol 11, 451–499.

Peyron, J. F., Verma, S., de Waal Malefyt, R., Sancho, J., Terhorst, C., & Spits, H. (1991). The CD45 protein tyrosine phosphatase is required for the completion of the activation program leading to lymphokine production in the Jurkat human T cell line. Int. Immunol. 3, 1357–1366.

Pingel, J. T., & Thomas, M. L. (1989). Evidence that the leukocyte-common antigen is required for antigen-induced T lymphocyte proliferation. Cell 58, 1055–1065.

Pingel, J. T., Cahir McFarland, E. D., & Thomas, M. L. (1994). Activation of CD45-deficient T cell clones by lectin mitogens but not anti-Thy-1. Int. Immunol. 6, 169–178.

Pot, D. A., Woodford, T. A., Remboutsika, E., Haun, R. S., & Dixon, J.E. (1991). Cloning, bacterial expression, purification, and characterization of the cytoplasmic domain of rat LAR, a receptor-like protein tyrosine phosphatase. J. Biol. Chem. 266, 19688–19696.

Powell, L. D., Sgroi, D., Sjoberg, E. R., Stamenkovic, I., & Varki, A. (1993). Natural ligands of the B cell adhesion molecule CD22β carry N-linked oligosaccarides with α-2,6-sialic acids that are required for recognition. J. Biol. Chem. 268, 7019–7027.

Prickett, T. C. R., & Hart, D. N. J. (1990). Anti-leucocyte common (CD45) antibodies inhibit dendritic cell stimulation of CD4 and CD8 T-lymphocyte proliferation. Immunology 69, 250–256.

Rao, V. V. N. G., Löffler, C., Sap, J., Schlessinger, J., & Hansmann, I. (1992). The gene for receptor-linked protein-tyrosine-phosphatase (PTPA) is assigned to human chromosome 20p12-pter by *in situ* hybridization (ISH and FISH). Genomics 13, 906–907.

Robb, B., Matthews, R. J., & Thomas, M. L. (Forthcoming). J. Immunol.

Saga, Y., Tung, J.-S., Shen, F.-W., Pancoast, T. C., & Boyse, E. A. (1988). Organization of the Ly-5 gene. Mol. Cell. Biol. 8, 4889–4895.

Sap, J., D'Eustachio, P., Givol, D., & Schlessinger, J. (1990). Cloning and expression of a widely expressed receptor tyrosine phosphatase. Proc. Natl. Acad. Sci. USA 87, 6112–6116.

Sap, J., Jiang, Y., Friedlander, D., Grumet, M., & Schlessinger, J. (1994). Receptor tyrosine phosphatase R-PTP-κ mediates homophilic binding. Mol. Cell. Biol. 14, 1–9.

Schepens, J., Zeeuwen, P., Wieringa, B., & Hendriks, W. (1992). Identification and typing of members of the protein-tyrosine phosphatase gene family expressed in mouse brain. Mol. Biol. Rep. 16, 241–248.

Schraven, B., Roux, M., Hutmacher, B., & Meuer, S. C. (1989). Triggering of the alternative pathway of human T cell activation involves members of the T200 family of glycoproteins. Eur. J. Immunol. 19, 397–403.

Schraven, B., Kirchgessner, H., Gaber, B., Samstag, Y., & Meuer, S. (1991). A functional complex is formed in human T lymphocytes between the protein tyrosine phosphatase CD45, the protein tyrosine kinase $p56^{lck}$ and pp32, a possible substrate. Eur. J. Immunol. 21, 2469–2477.

Schraven, B., Schirren, A., Kirchgessner, H., Siebert, B., & Meuer, S. C. (1992). Four CD45/$p56^{lck}$-associated phosphoproteins (pp29–pp32) undergo alterations in human T cell activation. Eur. J. Immunol. 22, 1857–1863.

Sgroi, D., Varki, A., Braesch-Andersen, S., & Stamenkovic, I. (1993). CD22, a B cell-specific immunglobulin superfamily member, is a sialic acid-binding lectin. J. Biol. Chem. 268, 7011–7018.

Shiroo, M., Goff, L., Biffen, M., Shivnan, E., & Alexander, D. (1992). CD45 tyrosine phosphatase-activated $p59^{fyn}$ couples the T cell antigen receptor to pathways of diacylglycerol production, protein kinase C activation and calcium influx. EMBO J. 11, 4887–4897.

Sieh, M., Bolen, J. B., & Weiss, A. (1993). CD45 specifically modulates binding of Lck to a phosphopeptide encompassing the negative regulatory tyrosine of Lck. EMBO J. 12, 315–321.

Songyang, Z., Shoelson, S. E., Chaudhuri, M., Gish, G., Pawson, T., Haser, W. G., King, F., Roberts, T., Ratnofsky, S., Lechleider, R. J., Neel, B. G., Birge, R. B., Fajardo, J. E., Chou, M. M., Hanafusa, H., Schaffhausen, B., & Cantley, L. C. (1993). SH2 domains recognize specific phosphopeptide sequences. Cell 72, 767–778.

Streuli, M., Krueger, N. X., Hall, L. R., Schlossman, S. F., & Saito, H. (1988). A new member of the immunoglobulin superfamily that has a cytoplasmic region homologous to the leukocyte common antigen. J. Exp. Med. 168, 1523–1530.

Streuli, M., Krueger, N. X., Tsai, A. Y. M., & Saito, H. (1989). A family of receptor-linked protein tyrosine phosphatases in humans and *Drosophila*. Proc. Natl. Acad. Sci. USA 86, 8698–8702.

Streuli, M., Krueger, N. X., Thai, T., Tang, M., & Saito, H. (1990). Distinct functional roles of the two intracellular phosphatase like domains of the receptor-linked protein tyrosine phosphatases LCA and LAR. EMBO J. 9, 2399–2407.

Streuli, M., Krueger, N. X., Ariniello, P. D., Tang, M., Munro, J. M., Blattler, W. A., Adler, D. A., Disteche, C. M., & Saito, H. (1992). Expression of the receptor-linked protein tyrosine phosphatase LAR: Proteolytic cleavage and shedding of the CAM-like extracellular region. EMBO J. 11, 897–907.

Thomas, M. L. (1989). The leukocyte common antigen family. Ann. Rev. Immunol. 7, 339–369.

Thomas, M. L., Barclay, A. N., Gagnon, J., & Williams, A. F. (1985). Evidence from cDNA clones that the rat leukocyte-common antigen (T200) spans the lipid bilayer and contains a cytoplasmic domain of 80,000 Mr. Cell 41, 83–93.

Thomas, M. L., Bowne, D. B., Cahir McFarland, E., et al. (1993). A diversity of protein tyrosine phosphatases expressed by T lymphocytes. In: Progress in Immunology VIII, (Gergely, J., Benczur, M., Erdei, A., eds.) pp. 213–219. Berlin, Springer Verlog.

Tian, S.-S., Tsoulfas, P., & Zinn, K. (1991). Three receptor-linked protein-tyrosine phosphatases are selectively expressed on central nervous system axons in the *Drosophila* embryo. Cell 67, 675–685.

Tonks, N. K., Charbonneau, H., Diltz, C. D., Fischer, E. H., & Walsh, K. A. (1988a). Demonstration that the leukocyte common antigen CD45 is a protein tyrosine phosphatase. Biochemistry 27, 8695–8701.

Tonks, N. K., Diltz, C. D., & Fischer, E. H. (1988b). Characterization of the major protein-tyrosine-phosphatases of human placenta. J. Biol. Chem. 263, 6731–6737.

Tonks, N. K., Diltz, C. D., & Fischer, E. H. (1990). CD45, an integral membrane protein tyrosine phosphatase. J. Biol. Chem. 265, 10674–10680.

Tsai, A. Y. M., Itoh, M., Streuli, M., Thai, T., & Saito, H. (1991). Isolation and characterization of temperature-sensitive and thermostable mutants of the human receptor-like protein tyrosine phosphatase LAR. J. Biol. Chem. 266, 10534–10543.

Tsukamoto, T., Takahashi, T., Ueda, R., Hibi, K., & Saito, H. (1992). Molecular analysis of the protein tyrosine phosphatase γ gene in human lung cancer cell lines. Cancer Research 52, 3506–3509.

Veillette, A., Bookman, M. A., Horak, E. M., Samelson, L. E., & Bolen, J. B. (1989). Signal transduction through the CD4 receptor involves the activation of the internal membrane tyrosine-protein kinase $p56^{lck}$. Nature 338, 257–259.

Volarevic, S., Niklinska, B. B., Burns, C. M., Yamada, H., June, C. H., Dumont, F. J., & Ashwell, J. D. (1992). The CD45 tyrosine phosphatase regulates phosphotyrosine homeostasis and its loss reveals a novel pattern of late T cell receptor-induced Ca^{2+} oscillations. J. Exp. Med. 176, 835–844.

Volarevic, S., Niklinska, B. B., Burns, C. M., June, C. H., Weissman, A. M., & Ashwell, J. D. (1993). Regulation of TCR signaling by CD45 lacking transmembrane and extracellular domains. Science 260, 541–543.

Wagner, N., Engel, P., & Tedder, T. F. (1993). Regulation of the tyrosine kinase-dependent adhesion pathway in human lymphocytes through CD45. J. Immunol. 150, 4887–4899.

Walton, K. M., Martell, K. J., Kwak, S. P., Dixon, J. E., & Largent, B. L. (1993). A novel receptor-type protein tyrosine phosphatase is expressed during neurogenesis in the olfactory neuroepithelium. Neuron 11, 387–400.

Wang, Y., & Pallen, C. (1991). The receptor-like protein tyrosine phosphatase HPTPα has two active catalytic domains with distinct substrate specificies. EMBO J. 10, 3231–3237.

Wang, Y., & Pallen, C. J. (1992). Expression and characterization of wild type, truncated, and mutant forms of the intracellular region of the receptor-like protein tyrosine phosphatase HPTPβ. J. Biol. Chem. 267, 16696–16702.

Wary, K. K., Lou, Z., Buchberg, A. M., Siracusa, L. D., Druck, T., LaForgia, S., & Kuebner, K. (1993). A homozygous deletion within the carbonic anhydrase-like domain of the *Ptprg* gene in murine L-cells. Cancer Res. 53, 1498–1502.

Weaver, C. T., Pingel, J. T., Nelson, J. O., & Thomas, M. L. (1991). $CD8^+$ T-cell clones deficient in the expression of the CD45 protein tyrosine phosphatase have impaired responses to T-cell receptor stimuli. Mol. Cell. Biol. 11, 4415–4422.

Wong, E. C. C., Mullersman, J. E., & Thomas, M. L. (1993). Leukocyte common antigen-related phosphatase (LRP) gene structure: Conservation of the genomic organization of transmembrane protein tyrosine phosphatases. Genomics 17, 33–38.

Yan, H., Grossman, A., Wang, H., D'Eustachio, P., Mossie, K., Musacchio, J. M., Silvennoinen, O., & Schlessinger, J. (1993). A novel receptor tyrosine phosphatase-σ that is highly expressed in the nervous system. J. Biol. Chem. 268, 24880–24886.

Yang, X., Seow, K. T., Bahri, S. M., Oon, S. H., & Chia, W. (1991). Two *Drosophila* receptor-like tyrosine phosphatase genes are expressed in a subset of developing axons and pioneer neurons in the embryo CNS. Cell 67, 661–673.

Yi, T., Cleveland, J. L., & Ihle, J. N. (1991). Identification of novel protein tyrosine phosphatases of hematopoietic cells by polymerase chain reaction amplification. Blood 78, 2222–2228.

Zheng, X. M., Wang, Y., & Pallen, C. J. (1992). Cell transformation and activation of $pp60^{c\text{-}src}$ by overexpression of a protein tyrosine phosphatase. Nature 359, 336–339.

REGULATION OF THE MAMMALIAN ADENYLYL CYCLASES

Roy J. Duhe, Andrew H. Dittman, Zhiliang Wu, and Daniel R. Storm

Biomembranes
Volume 6, pages 107–128.

ISBN: 1-55938-663-0.

I. INTRODUCTION

The mammalian adenylyl cyclases are a family of integral membrane proteins that catalyze the formation of cAMP. Cyclic AMP is an important intracellular messenger which regulates a variety of biochemical and physiological systems in animal cells including carbohydrate, lipid, protein, and nucleic acid metabolism, as well as synaptic transmission, ion channel function, muscle contraction, secretion, and transcription (reviewed by Krebs and Beavo, 1979; Nestler and Greengard, 1983; Nairn et al., 1985).

Adenylyl cyclases are regulated by a variety of stimulatory and inhibitory receptors coupled to the catalytic subunit through the guanyl nucleotide regulatory proteins, G_s and G_i. In addition, modulation of adenylyl cyclase activity by Ca^{2+} has been demonstrated in several tissues and it has been proposed that Ca^{2+}-sensitive adenylyl cyclases may provide coupling between the Ca^{2+} and cAMP signal transduction systems. Characterization of individual forms of mammalian adenylyl cyclases has been greatly facilitated by purification of the enzymes and isolation of cDNA clones encoding specific adenylyl cyclases. Specific adenylyl cyclases may be regulated by multiple effector molecules including Ca^{2+}, receptors coupled through G_s or G_i, and protein kinase C. The diversity of this enzyme system undoubtedly reflects the need for different mechanisms for regulation of cAMP levels in animal cells and the variety of physiological processes that are regulated by intracellular cAMP. This is illustrated by the neurospecific, type I calmodulin-sensitive adenylyl cyclase which may be important for neuroplasticity in the central nervous system. This review reports on recent developments in the characterization of adenylyl cyclases and focuses on the unique properties of the type I adenylyl cyclase.

II. IDENTIFICATION OF A FAMILY OF MAMMALIAN ADENYLYL CYCLASES

A. Separation of Calmodulin-Sensitive and Calmodulin-Insensitive Adenylyl Cyclases

The first evidence for the existence of distinct forms of adenylyl cyclases was the separation of calmodulin-sensitive and calmodulin-insensitive adenylyl cyclases from bovine brain using CaM-Sepharose affinity chromatography (Westcott et al., 1979). The calmodulin-sensitive adenylyl cyclase purified from bovine brain absorbed to CaM-Sepharose in the presence of Ca^{2+} and was stimulated by Ca^{2+} when reconstituted with calmodulin. Half-maximal stimulation of the enzyme occurred at approximately 80 nM free Ca^{2+}. The calmodulin-insensitive forms of adenylyl cyclase present in brain did not absorb to CaM-Sepharose in the presence or absence of Ca^{2+}, and were not stimulated by Ca^{2+} either in the presence or absence of calmodulin. Furthermore, polyclonal and monoclonal antibodies have been isolated that distinguished between the calmodulin-sensitive and calmodulin-insensitive adenylyl cyclases providing further evidence for separate adenylyl cyclases (Rosenberg and Storm, 1987; Mollner and Pfeuffer, 1991; Mollner et al., 1988).

The catalytic subunit of a calmodulin-sensitive adenylyl cyclase from brain was purified to homogeneity using calmodulin-Sepharose and forskolin-Sepharose affinity chromatography (Yeager et al., 1985; Smigel et al., 1986; Minocherhomjee et al., 1987). In general, it has been difficult to purify the mammalian adenylyl cyclases from membrane preparations because of the lability of catalytic activity and the small amount of these proteins present in membranes. Consequently, the purified proteins have not yielded significant structural information about the enzymes. Nevertheless, characterization of the purified calmodulin-sensitive adenylyl cyclase from brain indicated that it is a glycoprotein which interacts directly with calmodulin (Minocherhomjee et al., 1987). Furthermore, it can couple to β-adrenergic receptors through G_s (May et al., 1985; Rosenberg et al., 1987) and muscarinic receptors through G_i (Tota et al., 1990). The expression of the type I adenylyl cyclases in the insect/Baculovirus system and the development of new strategies for its purification from these cells offers great promise for significant progress in direct characterization of the protein (Taussig et al., 1993b).

B. Isolation of cDNA Clones for Mammalian Adenylyl Cyclases

Characterization of the adenylyl cyclases was significantly advanced by the isolation of a cDNA clone for the type I adenylyl cyclase from a bovine brain cDNA library (Krupinski et al., 1989) and cDNA clones for five additional adenylyl cyclases designated types II–VI (Bakalyar and Reed, 1990; Gao and Gilman, 1991; Yoshimura and Cooper, 1992; Feinstein et al., 1991; Ishikawa et al., 1992; Premont et al., 1992). Although these enzymes share considerable sequence homology and can be fit to a common membrane topology on the basis of hydropathy plots, they

Table 1. Properties of the Cloned Mammalian Adenylyl Cyclases

Type	*mRNA*	*Mr*	*Ca/CaM*	*β/γ*	*Tissue*
I	11.5	120 kD	Yes	Inhibition	Brain, Retina, Adren. Med.
II	4.1	119 kD	No	Stimulation	Brain, Lung Olf. Epith.
III	4.7	125 kD	Yes	Insensitive	Olf. Neurons, Brain, Heart, Adren., Lung, Retina
IV	3.5	117 kD	No	Stimulation	Brain, Kidney, Heart, Liver, Lung
V	5.7	119 kD	No	Insensitive	Heart, Brain, Liver, Kidney
VI	6.1	132 kD	Ca^{2+} Inhib.	Insensitive	Heart, Brain, Intestine, Kidney, Sk. Mus.

contain hypervariable regions which distinguish each enzyme. All six adenylyl cyclase catalytic subunits have a predicted M_r of approximately 120- to 130-kD and are proposed to contain 12 transmembrane sequences and two large cytoplasmic domains (Table 1). Although the topographical homology between the mammalian adenylyl cyclases and various ion channels suggested that the enzyme may have channel activity (Krupinski et al., 1989), there is no experimental evidence that any of these enzymes possess ion channel activity. All six enzymes are stimulated by forskolin, but differ in their sensitivity to calmodulin and the βγ complex of G-proteins. On the basis of sequence homologies, the six adenylyl cyclases can be divided into four subclasses comprised of type I, type III, type II and IV, and type V and VI (Katsushika et al., 1992; Premont et al., 1992).

III. REGULATION OF ADENYLYL CYCLASES BY G-COUPLING PROTEINS AND RECEPTORS

A. Coupling of Receptors to Adenylyl Cyclases Through G-Coupling Proteins

The adenylyl cyclases are multi-subunit protein complexes comprised of catalytic, hormone receptor, and guanine nucleotide-binding regulatory subunits, G_s and G_i (Pfeuffer and Helmreich, 1975; Orly and Schramm, 1976; Ross and Gilman, 1980). The guanyl nucleotide-stimulatory complex, G_s, binds GTP and its nonhydrolyzable analog, GppNHp. G_s is required for hormone, NaF, and guanine nucleotide stimulation of adenylyl cyclases whereas G_i mediates inhibition of the enzyme. G_s and G_i are both heterotrimers of α-, β-, and γ-subunits. The α-subunits are distinct for each G-protein complex, whereas β- and γ-subunits are similar for different G-proteins. Binding of GTP to the G_s or G_i α-subunit results in the dissociation of the βγ complex and activation of the α-subunits which may interact directly with the catalytic subunits of adenylyl cyclases. Hydrolysis of bound GTP causes the reassociation of Gα with the βγ complex (Gilman, 1984).

B. Regulation of Adenylyl Cyclases by α-Subunits of G-Proteins

It is well established that stimulatory receptors catalyze the exchange of GTP for GDP on G_s-α subunits with dissociation of the βγ complex. The activated G_s-α · GTP complexes can directly stimulate adenylyl cyclases (Gilman, 1987; Bourne et al., 1990; Simon et al., 1991). For example, the purified catalytic subunits of the type I and type II adenylyl cyclases can be directly stimulated by G_s-α · GTP (Taussig et al., 1993b). However, the mechanism(s) for receptor inhibition of adenylyl cyclases through G_i have not been clear until recently. Chen and Iyengar (1993) examined the effect of a mutant activated G_{i2}-α on types II, III, and VI adenylyl cyclases by cotransfection in COS-cells. Their data indicate that most, if not all adenylyl cyclases can be directly inhibited by G_i-α. Similarly, Taussig et al. (1993a) have shown that type-I and -V adenylyl cyclases expressed in Sf9 cell membranes are directly inhibited by G_i-α. The inhibition of type I adenylyl cyclase by G_i was much more evident when the enzyme was activated by forskolin or calmodulin. This is consistent with the data of Tota et al. (1990) who demonstrated that carbachol inhibition of type I adenylyl cyclase reconstituted with G_i and muscarinic receptors was most pronounced when the enzyme was stimulated by calmodulin.

C. Regulation of Adenylyl Cyclases by βγ-Subunits of G-Proteins

One of the most interesting properties of the mammalian adenylyl cyclases is their different sensitivities to the βγ complex from G-coupling proteins (Tang and Gilman, 1991; Federman et al., 1992). Although hormone regulation of adenylyl cyclases through G-proteins may involve significant contributions from the direct actions of G_s-α and G_i-α on the catalytic subunit of the enzymes, βγ can directly or indirectly modulate adenylyl cyclase activities. For example, Tang and Gilman (1991) demonstrated that addition of βγ to preparations of the type I adenylyl cyclase, stimulated by either G_s-α or calmodulin, inhibited adenylyl cyclase activity. In contrast, βγ enhanced the stimulation of the type II or type IV adenylyl cyclases by G_s-α. The other adenylyl cyclases described are apparently insensitive to βγ.

The mechanism for βγ regulation of adenylyl cyclases may be due to direct interactions with catalytic subunits of the adenylyl cyclases. Using purified adenylyl cyclases, it has been demonstrated that G_s-α-stimulated type I adenylyl cyclase is inhibited by βγ, whereas βγ stimulated the type II adenylyl cyclase (Taussig et al., 1993b). These investigators also saw modest, but significant, stimulation of the type II cyclase without G_s present. This strongly suggests that βγ can regulate adenylyl cyclases by direct interactions with catalytic subunits.

D. Stimulation of Adenylyl Cyclase Activities by Receptors that Normally Couple to Inhibition of the Enzymes Through G_i

Stimulation of adenylyl cyclases by βγ is a very important regulatory mechanism and it suggests that receptors coupled to G_i might stimulate certain types of adenylyl cyclases, providing that G_s-α is also active. For example, co-expression of mutationally active G_s-α with type II adenylyl cyclase in 293 cells resulted in stimulation of adenylyl cyclase by agonists, for example, α-2-adrenergic agonists, that normally act through inhibitory receptors (Federman et al., 1992). Stimulation of the type II adenylyl cyclase by α-2-adrenergic agonists was apparently mediated by the liberation of βγ from G_i-α. Similarly unexpected stimulation of adenylyl cyclase activity by inhibitory agonists has been found in membranes from rat olfactory bulb (Olinas and Onali, 1992). The muscarinic agonist, carbachol, normally inhibits adenylyl cyclase activity through M2 or M4 receptors coupled to adenylyl cyclase through G_i. However, carbachol stimulates adenylyl cyclase activity in membrane preparations from rat olfactory bulb. Carbachol stimulation is dependent upon GTP, sensitive to pertussis toxin, and independent of Ca^{2+} increases. Similarly, Baumgold et al. (1992) have discovered that stimulation of cAMP levels in SK-N-SH cells by carbachol is independent of Ca^{2+} increases and therefore, not due to Ca^{2+} stimulation of calmodulin-sensitive adenylyl cyclases. One possible interpretation of these data is that carbachol liberates βγ from G_i which, in turn, stimulates type II- or type IV-adenylyl cyclases in these cells.

E. Cross-Over From G_i Inhibition to G_s Stimulation of Adenylyl Cyclases

Muscarinic M4 acetylcholine receptors normally couple through G_i to inhibit adenylyl cyclases. Recent data suggest that M4 muscarinic receptors may undergo "cross-over" from G_i inhibition to G_s stimulation at high receptor density or when G_i is inhibited by pertussis toxin (Dittman et al., 1994). In these experiments, M4 receptors and specific isozymes of adenylyl cyclases were co-expressed in HEK-293 cells to characterize the mechanism(s) for M4 receptor regulation of adenylyl cyclases. The calmodulin-sensitive type I- and type III-adenylyl cyclases were chosen for this study because neither enzyme is stimulated by the βγ complex of G-coupling proteins. M4 receptors exhibited either inhibition or stimulation of type-I and -III adenylyl cyclases depending upon receptor density and agonist concentration. Inhibition of adenylyl cyclase was apparently due to M4 coupling through G_i since this effect was blocked by pertussis toxin.

Adenylyl cyclase stimulation through M4 receptors was not due to increases in intracellular Ca^{2+} since it was evident in isolated membranes in the absence of free Ca^{2+} and with whole cells preloaded with the Ca^{2+} chelator, BAPTA-AM. Stimulation of adenylyl cyclase activities by M4 receptors was apparently mediated via G_s since it was GTP-dependent, insensitive to pertussis toxin, and was not due to βγ stimulation. Synthetic peptides derived from a G-protein activating region of the M4 receptor mimicked the M4-mediated stimulation of adenylyl cyclase activity. These data demonstrated a novel mechanism for muscarinic regulation of adenylyl cyclases that apparently involves cross-over from inhibitory to stimulatory G-protein coupling.

What role does M4 receptor coupling play in cellular signal transduction? M4 activation of G_s may act directly to stimulate adenylyl cyclases. Alternatively, the dual G_s and G_i coupling properties of M4 might facilitate $\beta\gamma$ stimulation of type-II and -IV adenylyl cyclases. Since activated G_s-α is required for $\beta\gamma$ stimulation of type-II and -IV adenylyl cyclase, receptors such as M4 or $\alpha_{2C}10$, which can activate both G_s-α and stimulate $\beta\gamma$ release from G_i and G_o, may provide signal amplification through a single receptor.

IV. REGULATION OF ADENYLYL CYCLASE ACTIVITIES BY CALCIUM AND CALMODULIN

A. Stimulation of the Type I Adenylyl Cyclase by Ca^{2+} and CaM *In Vitro* and *In Vivo*

Expression of type I adenylyl cyclase activity in insect Sf9 cells (Tang et al., 1991) and kidney 293 cells (Choi et al., 1992a) has established that this enzyme is directly stimulated by calmodulin and Ca^{2+} *in vitro*. Calmodulin stimulated the type I enzyme activity in membranes from 293 cells at a half-maximal concentration of approximately 20 nM. Half-maximal stimulation of type I adenylyl cyclase occurs at approximately 50 nM free Ca^{2+} (Choi et al., 1992a) which is consistent with the Ca^{2+} sensitivity of the calmodulin-stimulated adenylyl cyclase isolated from bovine brain (Westcott et al., 1979). It has also been established that the type I adenylyl cyclase can couple increases in intracellular Ca^{2+} to cAMP production *in vivo* in cultured 293 cells (Choi et al., 1992b). In the presence of 2 mM extracellular Ca^{2+}, the intracellular cAMP levels of control 293 cells lacking type I adenylyl cyclase were unaffected by addition of the Ca^{2+} ionophore, A23187 (Figure 1a). In contrast, intracellular cAMP in 293 cells stably expressing the type I adenylyl cyclase increased approximately 16-fold with the addition of 10 μM A23187. The increase in intracellular cAMP stimulated by A23187 depended upon the concentration of extracellular Ca^{2+} (Figure 1b). Elevated cAMP was detectable within a few minutes after addition of 10 μM A23187 and 2 mM Ca^{2+} (Figure 1c). These data demonstrated that intracellular Ca^{2+} can stimulate the type I adenylyl cyclase activity *in vivo*. Furthermore, the type I adenylyl cyclase was also indirectly stimulated by muscarinic receptors by mobilization of intracellular free Ca^{2+} (Choi et al., 1992b).

Ca^{2+} stimulation of the type I adenylyl cyclase *in vivo* may be due to direct interactions of the enzyme with Ca^{2+} and calmodulin, or indirect mechanism involving stimulation of the enzyme by Ca^{2+}-activated protein kinases. Several point mutations within the calmodulin-binding domain of the enzyme were made to determine if the Ca^{2+} sensitivity of the enzyme can be modified by mutagenesis (Wu et al., 1993). The catalytic activities of the mutant enzymes were comparable to wild-type type I adenylyl cyclase. However, Ca^{2+} and calmodulin stimulation were abolished by substitution of Phe-503 with Arg-503. Stimulation of type I adenylyl cyclase activity *in vivo* by intracellular Ca^{2+} was also greatly diminished

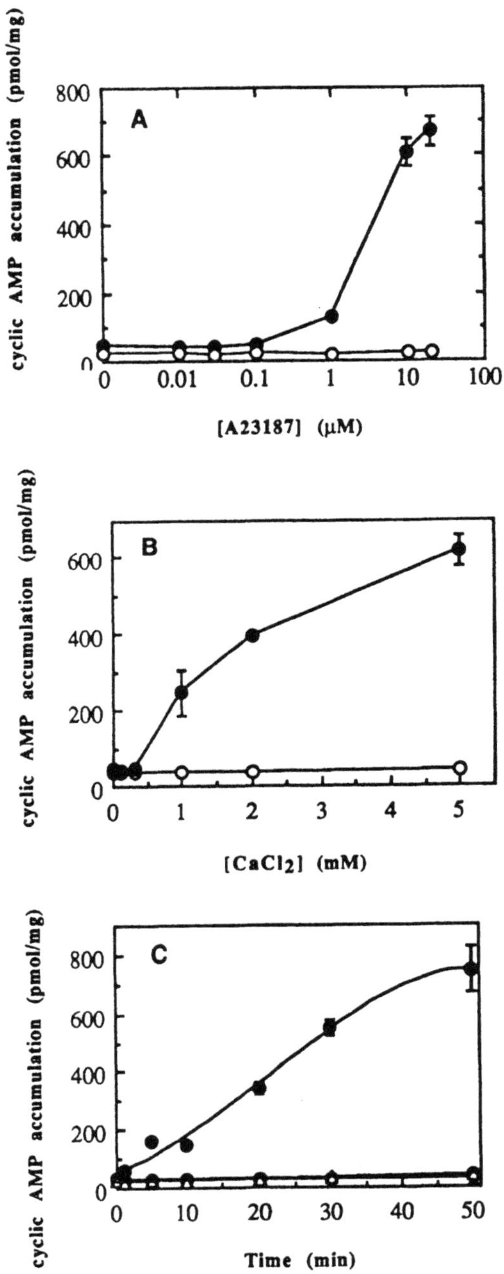
cyclic AMP accumulation (pmol/mg)
A
800
600
400
200
0
0
0.01
0.1
1
10
100
[A23187] (μM)
B
600
400
200
0
1
2
3
4
5
[CaCl2] (mM)
C
800
600
400
200
0
10
20
30
40
50
Time (min)

with the Arg-503 mutant indicating that Ca^{2+} stimulation of the enzyme *in vivo* is due primarily to direct interactions with calmodulin and Ca^{2+}.

B. Regulation of the CaM Sensitivity of the Type I Adenylyl Cyclase by Nitric Oxide

The calmodulin-binding domain of the type I adenylyl cyclase has recently been identified as an amino acid sequence (residues 495–522) that contains two cysteine residues (Vorherr et al., 1993; Wu et al., 1993). Therefore, we examined the effect of several sulfhydryl reagents on the calmodulin sensitivity of the enzyme. Treatment of membranes containing the type I adenylyl cyclase with limiting amounts of o-iodosobenzoate, which oxidizes vicinal sulfhydryls to disulfides, inhibited stimulation by Ca^{2+} without affecting basal adenylyl cyclase activity. Calmodulin stimulation of the enzyme was restored by treatment with dithiothreitol or glutathione, which reduce disulfides to free thiols. NO and sodium nitroprusside also reversibly inhibited CaM stimulation of the enzyme. The loss in calmodulin sensitivity caused by NO may be due to the oxidation one or more sets of vicinal thiols present in the enzyme.

C. Stimulation of the Type III Adenylyl Cyclase by Ca^{2+} and Calmodulin

Although the type III adenylyl cyclase is highly enriched in olfactory tissue (Bakalyar and Reed, 1990) the enzyme is also expressed in other tissues including brain and heart muscle (Xia et al., 1992). Until recently, the only documented calmodulin-sensitive mammalian adenylyl cyclase was the type I enzyme. However, the type III adenylyl cyclase is also stimulated by Ca^{2+} and calmodulin (Choi et al., 1992a). In contrast to the type I enzyme, type III adenylyl cyclase is not stimulated by calmodulin and Ca^{2+} in the absence of other effectors. However, calmodulin stimulates the type III enzyme activity in the presence of forskolin or GppNHp. Half-maximal stimulation of the type III adenylyl cyclase occurs at approximately 1 μM calmodulin and 5 μM free Ca^{2+}. This enzyme may allow Ca^{2+}

Figure 1. Ca^{2+} and A23187 stimulation of intracellular cAMP levels in 293 cells expressing the type I adenylyl cyclase. (**A**) Cultured 293 cells expressing the type I adenylyl cyclase (●) or control cells transfected with the CDM8 expression vector without the cyclase insert (○) were treated with varying concentrations of A23187 in the presence of 2 mM $CaCl_2$ for 30 minutes. (**B**) 293 cells expressing the type I adenylyl cyclase (●) or control cells (○) were treated with varying concentrations of $CaCl_2$ for 30 minutes in the presence of 10 μM A23187. (**C**) 293 cells expressing the type I adenylyl cyclase (●, Δ) or control cells lacking type I adenylyl cyclase (○, X) were treated for various periods of time with 2 mM $CaCl_2$ and 10 μM A23187 in the presence (●, ○) or absence (Δ, X) of 1 mM IBMX. (Taken from Choi et al., 1992b, with permission.)

amplification of cyclic AMP signals that are generated by receptor stimulation of adenylyl cyclase. For example, the existence of cAMP-gated ion channels in olfactory sensory neurons suggests that initial cAMP signals, generated through olfactory receptors coupled to adenylyl cyclase, may be further amplified by increases in intracellular Ca^{2+} through cAMP-gated ion channels.

D. Regulation of Adenylyl Cyclases by Protein Kinase C

Cross-talk between the cAMP and the phosphoinositide/calcium signal transduction pathways can occur by several different mechanisms. For example, muscarinic agonists can indirectly increase intracellular cAMP through mobilization of free Ca^{2+} which activates the type I calmodulin (CaM)[1]-sensitive adenylate cyclase (Choi et al., 1992b). Phorbol esters and other activators of protein kinase C (PKC) can also affect intracellular cAMP levels in various tissues and cultured cells. Activation of protein kinase C in intact cells can either facilitate (Bell and Brunton, 1987; Choi and Toscano, 1988; Johnson and Toews, 1990; Rozengurt et al., 1987; Quilliam et al., 1989) or inhibit (Bell and Brunton, 1987; Summers and Cronin, 1988; Bushfield et al., 1987; Dixon et al., 1988; Yamashita et al., 1988) cAMP accumulations caused by forskolin or receptor stimulation of adenylyl cyclases. The response of various cell types and membrane preparations to phorbol esters is complex and varied. For example, treatment of a rat osteosarcoma cell line with 12-O-tetradecanoylphorbol-13-acetate (TPA) increased parathyroid hormone-stimulated adenylyl cyclase activity and inhibited prostaglandin E2-responsive enzyme activity (Freyaldenhoven et al., 1992). Activation of PKC in PC12 cells increased the response of the cAMP generating systems, whereas PKC activation in NCB20 cells and NIH 3T3 caused inhibition of cAMP-generating systems (Gusovsky and Gutkind, 1991). These multiple and contradicting effects of phorbol esters on hormone-sensitive adenylyl cyclase can occur within the same cell, and may be due to multiple PKC phosphorylation sites within the adenylyl cyclase system, different forms of adenylyl cyclase present in various cells, or different isozymes of PKC present.

The availability of cDNA clones for different adenylyl cyclases has made it possible to examine the sensitivity of distinct adenylyl cyclases to direct or indirect regulation by activators of PKC in whole cells. The effect of phorbol esters on the activity of the type I- and type III-adenylyl cyclases in whole cells has been examined using stably transfected 293 cells expressing either enzyme (Choi et al., 1993). TPA markedly enhanced the forskolin responsiveness of the type I- and type III-adenylyl cyclases expressed in kidney 293 cells. The effect of TPA on the activity of the calmodulin-sensitive adenylyl cyclases was not mediated through increases in intracellular free calcium. Jacobowitz et al. (1993) have also examined the sensitivities of various adenylyl cyclases to phorbol esters by transient expression of the enzymes in 293 cells. Their data indicated that the type II adenylyl cyclase was particularly sensitive to phorbol esters and that types IV, V, and VI showed modest stimulations upon PMA treatment. The stimulation of adenylyl cyclases by phorbol esters suggests that activation of protein kinase C can elevate intracellular cAMP in animal cells.

V. PHYSIOLOGICAL FUNCTION OF THE TYPE I ADENYLYL CYCLASE AND ITS IMPLIED ROLE IN NEUROPLASTICITY

A. Physiological Role(s) of the Type I Adenylyl Cyclase

The type I adenylyl cyclase is not required for survival of mice; live mice disrupted in the gene for this enzyme have lived for at least five weeks after birth (Wu and Storm, unpublished observations). Although the physiological function of the type I adenylyl cyclase in brain is not known, it can function to link changes in intracellular free Ca^{2+} to increases in cAMP *in vivo* and thus couple the calcium and cAMP regulatory systems (Choi et al., 1992b). Data from a number of different laboratories have suggested that the type I calmodulin-sensitive adenylyl cyclase may be important for neuroplasticity. For example, the *Drosophila melanogaster* learning mutant, rutabaga, is deficient in calmodulin-sensitive adenylyl cyclase activity (Livingstone, 1985; Dudai, 1988; Dudai and Zvi, 1985; Livingstone et al., 1984). The gene for the type I adenylyl cyclase maps within a region on the X chromosome that includes the *rut* locus (Levin et al., 1992). Furthermore, the rutabaga mutant showed altered synaptic plasticity at neuromuscular junctions (Zhong and Wu, 1991). Feany (1990) has proposed that calcium responsiveness, rather than the overall cAMP synthesis may be the crucial component of adenylyl cyclase activity required for associative learning in *Drosophila*. On the basis of their work with *Aplysia*, it has been proposed that the calmodulin-sensitive adenylyl cyclase may play a key role for associative processes in brain (Kandel and Schwartz, 1982; Walters and Byrne, 1983; Klein et al., 1986; Greenberg et al., 1987). There is also evidence that neurotransmitter stimulation of adenylyl cyclase is important for long-term potentiation in the dentate gyrus, CA1 and CA3 regions of mammalian brain (Stanton and Sarvey, 1985; Hopkins and Johnston, 1988; Frey et al., 1993).

B. Tissue Distribution of the Type I Adenylyl Cyclase

Because of the implied role of the type I adenylyl cyclase in neuroplasticity, an extensive northern analysis using poly A+ selected RNA from various bovine tissues and murine cell lines was carried out (Xia et al., 1993). As discussed above, the complete coding cDNA sequences have been determined for six mammalian adenylyl cyclases. Although these proteins show considerable overall homology, they contain several hypervariable domains including a region in a cytoplasmic loop designated C1b and a transmembrane region designated M2 (Gao and Gilman, 1991). Probes specific for the type I adenylyl cyclase were made based upon cDNA sequence in the C1b and the M2 region. The specificity of these probes used for northern analysis is readily demonstrated by the negative signals obtained when these probes were used to analyze a variety of tissues that contain types II–VI adenylyl cyclases. In order to determine if mRNA for this enzyme is expressed in brain from other vertebrates, mRNA from whole mouse brain, rat brain cortex, bovine brain cortex, and human fetal brain cortex was analyzed by northern analysis using the type I adenylyl cyclase-specific probe. All four species showed a specific transcript at approximately 11.7 kb, consistent with the size of the type I enzyme mRNA in bovine brain. cDNA Clones for the human type I adenylyl cyclase have been obtained and the human enzyme showed very high homology to the brain enzyme (Villacres et al., 1993). The gene for the human enzyme has been mapped to the short arm of chromosome 7.

A number of bovine tissues were analyzed by northern analysis for type I specific mRNA using the bovine type I specific probe (Figure 2). The only bovine tissues showing a positive signal for type I enzyme mRNA were brain, retina, and adrenal medulla. Several cultured cell lines, including neuroblastoma cell N1E-115, neuroglio-hybridoma cell NG-108, rat glioma 36B-10 cell, and PC-12 cells were also analyzed for type I adenylyl cyclase mRNA using the bovine probe, 3C. None of these cell lines gave a positive signal with the 3C probe. These data indicated that mRNA for the type I adenylyl cyclase has a limited tissue expression and that the type I enzyme may be classified as a neural-specific adenylyl cyclase. The restricted expression of type I adenylyl cyclase mRNA in neural tissues contrasts sharply with that of the other mammalian enzymes which show fairly broad distribution in both neural and non-neural tissues.

C. Distribution of Type I Adenylyl Cyclase mRNA in Rat Retina

The presence of type I adenylyl cyclase mRNA in retina is consistent with the Ca^{2+} and CaM sensitivities of adenylyl cyclase activities reported for retina. For example, Gnegy et al. (1984) demonstrated that activation of retina adenylyl cyclase by calmodulin was maximal at 120 nM free Ca^{2+} with an apparent K_a for CaM stimulation of 67 nM. The concentrations of Ca^{2+} and CaM required for half-maximal stimulation of the type I adenylyl cyclase are 50 nM and 20 nM, respectively. The distribution of type I adenylyl cyclase mRNA in retina was

Figure 2. Northern analysis of the type I Ca^{2+}/CaM sensitive adenylyl cyclase using mRNA from various bovine tissues and cultured cells. Two micrograms of poly (A)+ selected RNA samples were electrophoresed to a 1.2% agarose/formaldehyde gel. The blots were hybridized with an α [^{32}P]dCTP labeled cDNA probe, 3C, that is specific for the bovine type I adenylyl cyclase. The 0.24–9.5 kb RNA ladder from BRL was used as the molecular weight standard. Poly (A)+ selected RNA was isolated from various bovine tissues. (Taken from Xia et al., 1993, with permission.)

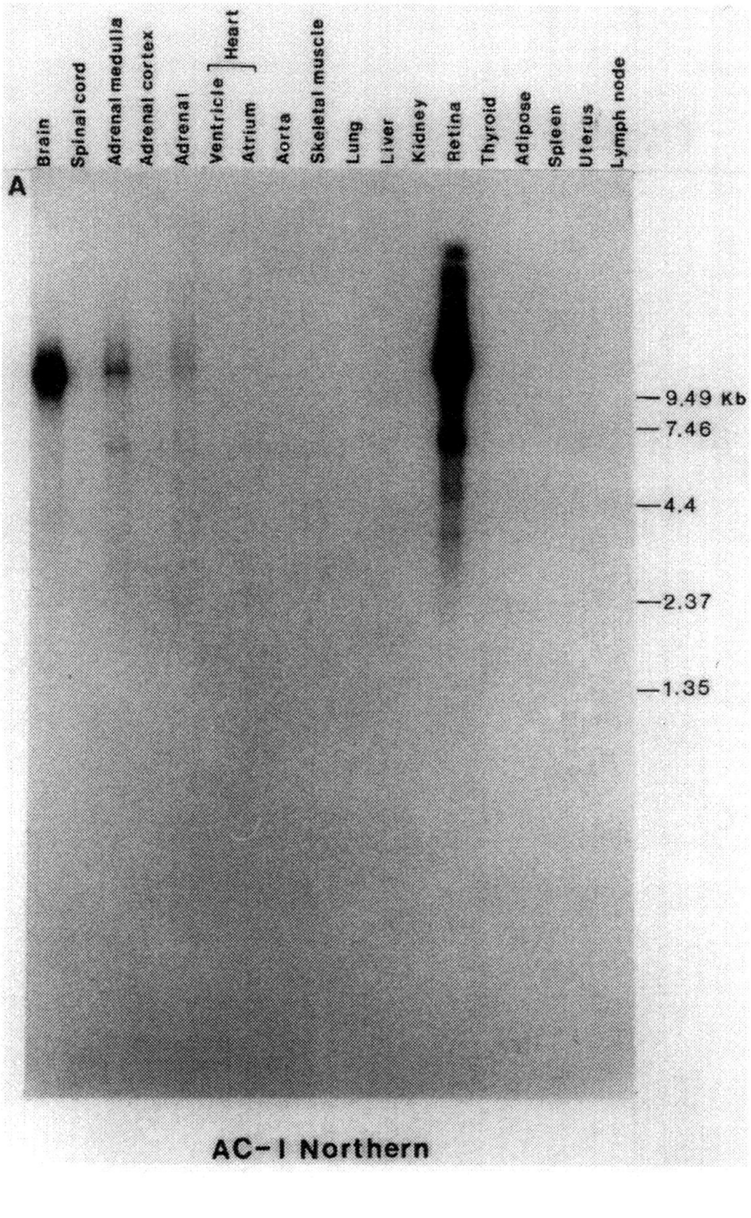
Brain
Spinal cord
Adrenal medulla
Adrenal cortex
Adrenal
Ventricle
Heart
Atrium
Aorta
Skeletal muscle
Lung
Liver
Kidney
Retina
Thyroid
Adipose
Spleen
Uterus
Lymph node
A
—9.49 Kb
—7.46
—4.4
—2.37
—1.35
AC-I Northern

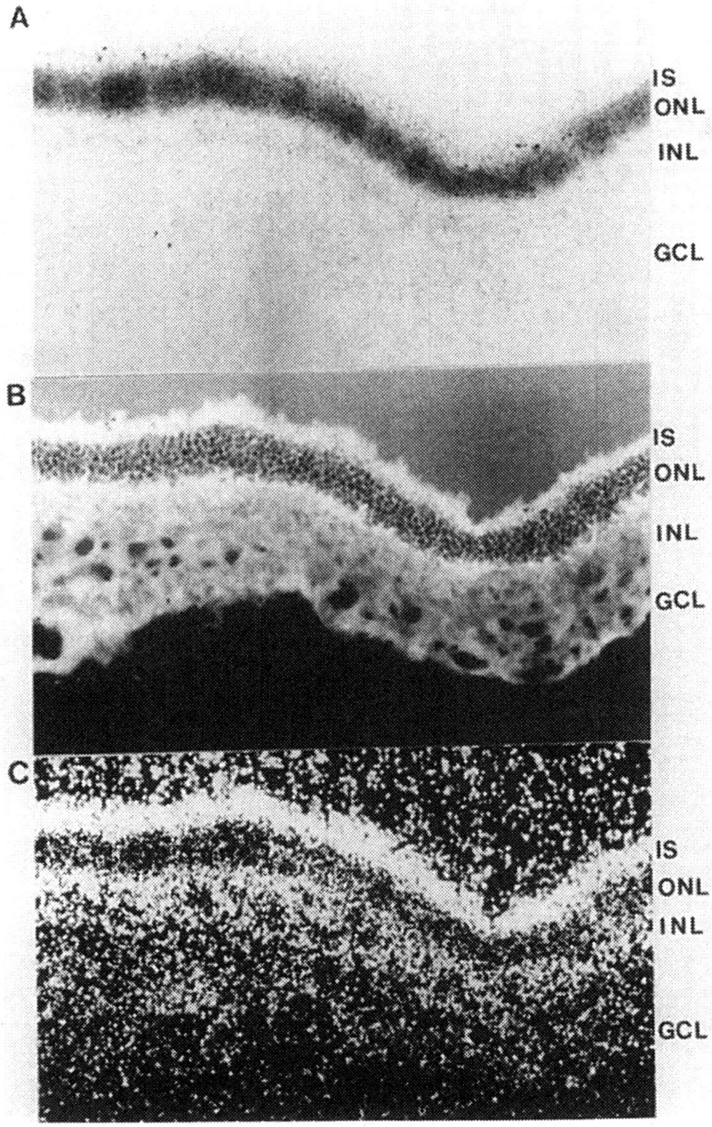
A
IS
ONL
INL
GCL
B
IS
ONL
INL
GCL
C
IS
ONL
INL
GCL

examined in more detail by *in situ* hybridization. Retina cross-sections from rat, rabbit, and bovine tissue were analyzed with a [^{35}S]-UTP labeled bovine riboprobe under conditions previously described (Xia et al., 1991). The neural retina was specifically labeled with the antisense riboprobe and the labeling was significantly reduced by addition of excess amount of unlabeled probe. mRNA for the type I adenylyl cyclase was detected in the inner segment layer of the photoreceptors (IS), and in all three nuclear layers of the neural retina (Figure 3). The outer nuclear layer (ONL) which contains rods and cones, the inner nuclear layer (INL) which contains horizontal cells, bipolar cells and amacrine cells, and the ganglion cell layer (GCL) all contained mRNA for type I adenylyl cyclase. The intensity of the labeling was strongest in the IS (the cytoplasm of photoreceptors) and the ONL.

The presence of type I adenylyl cyclase mRNA in the photoreceptor cells, suggests that coupling of Ca^{2+} to changes in cAMP may play an important regulatory role in photoreception or light-mediated cellular events. For example, elevated Ca^{2+} during stimulation of photoreceptor cells may increase intracellular cAMP through the type I adenylyl cyclase. Photoadaptation may be due, at least in part, to cAMP-dependent protein phosphorylations of channels, photoreceptors, or other regulatory molecules involved in phototransduction. While several transmitters modulate the activity of adenylyl cyclase in intact retina, or in retinal homogenates, most of the receptors are found in the inner retina.

D. Distribution of Type I Adenylyl Cyclase mRNA in Rat Brain

The distribution of mRNA encoding the type I adenylyl cyclase in rat brain was also examined by *in situ* hybridization (Figure 4; Xia et al., 1991). *In situ* hybridizations in adult rat brain revealed high levels of type I adenylyl cyclase mRNA in specific areas of brain associated with learning and memory, including the hippocampal formation, the neocortex, entorhinal cortex, cerebellum cortex, and the olfactory system. The dentate gyrus in the hippocampal formation showed very intense labeling which appeared to be associated with the granule cell layer. Moderately strong labeling was also evident in association with the pyramidal cells

Figure 3. *In situ* hybridization studies of the type I Ca^{2+}/CaM-sensitive adenylyl cyclase in bovine retina. Bovine retina eyecup cross-sections were hybridized with digoxigenin–dUTP labeled oligonucleotide probe, ZX5, derived from the 3′-noncoding sequence of the bovine type I adenylyl cyclase cDNA. The signals were detected using anti-digoxigenin antibody conjugated with alkaline phosphatase as described by the manufacturer (Boehringer Mannheim). Abbreviations: IS, inner segment layer of the photoreceptor cells; ONL, the outer nuclear layer which contains nuclei of rods and cones; INL, the inner nuclear layer which contains cell bodies of horizontal cells, bipolar cells, and amacrine cells; GCL, the ganglion cell layer. (Taken from Xia et al., 1993, with permission.)

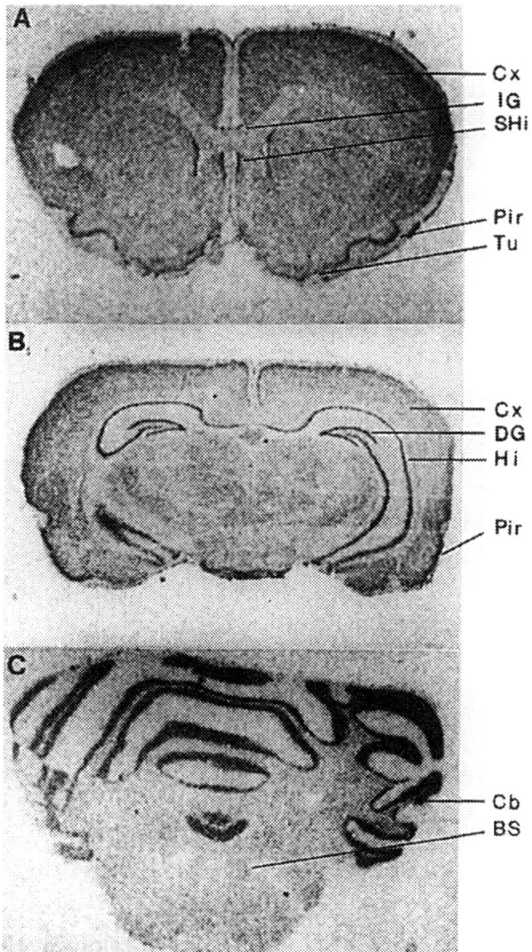

Figure 4. Autoradiographs of Coronal Rat Brain Sections Hybridized with [^{35}S]-Labeled Antisense Riboprobe Derived from the 3'-noncoding Sequence of the Type I CaM-Sensitive Adenylate Cyclase cDNA. Coronal rat brain sections were hybridized with [^{35}S]-labeled antisense riboprobe. Exposure time: 3 days. Cx, neocortex; DG, dentate gyrus; Hi, hippocampus; IG, indusium griseum; Pir, piriform cortex; SHi, septohippocampal nucleus; Tu, olfactory tubercle; Cb, cerebellum; BS, brain stem.

in CA1, CA2, and CA3 layers of the hippocampus. Similar patterns for the distribution of type I adenylyl cyclase mRNA in brain have been reported by Glatt and Snyder (1993). Messenger RNA for the type I adenylyl cyclase is not generally distributed throughout the brain, suggesting that the enzyme does not play a general regulatory role (e.g., in regulation of cell metabolism), and that it may be important for specific neuronal functions. Messenger RNA for the type I adenylyl cyclase is highly localized to specific regions of brain, including those areas that have showed long-term potentiation and have been implicated in learning and memory. Although these data do not define the function of the type I CaM-sensitive adenylyl cyclase, its mRNA distribution in these specific areas is consistent with the proposal that this enzyme may be important for learning and memory.

E. Biochemical Model for the Role of the Type I Adenylyl Cyclase in Neuroplasticity

The type I adenylyl cyclase is a neural-specific adenylyl cyclase with a highly restricted expression in mammalian brain. What is the special function of this enzyme in neurons and what is its molecular role in neuroplasticity? One model for

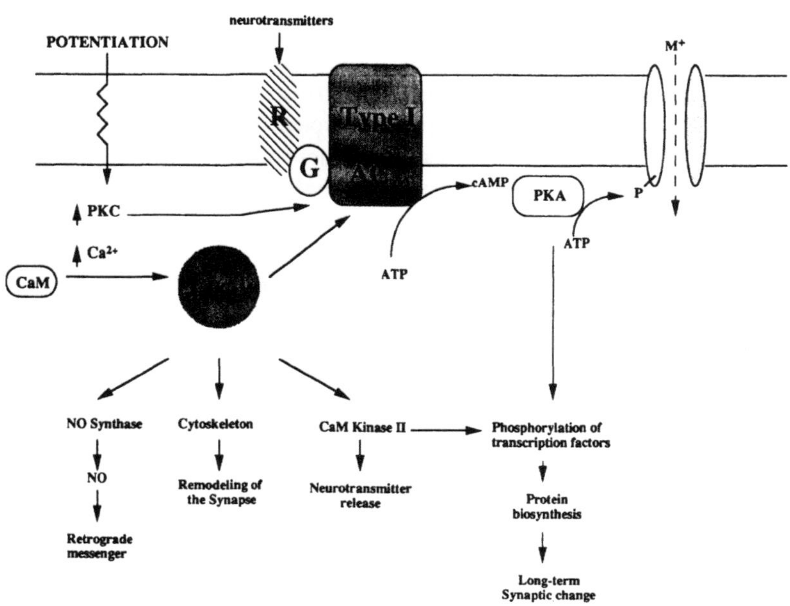

Figure 5. Model for the role of the type I adenylyl cyclase in neuroplasticity. During LTP, protein kinase C is activated and Ca^{2+} is mobilized. It is hypothesized that the type I adenylyl cyclase is activated during some forms of LTP by Ca^{2+}/calmodulin, protein kinase C, neurotransmitters, or a combination of these effectors. Activation of the type I adenylyl cyclase results in coupling of the Ca^{2+} and cAMP regulatory systems during LTP. Simultaneous or ordered activation of the Ca^{2+} and cAMP regulatory systems may be important for amplified cAMP signals, synergism between Ca^{2+} and cAMP activated kinases, and/or positive feedback regulation of Ca^{2+} channels by cAMP dependent kinase. Abbreviations: calmodulin; PKC, protein kinase C; PKA, cAMP-dependent protein kinase.

learning and memory that many neurobiologists have found useful is long-term potentiation (LTP). Although the molecular basis of synaptic plasticity is not known, we propose that some long-term changes in neurons and at synapses may require synergism between the cAMP and Ca^{2+} signal transduction systems (Figure 5). Furthermore, long-term changes suggest that transcriptional regulation may be required with synthesis of specific proteins required for long-term synaptic changes.

During LTP, protein kinase C is activated (reviewed by Linden and Routtenberg, 1989) and intracellular Ca^{2+} increases. We hypothesize that the type I adenylyl cyclase is activated during some forms of LTP because of increases in intracellular Ca^{2+}, release of neurotransmitters, activation of protein kinase C, or a combination of these events. Long-term changes in neurons may be due, at least in part, to cAMP control of transcription through cAMP responsive DNA elements such as CRE (reviewed by Mitchell and Tjian, 1989). The coupling of Ca^{2+} to cAMP increase by the type I adenylyl cyclase may produce a stronger or more persistent cAMP signal required for transcriptional control. Alternatively, the coupling of the Ca^{2+} and cAMP systems may result in simultaneous or sequentially ordered activation of CaM kinase II and the cAMP-dependent protein kinase. Convergence of the Ca^{2+}- and cAMP-dependent kinases on one or more transcriptional factors may be one of the key events in long-term adaptive responses in neurons. For example, the transcriptional factor, Creb, is phosphorylated by both CaM-dependent and cAMP-dependent kinases (Sheng et al., 1991). This mechanism may be limited to a subset of neurons in the central nervous system that contain the type I adenylyl cyclase.

VI. CONCLUSION

The mammalian adenylyl cyclases are complex membrane protein systems that play a pivotal role in regulation of animal cell functions. Differential regulation of adenylyl cyclases by G-α subunits, the $\beta\gamma$ complex of G-proteins, protein kinase C, and calmodulin provides a mechanism for cross-talk between different signal transduction systems. Synergistic stimulation of the type I- and III-adenylyl cyclases by calcium and G_s-coupled receptors may be an important form of cross-talk between signal transduction systems that perhaps plays a role in the visual and olfactory signal transduction systems. There has been considerable interest in the type I adenylyl cyclase because of its implied role in learning and memory. The limited expression of this neurospecific enzyme to areas of brain implicated in neuroplasticity, and the demonstration that it can couple increases in intracellular Ca^{2+} to elevations in cAMP *in vivo* provide important new insights concerning the physiological function of this enzyme. The Ca^{2+} and cAMP signal transduction systems may be directly coupled in neurons that contain the type I adenylyl cyclase, and this regulatory mechanism may be important for long-term adaptive responses in neurons.

REFERENCES

Bakalyar H. A., & Reed R. R. (1990). Identification of a specialized adenylyl cyclase that may mediate odorant detection. Science 250, 1403–1406.

Baumgold, J., Paek, R., & Fiskum, G. (1992). Calcium independence of phosphoinositide hydrolysis induced increase in cAMP accumulation in SK-N-SH human neuroblastoma cells. J. Neurochemistry 58, 1754–1759.

Bell, J. D., & Brunton L. L. (1987). Amer. J. Physiol. 252, E783–789.

Bourne, H. R., Sanders, D. A., & McCormick, F. (1990). The GTPase superfamily: A conserved switch for diverse cell functions. Nature 348, 125–132.

Bushfield, M., Hopple, S. L., Gibson, I. F., Murdoch, F. A., & MacIntyre, D. E. (1987). Effects of protein kinase C activation on human platelet cyclic AMP metabolism. FEBS Letters 222, 299–304.

Chen, J., & Iyengar, R. (1993). Inhibition of cloned adenylyl cyclases by mutant activated G_i-alpha & specific suppression of type 2 adenylyl cyclase by phorbol ester treatment. J. Biol. Chem. 268, 12253–12256.

Choi, E. J., & Toscano, W. A. Jr. (1988). Modulation of adenylate cyclase in human keratinocytes by protein kinase. C. J. Biol. Chem. 263, 17167–17172.

Choi, E. J., Xia, Z., & Storm, D. R. (1992a). Stimulation of the type III olfactory adenylyl cyclase by calcium and calmodulin. Biochemistry 31, 6492–6498.

Choi, E. J., Wong, S. T., Hinds, T. J., & Storm, D. R. (1992b). Calcium and muscarinic agonist stimulation of type I adenylyl cyclase in whole cells. J. Biol. Chem. 267, 12440–12442.

Choi, E. J., Wong, S. T., Dittman, A. H., & Storm, D. R. (1993). Phorbol ester stimulation of the type I and type III adenylyl cyclases in whole cells. Biochemistry 32, 1891–1894.

Dittman, A. H., Weber, J. P., Hinds, T. J., Choi, E. J., Migeon, J. C., Nathanson, N. M., & Storm, D. R. (1993). A novel mechanism for coupling of M4 muscarinic acetylcholine receptors to calmodulin sensitive adenylyl cyclases: Cross-over from G protein coupled inhibition to stimulation. Biochemistry 33, 943–951.

Dixon, B. S., Breckon, R., Burke, C., & Anderson, R. J. (1988). Phorbol esters inhibit adenylate cyclase activity in cultured collecting tubular cells. Amer. J. Physiol. 254, C183–191.

Dudai, Y. (1988). Neurogenetic dissection of learning and short-term memory in *Drosophila*. Annu. Rev. Neurosci. 11, 537–563.

Dudai, Y., & Zvi, S. (1985). Multiple defects in the activity of adenylyl cyclase from the *Drosophila* memory mutant rutabaga. J. Neurochem. 45, 355–364.

Duhe, R. J., Nielsen, M. D., Dittman, A. H., Villacres, E. C., Choi, E. J., & Storm, D. R. (1993). Oxidation of critical cysteine residues of type I adenylyl cyclase by o-iodosobenzoate and nitric oxide reversibly inhibits stimulation by calcium and calmodulin. J. Biol. Chem. 269, 7290–7296.

Feany, M. B. (1990). Rescue of the learning defect in dunce, a *Drosophila* learning mutant, by an allele of rutabaga, a second learning mutant. Proc. Natl. Acad. Sci. USA 87, 2795–2799.

Federman, A. D., Conklin, B. R., Schrader, K. A., Reed, R. R., & Bourne, H. R. (1992). Hormonal stimulation of adenylyl cyclase through G_i protein $\beta\gamma$ subunits. Nature 356, 159–161.

Feinstein, P. G., Schrader, A., Bakalyar, H. A., Tang, W. J., Krupinski, J., Gilman, A. G., & Reed, R. R. (1991). Molecular cloning and characterization of a Ca^{2+}/calmodulin insensitive adenylyl cyclase for rat brain. Proc. Natl. Acad. Sci. USA 88, 10173–10177.

Frey, U., Huang, Y. Y., & Kandel, E. R. (1993). Effects of cAMP stimulate a late stage of LTP in hippocampal CA1 neurons. Science 260, 1661–1664.

Freyaldenhoven, A. M., Gutierrez, G. E., Lifschitz, M. D., & Katz, M. S. (1992). Protein kinase C differentially modulates PTH- and PGE2-sensitive adenylate cyclase in osteoblast-like cells. Amer. J. Physiol. 262, E87–95.

Gao, B., & Gilman, A. (1991). Cloning and expression of a widely distributed (type IV) adenylyl cyclase. Proc. Natl. Acad. Sci. USA 88, 10178–10182.

Gilman, A. G. (1984). G proteins and dual control of adenylate cyclase. Cell 36, 577–579.

Gilman, A. G. (1987). Effects of protein kinase C activation on human platelet cyclic AMP metabolism. Annu. Rev. Biochem. 56, 615–649.

Glatt, C. E., & Snyder, S. H. (1993). Cloning and expression of an adenylyl cyclase localized to the corpus striatum. Nature 361, 536–538.

Gnegy, M., Muirhead, N., Roberts-Lewis, J. M., & Treisman, G. (1984). Calmodulin stimulates adenylyl cyclase activity and increases dopamine activation in bovine retina. J. Neurosci. 4, 2712–2717.

Greenberg, S. M., Castellucci, V. F., Bayley, H., & Schwartz, J. H. (1987). A molecular mechanism for long-term sensitization in *Aplysia*. Neuron 329: 62–65.

Gusovsky, F., & Gutkind, J. S. (1991). Selective effects of activation of protein kinase C isozymes on cyclic AMP accumulation. Mol. Pharm. 39, 124–129.

Han, P. L., Levin, L. R., Reed, R. R., & Davis, R. L. (1992). Preferential expression of the *Drosophila* rutabaga gene in mushroom bodies, neural centers for learning in insects. Neuron 9, 619–627.

Hawkins, R. D., Abrams, T. W., Carew, T. J., & Kandel, E. R. (1983). A cellular mechanism of classical conditioning in *Aplysia*: Activity dependent amplification of presynaptic facilitation. Science 219, 400–405.

Hopkins, W. F., & Johnston, D. (1988). Noradrenergic enhancement of long-term potentiation at mossy fiber synapses in the hippocampus. J. Neurophysiol. 59, 667–687.

Ishikawa, Y., Katsushika, S., Chen, L., Halnon, N. J., Kawabe, J., & Homcy, C. J. (1992). Isolation and characterization of a novel cardiac adenylylcyclase. J. Biol. Chem. 267, 13553–13557.

Jacobowitz, O., Chen, J., Premont, R. T., & Iyengar, R. (1993). Stimulation of specific types of G_s-stimulated adenylyl cyclases by phorbol ester treatment J. Biol. Chem. 268, 3829–3822.

Johnson, R. A., & Toews, M. L. (1990). Protein kinase C activators sensitize cyclic AMP accumulation by intact 1321N1 human astrocytoma cells. Mol. Pharm. 37, 296–303.

Kandel, E. R., & Schwartz, J. H. (1982). Molecular biology of learning: Modulation of transmitter release. Science 218, 433–443.

Katsushika, S., Chen, L., Kawabe, J. I., Nilakantan, R., Halnon, N. J., Homcy, C. J., & Ishikawa, Y. (1992). Cloning and characterization of a sixth adenylyl cyclase isoform: Type V & VI constitute a subgroup within the mammalian adenylyl cyclase family. Proc. Natl. Acad. Sci. USA 89, 8774–8778.

Klein, M., Hochner, B., & Kandel, E. R. (1986). Facilitatory transmitters and cAMP can modulate accommodation as well as transmitter release in *Aplysia* sensory neurons: Evidence for parallel processing in a single cell. Proc. Natl. Acad. Sci. USA 83, 7994–7998.

Krebs, E. G., & Beavo, J. A. (1979). Phosphorylation–dephosphorylation of enzymes. Annu. Rev. Biochem. 48, 923–959.

Krupinski, J., Coussen, F., Bakalyar, H. A., Tang, W. J., Feinstein, P. G., Orth, K., Slaughter, C., Reed, R. R., & Gilman, A. G. (1989). Adenylyl cyclase amino acid sequence: Possible channel- or transporter-like structure. Science 244, 1558–1564.

Levin, L. R., Han, P. L., Hwang, P. M., Feinstein, P. G., Davis, R. L., & Reed, R. R. (1992). The *Drosophila* learning and memory gene rutabaga encodes a Ca^{2+}/Calmodulin responsive adenylyl cyclase. Cell 68, 479–489.

Livingston, M. S. (1985). Genetic dissection of *Drosophila* adenylyl cyclase. Proc. Natl. Acad. Sci. USA 82, 5992–5996.

Livingston, M. S., Sziber, P. P., & Quinn, W. G. (1984). Loss of calcium calmodulin responsiveness in adenylyl cyclase of rutabaga, a *Drosophila* learning mutant. Cell 37, 205–215.

May, D., Ross, E., Gilman, A., & Smigel, M. (1985). Reconstitution of catecholamine-stimulated adenylate cyclase activity using three purified proteins. J. Biol. Chem. 260, 15829–15833.

Mollner, S., & Pfeuffer, T. (1988). Two different adenylate cyclases in brain distinguished by monoclonal antibodies. Eur. J. Biochem. 171, 265–271.

Mollner, S., Simmoteit, R., Palm, D., & Pfeuffer, T. (1991). Monoclonal antibodies against various forms of the adenylyl cyclase catalytic subunit and associated proteins. Eur. J. Biochem. 195, 280–286.

Nairn, A. C., Hemmings, H. C., & Greengard, P. (1985). Protein kinases in the brain. Annu. Rev. Biochem. 54, 931–976.

Nestler, E. J., & Greengard, P. (1983). Protein phosphorylation in the brain. Nature 305, 583–588.

Olinas, M. C., & Onali, P. (1992). Properties of muscarinic stimulated adenylate cyclase activity in rat olfactory bulb. J. Neurochemistry 58, 1723–1729.

Orly, J., & Schramm, M. (1976). Coupling of catecholamine receptor from one cell with adenylate cyclase from another cell by cell fusion. Proc. Natl Acad. Sci. USA 73, 4410–4414.

Pfeuffer, T., & Helmreich, E. (1975). Activation of pigeon erythrocyte membrane adenylate cyclase by guanylnucleotide analogues and separation of a nucleotide binding protein. J. Biol. Chem. 250, 867–876.

Premont, R. T., Chen, J., Ma, H. W., Ponnapalli, M., & Iyengar, R. (1992). Two members of a widely expressed subfamily of hormone stimulated adenylyl cyclases. Proc. Natl Acad. Sci. USA 89, 9809–9813.

Quilliam, L. A., Dobson, P. R., & Brown, B. L. (1989). Regulation of GH3 pituitary tumour-cell adenylate cyclase activity by activators of protein kinase C. Biochem. J. 262, 829–834.

Rosenberg, G. B., & Storm, D. R. (1987). Immunological distinction between calmodulin-sensitive and calmodulin-insensitive adenylate cyclases. J. Biol. Chem. 262, 7623–7628.

Rosenberg, G. B., Minocherhomjee, A., & Storm, D. R. (1987). Reconstitution of calmodulin-sensitive adenylate cyclase. Methods in Enzymology 139, 776–79.

Ross, E., & Gilman, A. (1980). Biochemical properties of hormone-sensitive adenylate cyclase. Ann. Rev. Biochem. 49, 533–564.

Rozengurt, E., Murray, M., Zachary, I., & Collins, M. (1987). Protein kinase C activation enhances cAMP accumulation in Swiss 3T3 cells: Inhibition by pertussis toxin. Proc. Natl. Acad. Sci. USA 84, 2282–2286.

Simon, M. I., Strathmann, M. P., & Gautam, N. (1991). Diversity of G proteins in signal transduction. Science 252, 802–806.

Stanton, P. K., & Sarvey, J. M. (1985). Depletion of norepinephrine, but not serotonin, reduces long-term potentiation in the dentate gyrus of rat hippocampal slices. J. Neurosci. 5, 2169–2176.

Summers, S. T., & Cronin, M. J. (1988). Phorbol esters induce two distinct changes in GH3 pituitary cell adenylate cyclase activity. Arch. Biochem. Biophys. 262, 12–18.

Tang, W. T., Krupinski, J., & Gilman, A. G. (1991). Expression and characterization of calmodulin activated (typeI) adenylyl cyclase. J. Biol. Chem. 266, 8595–8603.

Tang, W. J., & Gilman, A. G. (1991). Type specific regulation of adenylyl cyclase by G protein β/γ subunits. Science 254, 1500–1503.

Taussig, R., Iniguez-Lluhi, J. A., & Gilman, A. G. (1993a). Inhibition of adenylyl cyclase by G_i-α. Science 261, 218–221.

Taussig, R. T., Quarmby, L. R., & Gilman, A. G. (1993b). Regulation of purified type I and type III adenylyl cyclases by G protein β/γ subunits. J. Biol. Chem. 268, 9–12.

Tota, M. R., Xia, Z., Storm, D. R., & Schimerlik, M. I. (1990). Reconstitution of muscarinic receptor mediated inhibition of adenylyl cyclase. Mol. Pharm. 37, 950–957.

Vorherr, T., Knopfel, L., Hofmann, F., Mollner, S., Pfeuffer, T., & Carafoli, E. (1993). The calmodulin binding domain of nitric oxide synthase and adenylyl cyclase. Biochemistry 32, 6081–6088.

Walters, E. T., & Byrne, J. H. (1983). Associative conditioning of single sensory neurons suggests a cellular mechanism for learning. Science 219, 405–408.

Westcott, K. R., LaPorte, D. C., & Storm, D. R. (1979). Resolution of adenylate cyclase sensitive and insensitive to Ca^{2+} & calmodulin by calmodulin-Sepharose affinity chromatography. Proc. Nat. Acad. Sci., USA 76, 204–228.

Wu, Z., Wong, S. T., & Storm, D. R. (1993). Modification of the calcium and calmodulin sensitivity of the type I adenylyl cyclase by mutagenesis of its calmodulin binding domain. J. Biol. Chem. 268, 23766–23768.

Xia, Z., Choi, E. J., Wang, F., & Storm, D. R. (1992). The type III calcium/calmodulin sensitive adenylyl cyclase is not specific to olfactory sensory neurons. Neurosci. Letters 144, 169–173.

Xia, Z., Choi, E. J., Wang, F., Blazynski, C., & Storm, D. R. (1993). The type I calmodulin sensitive adenylyl cyclase is neural specific. J. Neurochem. 60, 305–311.

Xia, Z., Refsdal, C. D. R., Merchant, K. M., Dorsa, D. M., & Storm, D. R. (1991). Distinct patterns for the distribution of mRNA for the calmodulin-sensitive adenylate cyclase in rat brain: Expression in areas associated with learning and memory. Neuron 6, 431–443.

Yamashita, A., Kurokawa, T., Une, Y., & Ishibashi, S. (1988). Phorbol ester regulates stimulatory and inhibitory pathways of the hormone-sensitive adenylate cyclase system in rat reticulocytes. Eur. J. Pharmacol. 15, 167–175.

Yovell, Y., & Abrams, W. (1992). Temporal asymmetry in activation of *Aplysia* adenylyl cyclase by calcium and transmitter may explain temporal requirements of conditioning. Proc. Natl. Acad. Sci. USA 89, 6526–6530.

Yoshimura, M., & Cooper, D. M. (1992). Cloning and expression of a Ca^{2+}-inhibitable adenylyl cyclase from NCB-20 cells. Proc. Natl. Acad. Sci. USA 89, 6716–6720.

Yoshimura, M., & Cooper, D. M. (1993). Type-specific stimulation of adenylyl cyclase by protein kinase C. J. Biol Chem. 268, 4604–4607.

Zhong, Y., & Wu, C. F. (1991). Altered synaptic plasticity in *Drosophila* memory mutants with a defective cAMP cascade. Science 251, 198–201.

THE PROLACTIN/GROWTH HORMONE/CYTOKINE RECEPTOR SUPERFAMILY

P.A. Kelly, J. Finidori, M. Edery, and M.C. Postel-Vinay

Biomembranes
Volume 6, pages 129–145.

ISBN: 1-55938-663-0.

I. INTRODUCTION

Prolactin (PRL) and growth hormone (GH), along with placental lactogens (PL) form a family of hormones which result from duplication of an ancestral gene (Niall et al., 1971). A very broad spectrum of actions is associated with prolactin (more than 85 in various vertebrate species), whereas growth hormone is best known for its effects on the growth of skeletal and soft tissues, and for its metabolic functions.

All hormones act by first binding to a specific receptor. For PRL and GH receptors, this binding occurs at the level of the cell surface. Up until a few years ago, very little was known about the mechanism of action of prolactin and growth hormone, however, recent studies have demonstrated that one of the initial events in the action of these hormones is the activation of a tyrosine kinase, that in turn, phosphorylates other proteins including the receptor itself.

In the following sections, we will describe the expanded superfamily of PRL/GH/cytokine receptors, ligand-binding determinants of the receptors, and the role of receptor dimerization and tyrosine phosphorylation in signal transduction, and the newly discovered family of transcriptional activators utilized by the receptor superfamily.

II. IDENTIFICATION OF THE PRL/GH RECEPTOR GENE FAMILY

In 1987, the group of William Wood at Genentech purified to homogeneity and sequenced the GH receptor and binding protein (BP) in rabbit and cloned cDNAs encoding the GH receptor in rabbits and humans (Leung et al., 1987). Shortly thereafter, our group purified to homogeneity and sequenced the prolactin receptor in rats and cloned its cDNA (Boutin et al., 1988). The first form of the PRL receptor to be identified was termed *short*, because its cytoplasmic domain had only 57 amino acids, compared to the 350 amino acids of the GH receptor. Since that time, however, GH and PRL receptors have been identified and characterized from a number of different species. As can be seen in Figure 1, there are now *short* and *long* forms of both PRL and GH receptors: for GH receptors, the short-form is a binding protein (BP) that circulates in the blood, specifically binding GH; for PRL receptors, the short-form is membrane bound, representing a truncated version of the full-length, long-form of the receptor. Short and long forms of the PRL receptor are produced by alternative splicing of a single receptor gene in rat and mice (Davis and Linzer, 1989; Shirota et al., 1990). Finally, an intermediate form of receptor, missing 198 amino acids in the cytoplasmic domain, is found in the rat lymphoma cell line, Nb2, that is dependent upon prolactin for growth (Ali et al., 1991). This cell line is frequently used as a model to study the mechanism of action of prolactin.

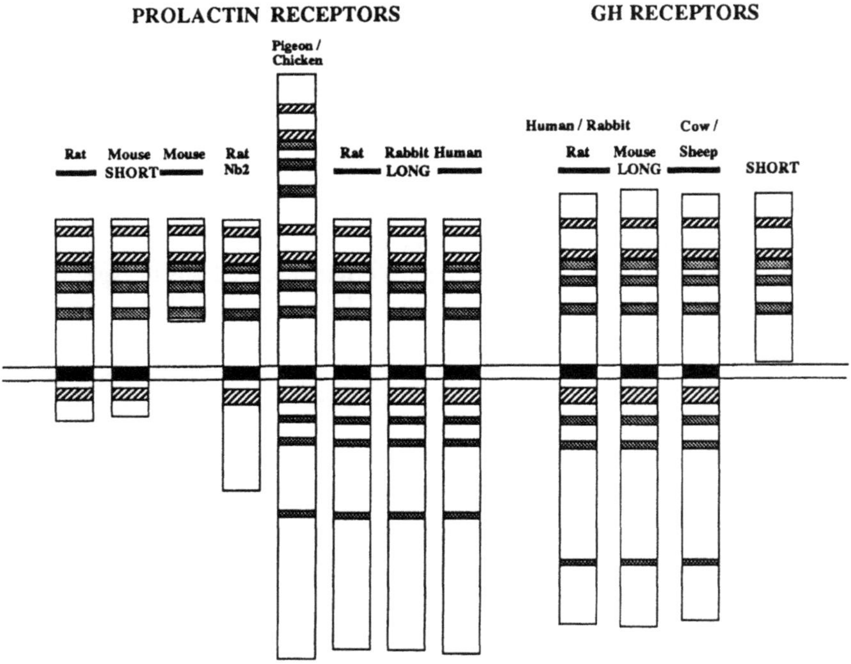

Figure 1. Representation of forms of prolactin (PRL) and growth hormone (GH) receptors found in different species. The short-form of the PRL receptor from rat and mouse, the intermediate Nb2 form, and the long-form of the PRL receptor in rat, rabbit, human, and birds (with the duplicated extracellular domain) are compared with the long and short (binding protein) forms of the GH receptor in human, rabbit, rat, mouse, and cow. The transmembrane domain is represented by a black box. Regions of high (~70%) amino acid identity are cross-hatched, and those of moderate (40–60%) are stippled.

III. ORIGIN AND POTENTIAL ACTIONS OF GH-BINDING PROTEINS

Two separate mechanisms have been proposed for the production of the GH-binding protein. In mice and rats, alternative splicing of a single primary transcript results in two distinct mRNAs. The 4.5 kb transcript encodes the full-length receptor, and the 1.2 kb form encodes a truncated receptor, in which the transmembrane region has been replaced by a short hydrophilic sequence (Baumbach et al., 1989; Smith et al., 1989). However, these murine species appear to represent the only examples of a separate transcript encoding the GHBP. For all other species having only a single mRNA transcript (~4.5 kb), the production of GH-binding protein appears to result from a second mechanism, specific proteolysis of the membrane receptor. Recently, we demonstrated that CHO cells stably transfected with a cDNA encoding the long-form of the rabbit GHR produced, in addition to the membrane-bound form of the receptor, high concentrations of soluble BP in the media (Sotiropoulos et al., 1993). When the cDNA encoding the GHR of the rat was stably expressed in CHO cells, only the membrane-bound form was observed. These observations, coupled with the fact that N-terminal sequence analysis of the GHBP was shown to be identical to the similar extracellular region of the GHR (Leung et al., 1987), suggest that proteolytic cleavage is the main mechanism of production of BP in most non-murine species.

Although a great deal is known about the origin of the GHBP, the specific role of such a binding protein is not appreciated. This protein could act as a reservoir for GH in the circulation. Decreased degradation and metabolic clearance of GH has been reported in a rat model when GH is bound to the BP (Baumann et al., 1987). Alternatively, the binding protein could serve to block GH actions, preventing further binding to membrane receptors.

IV. GROWTH HORMONE BINDING-PROTEIN IN HUMAN PLASMA

The GHBP present in human plasma binds the hormone with a relative high affinity ($5 \times 10^8\ M^{-1}$) and low capacity (Baumann et al., 1988; Tar et al., 1990). The affinity of the BP for the ligand is somewhat lower than that of the human GH receptor (Hocquette et al., 1990). The molecular weight of the GHBP is ~55,000, as evaluated by a number of different techniques. Human GH, when complexed to the BP in adult plasma, remains immunoreactive, thus RIA values should generally reflect total hGH concentrations. The proportion of human GH bound to the binding protein has been evaluated to be ~45%.

GHBP is measured by incubating serum plasma with [^{125}I]-hGH and different procedures for separating bound- and free-hormone are used, such as gel filtration (Baumann et al., 1988), HPLC (Tar et al., 1990), and dextran coated-charcoal (Amit et al., 1990). Recently, an "immunofunctional" assay and a radioimmunoassay involving specific antisera to the GHBP have been reported (Carlsson et al., 1991; Fairhall et al., 1992).

V. THE EXPANDED PRL/GH/CYTOKINE RECEPTOR SUPERFAMILY

The family that originally included GH and PRL receptors has grown to include receptors of a number of cytokines. Although the overall amino acid identity is low between members of this family, there is a significant (14–25%) identity over ~200 amino acids of the extracellular region of these receptors (Figure 2). Two characteristic features are seen in members of the superfamily: the first is the presence of two pairs of cysteines, almost always found in the N-terminal region of the molecule, which for the GH receptor, have been shown to be linked sequentially.

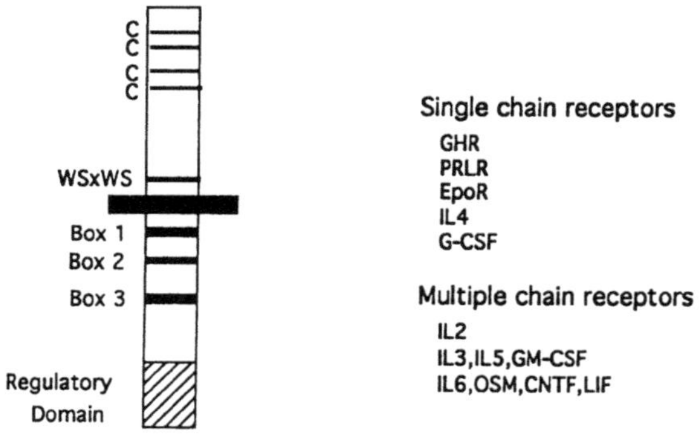

Figure 2. Common features shared between members of the PRL/GH/cytokine receptor superfamily: four cysteines in the extracellular domain, the tryptophan, serine, any amino acid, tryptophan, serine (WsXWs) motif, Box 1, Box 2, Box 3, and a potential C-terminal regulatory domain.

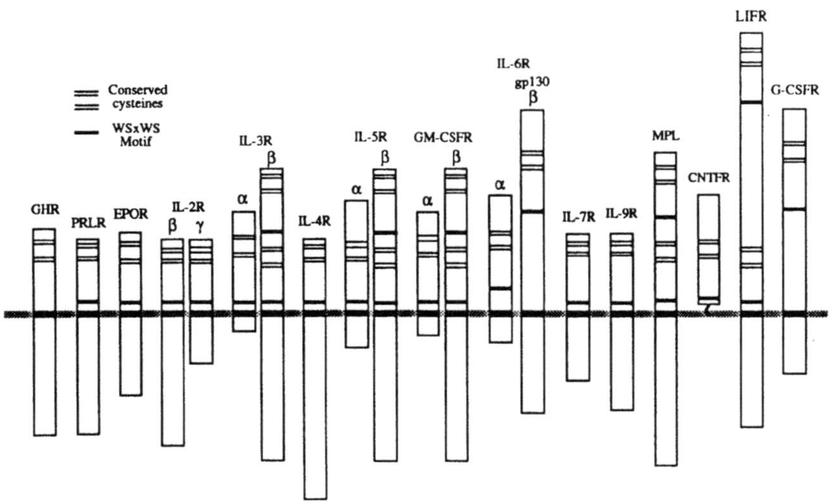

Figure 3. Representation of various members of the GH/PRL/cytokine receptor family. The abbreviations used are GHR: growth hormone receptor; PRLR: prolactin receptor; EPOR: erythropoietin receptor; IL-2R: interleukin-2 receptor; IL-3R: IL-3 receptor; IL-4R: IL-4 receptor; IL-5R: IL-5 receptor; GM-CSFR: granulocyte–macrophage colony stimulating factor; IL-6R: IL-6 receptor; gp130: glycoprotein of M_r 130,000 (or β-subunit of IL-6R, CNTFR, LIFR, or oncostatin M receptor); IL-7R: IL-7 receptor; IL-9R: IL-9 receptor; MPL: myeloproliferative leukemia virus or orphan receptor of unknown ligand; CNTFR: ciliary neurotrophic factor receptor; LIFR: leukocyte inhibitory factor receptor; G-CSFR: granulocyte colony stimulatory factor. The plasma membrane is indicated by a stippled rectangle. The transmembrane region is shown in black. The thin black lines indicate the conserved cysteines and the thick black lines the WSXWS motif (tryptophan, serine, any amino acid, tryptophan, serine). Several receptors are formed by subunits, indicated α, β, or γ.

In addition, near the C-terminal extremity of this homologous region, is a highly conserved WSXWS motif (tryptophan, serine, any amino acid, tryptophan, serine), which is found in all members, except the GH receptor, in which conservative substitutions occur. Finally, although there is very little conservation of primary sequence in the cytoplasmic domains, three regions, known as Box 1, Box 2, and Box 3, are found in many members of this family. Figure 3 shows the expanded GH/PRL/Cytokine receptor family. As can be seen, a number of receptors are formed by multiple subunits (IL-2, IL-3, IL-5, GM-CSF, IL-6, and LIF and CNTFR; Cosman, 1993).

VI. EXPRESSION OF TWO FORMS OF PRL RECEPTOR TRANSCRIPT

Two approaches have been employed to measure the expression of the short and long forms of the PRL receptor in the rat. We have recently developed a quantitative polymerase chain reaction (Q-PCR) in order to measure the absolute number of mRNA molecules encoding both forms of PRL receptor. A fundamental aspect of Q-PCR is the construction of an internal control. The control RNA is reverse transcribed and amplified along with sample RNA under conditions that allow parallel responses. The details involved in the Q-PCR have been described (Nagano and Kelly, 1994). Using this technique, it was possible to detect as few as 500 molecules of receptor transcript per μg of total RNA. Sixteen tissues of adult female rats at two stages of the estrous cycle (proestrus and diestrus I) and the mammary gland of 20-day pregnant and seven-day lactating rats were examined. Receptor transcripts were detected in all tissues, with values ranging from $\sim 10^3$ molecules per μg of total RNA in skeletal muscle to $\sim 10^7$ molecules per μg of total RNA in liver, choroid plexus, and ovary. Most tissues predominantly expressed the long-form transcript , although in the liver, the short-form is the major component, while in the thymus and kidney, both forms are expressed equally. These results indicate that PRL receptor mRNA is ubiquitously, but variably expressed in a tissue-specific manner and is clearly regulated by the hormonal environment associated with the stage of the estrous cycle, pregnancy, or lactation (Nagano and Kelly, 1994).

A second approach to measure the two receptor transcripts involves *in situ* hybridization. Probes specific to the intracellular domains of the short or long forms of PRL receptor were prepared. The specificity of the signals was controlled by competition with excess unlabeled homologous probes or by hybridization with heterologous probes. Moreover, some tissues showed no expression of either form of receptor mRNA, and thus served as controls. The methods employed have recently been described (Ouhtit et al., 1993a, 1993b). *In situ* hybridization offers a means of directly identifying cells that express the mRNA of interest. Such an approach is especially well suited to the identification of short- and long-form transcripts in various tissues of the rat. The surprising finding, agreeing well with the results of Q-PCR, is that almost all tissues, and most frequently specific cells within a tissue, express varying levels of PRL receptor transcripts. Since mRNA expression has been detected in many tissues for which PRL is not known to have an action, it will be important to pursue future studies to correlate PRL receptor gene expression (by Q-PCR, *in situ* hybridization, and western blot) with specific effects associated with each form of receptor (Ouhtit et al., 1993a, 1993b).

VII. BINDING DETERMINANTS OF GH, PRL, AND THEIR RECEPTORS

Growth hormone is a molecule formed by four α-helices, as originally shown by analysis of its crystal structure. Although there is little amino acid identity between GH and the various cytokines, it is thought that the four α-helix model will probably hold true for all ligands of this superfamily of receptors.

Homolog and alanine scanning mutagenesis of hGH originally localized a receptor-binding domain on the hGH molecule that involves two of the α-helices and the 54–74 loop region. Twelve residues were identified to form a patch on a two-dimensional structural model of hGH (Cunningham and Wells, 1991). A similar approach was used to identify the binding determinants of the extracellular region of the GH (Bass et al., 1991) and PRL (Rozakis-Adcock et al., 1992) receptor. The binding domain of the receptors involves a region of ~100 amino acids, including the first four cysteines commonly found in the receptor superfamily. This region has been shown in structural models to form seven anti-parallel β strands grouped in a β-sheet sandwich. Although such approaches provided important information about how the ligands and receptors interacted, a major advance was made when the crystal structure of growth hormone bound to its receptor was solved.

VIII. THREE-DIMENSIONAL STRUCTURE, RECEPTOR DIMERIZATION, AND ACTION

The three-dimensional crystal structure of the human GH-binding protein and hGH has confirmed that this complex forms a dimer with the ligand (De Vos et al., 1992). Before this work was completed, however, in a series of studies carried out by Kopchick's group, a second site of interaction on the growth hormone molecule was found in the third α-helix of bovine GH (Chen et al., 1991). These authors in fact mutated the glycine residue at position 119 into an arginine and established transgenic animals expressing this mutated gene. They expected to see giant mice, but to their surprise, found dwarfs. The authors reasoned that the first and fourth α-helices were directly interacting with the receptor, but that residues in the third α-helix interacted with an unknown transmembrane protein that was necessary for functional activity. We now know, of course, based on the outstanding studies of de Vos (De Vos et al., 1992), that the second protein is, in fact, a second receptor molecule, since the crystal structure studies, combined with the biochemical data by the group of Wells (Cunningham et al., 1991), demonstrate that the extracellular binding protein exists as an unusual homodimer consisting of two molecules of receptor and one molecule of ligand (2:1). Thus, there are two receptor sites on hGH (identified as sites 1 and 2). Both sites bind to the same region of the hGH receptor. A sequential complex appears to form with the receptor first binding to site 1, after which a second receptor binds to site 2, followed by an interaction between the receptor molecules themselves that maintains the dimer complex (Cunningham et al., 1991).

A group at Genentech prepared a chimeric protein, consisting of the extracellular domain of the hGH receptor and the transmembrane and cytoplasmic domains of the G-CSF receptor. Stable transfection of this chimeric cDNA into FDC-P1 cells resulted in a biological test capable of measuring GH activity. Human GH mutants, with a reduced affinity for site 1, had a greatly reduced ability to stimulate proliferation of FDC-P1 cells containing the GH/G-CSF hybrid receptor compared to wild-type hGH. A mutant in site 2 of hGH (G120R, corresponding to glycine 119 of bGH) that was fully capable of binding to the receptor, failed to activate proliferation, and antagonized hGH stimulation, confirming the sequential two-site model of GH action (Fuh et al., 1992).

There is strong evidence that activation of PRL receptors also follows a similar mechanism involving dimerization. Monoclonal antibodies to the rat PRL receptor which are able to form receptor dimers have previously been shown to be partial agonists (Elberg et al., 1990). More recently, using adjusted concentrations of second antibody to cross-link receptors, we have been able to show that the same monoclonal antibodies are able to transduce full functional activity. In addition, monovalent Fab fragments which bind receptors are devoid of activity, but addition of a second antibody restores the functional capacity (Rui et al., 1994). In addition, the soluble, extracellular domain of the rat PRL receptor has been shown to form 2:1 dimers with ovine PRL (Hooper et al., 1993), identical to the complex formation between hGH and its binding protein (Cunningham et al., 1991). Finally, G120R, which binds to the PRL receptor in the Nb2 lymphoma cell line, is able to block cell proliferation induced by PRL (Fuh et al., 1993). Final confirmation of the two-site model for prolactin and dimerization of the receptor, however, must await three-dimensional crystallography.

IX. FUNCTIONAL ACTIVITY OF PRL AND GH RECEPTORS

We previously developed an assay to measure the functional activity of transfected forms of the PRL receptor (Lesueur et al., 1991). This assay involves the co-transfection of a prolactin-responsive gene, such as ovine β-lactoglobulin or rat β-casein, coupled to a reporter gene, chloramphenicol acetyltransferase (CAT). Chinese hamster ovary (CHO) cells are transiently transfected with the PRL receptor cDNA and the promoter/CAT fusion reporter gene. The transfected cells respond to prolactin in the incubation media as measured by production of acetylated forms of chloramphenicol. More recently, we developed a similar assay to measure the functional activity of GH receptors, using a fusion gene consisting of either ovine β-lactoglobulin/CAT or the serine protease inhibitor (SPI) 2.1/CAT (Goujon et al., 1994). CHO cells transiently transfected with either of these constructs and the wild-type GH receptor respond to GH in the incubation media. We have recently started using the reporter gene, luciferase, for both PRL and GH functional assays, which increases the sensitivity and speed of the test. The advantage of such functional assays is that they use transient transfection, and thus are well-adapted to evaluate the cytoplasmic regions of the receptor required for the hormonal response. Other groups have developed functional tests of GH receptors expressed in CHO cells to evaluate effects on protein synthesis (Emtner et al., 1990; Fiddes et al., 1992), activation of MAP kinases (Campbell et al., 1992; Möller et al., 1992), and mitogenesis (Möller et al., 1992). Transcriptional effects of GH also have been evaluated by transfecting the GH receptor cDNA into rat insulinoma cells (Moldrup et al., 1991) or in Buffalo rat liver cells (Francis et al., 1993), and measuring insulin or lipoprotein lipase gene activation, respectively.

X. ROLE OF BOX 1 IN SIGNAL TRANSDUCTION

Truncation and deletion mutants of GH receptors were prepared and expressed in CHO cells. While the presence of ~50% of the cytoplasmic domain is sufficient for

GHR	I	L	**P**	P	V	**P**	V	**P**
PRLR	I	F	**P**	P	V	**P**	G	**P**
GP130	I	W	**P**	N	V	**P**	D	**P**
G CSF	I	W	**P**	S	V	**P**	D	**P**
EPO R	I	W	**P**	G	I	**P**	S	**P**
LIFR	F	Y	**P**	D	I	**P**	D	**P**
IL3 β	W	E	E	K	I	P	N	**P**
IL3 α	Q	R	L	F	P	R	I	**P**
GM CSF α	Q	R	L	F	P	P	V	**P**
IL5 α	I	R	L	F	P	P	V	**P**
IL2 β	I	K	C	N	T	P	D	**P**
IL4	W	W	D	Q	I	P	T	**P**

Figure 4. Amino acid identity of Box 1 regions of various members of the PRL/GH/cytokine receptor family.

full activity of the PRL receptor (Edery et al., 1994), a similar mutant of the GH receptor is inactive (Goujon et al., 1994). A 25 amino acid region just inside the transmembrane domain is highly conserved between GH and PRL receptors. Because of this, we originally proposed that this region may be important in the process of signal transduction (Boutin et al., 1988). In fact, deletion of this juxtamembrane domain, either in PRL (Edery et al., 1994) or GH (Goujon et al., 1994) receptors leads to the complete loss of PRL- or GH-stimulated activity. A more restrictive region within the juxtamembrane domain consisting of 8 amino acids, known as Box 1 (Figure 4), has been identified in several members of the cytokine receptor family. As shown in Figure 5, deletion and alanine scanning

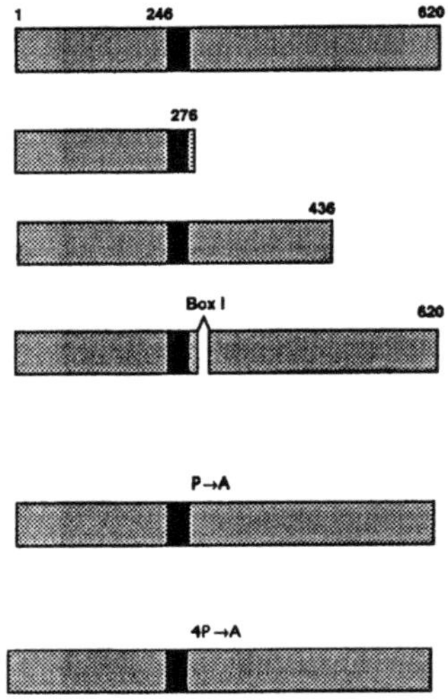

	Fold-induction of		^{125}I hGH binding to cells	
	Spi/CAT Activity	β-lacto/CAT Activity	K_d (nM)	sites/cell x 10^{-3}
Wild Type	3.7 ± 0.3	4.3 ± 0.3	0.29	143
T276	1.1 ± 0.1	1.3 ± 0.1	0.22	67
T436	1.3 ± 0.2	1.2 ± 0.1	0.32	238
Δ279-293	1.0 ± 0.1	0.9 ± 0.2	0.33	71
P282A	3.3 ± 0.4	ND	0.34	156
P283A	4.4 ± 0.8	ND	0.23	168
P285A	3.9 ± 0.4	ND	0.24	207
P287A	3.3 ± 0.3	ND	0.19	168
4P/A	1.4 ± 0.1	ND	0.23	115

Figure 5. Functional activity of wild-type and mutant forms of the GHR. The wild-type receptor consists of 620 amino acids with a single transmembrane domain region (black). Truncated mutants (T276 and T436) have 6- and 166-amino acids in their cytoplasmic domain, respectively. The mutant Δ279–293 has 15 amino acids deleted in the cytoplasmic domain. The substitution mutants have one proline mutated to alanine in the Box 1, the 4P mutant has four prolines (282, 283, 285, and 287) all changed to alanine. The functional activity (fold-induction of CAT activity) of each mutant is shown. Results represent the means ± SEM of 5 to 14 independent experiments. All mutants were tested in the Spi/CAT functional test. The mutated forms T276, T436, and Δ279–293 were tested in the β-lactoglobulin/CAT test. Characteristics of cell surface binding of [^{125}I] hGH to wild-type and mutant forms are shown. ND = not determined (from Goujon et al., 1994).

mutagenesis has confirmed that prolines of Box 1 are essential for the process of signal transduction for both receptors (Goujon et al., 1994). The hydrophobic amino acids (ILV) at the N-terminal end of Box 1 of the GH receptor are also essential for functional activity (data not shown).

In another functional assay which utilizes the full-length GH receptor transfected into FDC-P1 cells and is based on the original assay developed with the GH/G-CSF receptor hybrid, GH- induced cell proliferation occurs in the absence of IL-3 (Colosi et al., 1993). Interestingly, these authors found that as little as 54 amino acids in the cytoplasmic domain were able to transmit a positive proliferative signal. Thus, only the juxtamembrane region containing Box 1 and a few downstream residues were necessary for the stimulation of cell proliferation, but it appears that this is insufficient to stimulate GH-induced gene transcription (Goujon et al., 1994).

XI. PRL- AND GH-INDUCED TYROSINE PHOSPHORYLATION

Several years ago, Carter-Su's group reported that growth hormone stimulated the phosphorylation of a protein with a molecular weight of ~120,000 in a number of different cell systems. Originally, it was thought that the pp120 represented the GH receptor itself (Foster et al., 1988; Carter-Su et al., 1989). More recently, however, studies have clearly shown that an associated protein, and not the receptor, is the primary and initial tyrosine-phosphorylated protein (Wang et al., 1993). In addition, we and others have demonstrated, using CHO cells stably expressing the rabbit GH receptor, that at least three tyrosine-phosphorylated proteins are induced following stimulation with GH. The receptor itself is also phosphorylated, but the degree of phosphorylation appears to depend on the cell system used.

Using Nb2 cells, we and others have demonstrated the rapid stimulation of tyrosine kinase activity by prolactin (Rillema et al., 1992; Rui et al., 1992). We have identified at least 3 tyrosine-phosphorylated proteins (pp120, pp97, and pp42) induced by lactogenic hormones. Phosphorylation of p120 is maximal following incubation of cells with prolactin for 1 min. Peak levels of pp97 and pp40 occur at somewhat later periods. The 42–44 kD protein induced by both GH and PRL appears to be MAP kinase, a protein frequently involved in proliferation, and already shown to be activated by GH (Campbell et al., 1992; Möller et al., 1992).

The functional role of the phosphorylated tyrosines in the receptor molecules has not yet been determined, but it is reasonable to assume that they interact with proteins containing SH2 domains, as has been shown for a number of growth factor receptors.

XII. JAK2, THE TYROSINE KINASE INVOLVED IN PRL AND GH ACTION

Although neither the PRL nor the GH receptor contain a consensus sequence for ATP/GTP binding, nor a kinase domain, a major advance in the field was made by

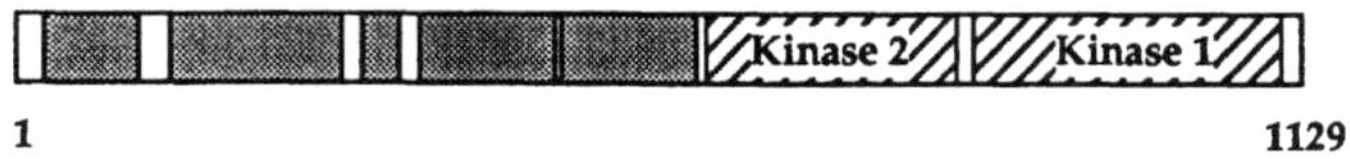

- **3 Family Members: JAK1, JAK2, & Tyk2**
- **2 Kinase domains**
- **50 % amino acid identity between family members**
- **Cytoplasmic localization**

Figure 6. The JAK family of tyrosine kinases, including JAK1, JAK2, and Tyk2. The two kinase domains are shown cross-hatched and the other regions of homology are stippled.

the identification of JAK2 as a growth hormone receptor-associated tyrosine kinase. JAK2 is a member of a family that also includes JAK1 and Tyk2 (Figure 6). All these proteins share the unusual feature of having two kinase domains. Complementary DNAs encoding these kinases were originally identified a few years ago (Firmbach-Kraft et al., 1990; Wilks et al., 1991; Harpur et al., 1992), although it was not known how they were activated. Stimulation of various cells expressing the GH receptor induced tyrosine phosphorylation of a protein with a M_r of 130,000, that could be immunoprecipitated with an antibody specific to JAK2 (Argetsinger et al., 1993). Erythropoietin is also known to activate rapid tyrosine phosphorylation of a similar sized protein, and this kinase has been shown also to be JAK2 (Witthuhn et al., 1993). In addition to the phosphorylation of JAK2, the GH and erythropoietin receptors themselves are also phosphorylated. Using GH or erythropoietin receptor mutants, a membrane-proximal region of the cytoplasmic domain was shown to be important for biological activity, similar to the results presented above for GH on the role of the juxtamembrane region. JAK2 has been shown to be the kinase that couples to the IL-3 receptor (Silvennoinen et al., 1993a), and the PRL receptor (Lebrun et al., 1994; Rui et al., 1994) and is most certainly implicated for GM-CSF, G-CSF, and IFNγ receptors, as the first event in the process of signal transduction. This is probably only the first step, however, in the signal transduction process that may involve other kinases, phospholipase C-γ (PLC-γ), diacylglycerol (DAG), and several effector proteins (Figure 7).

XIII. STAT FAMILY OF TRANSCRIPTIONAL ACTIVATORS

A more direct path to the activation of transcription of specific genes appears to involve the phosphorylation of a cytoplasmic protein known as p91 or STAT (signal transducer and activator of transcription). Interferon α/β has been shown to stimulate the transcription of specific target genes through a multimeric complex known

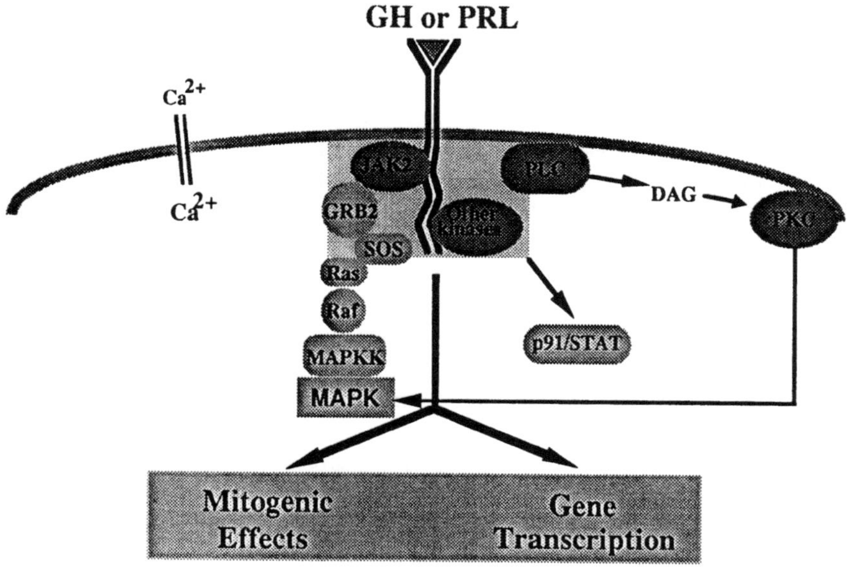

Figure 7. Schematic diagram representing potential pathways involved in PRL- and GH-signal transduction.

as interferon-stimulated gene factor-3 (ISGF-3). This factor consists of the cytoplasmic proteins p91/84 and p113, which were originally thought to be phosphorylated in response to INF-α/β by Tyk2. ISGF-3 once activated, combines with the cytoplasmic protein, p48, and this complex migrates to the nucleus to activate transcription (Fu, 1992; David and Larner, 1992; Velazquez et al., 1992; Schindler et al., 1992). Interferon-γ phosphorylates only one subunit (p91) of ISGF-3, probably via JAK1 or JAK2 (Shuai et al., 1992). More recently, it has been shown that GH also phosphorylates and activates p91/STAT (Meyer et al., 1994).

One can now speak of the STAT family of transcriptional activators. Darnell's group has recently suggested a nomenclature for these activators: STAT-1α and -1β are the 91- and 84-kD proteins, respectively, and STAT2 is the 113 kD protein. STAT1 (α and β) and STAT2 are phosphorylated by IFN-α, whereas IFN-γ leads to phosphorylation of STAT1 (α and β), but not STAT2. Following stimulation by IFN-γ, STAT1 (α and β) forms a homodimer that directly binds DNA. This homodimer is formed by intermolecular interactions of phosphotyrosine residues with SH2 domains in STAT1 (Shuai et al., 1994). IFN-α apparently also promotes interactions of STAT1 with STAT2 via tyrosine phosphorylation (Schindler et al., 1992; Shuai et al., 1992).

The exciting discovery that polypeptide ligands such as epidermal growth factor (EGF), platelet-derived growth factor (PDGF), colony-stimulating factor 1 (CSF-

1), IL-10, CNTF, and probably others promote tyrosine phosphorylation of STAT1 (Larner et al., 1993; Ruff-Jamison et al., 1993; Sadowski et al., 1993; Shuai et al., 1993; Silvennoinen et al., 1993b) raises the important question of how transcriptional specificity is maintained when cells are stimulated with different ligands. Since different STATs share ~40% amino acid identity, it was reasonable to assume that many of the previously mentioned ligands probably induce tyrosine phosphorylation of similar, but distinct DNA-binding proteins. In fact, a 92 kD protein, termed STAT3 has recently been identified (Zhong et al., 1994). A STAT4 clone that is related, but distinct from other STATs has also been found. The formation of homo- and heterodimers of these STAT proteins appears to be one of the ways in which transcriptional specificity is achieved. Equally important is the ability of STAT proteins to interact with proteins outside the family (e.g., p48 of ISGF-3), which greatly expands the potential to form and deliver transcriptional complexes to the nucleus.

Thus, in addition to the now classical pathway involving activation of an intrinsic tyrosine kinase binding to SH2-containing proteins, activation of MAP kinase and the transcription factor AP1, cytokines, and growth factors also lead to the phosphorylation of a specific tyrosine of STAT proteins, which is able to form homo- and heterodimers, and to activate transcription by the direct interaction with response elements on target DNA.

It will thus be most interesting to investigate how the STAT family of transcriptional activators are able to regulate the transcriptional effects of growth hormone and prolactin. Are homo- or heterodimers always induced, or will it be necessary to invoke the interaction of non-STAT proteins in the transcriptional machinery?

REFERENCES

Ali, S., Pellegrini, I., & Kelly, P. A. (1991). A prolactin dependent cell immune cell line (Nb2) expresses a mutant form of prolactin receptor. J. Biol. Chem. 266, 20110–20117.

Amit, T., Barkey, R. J., Youdim, M. B. H., & Hochberg, Z. (1990). A new and convenient assay of growth hormone-binding protein activity in human serum. J. Clin. Endocrinol. Metab. 71, 474–479.

Argetsinger, L. S., Campbell G. S., Yang, X., Witthuhn, B. A., Silvennoinen, O., Ihle, J. N., & Carter-Su, C. (1993). Identification of JAK2 as a growth hormone receptor-associated tyrosine kinase. Cell 74, 237–244.

Bass, S. H., Mulkerrin, M. G., & Wells, J. A. (1991). A systematic mutational analysis of hormone-binding determinants in the human growth hormone receptor. Proc. Natl.Acad. Sci. USA 88, 4498–4502.

Baumann, G., Amburn, K. D., & Buchanan, T. A. (1987). The effect of circulating on metabolic clearance, distribution, and degradation of human growth hormone-binding protein. J. Clin. Endocrinol. Metab. 65, 657–660.

Baumann, G., Amburn, K., & Shaw, M. (1988). The circulating growth hormone-binding protein complex: a major constituent of plasma GH in man. Endocrinology 122, 976–984.

Baumbach, W. R., Horner, D. L., & Logan, J. S. (1989). The growth hormone-binding protein in rat serum is an alternatively spliced form of the rat growth hormone receptor. Genes and Dev. 3, 1199–1205.

Boutin, J. M., Jolicoeur, C., Okamura, H., Gagnon J., Edery, M., Shirota, M., Banville, D., Dusanter-Fourt, I., Djiane, J., & Kelly, P. A. (1988). Cloning and expression of the rat PRL receptor, a member of the GH/PRL receptor gene family. Cell 53, 69–77.

Campbell, G. S., Pang, L., Miyasaka, T., Saltiel, A.R., & Carter-Su, C. (1992). Stimulation by growth hormone of MAP kinase activity in 3T3-F442A fibroblasts. J. Biol. Chem. 267, 6074–6080.

Carlsson, L. M. S., Rowland, A. M., Clark, R. G., Gesundheit, N., & Wong, W. L. T. (1991). Ligand-mediated immunofunctional assay for quantitation of growth hormone-binding protein in human blood. J. Clin. Endocrinol. Metab. 73, 1216–1223.

Carter-Su, C., Stubbart, J. R., Wang, X., Stred, S. E., Argetsinger, L. S., & Shafer, J. A. (1989). Phosphorylation of highly purified growth hormone receptors by a growth hormone receptor-associated tyrosine kinase. J. Biol. Chem. 264, 18654–18661.

Chen, W. Y., Wight, D. C., Mehta, B. V., Wagner, T. E., & Kopchick, J. J. (1991). Glycine 119 of bovine growth hormone is critical for growth-promoting activity. Mol. Endocrinol. 5, 1845–1852.

Colosi, P., Wong, K., Leong, S. R., & Wood, W. I. (1993). Mutational analysis of the intracellular domain of the human growth hormone receptor. J. Biol. Chem. 268, 12617–12623.

Cosman, D. (1993). The hematopoietin receptor superfamily. Cytokine 5, 95–106.

Cunningham, B. C., & Wells, J. A. (1991). Rational design of receptor-specific variants of human growth hormone. Proc. Natl. Acad. Sci. USA 88, 3407–3411.

Cunningham, B. C., Ultsch, M., De Vos, A. M., Mulkerrin, M. G., Clauser, K. R., & Wells, J. A. (1991). Dimerization of the extracellular domain of the human growth hormone receptor by a single hormone molecule. Science 254, 821–825.

David, M., & Larner, A. C. (1992). Activation of transcription factors by interferon-α in a cell-free system. Science 257, 813–815.

Davis, J. A., & Linzer, D. I. H. (1989). Expression of multiple forms of the prolactin receptor in mouse liver. Mol. Endocrinol. 3, 674–680.

De Vos, A. M., Ultsch, M., & Kossiakoff, A. A. (1992). Human growth hormone and extracellular domain of its receptor: Crystal structure of the complex. Science 255, 306–312.

Edery, M., Levi-Meyrueis, C., Paly, J., Kelly, P. A., & Djiane, J. (1994). A limited cytoplasmic region of the prolactin receptor proximal to the transmembrane domain is critical for signal transduction. Mol. Cell Endocrinol. 102, 39–44.

Elberg, G., Kelly, P. A, Djiane, J., Binder, L., & Gertler, A. (1990). Mitogenic and binding properties of monoclonal antibodies to the prolactin receptor in NB2 rat lymphoma cells: Selective enhancement by anti-mouse IgG. J. Biol. Chem. 265, 14770–14776.

Emtner, M., Mathews, L. S., & Norstedt, G. (1990). Growth hormone (GH) stimulates protein synthesis in cells transfected with GH receptor complementary DNA. Mol. Endocrinol. 4, 2014–2020.

Fairhall, K. M., Carmignac, D. F., & Robinson, I. C. A. F. (1992). Growth hormone (GH) binding protein and GH interactions *in vivo* in the guinea pig. Endocrinology 131, 1963–1969.

Fiddes, R. J., Brandon, M. R., & Adams, T. E. (1992). Functional expression of an ovine growth hormone receptor in transfected Chinese hamster ovary cells. Mol. Cell. Endocrinol. 86, 37–47.

Firmbach-Kraft, I., Byers, M., Shows, T., Dalla-Favera, R., & Krolewski, J. J. (1990). Tyk2, prototype of a novel class of non-receptor tyrosine kinase genes. Oncogene 5, 1329–1336.

Foster, C. M., Shafer, J. A., Rozsa, F. W., Wang, X., Lewis, S. D., Renken, D. A., Natale, J. E., Schwartz, J., & Carter-Su, C. (1988). Growth hormone promoted tyrosyl phosphorylation of growth hormone receptors in murine 3T3-F442A fibroblasts and adipocytes. Biochemistry 27, 326–334.

Francis, S. M., Enerbäck, S., Möller, C., Enberg, B., & Norstedt, G. (1993). A novel *in vitro* model for studying signal transduction and gene regulation via the growth hormone receptor. Mol. Endocrinol. 7, 972–978.

Fu, X. Y. (1992). A transcription factor with SH2 and SHE domains is directly activated by an interferon α-induced cytoplasmic protein tyrosine kinase(s). Cell 70, 323–335.

Fuh, G., Cunningham, B. C., Fukunaga, R., Nagata, S., Goeddel, D. V., & Wells, J. A. (1992). Rational design of potent antagonists to the human growth hormone receptor. Science 256, 1677–1680.

Fuh, G., Colosi, P., Wood, W. I., & Wells, J. A. (1993). Mechanism based design of prolactin receptor antagonists. J. Biol. Chem. 268, 5376–5381.

Goujon, L., Allevato, G., Simonin, G., Paquereau, L., Le Cam, A., Clark, J., Nielsen, J. H., Djiane, J., Postel-Vinay, M. C., Edery, M., & Kelly, P. A. (1994). Cytoplasmic domains of the growth hormone receptor necessary for signal transduction. Proc. Natl. Acad. Sci. USA 91, 957–961.

Harpur, A. G., Andres, A. C., Ziemiecki, A., Aston, R. R., & Wilks, A. F. (1992). JAK2, a third member of the JAK family of protein tyrosine kinases. Oncogene 7, 1347–1353.

Hocquette, J. F., Postel-Vinay, M. C., Djiane, J., Tar, A., & Kelly, P. A. (1990). Human liver growth hormone receptor and plasma binding protein: Characterization and partial purification. Endocrinology 127, 1665–1672.

Hooper, K. P., Padmanabhan, R., & Ebner, K. E. (1993). Expression of the extracellular domain of the rat liver prolactin receptor and its interaction with prolactin. J. Biol. Chem. 268, 22347–22352.

Larner, A. C., David, M., Feldman, G. M., Igarashi, K. I., Hackett, R. H., Webb, D. S. A., Sweitzer, S. M., Petricoin, E. F., & Finbloc, D. S. (1993). Tyrosine phosphorylation of DNA binding protein by multiple cytokines. Science 261, 1730–1733.

Lebrun, J. J., Ali, S., Sofer, L., Ulrich, A., & Kelly, P. A. (1994). Prolactin induced proliferation of Nb2 cells involves tyrosine phosphorylation of the prolactin receptor and its associated tyrosine kinase. J. Biol. Chem. 269, 14021–14026.

Lesueur, L., Edery, M., Ali, S., Paly, J., Kelly, P. A., & Djiane, J. (1991). Comparison of long and short form of the prolactin receptor on prolactin-induced milk protein gene transcription. Proc. Natl. Acad. Sci. USA 88, 824–828.

Leung, D. W., Spencer, S. A., Cachianes, G., Hammonds, R. G., Collins, C., Henzel, W. J., Barnard, R., Waters, M. J., & Wood, W. I. (1987). Growth hormone receptor and serum binding protein: purification, cloning and expression. Nature 330, 537–543, 1987.

Meyer, D. J., Campbell, G. S., Cochran, B. H., Argetsinger, L. S., Larner, A. C., Finbloom, D. S., Carter-Su, C., & Schwartz, J. (1994). Growth hormone induces a DNA binding factor related to the interferon-stimulated 91-kDa transcription factor. J. Biol. Chem. 269, 4701–4704.

Moldrup, A., Allevato, G., Dyrberg, T., Nielsen, J. H., & Billestrup, N. (1991). Growth hormone action in rat insulinoma cells expression truncated growth hormone receptors. J. Biol. Chem. 266, 17441–17445.

Möller, C., Hansson, A., Enberg, B., Lobie, P. E., & Norstedt, G. (1992). Growth hormone (GH) induction of tyrosine phosphorylation and activation of mitogen-activated protein kinases in cells transfected with rat GH receptor cDNA. J. Biol. Chem. 267, 23403–23408.

Nagano, M., & Kelly, P. A. (1994). Tissue distribution and regulation of prolactin receptor gene expression: Quantitative analysis by polymerase chain reaction. J. Biol. Chem. 269, 13337–13345.

Niall, H. D., Hogan, M. L., Sayer, R., Rosenblum, I. Y., & Greenwood, F. C. (1971). Sequences of pituitary and placental lactogenic and growth hormones. Evolution from a primordial peptide by gene duplication. Proc. Natl. Acad. Sci. USA 68, 866–869.

Ouhtit, A., Morel, G., & Kelly, P. A. (1993a). Visualization of gene expression of short and long forms of prolactin receptor in the rat. Endocrinology 133, 135.

Ouhtit, A., Morel, G., & Kelly, P. A. (1993b). Visualization of gene expression of short and long forms of prolactin receptor in rat reproductive tissues. Biol. Reprod. 49, 528.

Rillema, J. A., Campbell, G. S., Lawson, D. M., & Carter-Su, C. (1992). Evidence for a rapid stimulation of tyrosine kinase activity by prolactin in Nb2 rat lymphoma cells. Endocrinology 131, 973–975.

Rozakis-Adcock, M., & Kelly, P. A. (1992). Identification of ligand binding determinants of the prolactin receptor. J. Biol. Chem. 267, 7428–7433.

Ruff-Jamison, S., Chen, K., & Cohen, S. (1993). Induction by EGF and interferon-γ of tyrosine phosphorylated DNA binding proteins in mouse liver nuclei. Science 261, 1733–1736.

Rui, H., Djeu, J. Y., Evans, G. A., Kelly, P. A., & Farrar, W. L. (1992). Prolactin receptor triggering: Evidence for rapid tyrosine kinase activation. J. Biol. Chem. 267, 24076–24081.

Rui, H., Kirken, R. A., & Farrar, W. L. (1994). Activation of receptor-associated tyrosine kinase JAK2 by prolactin. J. Biol. Chem. 269, 5364–5368.

Rui, H., Lebrun, J. J., Kirken, R. A., Kelly, P. A., & Farrar, W. L. (1994). JAK2 activation and cell proliferation induced by antibody-mediated prolactin receptor dimerization. Endocrinology 135, 1299–1306.

Sadowski, H. B., Shuai, K., Darnell, J. E., Jr., & Gilman, M. Z. (1993). A common nuclear signal transduction pathway activated by growth factor and cytokine receptors. Science 261, 1739–1744.

Schindler, C., Shuai, K., Prezioso, V. R., & Darnell, J. E., Jr. (1992). Interferon-dependent tyrosine phosphorylation of a latent cytoplasmic transcription factor. Science 257, 809–812.

Shirota, M., Banville, D., Ali, S., Jolicoeur, C., Boutin, J. M., Edery, M., Djiane, J., & Kelly, P. A. (1990). Expression of two forms of the prolactin receptor rat ovary and liver. Mol. Endocrinol. 4, 1136–1143.

Shuai, K., Schindler, C., Prezioso, V. R., & Darnell, J. E., Jr. (1992). Activation of transcription by IFN-γ: Tyrosine phosphorylation of a 91-kD DNA binding protein. Science 258, 1808–1812.

Shuai, K., Stark, G. R., Kerr, I. M., & Darnell, J. E., Jr. (1993). A single phosphotyrosine residue of stat 91 required for gene activation by interferon-γ. Science 261, 1744–1746.

Shuai, K., Horvath, C. M., Tsai Huang, L. H., Qureshi, S. A., Cowburn, D., & Darnell, J. E., Jr. (1994). Interferon activation of the transcription factor stat91 involves dimerization through SH2-phosphotyrosyl peptide interactions. Cell 76, 821–828.

Silvennoinen, O., Zitthuhn, B. A., Quelle, F. W., Cleveland, J. L., Yi, T., & Ihle, J. A. (1993a). Structure of the murine Jak2 protein-tyrosine kinase and its role in interleukin 3 signal transduction. Proc. Natl. Acad. Sci. USA 90, 8429–8433.

Silvennoinen, O., Schindler, C., Schlessinger, J., & Levy, D. E. (1993b). Ras-independent growth factor signaling by transcription factor tyrosine phosphorylation. Science 261, 1736–1739.

Smith, W. C., Kuniyoshi, J., & Talamantes, F. (1989). Mouse serum growth hormone (GH) binding protein has GH receptor extracellular and substituted transmembrane domains. Mol. Endocrinol. 3, 984–990.

Sotiropoulos, A., Goujon, L., Simonin, G., Kelly, P. A., Postel-Vinay, M. C., & Finidori, J. (1993). Evidence for generation of the growth hormone-binding protein through proteolysis of the growth hormone membrane receptor. Endocrinology 132, 1863–1865.

Tar, A., Hocquette, J. F., Souberbielle, J. C., Clot, J. P., Brauner, R., & Postel-Vinay, M. C. (1990). Evaluation of the growth hormone-binding proteins in human plasma using HPLC-gel filtration. J. Clin. Endocrinol. Metab. 71, 1202–1207.

Velazquez, L., Fellous, M., Stark, G. R., & Pellegrini, S. (1992). A protein tyrosine kinase in the interferon α/β signaling pathway. Cell 70, 313–322.

Wang, X., Möller, C., Norstedt, G., & Carter-Su, C. (1993). Growth hormone-promoted tyrosyl phosphorylation of a 121-kDa growth hormone receptor-associated protein. J. Biol. Chem. 268, 3573–3579.

Wilks, A. F., Harpur, A. G., Kurban, R. R., Ralph, S. J., Zurcher, G., & Ziemiecki, A. (1991). Two novel protein-tyrosine kinases, each with a second phosphotransferase-related catalytic domain, define a new class of protein kinase. Mol. Cell. Biol. 11, 2057–2065.

Witthuhn, B. A., Quelle, F. W., Silvennoinen, O., Yi, T., Tang, B., Miura, O., & Ihle, J. N. (1993). JAK2 associates with the erythropoietin receptor and is tyrosine phosphorylated and activated following stimulation with erythropoietin. Cell 74, 227–236.

Zhong, Z., Wen, Z., & Darnell, J. E., Jr. (1994). Stat3: A stat family member activated by tyrosine phosphorylation in response to epedermal growth factor and interleukin-6. Science 264, 95–98.

INTERLEUKIN-1 RECEPTORS

Steven K. Dower and John E. Sims

I. INTRODUCTION

Infection and injury set in train a variety of local and systemic responses, the overall consequences of which are to activate and coordinate a disparate collection of systems that function to repair tissue damage, to limit the action of and ultimately destroy pathogens, and to replace blood elements lost during trauma. These systems include the antigen-specific arm of the immune response, local connective tissue and vascular elements at the site of injury, central nervous system elements that regulate body temperature (the hypothalamus) and innervation of key organs, the liver, and the bone marrow. A key role in the integration of immune and inflammatory responses is played by a group of soluble mediators comprising growth factors, colony stimulating factors, and cytokines. These polypeptide mediators, released

Biomembranes
Volume 6, pages 147–175.

ISBN: 1-55938-663-0.

locally, act both on the surrounding tissue at the site of injury (autocrine and paracrine actions) and systemically (endocrine actions) to bring all the physiological elements required for an integrated physiological response into play.

The cytokine interleukin-1 (IL-1) was one of the earliest identified endogenous mediators of inflammation (Dinarello, 1984, 1989). IL-1 induces a wide-range of biological responses including, fever, cartilage breakdown, bone resorption, thymocyte proliferation, activation of T- and B-lymphocytes, induction of acute phase protein synthesis from hepatocytes, fibroblast proliferation, and differentiation and proliferation of bone marrow stem cells. It is now known that there are two distantly related polypeptides (IL-1α and IL-1β), each of which appears to possess all of the biological activities previously ascribed to IL-1 (March et al., 1985; Oppenheim et al., 1986). Both IL-1s are synthesized as precursors of approximately 270 residues, lacking a hydrophobic signal peptide, and processed prior to or during release from cells to yield C-terminal fragments of approximately 160 residues which constitute the mature forms of the hormones. The nature of the processing and secretion pathway remains largely obscure, although it appears that the processing of IL-1β is performed by a specific protease, IL-1β converting enzyme (ICE), that is co-induced with the IL-1β precursor (Black et al., 1989a, 1989b; Cerretti et al., 1992; Thornberry et al., 1992). In keeping with the lack of a hydrophobic leader sequence, immuno-cytochemical localization showed that intracellular IL-1 is distributed throughout the cytoplasm, rather than being associated with any particular organelle (Singer et al., 1988). A third from of interleukin-1 has been identified and characterized by cDNA cloning (Eisenberg et al., 1990, 1991; Hannum et al., 1990). This form of IL-1, (IL-1ra), appears to be a pure IL-1 antagonist showing a binding affinity for IL-1 receptors that is similar to those possessed by the two agonist IL-1s, but no capacity to induce biological responses (Hannum et al., 1990). Unlike the agonist forms, IL-1ra possesses a conventional hydrophobic leader sequence, and is therefore, presumably secreted from activated macrophages, the major cellular source, by the normal Golgi-mediated pathway. This view is supported by the finding that IL-1ra, unlike the other forms of IL-1, exists as a mixture of forms that differ in oligosaccharide content (Hannum et al., 1990). Recently, an intracellular form of IL-1ra (icIL-1ra) lacking a substantial portion of the hydrophobic leader sequence has been identified in keratinocytes (Haskill et al., 1991).

IL-1 receptors were first identified and characterized on an IL-1 responsive T-cell line (Dower et al., 1985). The receptor was found to be an M_r 80,000 cell surface protein present at low copy number (ca. 100–1000 sites/cell). Subsequent studies revealed that IL-1 receptors are widely distributed at low levels on cells of many different lineages (Dower et al., 1990, 1992b). It was also established that IL-1α and IL-1β bound to a common receptor with similar affinities on all cell types tested (Bird and Saklatvala, 1986; Dower et al., 1986; Kilian et al., 1986). These findings were entirely in keeping with the wide-range of cellular targets and biological activities previously reported for IL-1 and the complete overlap in activities seen for IL-1α and IL-1β (Dinarello, 1989). Isolation of cDNA clones encoding both mouse and human IL-1 receptors confirmed the original observations that the IL-1 receptor was an M_r 80,000 integral membrane protein which bound both IL-1α and IL-1β, and showed that it was a single-chain molecule, with an extracellular region structurally related to immunoglobulin, and a moderate sized cytoplasmic domain (213 residues; Sims et al., 1988, 1989).

Studies on B cell and monocytic cell lines showed that the IL-1 receptor on these cells was M_r 60,000, considerably smaller than the receptor originally described on T cells (Matsushima et al., 1986; Bomsztyk et al., 1989; Kilian et al., 1990; Spriggs et al., 1990, 1992b). Two studies have shown that this smaller receptor is the product of a distinct gene from the M_r 80,000 receptor (Bomsztyk et al., 1989; Chizzonite

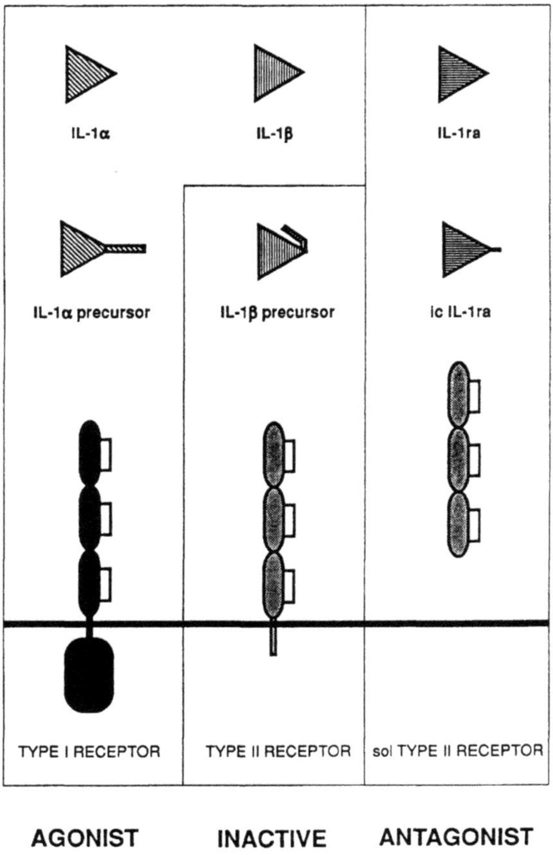

Figure 1. The components of the IL-1 system. The ligands are indicated by triangles, the immunoglobulin domains of the receptors by ovals, and the postulated inter-chain disulphide bonds by square brackets.

et al., 1989). These two receptor forms were originally thought to be expressed in a lineage-specific fashion; T cells and connective tissue cells expressing p80, and B-cells, monocytes, and granulocytes expressing p60. However, subsequent studies have revealed that both receptors are more widely distributed than at first supposed, and indeed, can be co-expressed on cells of a variety of lineages (Dripps et al., 1991; McMahan et al., 1991; Dower et al., 1992b; Stylianou et al., 1992). The two molecules are now termed type I (p80) and type II (p60) IL-1 receptors. cDNA clones for the type II IL-1 receptor have been isolated from human and murine B cell lines (McMahan et al., 1991). The type II receptor has an extracellular region that is immunoglobulin-like, the same overall size as the type I receptor (329 residues) and clearly related to it. The type II receptor has a small (29 residues) cytoplasmic region accounting for the size difference between the two receptor types. The overall structures of the three forms of IL-1 and the two receptors are summarized in Figure 1.

In addition to the integral membrane-bound forms of IL-1 receptor, interest has recently centered on the finding that substantial amounts of soluble type II receptor and smaller amounts of soluble type I receptor can be found both in supernatants from cell lines expressing IL-1 receptors on the cell surface, and in such *in vivo* sources as synovial fluid or serum (Giri et al., 1990; Symons et al., 1990, 1991). In addition, the levels of soluble type II receptor are elevated in animals or patients during inflammatory responses (Giri et al., 1994). The mechanism by which the soluble forms are generated seems to be proteolytic processing of the membrane-associated forms rather than alternate splicing of the initial transcript from the receptor genes. Two recent studies in which virally encoded soluble type II IL-1 receptors were identified and found to play a role in viral infection *in vivo* support the notion that soluble type II receptors play a relevant regulatory role in the IL-1 system (Alcami and Smith, 1992; Spriggs et al., 1992a). No such role has been defined for soluble type I receptors.

II. BINDING OF IL-1s TO IL-1 RECEPTORS

Binding of IL-1s (IL-1α and IL-1β) to cell surface receptors has been extensively documented in the literature (Dower et al., 1992a, 1992b). Figure 2 shows a summary of the binding properties of the three forms of IL-1 to both type I and type II receptors, generated by a relatively new method using the Pharmacia Biacore. The affinity values derived from the data are summarized in Table 1, along with average values for the cell surface forms from many studies in the literature. Both types of IL-1 receptor bind both agonist forms of human IL-1. Type I receptors exhibit an affinity for human IL-1α in the range of 10^9–10^{10} M^{-1} with a similar affinity for IL-1β. The affinity of this receptor for human IL-1ra is higher than that for either of the agonist forms. This, however, only partially characterizes the difference between the binding of IL-1ra and the agonist forms, since the dissociation rate of the complex IL-1RI/IL-1ra is so slow as to render the binding effectively

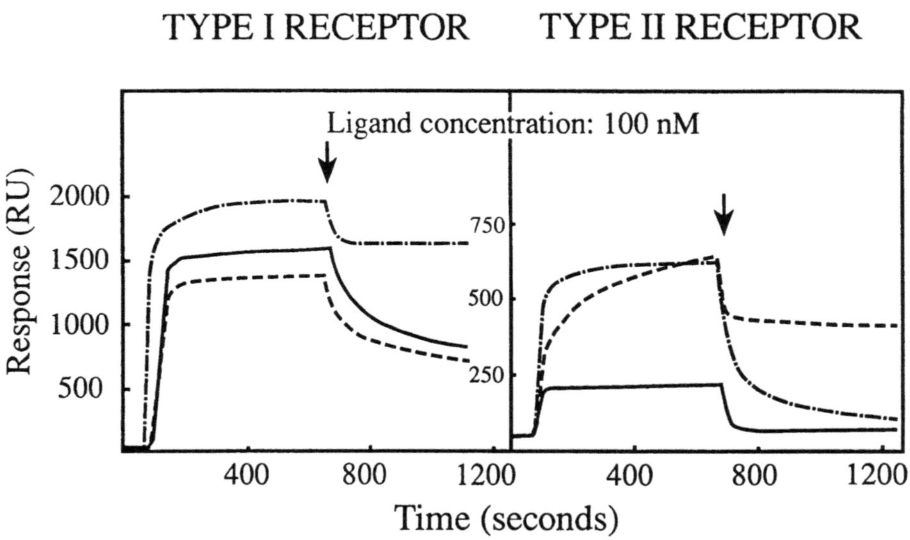

Figure 2. Binding Interactions in the IL-1 System. Kinetics of human IL-1α, IL-1β, and IL-1ra binding to soluble human IL-1RI and IL-1RII. Data were collected on a Pharmacia Biacore instrument using NHS/EDC chemistry to immobilize soluble human IL-1RI and IL-1RII on the chip surface. One thousand resonance units on the Y-axis represent approximately 1 ng/mm^2 bound. The curves are IL-1α (solid line), IL-1β (dashed line), and IL-1ra (dot–dash line). The arrow-head represents the point at which the injected solution was switched from buffer containing 1000 nM ligand to buffer alone, and hence dissociation of IL-1/IL-1R complexes begins.

Table 1. Binding Interactions in the IL-1 System[a]

	Affinity (M^{-1}) *for:*	
Receptor Form	*Type I Receptor*	*Type II Receptor*
IL-1α cell surface	$2–10 \times 10^9$	0.03×10^9
IL-1α soluble	$1–2 \times 10^9$	0.1×10^9
IL-1β cell surface	$1–10 \times 10^9$	$2–20 \times 10^9$
IL-1β soluble	$1–10 \times 10^9$	$2–6 \times 10^9$
IL-1ra cell surface	$10–20 \times 10^9$	0.03×10^9
IL-1ra soluble	$3–10 \times 10^9$	$0–0.02 \times 10^9$

Note: [a]The cell surface binding data are based on the literature, primarily on the data in McMahan et al. (1991) and Slack et al. (1993). The soluble receptor values are based on Biacore measurements as shown in Figure 2.

irreversible (Figure 2). The type II receptor shows an approximately 10–100 fold lower affinity for IL-1α than the type I IL-1 receptor and binds IL-1ra very weakly also; the affinity for IL-1β is, by contrast, similar to that of the type I receptor. The type II receptor may, therefore, be thought of as a relatively IL-1β specific binding protein.

One area of unresolved controversy is the nature and functional significance of the two affinity classes of IL-1 receptors that are found on some cell types, particularly some T cell lines or helper T cell clones. Specifically, it is possible to detect a small population of IL-1 binding sites with an affinity in the range of 10^{11} M^{-1}–10^{12} M^{-1} for both IL-1α and IL-1β on cells expressing predominantly or exclusively type I receptors (see for example Lowenthal and MacDonald, 1986; Sims et al., 1989; Dower et al., 1992a). In addition, cells expressing type II IL-1 receptor generally show two or more classes of binding sites for [^{125}I]IL-1α, but not [^{125}I]IL-1β (Benjamin and Dower, 1990; Bensimon et al., 1989; Benjamin et al., 1990). One possibility raised when it was found that the type II receptor had a small cytoplasmic region and that all cells expressing type II receptors also express low levels of type I receptor was that these two receptors are subunits of a heterodimeric (or oligomeric) receptor complex (Slack et al., 1993). This view seemed possible by comparison with many other cytokine receptor systems, for example those for IL-2/IL-4/IL-7/IL-15 (Hatakeyama et al., 1985, 1989; Cosman et al., 1990; Takeshita et al., 1992) or LIF/Oncostatin-M/IL-6/CNTF (Gearing and Ziegler, 1993; Hibi et al., 1990; Gearing et al., 1992a, 1992b) where the functional receptor complexes are assembled from two or more different cell surface subunits. In several instances, for example IL-6 and CNTF receptors, one chain lacks any substantial sized cytoplasmic region and acts as a binding protein recruiting cytokine for binding to other chains that have large cytoplasmic regions and transduce signals. The association of these chains also creates high-affinity binding sites. Thus, by analogy, on cells expressing both IL-1 receptors, low-affinity sites would be composed of the predominant chain, and complexes containing both type I and type II receptors would be the high-affinity receptors. However, an extensive series of experiments designed to test this hypothesis showed that it is unlikely to be correct, since IL-1 cannot bind to a type I receptor and a type II receptor simultaneously (Slack et al., 1993). Thus, when a cell expresses both types of receptors on the cell surface, only complexes of the structure IL-1RI/IL-1 and IL-1RII/IL-1, but not complexes of the structure IL-1RI/IL-1/IL-1RII can be formed. Finally, binding experiments with soluble receptor ectodomains (see Figure 2) usually show both high- and low-affinity components suggesting that this behavior is intrinsic to the interacting pairs of proteins rather than a function of some higher order molecular complex formation (Slack et al., 1993).

The functional significance of the high-affinity sites remains unclear, since a number of cell lines can respond to IL-1 at fm concentrations, far below the reported dissociation constant reported even for such high-affinity receptors (ca. 1 pm at the lowest; Orencole and Dinarello, 1989; Dower et al., 1986a; Rosoff et al., 1988), and the numbers of such sites have always been reported to be very low (<100 sites/cell). Further, cells bearing no detectable high-affinity receptors can respond to IL-1 at sub-picomolar concentrations (Rosoff et al., 1988). We have shown elsewhere (Dower et al., 1992a) that at concentrations more than ten-fold lower than the lowest measured K_d in a system, since for all K_{di}, K_{di},x L<<1 (when L is the ligand concentration), the binding dose–response curve is linear and can be written:

$$\text{Bound} = \sum N_i K_{di} L \tag{1}$$

The amount of ligand bound in this concentration range can thus be influenced as much by changes in the number of low-affinity sites as by the presence of a small number of high-affinity sites. Indeed it appears, as we had originally suggested (Dower et al., 1986a), that on many cell types, as few as 1–10 molecules of IL-1 bound/cell will elicit biological responses. Finally, it is useful to point out that while as a general principle of pharmacology, it is expected that for any given agonist, a class of receptors exists with a K_d close to the mid-point of the biological dose response curve, IL-1 may not follow this rule since its biological efficacy renders affinity to some extent irrelevant. More specifically, at concentrations in the picomolar to femtomolar range, where IL-1 dose–response curves often plateau, calculations based on the design of several standard bioassays show that even if *all* the IL-1 present binds to cells, there will only be a few molecules bound/cell. Hence, it is stoichiometry rather than concentration or affinity that likely determines the extent of the biological response. There has been less attention focused on the heterogeneity in IL-1α binding to B-cells in attempts to correlate binding with biological response, as for these cells, the lower affinity sites show an affinity constant of approximately 10^8 M^{-1} (Dower et al., 1986a; Benjamin et al., 1990; Slack et al., 1993) and the higher affinity sites have an affinity constant of approximately $0.5–1 \times 10^{10}$ M^{-1} (Table 1).

Whatever the functional significance of high and low affinity IL-1 receptors may be, there is no doubt that such receptor heterogeneity exists, and it is therefore, of interest to explore the structural basis of the binding of IL-1 to cells and how this bears on IL-1 receptor heterogeneity. Transfection of CVI-EBNA cells with cDNAs encoding murine or human type I- or type II-receptor leads to the expression of a large number of IL-1 binding sites which bind both IL-1α and IL-1β (McMahan et al., 1991; Slack et al., 1993). The binding properties of the recombinant receptors are similar to those observed for cells expressing the receptor naturally (McMahan et al., 1991; Slack et al., 1993). These data suggest that for both type I and type II receptors a single polypeptide, the product of the relevant cDNA is sufficient to reconstitute IL-1 binding in cells. Further, as illustrated in Figure 2, the truncated extracellular region of both types of IL-1 receptor, expressed as soluble proteins and tested in a cell-free binding assay, reproduces the binding properties of the entire receptor in cells or in detergent extracts (Urdal et al., 1988; Dower et al., 1989; Slack et al., 1989, 1993). Taken as a whole, the data that we have generated using both soluble and full-length recombinant IL-1 receptors suggest that the product of a single cDNA, expressed as a single glycosylated polypeptide, can reproduce most, if not all, the binding phenomena observed regarding the interaction of IL-1α and IL-1β with cells.

In order to examine the structural basis of ligand binding by the type I receptor, we exploited the immunoglobulin-like nature of the receptor by deleting each of the three domains in turn and testing each of the mutants, ΔD1 (lacking domain 1), ΔD2 (lacking domain 2), and ΔD3 (lacking domain 3), for IL-1 binding function after transient expression in COS cells (Dower and Sims, 1990b). In order to rule out the potential artifact that deletion of domains prevented expression of the receptor at the cell surface, we used a series of antibodies directed against the extracellular region to test for cell surface expression of receptor protein. The results from these experiments, reported in detail elsewhere (Dower and Simms, 1990), suggested that deletion of any of the three domains from the murine type I IL-1 receptor generated a mutant receptor protein that had lost all detectable IL-1 binding activity, but had retained sufficient structural integrity to bind at least one monoclonal antibody.

The ligand-binding data suggest that all three domains of the type I IL-1 receptor are required, either to obtain correctly folded receptor, to form an IL-1 binding site, or both. The overall structural resemblance between the type I and type II receptors and the observation that they bind to overlapping sites on the IL-1 molecules would suggest that the interactions of IL-1 and type II receptor would be similar to those with the type I receptor, however, we have not tested this directly.

III. SIGNALING THROUGH IL-1 RECEPTORS

The roles of the two IL-1 receptors in transmitting signals to cells have been defined in a series of recent studies (Stylianou et al., 1992; Colotta et al., 1993b, 1994; Sims et al., 1993). As outlined above, the type I and type II receptors do not form a receptor complex even when co-expressed on the same cell. Thus, since each interacts with IL-1 separately, it is possible that each could form a separate receptor system coupling to different signaling pathways. By using receptor isotype-specific blocking monoclonal antibodies, we were able to show that all IL-1 responses in several cell types could be blocked by antibodies directed to the type I receptor (Bird et al., 1992a; Sims et al., 1993). By contrast, antibodies directed to the type II receptor had no inhibitory effect on IL-1 action, even on cells that expressed predominantly type II receptors on the cell surface, as determined either by cross-linking or inhibition of [^{125}I]IL-1 binding. For example, neither IL-1 induction of IL-8 in the monocytic cell line THP-1 (Sims et al., 1993), nor activation of the MAP kinase pathway in the hepatoma cell line, HEPG2 (Bird et al., 1992a; Bird, T.A., unpublished data), can be blocked by anti-type II receptor antibodies, but in both instances, anti-type I receptor antibodies, which block only a small fraction of IL-1 binding to either cell type, block IL-1 responses completely. In another

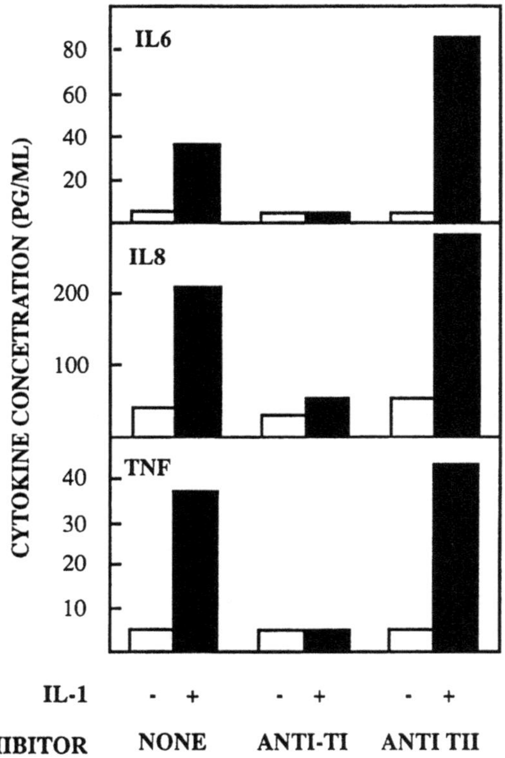

Figure 3. Primary Human Neutrophils Respond to IL-1 through type I but not type II IL-1 receptors. Human neutrophils isolated from whole blood were cultured with or without IL-1 for 24 hours, in the presence of no antibody, IL-1RIM4 or IL-1RIIM22 at 100 μg/ml. Subsequently, supernatants were harvested and assayed for the presence of IL-6, IL-8, or TNF by using specific ELISA assays in each case.

example, shown in Figure 3, induction of cytokine production by human neutrophils *in vitro* is not blocked by anti-type II receptor antibodies, but is completely blocked by anti-type I receptor antibodies. The type II receptor thus plays no role in transmitting IL-1 signals and interest is focused on the type I receptor in terms of understanding this process.

Molecular cloning techniques have led to the classification of many receptors into relatively few general types based on primary sequence motifs. The first of these comprises those receptors whose intracellular domains function as protein tyrosine kinases and share sequence homologies typical of protein kinases and protein tyrosine kinases in particular; this group is exemplified by the receptors for epidermal growth factor (Downward et al., 1984; Ullrich et al., 1984), insulin (Ullrich et al., 1985), and platelet-derived growth factor (Yarden et al., 1986). A second major group is formed by receptors which couple to heterotrimeric GTP-binding proteins (G-proteins); typically, these receptors have seven membrane-spanning domains, and are hence termed serpentine receptors (Roth et al., 1991). This large family of molecules, of which the β-adrenergic receptor is considered prototypic, are all related at the primary sequence level. A third large group of receptors, the hemopoetin receptors, were originally grouped together on the basis of a unique short motif in the extracellular region (WSXWS) (Cosman, 1993; Kishimoto, 1992; Gearing and Ziegler, 1993), and have been recently found to share a common general signaling mechanism via interaction with cytoplasmic protein tyrosine kinases of the JAK family (Witthuhn et al., 1993; Darnell et al., 1994; Johnston et al., 1994; Narazaki et al., 1994; Nicholson et al., 1994; Stahl et al., 1994), likely by direct association of these molecules with specific sequence motifs in the cytoplasmic region.

The cytoplasmic region of the type I receptor, a structure of approximately 200 residues in length, shows significant sequence similarity with regions from several other cloned proteins in the data bases, but not with any of the large, well characterized receptor families, nor with protein kinases (Hanks et al., 1988). Proteins with clear homology include, TOLL (Gay and Keith, 1991; Anderson et al., 1985a,b), a *Drosophila* cell surface protein that was identified as the product of a gene involved in the establishment of dorsal–ventral polarity in the developing embryo. TOLL homologs from other insect species have recently been identified (Yamagata et al., 1994). A second *Drosophila* gene product (eighteen-wheeler) which possesses a cytoplasmic region with sequence similarity to the IL-1 receptor cytoplasmic region, has also been recently cloned. In addition, two mammalian proteins have been recently identified that contain regions resembling the type I IL-1 receptor cytoplasmic domain. The first of these, RSC786, a mammalian cell surface protein, was isolated from a library made from the human myelomonocytic cell line, KG-1; the RSC786 cDNA was selected at random as a candidate for sequence analysis. The second clone, MyD88, was isolated as a transcript induced by IL-6 treatment of myeloid precursor cells (Hultmark, 1994; Lord et al., 1990; Yamagata et al., 1994). Unlike the other molecules in the family, MyD88 does not possess a sequence predicted to act as a membrane anchor, nor does it have a signal peptide. It appears, therefore, to be an intracellular protein. The last member of this family of molecules, ST2 (Yanagisawa et al., 1993; Bergers et al., 1994), was originally identified as a soluble homolog of the murine type I IL-1 receptor extracellular ligand-binding region, induced by serum stimulation of 3T3 cells (Tominaga, 1989). More recently, cDNAs have been isolated for murine and rat ST2, the rat homolog being termed Fit-1 (Bergers et al., 1994), which encode integral membrane proteins having extracellular and cytoplasmic regions homologous to type I IL-1 receptor. Despite the fact that ST2/Fit-1 is as similar to type I or type II IL-1 receptors in its N-terminal extracellular region as type I IL-1 receptors are to each other (ca. 27% identity at the amino acid level for any of the three pairs), it is not a type III IL-1 receptor since it does not bind IL-1α, IL-1β, or IL-1ra (Gayle et al., 1996).

While all of the molecules which contain the type I IL-1 receptor cytoplasmic domain motif are clearly related in this region, they show extensive diversity in the rest of the sequences. Thus, the extracellular, putative ligand binding, domains of type I IL-1 receptor, TOLL, eighteen-wheeler, and RSC786 are only distantly related to one another. TOLL and eighteen-wheeler have large extracellular regions that do not have any close similarity to any other sequences in the data bases. Both, however, possess a number of repeats of a 24-residue leucine rich repeat (LRR) that is found in a diverse group of proteins including proteoglycans such as decorin and biglycan, platelet GP1b, chaoptin, and adenylate cyclase (Bourdon, 1990). RSC786 also contains 12 of these LRR 24-residue repeats, and the homology spanning the first 300 residues of the RSC786 extracellular region beginning at residue 10 of the predicted mature sequence is most closely related to decorin and biglycan. However, it lacks the sequence at the N-terminus that contains the glycosaminoglycan addition sites in the proteoglycan core proteins, and would not, therefore, be predicted to be a cell surface proteoglycan (Bourdon, 1990). The intracellular molecule, MyD88, contains, in addition to the IL-1 receptor cytoplasmic domain homology region, an N-terminal extension which appears to be unique when analyzed against the GENBANK and Swiss Protein databases. This family is thus composed of molecules which have arisen by modular recombination, in which the IL-1 receptor-like cytoplasmic domain has become associated a set with unrelated domains to produce a series of hybrid molecules. In agreement with this idea, the region of the type I IL-1 receptor that is homologous to the rest of the family is encoded on two exons, thus making it more likely that after gene duplication, productive recombination events involving this region of the molecule would occur. The functional diversity of the family is illustrated by the finding that the ligand for TOLL, the product of the spatzle gene, is completely unrelated to the three forms of IL-1. This underscores the obvious point that TOLL is not the insect type I IL-1 receptor homolog. The structures of the family of IL-1 receptor-related molecules are summarized in diagrammatic form in Figure 4.

The shared cytoplasmic region found in the receptor family members suggests that all of these molecules couple through this region to a common set of signaling pathways. One would presume, in this context, that MyD88 may lie down-stream of other signaling systems which, by acting on the N-terminal region, feed into the same set of pathways as the other family members. One might speculate, for example, that MyD88 is the cytoplasmic region for an as yet unidentified transmembrane receptor, or that it can be phosphorylated by one or more protein kinases to be activated. We are currently investigating these possibilities.

The best guide to the nature of at least one of the pathways which this family of molecules may activate comes from the genetic analysis of the TOLL pathway in *Drosophila melanogaster*. Toll is one of a set of genes that control the establishment of dorsal–ventral polarity during embryogenesis. The others are EASTER (Jin and Anderson, 1990; Chasan et al., 1992), SPATZLE (Morisato and Anderson, 1994; Schneider et al., 1994), (TOLL), TUBE (Hecht and Anderson, 1993), PELLE

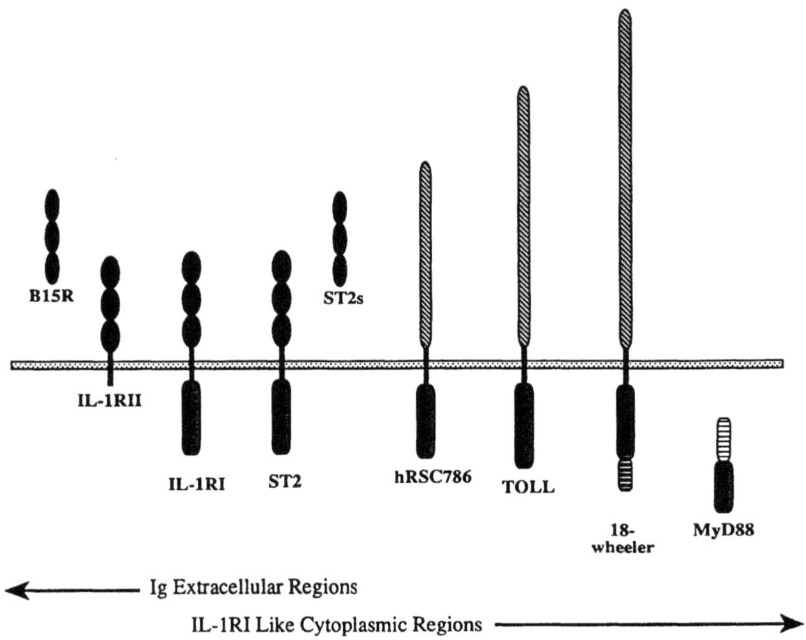

Figure 4. Schematic Illustration of IL-1 Receptors and Related Molecules. The stippled horizontal line indicates the plasma membrane with regions below it being cytoplasmic. Those domains showing detectable shared homology are shaded as follows: solid black—IL-1 receptor extracellular/immunoglobulin C2 domain-like; dark gray stippling—IL-1 type I receptor cytoplasmic region-like; diagonal cross-hatching—containing LRR repeats.

(Shelton and Wasserman, 1993), CACTUS (Geisler et al., 1992; Kubota et al., 1993; Gillespie and Wasserman, 1994) and DORSAL (Steward, 1987; Ghosh et al., 1990; Norris and Manley, 1992): they are listed in functional order. The Easter gene encodes a protease which acts on Spatzle, activating it. Spatzle is the ligand for Toll. The products of the Tube and Pelle genes encode cytoplasmic proteins which lie down-stream of Toll. cDNAs have been isolated for both of these; pelle is a protein serine/threonine kinase of the raf/mos family. It is not *Drosophila* raf (D-raf) (Shelton and Wasserman, 1993; Tsuda et al., 1993). The sequence of the tube protein is unique at this point and offers no clues as to its function. Finally, CACTUS is an IκB like molecule, and DORSAL is a member of the rel/NFκB family of transcription factors. The functional parallels between family members are underlined by the extensively documented observation that IL-1 is a potent activator of rel family transcription factors (e.g., Leung et al., 1994).

While the shared cytoplasmic region structure suggests that all of these proteins share common proximal interactions, little is known about the nature of these interactions. It has been clearly established that IL-1 is a good activator of several protein serine–threonine kinases in the central MAP kinase or ERK pathway (Bird et al., 1990, 1992, 1994, 1995). These include, in certain cells, both the p44 and p42 forms of MAP kinase and the down-stream protein kinase that phosphorylates the small heat shock protein, hsp27 (Guesdon and Saklatvale, 1991; Kaur et al., 1989; Guesdon et al., 1990, Bird et al., 1994). This protein kinase has also been termed MAPKAP kinase II (Bird et al., 1994). IL-1 has also been recently shown to activate the p54 form of MAP kinase or stress-activated protein kinase (SAPK) (Kyriakis et al., 1991, 1994; Bird et al., 1995). However, it is not known how IL-1 activates steps earlier in this pathway. It does not, for example, activate ras and is a poor activator of raf (Bird et al., 1995), the protein serine–threonine kinase that has been suggested to be a major form of MEK-kinase (MEKK) (Kyriakis et al., 1992). This is consistent with a number of recent reports that ras is a major regulator of raf activity (Cook and McCormick, 1993; Johnson and Vaillancourt, 1994; Wood et al., 1992; Moodie et al., 1993; Vojtek et al., 1993; Warne et al., 1993; Zhang et al., 1993; Chuang et al., 1994; Hallberg et al., 1994; Leevers et al., 1994; Stokoe et al., 1994). In addition, while there are some reports in the literature of IL-1 action being partially inhibited by tyrosine kinase inhibitors (Coyne et al., 1990; Katabami et al., 1993; Marczin et al., 1993; Corbett et al., 1994), for example, tyrphostins (Munoz et al., 1991, 1992) or herbimycin A (Iwasaki et al., 1992), there is no evidence for increases in tyrosine phosphorylation of proteins in IL-1 treated cells, with the exception of increases in MAP kinase phosphorylation on tyrosine residues (Bird et al., 1992, 1995). This, however, is a special case as MEK is a dual specificity kinase that activates MAP kinases by phosphorylation on a pair of residues T/S–X–Y (Payne et al., 1991).

It would seem reasonable to suppose that the mechanism delineated for ras/raf may be a specific example of a more general phenomenon in which other members of the small GTP-binding protein family (ras, rac-1, rho-A, rho-B, rho-C, etc.) act to regulate other members of the raf/mos family of protein serine/threonine kinases (see for example Manser et al., 1994). It is possible that pelle activity is dependent on one or more small GTP-binding proteins. It is also possible that IL-1 receptor and related proteins act through a small GTPase other than ras to activate an as yet uncharacterized raf-related protein kinase that is the major MEKK activity which triggers MAP kinase activation in response to IL-1. How such a putative early step would be triggered by IL-1 binding to receptor is unclear. The major established pathway to ras activation is from receptors with endogenous tyrosine kinase activity, which upon auto-phosphorylation bind GRB-2 and activate the ras GTP/GDP exchange factor SOS, thus activating ras (Cook and McCormick, 1993; Lowenstein et al., 1992). Since there is no evidence for rapid protein tyrosine phosphorylation induced by IL-1, it would seem unlikely that this or an analogous pathway involving other SH2-domain containing proteins is involved in IL-1 action. Indeed, recent experiments from our laboratory and another group, suggest that there is an IL-1 activated phosphorylation event triggered in a complex involving the type I receptor, in IL-1 responsive cells (Martin et al., 1994; Eriksson, A., Bird, T.A., and Dower, S.K., unpublished observations). The major substrate, a M_r 65,000 protein distinct from the receptor, is phosphorylated on threonine residues when complexes immunoprecipitated from IL-1-treated EL4 cells with a non-blocking anti-type I IL-1 receptor antibody are incubated with labeled ATP *in vitro*. There is a 2–5-fold stimulation of the *in vitro* activity when IL-1-treated cells are compared with untreated cells. These data are consistent with the notion that if there is a multisubunit IL-1 receptor signaling complex, that any protein kinase it contains is a serine/threonine kinase. It is also clear, based on the sequence of the receptor cytoplasmic region, that this kinase is not the receptor itself.

Finally it is also not clear how IL-1 binding activates the receptor. A wide variety of evidence suggests that lateral aggregation of receptors in the plane of the membrane is a universal mechanism for triggering transmembrane signals. To date, no convincing evidence has been generated that IL-1 cross-links or aggregates type I receptors when it binds (Dower et al., 1985, 1992b; Slack et al., 1993; Sims et al., 1994). Thus, none of the anti-receptor antibodies are agonists, no evidence for an IL-1 induced increase in receptor dimers as detected by chemical cross-linking has been found, nor can IL-1 be shown to cause dimerization of soluble IL-1 receptors (Dower et al., 1989), nor are the binding and kinetic properties of IL-1 receptors indicative of multivalent binding. Nevertheless, it seems reasonable to suppose that receptor aggregation triggered by ligand is the proximal signal, for not only is this widely established as the mechanism in other systems, but such a mechanism would provide a straight-forward explanation for the mode of action of IL1ra, which would fail to cross-link receptors (Ju et al., 1991).

One broader issue that is as yet unresolved is the mechanism by which IL-1 receptor and the related molecules couple to NFkB activation and how this relates to the control of NFkB activity by other agonists. The general mechanism of regulation of rel family transcription factor activity is well understood in outline (Liou and Baltimore, 1993). The DNA binding and transactivation components (for example p65 or c-rel) are held in an inactive state by binding to inhibitory subunits (for example IκBα). Activation of the complexes involves modification of the inhibitory component, likely serine or threonine phosphorylation (Beg and Baldwin, 1993; Beg et al., 1993), the complex dissociates releasing the transcription factor, and the free-modified inhibitor is rapidly degraded (Beg et al., 1993; Henkel et al., 1993). The identity of the protein kinase(s) that mediate this in intact cells has not been established. Whether all upstream pathways converge on a common kinase or whether many enzymes can produce the same effect is not known. One obvious possibility is that pelle is the IκB kinase in *Drosophila* (Gillespie and Wasserman, 1994), and that there is a mammalian homolog of pelle which serves an a analogous function. Several years ago, Mizel and collaborators showed that addition of protein kinase C or protein kinase A preparations to cell extracts was sufficient to produce detectable NFκB activity *in vitro* as measured by a gel-shift assay (Shirakawa and Mizel, 1989; Shirakawa et al., 1989). It was not clear, however, from these experiments whether the action of the particular kinase preparations used was directly on one or more IκBs or whether other protein kinases in the assays were being activated by the added enzymes. We have found that there is some functional separation between the MAP kinase pathway and the activation of rel family transcription factors, since inhibitors of MAP kinases do not block IL-1 activation of NFκB as measured by gel-shift assay (Bird et al., 1992a). This suggests that none of the ERKs and no downstream enzymes such as MAPKAP kinase or S6 kinase, are involved in NFκB activation.

Clearly, much remains to be done to elicit the nature of IL-1 signaling, in particular with respect to the receptor-proximal events. Nevertheless, the finding of what is undoubtedly the beginning of a family of molecules, which likely use common elements to effect cellular signaling, should allow us to define which regions of the IL-1 receptor are critical for function and to begin to understand the mechanism of action of these signaling molecules. These findings also suggest that the mechanism, whatever it may prove to be, is likely of wider significance than just the mediation of IL-1 action. Some initial steps have been made in this direction. Thus, based on a comparison of IL-1R type I and Toll alone, Heguy and collaborators were able to note that a short motif close to the C-terminus of the IL-1 receptor, $RFWKX_1X_2RY(X_1 = K$ or N, $X_2 = V$ or L), in all four cloned IL-1 receptors, and WFWDKLRF in Toll, was particularly strongly conserved and that mutations in this region abolished signaling function in the receptor (Heguy et al., 1992). A later mutagenesis study by Matsushima and collaborators also identified this region as critical for induction of IL-8 gene transcription by IL-1 (Kuno et al., 1993). With the availability of a larger data set of sequences, it should be possible to dissect the structure–function relationships for this family of molecules and determine, for example, whether coupling to the NFκB and MAP kinase pathways is mediated by common or distinct sites on the receptor.

IV. THE PHYSIOLOGICAL ROLES OF THE COMPONENTS OF THE IL-1 SYSTEM

In the introduction, we outlined the ligands and receptors that form the IL-1 system. There are three distinct ligand genes encoding IL-1α, IL-1β, and IL-1ra. Each of these genes gives rise to two different forms of ligand. The genes for IL-1α and IL-1β are each initially transcribed to yield mRNA encoding a single form of protein, a precursor of the mature form of IL-1. Neither form of IL-1 precursor has an N-terminal signal peptide, and thus neither is secreted. The precursor of IL-1β is incapable of binding to type I receptors and has no biological activity, the precursor of IL-1α binds to the receptor and is biologically active (Mosley et al., 1987). The precursor form of IL-1β (pro-IL-1β) is processed to the mature form by a specific protease, the IL-1 converting enzyme (ICE) (Kronheim et al., 1992;

Thornberry et al., 1992). No specific enzyme has been identified for the processing of IL-1α. However, calpain has been suggested to play a major role in IL-1α processing in intact cells (Kobayashi et al., 1990). The gene encoding IL-1ra can generate two different forms of protein since the initial transcript give rise to two different forms of mRNA by alternative splicing (Hannum et al., 1990; Haskill et al., 1991). One form encodes a form of IL-1ra with a signal peptide, and this protein is thus secreted by the conventional pathway. The second form of mRNA lacks sequences that correspond to an exon encoding the extreme N-terminus of the precursor, and hence contains a defective signal peptide. It is not secreted, but like the pro-forms of IL-1α and IL-1β, accumulates in the cytoplasm (Haskill et al., 1991).

As discussed previously, all the experimental data that we have accumulated to date suggest that the type II IL-1 receptor does not mediate the transmission of IL-1 signals across the plasma membrane of cells. Nevertheless, the only mRNAs found to date encode an integral membrane form of this protein, and it binds IL-1α and IL-1β with high affinity. However, a variety of lines of evidence suggest that the type II receptor does not function as receptor in the conventional meaning of the term, but is rather an inhibitor of IL-1 responses, binding IL-1α and IL-1β to form complexes that cannot bind to the type I receptor and trigger signals.

First, we have shown that soluble recombinant type II IL-1 receptor will completely block binding of IL-1α and IL-1β to cells expressing type I IL-1 receptor (Slack et al., 1993). It is, as the affinity constants measured by direct binding would predict (Table 1), a much more effective inhibitor of IL-1β binding than of IL-1α binding.

Second, soluble type II IL-1 receptor will also block biological responses to IL-1 (Giri et al., 1990, 1994; Symons et al., 1991; Colotta et al., 1993, 1994). Interestingly, while it is a good inhibitor of IL-1β action, it shows virtually no capacity to block IL-1α action. The difference between the potency of the type II receptor as an IL-1β antagonist and its potency as an IL-1α antagonist is more marked than would be predicted on the basis of affinity alone, and is presumably also affected by the much more rapid dissociation of IL-1α than IL-1β from complexes with the type II receptor. Because of the rapid off-rate of IL-1α, it seems likely than in the presence of cells which bind, respond to, and internalize IL-1, that IL-1 consumed by the cells would be rapidly replaced due to dissociation of the IL-1α/ soluble type II receptor complexes. These data thus suggest that soluble type II receptor may be a specific antagonist for IL-1β. By contrast, IL-1ra would be an antagonist for both IL-1α and IL-1β since it binds irreversibly to the signaling type I receptor and does not trigger it, thus blocking the capacity of a cell to respond to IL-1.

A third line of evidence that points to a physiological role of type II receptor as a IL-1 antagonist, is the finding that a variety of signals will trigger rapid shedding of the receptor from cells due to proteolysis of the integral membrane form of the receptor (Giri et al., 1994). Figure 5 shows an example of this in which CB23 cells, a human B lymphoma line that expresses 3–10,000 type II receptors and 10–100

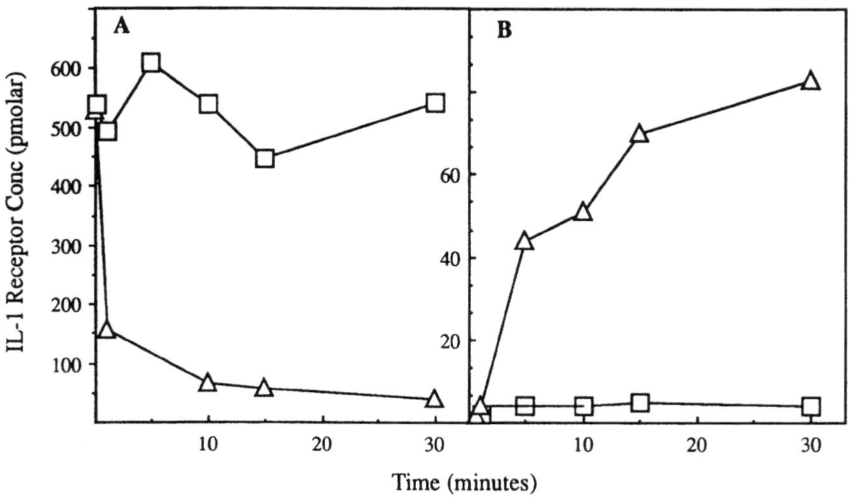

Figure 5. Phorbol Ester treatment Induces Rapid Shedding of IL-1R Type II from Cells. CB23 human B-lymphoma cells in culture were treated with 100 nM phorbol myrystyl acetate (TPA) for the indicated times (open triangles) or with vehicle (DMSO) (open squares). (**A**) Receptor expression on cells measured by binding of radiolabeled IL-1. The data are expressed as cell surface receptor concentration in the culture by correcting for the cell concentration to allow direct comparison with the soluble receptor data. (**B**) Soluble receptor concentration in the culture. The receptor concentration was measured using a soluble receptor direct binding assay (Dower et al., 1989; Slack et al., 1989, 1993) in which supernatants were adsorbed to anti-type II receptor coated plates, the antibody being a non-blocking monoclonal antibody. The plates were subsequently washed and incubated with radiolabeled IL-1. The assay, therefore, detects only ligand binding active soluble type II IL-1 receptor.

type I receptors, loses >90% of its cell surface receptors when treated with TPA which directly activates several isoforms of protein kinase C. A significant amount of soluble type II IL-1 receptor appears in the medium in the same time-frame. It is also clear that cell lines that express type II receptor on their cell surface constitutively shed a soluble form of the protein into the medium. In this respect, the type II IL-1 system resembles the TNF receptors, the IL-2Rα chain, the LIF receptor, and many others which are shed from most, if not all, cell types that express them as integral membrane forms. While this phenomenon is wide-spread, the molecular mechanism has not been resolved in detail for any system. The most thorough studies have been done with TNF receptors and suggest that the cleavage generally occurs within a short stretch of sequence close to the membrane. Mutagenic analysis of this region, however, suggested that the susceptibility to cleavage was relatively insensitive to which residues were present in this region, with one notable exception: introduction of prolines in this stretch led to a marked reduction in shedding. This observation suggests that the susceptibility may be a function of the overall conformation, possibly a stretch of α-helix and that resistance conferred by prolines is due to disruption of the helix. We have no such detailed information for IL-1R type II.

A fourth line of evidence for the role of type II receptor comes from the observation that a number of genomes of mammalian viruses (for example the WR strain of vaccinia virus) contain open reading frames that are sufficiently closely related in primary sequence to IL-1RII that it is highly likely that they have been acquired from host cells by recombination (Alcami and Smith, 1992; Spriggs et al., 1992a). Significantly, these open reading frames lack sequences encoding membrane-spanning regions and are, hence, predicted to encode soluble proteins. It has been shown that virally infected cells shed soluble IL-1β binding proteins into the medium, and when the vaccinia gene (B15R, identified by sequence similarity) was cloned out of the viral genome and expressed in baculovirus, it was shown to encode an IL-1 binding protein. Recombinant mutant virus in which the B15R gene was deleted showed significantly modified host pathology in mice, suggesting that the gene is expressed in the animals and that the gene product inhibits host immune and inflammatory responses. Finally, it should be pointed out that this is a particular example of a wide-spread phenomenon, namely the acquisition by viruses of genes encoding negative regulators of host immune and inflammatory responses (for a review see Gooding, 1992).

That the endogenous IL-1 receptor may play a similar role *in vivo* is suggested by recent findings that serum and synovial fluid contain soluble IL-1 receptors and these levels of detectable receptor increase when active inflammation is ongoing, for example, in septic shock or arthritis (Giri et al., 1994). One additional piece of circumstantial evidence, that is at least consistent with this picture, is the observation that the type II receptor binds IL-1ra weakly in man and shows no detectable affinity in mouse and rat (Dripps et al., 1991; McMahan et al., 1991). While teleological, it is reasonable that the two antagonist components of the IL-1 system fail to mutually neutralize.

Some of these themes were brought together recently in a study by Colotta and collaborators (Colotta et al., 1993a) in which it was shown that a well-documented activity of IL-1, that it will maintain survival of human peripheral blood neutrophils *in vitro*, can be blocked by IL-4. IL-4 will not inhibit the same biological endpoint when GMCSF is used to maintain the cells. The investigators showed that the mechanism by which IL-4 specifically antagonizes IL-1β is by inducing the synthesis and shedding of type II IL-1 receptor. The study by Colotta et al. thus demonstrates for the first time, that the antagonist type II receptor can act as a integral component of the cytokine networks that regulate activation, differentiation, and survival of the cellular elements of the immune/inflammatory system. This type of mechanism may also account, in part, for the immuno-suppressive action of steroids, since it well documented that dexamethasone, for example, will cause a large up-regulation in type II IL-1 receptor expression in myeloid lineage cells and cell lines (Akahoshi et al., 1988).

While there is thus a substantial and growing body of evidence defining the negative regulatory role of type II IL-1 receptor, there is, surprisingly, much less evidence consistent with IL-1ra filling an analogous role *in vivo*. It is certainly true that IL-1ra will block the receptor binding and biological activity of both IL-1α and IL-1β in any system where it has been tested, and the list of such systems is a long one. However, *in vitro* it requires in general >100-fold molar excess of IL-1ra to block IL-1α or IL-1β activity, (Arend et al., 1990, 1991; Firestein et al., 1994) despite the fact that the affinities of the three forms of IL-1 are comparable and indeed that IL-1ra, unlike IL-1α or IL-1β, binds to type I receptors irreversibly. *In vivo*, the data are more skewed and very large doses of IL-1ra are required to inhibit biological responses mediated by IL-1 in experimental models (Ohlsson et al., 1990; Wakabayashi et al., 1991). For example, a dose of 100 mg/kg body weight was required to reduce mortality to ≤ 10% in a group of rabbits after endotoxin treatment: in the control group receiving endotoxin plus saline, 85% of the animals died (Ohlsson et al., 1990). The levels of circulating IL-1ra found in such animals far exceed the levels of endogenous IL-1ra found in serum or other sources (Arend et al., 1990, 1991; Firestein et al., 1994), suggesting that it is unlikely that the endogenous capacity to produce receptor antagonist results in levels capable of significantly blocking systemic IL-1 action. Admittedly, in many instances samples are taken from patients and animals where a pathological acute or chronic inflammatory response is ongoing, and one might argue that one likely cause of this is that the IL-1/IL-1ra balance has been perturbed in favor of IL-1 (Firestein et al., 1994). It is certainly possible, therefore, that during normal physiological homeostasis, IL-1ra does play a role in dampening IL-1 responses. In addition, searches of the sequence databases with the ILra sequence from several species fail to detect any significantly homologous viral open reading frames. On the basis of this finding, it seems reasonable to suggest that during infection by such a hypothetical virus it would not be possible to produce enough IL-1ra to interfere effectively with host responses and hence give the virus a selective advantage.

As discussed previously in this chapter, the underlying cause of the surprisingly weak efficacy of IL-1ra lies in the unusual hyper-responsiveness of the IL-1 receptor signaling to ligand binding. T-lymphocyte lineage cells are seen to be extreme in this respect, with several reports of cell lines responding to sub-picomolar–femtomolar IL-1 concentrations, even when the cells express relatively low levels (50–500 sites/cell) of receptors (Orencole and Dinarello, 1989; Dower et al., 1986a; Rosoff et al., 1988; Savage et al., 1989). Analysis of such data suggest that, in some instances, a single molecule of IL-1 bound to a cell will trigger that cell to respond (Dower et al., 1986; Curtis et al., 1989). Clearly, in the face of ongoing receptor synthesis and turnover, even an antagonist with such favorable binding properties as IL-1ra will be relatively inefficient at blocking agonist action, when the agonist/receptor system is so sensitive. One can only speculate as to why the IL-1 system seems to have evolved toward the logical limit of one molecule/cell triggering the signal pathway. One reasonable suggestion would be that the existence of the receptor antagonist itself has resulted in the present nature of the system. More specifically, if *in vivo* "IL-1" is always a mixture of IL-1/IL-1ra, then the relationship between receptor occupancy by ligand and biological response is lost, and there would be no selective pressure to prevent the system from tending towards increasing efficiency of receptor action and an increasing ratio of IL-1ra to IL-1 production. One would, therefore, argue that at present "real IL-1" is the mixture of IL-1 with the receptor antagonist and that the function of the "antagonist" is not to block IL-1 action, but to provide a mechanism by which the specific activity of IL-1, that is units/mg, can be varied continuously over a wide range.

In summary, the six genes that encode the IL-1 system, three for the ligands, two for the receptors, and one for ICE, give rise to six different forms of IL-1, and three or possibly four different forms of receptor—if soluble type I receptors are produced and sub-serve some distinct function. Our current understanding of the ways in which these proteins function suggests that the internal regulation of the system is complex and is centered on exquisitely regulating the extent to which IL-1/IL-1RI complexes are formed at the cell surface. The precise regulation of the level of this complex is critical because of the potent signaling activity that it possesses, and this, in turn, may well have arisen by virtue of the mechanism of regulation that this particular cytokine system happened to have arrived at some point in the past.

V. CONCLUSIONS

Since its initial description as an endogenous pyrogen, a large body of data has accumulated implicating interleukin-1 as a central mediator of inflammation. The studies we have summarized in the preceding sections illustrate that the IL-1 system is highly internally regulated, being comprised not only of agonist components (pro-IL-1α, IL-1α, IL-1β, ICE, and IL-1RI), but also antagonists (icIL-1ra, IL-1ra, and IL-1RII). The IL-1 system does not, of course, exist in a functional vacuum, but is part of a more extensive network of regulatory factors, cell adhesion molecules, lipid mediators, and so forth; many of these act on the IL-1 system. Thus, for example, PDGF up-regulates and TGFβ down-regulates IL-1RI gene expression. Nevertheless, the IL-1 system shows an unusual degree of internal regulation, containing both an inhibitory soluble receptor and a natural antagonist form of the hormone. Both of these negative regulators seem to be functional *in vivo*, and can also inhibit inflammation when administered exogenously in animal models. The molecular physiology of the IL-1 system and, indeed, that of inflammation in general is poorly understood. Thus, we have little clear understanding of how the members of the ever-growing list of molecular components identified as being "involved" in regulation of inflammation work in concert, nor which are the most relevant in different diseases. It is not yet clear whether IL-1ra is an antagonist or serves as a "diluent" to modulate the specific biological activity of IL-1α and IL-1β. By the same token, it is not clear whether sIL-1RII always acts as an antagonist or whether under some circumstances it might serve as carrier and agonist for IL-1, particularly IL-1α which can dissociate rapidly from complexes with type II receptor.

The molecular details of the signaling pathways from the type I receptor are still obscure, but the existence of a growing group of molecules that share the type I receptor cytoplasmic region will accelerate the pace at which our understanding of the immediate interactions of the receptor grows. Clearly, interest remains high in the IL-1 system because a large body of data implicate dis-regulation of the production of IL-1 as being a central process in the molecular pathogenesis of a wide-range of inflammatory diseases. It is clear that when expressed in recombinant form, both IL-1ra and soluble type I IL-1 receptor can suppress IL-1 action in a wide variety of *in vitro* and *in vivo* model systems. The data that have been generated so far support the potential value of these two recombinant proteins and IL-1 antagonists in general in the treatment of such diseases, and suggest the possibility that developing an understanding of the molecular basis of IL-1 signaling may enable us to develop a new generation of therapeutic agents that block interactions in these pathways.

REFERENCES

Akahoshi, T., Oppenheim, J. J., & Matsushima, K. (1988). Induction of high-affinity interleukin 1 receptor on human peripheral blood lymphocytes by glucocorticoid hormones. J. Exp. Med. 167, 924–936.

Alcami, A., & Smith, G. L. (1992). A soluble receptor for interleukin-1 β encoded by vaccinia virus: A novel mechanism of virus modulation of the host response to infection. Cell 71, 153–167.

Anderson, K. V., Bokla, L., & Nusslein-Volhard, C. (1985a). Establishment of dorsal–ventral polarity in the *Drosophila* embryo: The induction of polarity by the Toll gene product. Cell 42, 791–798.

Anderson, K. V., Jurgens, G., & Nusslein-Volhard, C. (1985b). Establishment of dorsal–ventral polarity in the *Drosophila* embryo: Genetic studies on the role of the Toll gene product. Cell 42, 779–789.

Arend, W. P., Welgus, H. G., Thompson, R. C., & Eisenberg, S. P. (1990). Biological properties of recombinant human monocyte-derived interleukin 1 receptor antagonist. J. Clin. Invest. 85, 1694–1697.

Arend, W. P., Malyak, M., Bigler, C. F., Smith, M. J., & Janson, R. W. (1991). The biological role of naturally-occurring cytokine inhibitors. Br. J. Rheumatol. 2, 49–52.

Beg, A. A., & Baldwin, A. J. (1993). The IκB proteins: Multifunctional regulators of Rel/NF-κB transcription factors. Genes Dev. 7, 2064–2070.

Beg, A. A., Finco, T. S., Nantermet, P. V., & Baldwin, A. J. (1993). Tumor necrosis factor and interleukin-1 lead to phosphorylation and loss of IκBα: A mechanism for NF-kappa B activation. Mol. Cell. Biol. 13, 3301–3310.

Benjamin, D., & Dower, S. K. (1990). B-cells express two types of IL-1 receptor. Blood 75, 2017–2023.

Benjamin, D., Wormsley, S., & Dower, S. K. (1990). Heterogeneity in interleukin (IL)-1 receptors expressed on human B cell lines. Differences in the molecular properties of IL-1α and IL-1β binding sites. J. Biol. Chem. 265, 9943–9951.

Bensimon, C., Wakasugi, N., Tagaya, Y., Takakura, K., Yodoi, J., Tursz, T., & Wakasugi, H. (1989). Two distinct affinity binding site for IL-1 on human cell lines. J. Immunol. 143, 1168–1174.

Bergers, G., Reikerstorfer, A., Braselmann, S., Graninger, P., & Busslinger, M. (1994). Alternative promoter usage of the Fos-responsive gene Fit-1 generates mRNA isoforms coding for either secreted or membrane-bound proteins related to the IL-1 receptor. Embo. J. 13, 1176–1188.

Bird, T. A., & Saklatvala, J. (1986). Identification of a common class of high affinity receptors for both types of porcine interleukin-1 on connective tissue cells. Nature 324, 263–266.

Bird, T. A., & Saklatvala, J. (1990). Down-modulation of epidermal growth factor receptor affinity in fibroblasts treated with interleukin-1 or tumor necrosis factor is associated with phosphorylation at a site other than threonine 654. J. Biol. Chem. 265, 235–240.

Bird, T. A., Schule, H. D., Delaney, P. B., Sims, J. E., Thoma, B., & Dower, S. K. (1992a). Evidence that MAP (mitogen-activated protein) kinase activation may be a necessary but not sufficient signal for a restricted subset of responses in IL-1-treated epidermoid cells. Cytokine 4, 429–440.

Bird, T. A., Sleath, P. R., deRoos, P. C., Dower, S. K., & Virca, G. D. (1992b). Interleukin-1 represents a new modality for the activation of extracellular signal-regulated kinases/microtubule-associated protein- 2 kinases. J. Biol. Chem. 267, 2065–2072.

Bird, T. A., Schule, H. D., Delaney, P., de Roos, P., Sleath, P., Dower, S. K., & Virca, G. D. (1994a). The interleukin-1-stimulated protein kinase that phosphorylates heat shock protein hsp27 is activated by MAP kinase. FEBS Lett. 338, 31–36.

Bird, T. A., Kyriakis, J. M., Tyshler, L., Gayle, M., Milne, A., & Virca, G. D. (1994b). Interleukin-1 activates p54 MAP kinase/stress activated protein kinase by a pathway which is independent of p21ras, RAF-1, and MEK. J. Biol. Chem. 269, 31838–31844.

Black, R. A., Kronheim, S. R., Merriam, J. E., March, C. J., & Hopp, T. P. (1989a). A pre-aspartate protease from human leukocytes that cleaves pro-interleukin-1β. J. Biol. Chem. 264, 5323–5326.

Black, R. A., Kronheim, S. R., & Sleath, P. R. (1989b). Activation of IL-1β by a co-induced protease. FEBS Lett. 247, 386–390.

Bomsztyk, K., Sims, J. E., Stanton, T. H., Slack, J., McMahan, C. J., Valentine, M. A., & Dower, S. K. (1989). Evidence for different interleukin 1 receptors in murine B- and T- cell lines. Proc. Natl. Acad. Sci. USA 86, 8034–8038.

Bourdon, M. A. (1990). Structure and Role of Cloned Small Proteoglycans. In: *Extracellular Matrix Genes* L. J. a. B. (Sandell, C.D., ed.) pp. 157–174. Academic Press Inc., San Diego.

Cerretti, D. P., Kozlosky, C. J., Mosley, B., Nelson, N., Van Ness, K., Greenstreet, T. A., March, C. J., Kronheim, S. R., Druck, T., Cannizzaro, L. A., Heubuer, K., & Blade, R. A. (1992). Molecular cloning of the interleukin-1β converting enzyme. Science 256, 97–100.

Chasan, R., Jin, Y., & Anderson, K. V. (1992). Activation of the easter zymogen is regulated by five other genes to define dorsal–ventral polarity in the *Drosophila* embryo. Development 115, 607–616.

Chizzonite, R., Truitt, T., Kilian, P. L., Stern, A. S., Nunues, P., Parker, K. P., Kaffka, K. L., Chua, A. O., Lugg, D. K., & Gubler, U. (1989). Two high affinity interleukin-1 receptors represent separate gene products. Proc. Natl. Acad. Sci. USA 86, 8029–8033.

Chuang, E., Barnard, D., Hettich, L., Zhang, X. F., Avruch, J., & Marshall, M. S. (1994). Critical binding and regulatory interactions between Ras and Raf occur through a small, stable N-terminal domain of Raf and specific Ras effector residues. Mol. Cell. Biol. 14, 5318–5325.

Colotta, F., Re, F., Muzio, M., Bertini, R., Polentarutti, N., Sironi, M., Giri, J. G., Dower, S. K., Sims, J. E., & Mantovani, A. (1993a). Interleukin-1 type II receptor: A decoy target for IL-1 that is regulated by IL-4. Science 261, 472–475.

Colotta, F., Sironi, M., Borre, A., Pollicino, T., Bernasconi, S., Boraschi, D., & Mantovani, A. (1993b). Type II interleukin-1 receptor is not expressed in cultured endothelial cells and is not involved in endothelial cell activation. Blood 81, 1347–1351.

Colotta, F., Re, F., Muzio, M., Polentarutti, N., Minty, A., Caput, D., Ferrara, P., & Mantovani, A. (1994). Interleukin-13 induces expression and release of interleukin-1 decoy receptor in human polymorphonuclear cells. J. Biol. Chem. 269, 12403–12406.

Cook, S. J., & McCormick, F. (1993). Inhibition by cAMP of Ras-dependent activation of Raf. Science 262, 1069–1072.

Corbett, J. A., Kwon, G., Misko, T. P., Rodi, C. P., & McDaniel, M. L. (1994). Tyrosine kinase involvement in IL-1β-induced expression of iNOS by β-cells purified from islets of Langerhans. Am. J. Physiol. 267, C48–54.

Cosman, D. (1993). The hematopoietin receptor superfamily. Cytokine 5, 95–106.

Cosman, D., Lyman, S., Izerda, R., Beckmann, M. P., Park, L. S., Goodwin, R., & March, C. (1990). A new cytokine receptor superfamily. Trends. Biochem. Sci. 15, 265–270.

Coyne, D. W., & Morrison, A. R. (1990). Effect of the tyrosine kinase inhibitor, genistein, on interleukin-1 stimulated PGE2 production in mesangial cells. Biochem. Biophys. Res. Commun. 173, 718–724.

Curtis, B. M., Gallis, B., Overell, R. O., McMahan, C. J., DeRoos, P., Ireland, R., Eisenman, J., Dower, S. K., & Sims, J. E. (1989). T cell interleukin-1 receptor expressed in Chinese hamster ovary cells regulates functional responses to interleukin-1. Proc. Natl. Acad. Sci. USA 86, 3045–3049.

Darnell, J., Jr., Kerr, I. M., & Stark, G. R. (1994). Jak-STAT pathways and transcriptional activation in response to IFNs and other extracellular signaling proteins. Science 264, 1415–1421.

Dinarello, C. A. (1984). Interleukin-1. Rev. Infect. Dis. 6, 51–95.

Dinarello, C. A. (1989). Interleukin-1 and its biologically related cytokines. Adv. Immunol. 44, 153–205.

Dower, S. K., & Sims, J. E. (1990a). Molecular characterisation of cytokine receptors. Ann. Rheum. Dis. 49, 452–459.

Dower, S. K., & Sims, J. E. (1990b). Structure and function of murine and human interleukin-1 receptors. In: *Cellular and Molecular Mechanisms of Inflammation* (Cochrane, C. G., & Gimbrone, M., eds.) pp. 137–172. Academic Press Inc., San Diego.

Dower, S. K., & Sims, J. E. (1992). Interleukin-1 receptor. In: *Encyclopedia of Immunology* (Roitt, I. M., & Delves, P. J., eds.) pp. 901–903. Academic Press Inc., San Diego.

Dower, S. K., Kronheim, S., March, C. J., Hopp, T., Conlon, P. J., Gillis, S., & Urdal, D. L. (1985). Detection and characterisation of high affinity plasma membrane receptors for Interleukin-1. J. Exp. Med. 162, 501–515.

Dower, S. K., Call, S. M., Gillis, S., & Urdal, D. L. (1986a). Similarity between the interleukin-1 receptors on a murine T-lymphoma cell line and on a murine fibroblast cell line. Proc. Natl. Acad. Sci. USA 83, 1060–1065.

Dower, S. K., Kronheim, S., Hopp, T. P., Cantrell, M., Deeley, M., Henney, C. S., Gillis, S., & Urdal, D. L. (1986b). The cell surface receptors for interleukin-1α and interleukin-1β are identical. Nature 324, 266–268.

Dower, S. K., Wignall, J., Schooley, K., McMahan, C. J., Jackson, J., Prickett, K. S., Lupton, S., Cosman, D., & Sims, J. E. (1989). Retention of ligand binding activity by the extracellular domain of the IL-1 receptor. J. Immunol. 142, 4314–4320.

Dower, S. K., Fanslow, W. C., Slack, J., Bird, T. A., & Sims, J. E. (1992a). Interleukin-1 receptors. In: *Therapeutic Proteins: Pharmacokinetics and Pharmacodynamics* (Kung, A. H. C., Baughman, R. A., & Larrick, J. W., eds.) pp. 201–237. W.H. Freeman and Co.: New York.

Dower, S. K., Sims, J. E., Cerretti, D. P., & Bird, T. A. (1992b). The interleukin-1 system: receptors, ligands and signals. In: *Interleukins: Molecular Biology and Immunology* (Ishizaka, K., Lachman, P. J., Lerner, R., & Waksman, B. H., eds.) pp. 33–64. Basel, S. Karger.

Dower, S. K., Bird, T. A., & Sims, J. E. (1992c). Interleukin 1. Adv. Neuroimmunol. 2, 1–16.

Downward, J., Yarden, Y., Mayes, E., Scrace, G., Totty, N., Stockwell, P., Ullrich, A., Schlessinger, J., & Waterfield, M. D. (1984). Close similarity of epidermal growth factor receptor and v-erb-B oncogene protein sequences. Nature 307, 521–527.

Dripps, D. J., Brandhuber, B. J., Thompson, R. C., & Eisenberg, S. P. (1991). Interleukin-1 (IL-1) receptor antagonist binds to the 80-kDa IL-1 receptor but does not initiate IL-1 signal transduction. J. Biol. Chem. 266, 10331–10336.

Eisenberg, S. P., Evans, R. J., Arend, W. P., Verderber, E., Brewer, M. T., Hannum, C. H., & Thompson, R. C. (1990). Primary structure and functional expression from complemenatary DNA of a human interleukin-1 receptor antagonist. Nature 343, 341–345.

Eisenberg, S. P., Brewer, M. T., Verderber, E., Heimdal, P., Brandhuber, B. J., & Thompson, R. C. (1991). Interleukin 1 receptor antagonist is a member of the gene family: Evolution of a cytokine control mechanism. Proc. Natl. Acad. Sci. USA 88, 5232–5236.

Firestein, G. S., Boyle, D. L., Yu, C., Paine, M. M., Whisenand, T. D., Zvaifler, N. J., & Arend, W. P. (1994). Synovial interleukin-1 receptor antagonist and interleukin-1 balance in rheumatoid arthritis. Arthritis. Rheum. 37, 644–652.

Gay, N., & Keith, F. (1991). *Drosphila* Toll and the IL-1 receptor. Nature 351, 355–356.

Gayle, M. A., Slack, J. L., Bonnert, T. P., Renshaw, B. R., Sonoda, G., Taguchi, T., Testa, J. R., Dower, S. K., & Sims, J. E. (1996). Cloning of a putative ligand for the T1/ST2 receptor. J. Biol. Chem. 271(10), 5784–5789.

Gearing, D. P., & Ziegler, S. F. (1993). The hematopoietic growth factor receptor family. Curr. Opin. Hematol. 1993, 19–25.

Gearing, D. P., Thut, C. J., VandenBos, T., Gimpel, S. D., Delaney, P. B., King, J., Price, V., Larsen, A., Cosman, D., & Beckmann, M. P. (1992a). Expression cloning of a receptor for the leukaemia inhibitory factor. In: *Molecular Biology of Haematopoiesis. Proceedings of the International Conference on Molecular Biology of Haematopoiesis, July 1991, Innsbruck, Austria* (Abraham, N. G., Konwalinka, G., Marks, P., Sachs, L., & Tavassoli, M., eds.) pp. 131–136. Itercept Ltd., Andover.

Gearing, D. P., VandenBos, T., Beckmann, M. P., Thut, C. J., Comeau, M. R., Mosley, B., & Ziegler, S. F. (1992b). Reconstitution of high affinity leukaemia inhibitory factor (LIF) receptors in haemopoietic cells transfected with the cloned human LIF receptor. In: *Polyfunctional Cytokines: IL-6 and LIF* (Bock, G. R., Marsh, J., & Widdows, K., eds.) pp. 245–255. John Wiley, & Sons, Chichester.

Geisler, R., Bergmann, A., Hiromi, Y., & Nusslein-Volhard, C. (1992). Cactus, a gene involved in dorsoventral pattern formation of *Drosophila*, is related to the IκB gene family of vertebrates. Cell 71, 613–621.

Ghosh, S., Gifford, A. M., Riviere, L. R., Tempst, P., Nolan, G. P., & Baltimore, D. (1990). Cloning of the p50 DNA binding subunit of NF-κ B: Homology to rel and dorsal. Cell 62, 1019–1029.

Gillespie, S. K., & Wasserman, S. A. (1994). Dorsal, a *Drosophila* Rel-like protein, is phosphorylated upon activation of the transmembrane protein Toll. Mol. Cell. Biol. 14, 3559–3568.

Giri, J. G., Newton, R. C., & Horuk, R. (1990). Identification of soluble interleukin-1 binding protein in cell free supernatants. Evidence for soluble interleukin-1 receptor. J. Biol. Chem. 265, 17416–17419.

Giri, J. G., Slack, J., Jackson, J., Gayle, M., McMahan, C. E., Dower, S. K., & Sims, J. E. (1994). Proinflammatory mediators induce rapid release of a soluble IL-1 binding form of the type II IL-1 receptor both *in vitro* and *in vivo*. J. Immunol. 153, 5802–5809.

Gooding, L. R. (1992). Virus proteins that counteract host immune defenses. Cell 71, 5–7.

Guesdon, F., & Saklatvala, J. (1991). Identification of a cytoplasmic protein kinase regulated by IL-1 that phosphorylates the small heat shock protein, hsp27. J. Biol. Chem. 266, 14343–14352.

Guesdon, F., Freshney, N., Waller, R. J., Rawlinson, L., & Saklatvala, J. (1990). Interleukin 1 and tumor necrosis factor stimulate two novel protein kinases that phosphorylate the heat shock protein hsp27 and β-casein. J. Biol. Chem. 265, 3146–3152.

Hallberg, B., Rayter, S. I., & Downward, J. (1994). Interaction of Ras and Raf in intact mammalian cells upon extracellular stimulation. J. Biol. Chem. 269, 3913–3916.

Hanks, S. K., Quinn, A. M., & Hunter, T. (1988). The protein kinase family: Conserved features and deduced phylogeny of the catalytic domains. Science 241, 42–52.

Hannum, C. H., Wilcox, C. J., Arend, W. P., Joslin, F. G., Dripps, D. J., Heimdal, P. L., Armes, L. G., Sommer, A., Eisenberg, S. P., & Thompson, R. C. (1990). Interleukin-1 receptor antagonist activity of a human IL-1 receptor inhibitor. Nature 343, 336–340.

Haskill, S., Martin, G., Van Le, L., Morris, J., Peace, A., Bigler, C., Jaffe, G. J., Hammerberg, C., Sporn, S., Fong, S., Arend, W. P., & Ralph, P. (1991). cDNA cloning of an intracellular form of the human interleukin-1 receptor antagonist associated with epithelium. Proc. Natl. Acad. Sci. USA 88, 3681–3685.

Hatakeyama, M., Minamoto, S., Uchiyama, T., Hardy, R. R., Yamada, G., & Taniguchi, T. (1985). Reconstitution of functional receptor for human interleukin-2 in mouse cells. Nature 318, 467–470.

Hatakeyama, M., Tsudo, M., Minamoto, S., Kono, T., Doi, T., Miyata, T., Miyasaka, M., & Taniguchi, T. (1989). Interleukin-2 receptor beta chain gene: Generation of three receptor forms by cloned human α and β chain cDNA's. Science 244, 551–556.

Hecht, P. M., & Anderson, K. V. (1993). Genetic characterization of tube and pelle, genes required for signaling between Toll and dorsal in the specification of the dorsal–ventral pattern of the *Drosophila* embryo. Genetics 135, 405–417.

Heguy, A., Baldari, C. T., Macchia, G., Telford, J. L., & Melli, M. (1992). Amino acids conserved in interleukin-1 receptors (IL-1Rs) and the *Drosophila* toll protein are essential for IL-1R signal transduction. J. Biol. Chem. 267, 2605–2609.

Henkel, T., Machleidt, T., Alkalay, I., Kronke, M., Ben, N. Y., & Baeuerle, P. A. (1993). Rapid proteolysis of IκB-α is necessary for activation of transcription factor NF-κB. Nature 365, 182–185.

Hibi, M., Murakami, M., Saito, M., Hirano, T., Taga, T., & Kishimoto, T. (1990). Molecular cloning and expression of an IL-6 signal transducer, gp130. Cell 63, 1149–1157.

Hultmark, D. (1994). Macrophage differentiation marker MyD88 is a member of the Toll/IL-1 receptor family. Biochem. Biophys. Res. Commun. 199, 144–146.

Iwasaki, T., Uehara, Y., Graves, L., Rachie, N., & Bomsztyk, K. (1992). Herbimycin A blocks IL-1-induced NF-κB DNA-binding activity in lymphoid cell lines. FEBS Lett. 298, 240–244.

Jin, Y. S., & Anderson, K. V. (1990). Dominant and recessive alleles of the *Drosophila* easter gene are point mutations at conserved sites in the serine protease catalytic domain. Cell 60, 873–881.

Johnson, G. L., & Vaillancourt, R. R. (1994). Sequential protein kinase reactions controlling cell growth and differentiation. Curr. Opin. Cell. Biol. 6, 230–238.

Johnston, J. A., Kawamura, M., Kirken, R. A., Chen, Y. Q., Blake, T. B., Shibuya, K., Ortaldo, J. R., McVicar, D. W., & O'Shea, J. J. (1994). Phosphorylation and activation of the Jak-3 Janus kinase in response to interleukin-2. Nature 370, 151–153.

Ju, G., Labriola-Thompkins, E., Campen, C. A., Benjamin, W. R., Karas, J., Plocinski, J., Biondi, D., Kaffka, K. L., Kilian, P. L., Eisenberg, S. P., & Evans, R. J. (1991). Conversion of the interleukin-1 receptor antagonist to an agonist by site specific mutagenesis. Proc. Natl. Acad. Sci. USA 88, 2658–2662.

Katabami, T., Shimizu, M., Okano, K., Yano, Y., Nemoto, K., Ogura, M., Tsukamoto, T., Suzuki, S., Ohira, K., & Yamada, Y. (1993). Intracellular signal transduction for interleukin-1 β-induced endothelin production in human umbilical vein endothelial cells. Science 259, 519–522.

Kaur, P., Welch, W. J., & Saklatvala, J. (1989). Interleukin-1 and tumour necrosis factor increase protein phosphorylation of the small heat-shock protein. FEBS Lett. 258, 269–273.

Kilian, P. L., Kaffka, K. L., Stern, A. S., Woehle, D., Benjamin, W. R., Dechiara, T. M., Gubler, U., Farrar, J. J., Mizel, S. B., & Lomedico, P. T. (1986). Interleukin-1α and Interleukin-1β bind to the same receptor on T cells. J. Immunol. 136, 4509–4513.

Kilian, P. L., Stern, A. S., Gubler, U., & Chizzonite, R. A. (1990). Characterization of mouse IL-1 receptors: Evidence for IL-1 receptor heterogeneity. Prog. Clin. Biol. Res. 349, 251–257.

Kishimoto, T. (1992). *Interleukins: Molecular Biology and Immunology*. Karger Basel, Munchen, Paris and London.

Kobayashi, Y., Yamamoto, K., Saido, T., Kawasaki, H., Oppenheim, J. J., & Matsushima, K. (1990). Identification of calcium-activated neutral protease as a processing enzyme of human interleukin 1 α. Proc. Natl. Acad. Sci. USA 87, 5548–5552.

Kronheim, S. R., Mumma, A., Greenstreet, T., Glackin, P. J., Van Ness, K., March, C. J., & Black, R. A. (1992). Purification of interleukin-1β converting enzyme, the protease that cleaves the interleukin-1β precursor. Arch. Biochem. Biophys. 296, 698.

Kubota, K., Keith, F. J., & Gay, N. J. (1993). Relocalization of *Drosophila* dorsal protein can be induced by a rise in cytoplasmic calcium concentration and the expression of constitutively active but not wild-type Toll receptors. Biochem. J. 296, 497–503.

Kuno, K., Okamoto, S., Hirose, K., Murakami, S., & Matsushima, K. (1993). Structure and function of the intracellular portion of the mouse interleukin 1 receptor (type I). Determining the essential region for transducing signals to activate the interleukin 8 gene. J. Biol. Chem. 268, 13510–13518.

Kyriakis, J. M., Brautigan, D. L., Ingebritsen, T. S., & Avruch, J. (1991). pp54 microtubule-associated protein-2 kinase requires both tyrosine and serine/threonine phosphorylation for activity. J. Biol. Chem. 266, 10043–10046.

Kyriakis, J. M., App, H., Zhang, X. F., Banerjee, P., Brautigan, D. L., Rapp, U. R., & Avruch, J. (1992). Raf-1 activates MAP kinase-kinase. Nature 358, 417–421.

Kyriakis, J. M., Banerjee, P., Nikolakaki, E., Dai, T., Rubie, E. A., Ahmad, M. F., Avruch, J., & Woodgett, J. R. (1994). The stress-activated protein kinase subfamily of c-Jun kinases. Nature 369, 156–160.

Leevers, S. J., Paterson, H. F., & Marshall, C. J. (1994). Requirement for Ras in Raf activation is overcome by targeting Raf to the plasma membrane. Nature 369, 411–414.

Leung, K., Betts, J. C., Xu, L., & Nabel, G. J. (1994). The cytoplasmic domain of the interleukin-1 receptor is required for nuclear factor-κ B signal transduction. J. Biol. Chem. 269, 1579–1582.

Liou, H. C., & Baltimore, D. (1993). Regulation of the NF-κ B/rel transcription factor and I κ B inhibitor system. Curr. Opin. Cell. Biol. 5, 477–487.

Lord, K. A., Hoffman-Liebermann, B., & Liebermann, D. A. (1990). Nucleotide sequence and expression of a cDNA encoding MyD88, a novel myeloid differentiation primary response gene induced by IL6. Oncogene 5, 1095–1097.

Lowenstein, E. J., Daly, R. J., Batzer, A. G., Li, W., Margolis, B., Lammers, R., Ullrich, A., Skolnik, E. Y., Bar-Sagi, D., & Schlessinger, J. (1992). The SH2 and SH3 domain-containing protein GRB2 links receptor tyrosine kinases to ras signaling. Cell 70, 431–442.

Lowenthal, J. W., & MacDonald, H. R. (1986). Binding and internalisation of interleukin-1 by T cells; direct evidence for high- and low-affinity classes of interleukin-1 receptor. J. Exp. Med. 164, 1060–1074.

Manser, E., Leung, T., Salihuddin, H., Zhao, Z. S., & Lim, L. (1994). A brain serine/threonine protein kinase activated by Cdc42 and Rac1. Nature 367, 40–46.

March, C. J., Mosley, B., Larsen, A., Cerretti, D. P., Braedt, G., Price, V., Gillis, S., Henney, C. S., Kronheim, S., Grabstein, K., Conlon, P. J., Hopp, T. P., & Cosman, D. C. (1985). Cloning, sequence and expression of two distinct human interleukin-1 complementary DNAs. Nature 315, 641–647.

Marczin, N., Papapetropoulos, A., & Catravas, J. D. (1993). Tyrosine kinase inhibitors suppress endotoxin- and IL-1β-induced NO synthesis in aortic smooth muscle cells. Am. J. Physiol. 265, H1014–1018.

Martin, M., Bol, G. F., Eriksson, A., Resch, K., & Brigelius-Flohe, R. (1994). Interleukin-1-induced activation of a protein kinase co-precipitating with the type I interleukin-1 receptor in T cells. Eur. J. Immunol. 24, 1566–1571.

Matsushima, K., Akahoshi, T., Yamada, M., Furutani, Y., & Oppenheim, J. J. (1986). Properties of a specific interleukin-1 receptor on human Epstein–Barr virus transformed B lymphocytes: Identity of the receptors for IL-1α and IL-1β. J. Immunol. 136, 4496–4508.

McMahan, C. J., Slack, J. L., Mosley, B., Cosman, D., Lupton, S. D., Brunton, L. L., Grubin, C. E., Huebner, K., Croce, C. M., Cannizzano, L. A., Benjamin, D., Dower, S. K., Spriggs, M. K., & Sims, J. E. (1991). A novel IL-1 receptor, cloned from B cells by mammalian expression is expressed on many cell types. EMBO J. 10, 2821–2832.

Moodie, S. A., Willumsen, B. M., Weber, M. J., & Wolfman, A. (1993). Complexes of Ras.GTP with Raf-1 and mitogen-activated protein kinase kinase. Science 260, 1658–1661.

Morisato, D., & Anderson, K. V. (1994). The spatzle gene encodes a component of the extracellular signaling pathway establishing the dorsal–ventral pattern of the *Drosophila* embryo. Cell 76, 677–688.

Mosley, B., Urdal, D. L., Prickett, K., Larsen, A., Cosman, D., Conlon, P. J., Gillis, S., & Dower, S. K. (1987). The IL-1 receptor binds the human interleukin-1α precursor but not the human interleukin-1β precursor. J. Biol. Chem. 262, 2941–2944.

Munoz, E., Zubiaga, A. M., Sims, J. E., & Huber, B. T. (1991). IL-1 signal transduction pathways. I. Two functional IL-1 receptors are expressed in T cells. J. Immunol. 146, 136–143.

Munoz, E., Zubiaga, A., Huang, C., & Huber, B. T. (1992). Interleukin-1 induces protein tyrosine phosphorylation in T cells. Eur. J. Immunol. 22, 1391–1396.

Narazaki, M., Witthuhn, B. A., Yoshida, K., Silvennoinen, O., Yasukawa, K., Ihle, J. N., Kishimoto, T., & Taga, T. (1994). Activation of JAK2 kinase mediated by the interleukin 6 signal transducer gp130. Proc. Natl. Acad. Sci. USA 91, 2285–2289.

Nicholson, S. E., Oates, A. C., Harpur, A. G., Ziemiecki, A., Wilks, A. F., & Layton, J. E. (1994). Tyrosine kinase JAK1 is associated with the granulocyte-colony-stimulating factor receptor and both become tyrosine-phosphorylated after receptor activation. Proc. Natl. Acad. Sci. USA 91, 2985–2988.

Norris, J. L., & Manley, J. L. (1992). Selective nuclear transport of the *Drosophila* morphogen dorsal can be established by a signaling pathway involving the transmembrane protein Toll and protein kinase A. Genes Dev. 6, 1654–1667.

Ohlsson, K., Bjork, P., Bergenfeldt, M., Hageman, R., & Thompson, R. C. (1990). Interleukin-1 receptor antagonist reduces mortality from endotoxin shock. Nature 348, 550–551.

Oppenheim, J. J., Kovacs, E. J., Matsushima, K., & Durum, S. K. (1986). There is more than one interleukin-1. Immunol. Today 7, 45–55.

Orencole, S. F., & Dinarello, C. A. (1989). Characterisation of a subclone (D10S) of the D10.G4.1 helper T-cell line which proliferates to attomolar concentrations of interleukin-1 in the absence of mitogens. Cytokine 1, 14–19.

Payne, D. M., Rossomando, A. J., Martino, P., Erickson, A. K., Her, J. H., Shabanowitz, J., Hunt, D. F., Weber, M. J., & Sturgill, T. W. (1991). Identification of the regulatory phosphorylation sites in pp42/mitogen-activated protein kinase (MAP kinase). EMBO J. 10, 885–892.

Rosoff, P. M., Savage, N., & Dinarello, C. A. (1988). Interleukin-1 stimulates diacylglycerol production in T lymphocytes by a novel mechanism. Cell 54, 8–81.

Roth, N. S., Lefkowitz, R. J., & Caron, M. G. (1991). Structure and function of the adrenergic receptor family. Adv. Exp. Med. Biol. 308, 223–238.

Savage, N., Puren, A. J., Orencole, S. F., Ikejima, T., Clark, B. D., & Dinarello, C. A. (1989). Studies on IL-1 receptors on D10s T helper cells: Demonstration of two molecularly and antigenically distinct IL-1 binding proteins. Cytokine 1, 23–25.

Schneider, D. S., Jin, Y., Morisato, D., & Anderson, K. V. (1994). A processed form of the Spatzle protein defines dorsal–ventral polarity in the *Drosophila* embryo. Development 120, 1243–1250.

Shelton, C. A., & Wasserman, S. A. (1993). pelle encodes a protein kinase required to establish dorsoventral polarity in the *Drosophila* embryo. Cell 72, 515–525.

Shirakawa, F., & Mizel, S. B. (1989). *In vitro* activation and nuclear translocation of NFκB catalyzed by cyclic-AMP dependant protein kinase and protein kinase C. Mol. Cell. Biol. 9, 2424–2430.

Shirakawa, F., Chedid, M., Suttles, J., Pollok, B., & Mizel, S. B. (1989). Interleukin-1 and cAMP induce κ immunoglobulin light chain expression via activation of an NF-κB-like DNA binding protein. Mol. Cell. Biol. 9, 959–964.

Sims, J. E., March, C. J., Cosman, D., Widmer, M. B., MacDonald, H. R., McMahan, C. J., Grubin, C. E., Wignall, J. M., Call, S. M., Friend, D., Alpert, A. R., Gillis, S., Urdal, D. L., & Dower, S. K. (1988). cDNA expression cloning of the IL-1 receptor, a member of the immunoglobulin superfamily. Science 241, 585–589.

Sims, J. E., Acres, R. B., Grubin, C. E., McMahan, C. J., Wignall, J. M., March, C. J., & Dower, S. K. (1989). Cloning of the interleukin-1 receptor from human T-cells. Proc. Natl. Acad. Sci. USA 86, 8946–8950.

Sims, J. E., Gayle, M. A., Slack, J. L., Alderson, M. R., Bird, T. A., Giri, J. G., Colotta, F., Re, F., Mantovani, A., Shanebeck, K., Grabstein, K. H., & Dower, S. K. (1993). Interleukin 1 signaling occurs exclusively via the type I receptor. Proc. Natl. Acad. Sci. USA 90, 6155–6159.

Sims, J. E., Giri, J. G., & Dower, S. K. (1994). The two interleukin-1 receptors play different roles in IL-1 actions. Clin. Immunol. Immunopathol. 72, 9–14.

Singer, I. I., Scott, S., Hall, G.L., Limjuco, G., Chin, J., & Schmidt, J.A. (1988). Interleukin-1β is localized in the cytoplasmic ground substance but is largely absent from the Golgi apparatus and plasma membranes of stimulated human monocytes. J. Exp. Med. 167, 389–407.

Slack, J., Sims, J. E., Pitt, A. M., & Dower, S. K. (1989). Application of the Mutiscreen system to cytokine radioreceptor assays. BioTechniques 7, 1132–1139.

Slack, J., McMahan, C. J., Waugh, S., Schooley, K., Spriggs, M. K., Sims, J. E., & Dower, S. K. (1993). Independent binding of interleukin-1α and interleukin-1β to type I and type II IL-1 receptors. J. Biol. Chem. 268, 2513–2524.

Spriggs, M. K., Lioubin, P. J., Slack, J., Dower, S. K., Janas, U., Cosman, D., Sims, J. E., & Bauer, J. (1990). Induction of IL-1 receptor on monocytic cells. Evidence that the IL-1 receptor is not encoded by a T-cell type IL-1 receptor message. J. Biol. Chem. 265, 22499–22505.

Spriggs, M. K., Hruby, D. E., Maliszewski, C. R., Pickup, D. J., Sims, J. E., Buller, R. M., & VanSlyke, J. (1992a). Vaccinia and cowpox viruses encode a novel secreted interleukin-1- binding protein. Cell 71, 145–152.

Spriggs, M. K., Nevens, P. J., Grabstein, K., Dower, S. K., Cosman, D., Armitage, R. J., McMahan, C. J., & Sims, J. E. (1992b). Molecular characterization of the interleukin-1 receptor (IL-1R) on monocytes and polymorphonuclear cells. Cytokine 4, 90–95.

Stahl, N., Boulton, T. G., Farruggella, T., Ip, N. Y., Davis, S., Witthuhn, B. A., Quelle, F. W., Silvennoinen, O., Barbieri, G., Pellegrini, S., & et al. (1994). Association and activation of Jak–Tyk kinases by CNTF–LIF–OSM–IL-6β receptor components. Science 263, 92–95.

Steward, R. (1987). Dorsal, an embryonic polarity gene in *Drosophila*, is homologous to the vertebrate proto-oncogene, c-rel. Science 238, 692–694.

Stokoe, D., Macdonald, S. G., Cadwallader, K., Symons, M., & Hancock, J. F. (1994). Activation of Raf as a result of recruitment to the plasma membrane. Science 264, 1463–1467.

Stylianou, E., O'Neill, L. A., Rawlinson, L., Edbrooke, M. R., Woo, P., & Saklatvala, J. (1992). Interleukin 1 induces NF-κB through its type I but not its type II receptor in lymphocytes. J. Biol. Chem. 267, 15836–15841.

Symons, J. A., Eastgate, J. A., & Duff, G. W. (1990). A soluble binding protein specific for IL-1β is produced by activated mononuclear cells. Cytokine 2, 190–194.

Symons, J. A., Eastgate, J. A., & Duff, G. W. (1991). Purification and characterization of a novel soluble receptor for interleukin 1. J. Exp. Med. 174, 1251–1254.

Takeshita, T., Asao, H., Ohtani, K., Ishii, N., Kumaki, S., Tanaka, N., Munakata, H., Nakamura, M., & Sugamura, K. (1992). Cloning of the γ Chain of the human IL-2 Receptor. Science 257, 379–382.

Thornberry, N. A., Bull, H. G., Calaycay, J. R., Chapman, K. T., Howard, A. D., Kostura, M. J., Miller, D. K., Molineaux, S. M., Weidner, J. R., Aunins, J., Elliston, K. O., Ayala, J. M., Casano, F. J., Chin, J., Ding, G. J.-F., Egger, L. A., Gaffney, E. P., Limjuco, G., Palyha, O., Raju, S. M., Rolando, A. M., Salley, J. P., Yamin, T. T., Lee, T. D., Shiveley, J. E., MacCross, M., Mumford, R. A., Schmidt, J. A., & Tocci, M. J. (1992). A novel heterodimeric cysteine protease is required for interleukin-1β processing in monocytes. Nature 356, 768–774.

Tominaga, S. (1989). A putative protein of a growth specific cDNA from BALB/c-3T3 cells is highly similar to the extracellular portion of mouse interleukin-1 receptor. FEBS Lett. 258, 301–304.

Tsuda, L., Inoue, Y. H., Yoo, M. A., Mizuno, M., Hata, M., Lim, Y. M., Adachi-Yamada, T., Ryo, H., Masamune, Y., & Nishida, Y. (1993). A protein kinase similar to MAP kinase activator acts downstream of the raf kinase in *Drosophila*. Cell 72, 407–414.

Ullrich, A., Coussens, L., Hayflick, J. S., Dull, T. J., Gray, A., Tam, A. W., Lee, J., Yarden, Y., Libermann, T. A., Schlessinger, J., Downward, J., Mayes, E. L. V., Whittle, N., Waterfeld, M. D., & Seebury,

P. H. (1984). Human epidermal growth factor receptor cDNA sequence and aberrant expression of the amplified gene in A431 epidermoid carcinoma cells. Nature 309, 418–425.

Ullrich, A., Bell, J. R., Chen, E. Y., Herrera, R., Petruzzelli, L. M., Dull, T. J., Gray, A., Coussens, L., Liao, Y. C., Tsubokawa, M., Mason, A., Seebury, P. H., Greenfeld, C., Rosen, O. H., & Ramcheuleu, J. (1985). Human insulin receptor and its relationship to the tyrosine kinase family of oncogenes. Nature 313, 756–761.

Urdal, D. L., Call, S. M., Jackson, J. J., & Dower, S. K. (1988). Affinity purification and chemical analysis of the interleukin-1 receptor. J. Biol. Chem. 263, 2870–2877.

Vojtek, A. B., Hollenberg, S. M., & Cooper, J. A. (1993). Mammalian Ras interacts directly with the serine/threonine kinase Raf. Cell 74, 205–214.

Wakabayashi, G., Gelfand, J. A., Burke, J. F., Thompson, R. C., & Dinarello, C. A. (1991). A specific receptor antagonist for interleukin-1 prevents *Escherichia coli*-induced shock in rabbits. FASEB J. 5, 338–343.

Warne, P. H., Viciana, P. R., & Downward, J. (1993). Direct interaction of Ras and the amino-terminal region of Raf-1 *in vitro*. Nature 364, 352–355.

Witthuhn, B. A., Quelle, F. W., Silvennoinen, O., Yi, T., Tang, B., Miura, O., & Ihle, J. N. (1993). JAK2 associates with the erythropoietin receptor and is tyrosine phosphorylated and activated following stimulation with erythropoietin. Cell 74, 227–236.

Wood, K. W., Sarnecki, C., Roberts, T. M., & Blenis, J. (1992). ras mediates nerve growth factor receptor modulation of three signal-transducing protein kinases: MAP kinase, Raf-1, and RSK. Cell 68, 1041–1050.

Yamagata, M., Merlie, J. P., & Sanes, J. R. (1994). Interspecific comparisons reveal conserved features of the *Drosophila* Toll protein. Gene 139, 223–228.

Yanagisawa, K., Takagi, T., Tsukamoto, T., Tetsuka, T., & Tominaga, S. (1993). Presence of a novel primary response gene ST2L, encoding a product highly similar to the interleukin 1 receptor type 1. FEBS Lett. 318, 83–87.

Yarden, Y., Escobedo, J. A., Kuang, W. J., Yang, F. T., Daniel, T. O., Tremble, P. M., Chen, E. Y., Ando, M. E., Harkins, R. N., Francke, U., Fried, V. A., Ullrich, A., & Wiliams, L. T. (1986). Structure of the receptor for platelet-derived growth factor helps define a family of closely related growth factor receptors. Nature 323, 226–232.

Zhang, X. F., Settleman, J., Kyriakis, J. M., Takeuchi-Suzuki, E., Elledge, S. J., Marshall, M. S., Bruder, J. T., Rapp, U. R., & Avruch, J. (1993). Normal and oncogenic p21ras proteins bind to the amino-terminal regulatory domain of c-Raf-1. Nature 364, 308–313.

NERVE GROWTH FACTOR RECEPTORS

Ralph A. Bradshaw and Hubert Hondermarck

I. INTRODUCTION

Nerve growth factor (NGF) is a member of a large group of secreted substances that function through cell-surface receptors (Sporn and Roberts, 1990). As such, these substances can be generally designated as hormones (Bradshaw and Ruben, 1980), although it is often helpful to subdivide them, usually along functional lines.

Biomembranes
Volume 6, pages 177–196.

ISBN: 1-55938-663-0.

Thus, designations such as tissue growth factor, neurotrophic factor, and cytokine provide useful categorizations (Pimental, 1994). However, the activities that separate these classes are often obscure and the categories can be substantially overlapping. Therefore, it is always useful to remember that the overall purpose of hormones, as a class, is to provide intercellular communication (even when the cells are of the same type); along with neural transmission, this is an essential feature of polycellular organisms (although the origins of this phenomenon can clearly be traced to unicellular species; Luporini et al., 1994).

A. Nerve Growth Factor: Functional Properties

NGF is primarily classed as a neurotrophic factor. It was initially identified through its trophic stimulation of sympathetic nerves and was subsequently shown to interact in a similar fashion with a subset of dorsal root sensory neurons (Levi-Montalcini and Angeletti, 1968). Somewhat later, cholinergic neurons of the central nervous system were also found to be responsive to NGF, albeit the nature of the response is not entirely the same as with the peripheral neurons (Longo et al., 1993). In all cases, there is a clear stimulation of cellular metabolism with defined changes in specific gene expression (Altin and Bradshaw, 1993). In addition, neurite proliferation can be demonstrated in culture and *in situ* with the peripheral neuronal targets and *in situ* with some CNS tracks (Bothwell, 1995). It is also clear that it prevents neuronal cell death by suppressing apoptosis (Deckwerth and Johnson, 1993). Quite distinct from its neuronal functions, NGF has also been shown to interact with a variety of non-neuronal cell types (Bothwell, 1995). These include mast cells, keratinocytes, Sertoli cells, and pancreatic islet cells among others. Although these observations are less well defined in terms of their physiological relevance, they may be no less important than the neuronal interactions.

B. Nerve Growth Factor: Chemical Properties

NGF, like many other hormones and growth factors, is a member of a well-defined family of similar neurotrophic factors, now called the neurotrophins. These include brain-derived neurotrophic factor (BDNF), neurotrophin 3 (NT3), and neurotrophin 4/5 (NT4/5) (Bradshaw et al., 1993; Figure 1). The activities of the other neurotrophins have only been primarily defined in terms of neuronal responses, but they too have non-neuronal targets and more may eventually be described (Bothwell, 1995). The family shares approximately 50% amino acid sequence similarity and clearly enjoys a homologous relationship (Bradshaw et al., 1994). Interestingly, this group is a part of a larger superfamily that has been defined on the basis of the principal structural motif characterizing the three-dimensional structure of NGF (and, by prediction, the other neurotrophins as well; Murray-Rust et al., 1993). This structural unit is characterized by a disulfide knot in which the three intrachain links have a well-defined relationship. Two of the disulfides form a loop through which

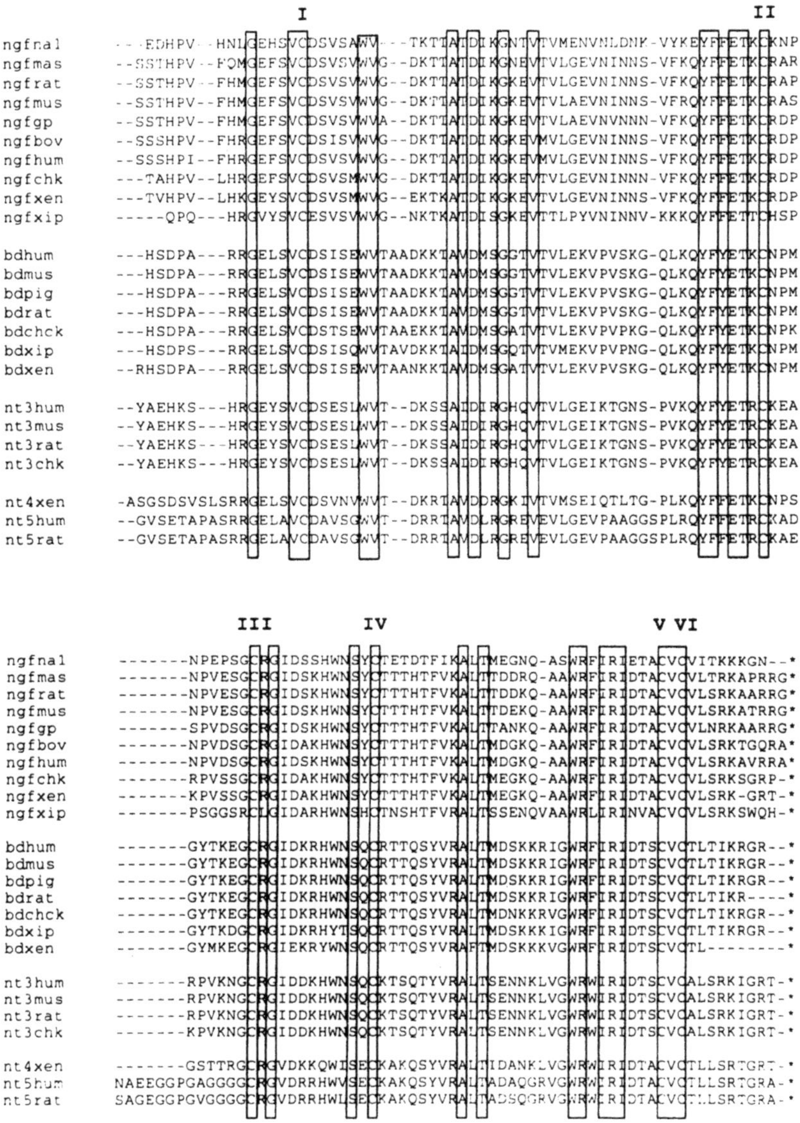

Figure 1. Sequence alignment of neurotrophins. Data extracted from the OWL database version 23.0 (Bleasby and Wotten, 1990). Residues conserved through all the sequences listed here are boxed; the Cys residues which form the disulfide bond cluster are numbered I–VI. Sequence codes and literature citations for the neurotrophins are cited in Bradshaw et al., 1994. (Used with permission from Cambridge University Press.)

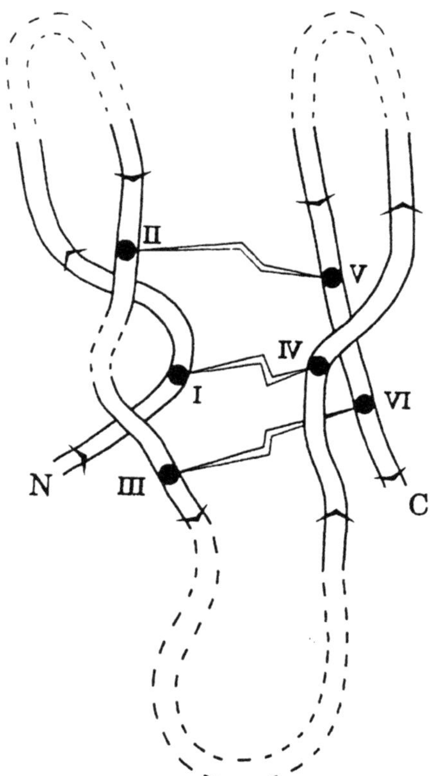

Figure 2. Schematic diagram of the disulfide knot topology. Arrows indicate the directions of the strands and dotted lines show links of variable length. The half-cystine residues involved in the knot are numbered I to VI in order of their occurrence in the sequence. (Taken from Murray-Rust et al., 1993, with permission.)

the third passes. In addition, all of the members of this family show extensions of four strands of β-pleated sheet which project in the same direction away from the cystine knot (Figure 2). This motif has now been found to occur in the family of transforming growth factor-β (TGFβ), platelet-derived growth factor (PDGF) and, most recently, the glycopeptide hormones (Lapthorn et al., 1994). It also occurs as a structural domain in some membrane-bound proteins, but the significance of this observation is not yet known. Clearly, this has been an important motif in the development of a large number of extracellular signaling substances.

The three-dimensional structure of NGF (McDonald et al., 1991), as noted above, has provided important insights into the structure/function properties of the molecule. In addition to revealing the fundamental structural core (the cystine knot–β-sheet motif), it has identified the organization of the subunits in the noncovalently-linked homodimer that characterizes the active entity (Angeletti et al., 1971) and has provided important clues with respect to residues and other structural elements that interact with the various NGF receptors (Bradshaw et al., 1994). Most importantly, studies using site-directed mutagenesis and chemical modification have pinpointed several regions of the molecule which, in 3-dimensional juxtaposition, form a face that is likely to contain the main sites of contact between NGF and its receptors. Interestingly, the best definition of these binding sites suggests that residues from both polypeptides contribute to interaction with a single receptor molecule (Ibanez et al., 1993). Like many hormones and growth factors, it is not known whether more than one ligand binds to the receptor dimer, which is required for the initiation of response (Ullrich and Schlessinger, 1990; see following). As the active form of NGF is a dimer, it appears likely that it interacts in a symmetrical fashion with two receptor polypeptides simultaneously ensuring the dimeric structure necessary for the initiation of signal transduction (Jing et al., 1992). This may also apply to its interaction with the low molecular weight receptor as well (see following).

II. NGF RECEPTORS

Over the past several years, it has become increasingly clear that there are two major signal transduction mechanisms employed by hormones and other extracellular messengers, albeit that there are many variations on each theme. One system utilizes receptors that interface with G-proteins (proteins that bind GTP), which in turn are linked to various effector molecules that produce the initial intracellular signal (Bourne et al., 1991). In contrast to this group are the substances that activate receptor-associated kinase effectors (Ullrich and Schlessinger, 1990). These are usually tyrosine-specific and, in some cases, they are found as the intracellular domain of the receptor, whereas in other cases, they are separate entities that must associate with the receptor (ligand binding) molecule to produce the initial tyrosine phosphorylation events (Fantl et al., 1993; Stahl et al., 1995). NGF belongs to the first category (usual denoted receptor tyrosine kinases or RTK), that is, it stimulates tyrosine phosphorylation events through the Trk family of receptors. As such, it can induce a number of intracellular signal cascades involving a variety of secondary effector molecules. The overall importance of these to the physiological responses of NGF are described in a later section. It is important to note that the involvement of G-protein mechanisms with NGF have not been rigorously eliminated and it has been suggested that another receptor form (termed low molecular weight receptor or low-affinity receptor and abbreviated, LNGFR or p75) has a consensus binding sequence for such molecules (Feinstein and Larhammar, 1990). However, there is otherwise no compelling evidence to suggest that the G-protein-connected effector systems are directly responsible for NGF transmission at the receptor level.

In the ensuing sections, detailed descriptions of the two major receptor classes for NGF (LNGFR and the Trk family) are given. In addition, possible interaction between these two receptor types are discussed and the relationship of each to the signal transduction pathways known to be important for NGF responses are described. It should be emphasized that although NGF functions in a wide variety of neuronal and non-neuronal targets, information regarding signal transduction mechanisms arising from either or both receptor groups are relatively limited and are primarily based on the tissue culture paradigm, the rat PC12 cell (Greene and Tischler, 1982).

A. Low Molecular Weight NGF Receptor (LNGFR, p75)

The low molecular weight NGF receptor was the first to be structurally defined and generally occurs in much higher concentrations than the corresponding Trk receptors. It is also found on NGF non-responsive cells, such as Schwann and melanoma cells, that lack any Trk receptors. As such, it was the principal entity recognized in binding experiments, which as in most hormone systems, provided the first molecular information regarding NGF receptors, and this abundance directly lead to structural characterization through cloning experiments (Johnson et al., 1986; Radeke et al., 1987). However, both cross-linking and initial isolation experiments that identified TrkA clearly preceded the characterization of LNGFR (Kouchalakos and Bradshaw, 1986; Massagué et al., 1981).

The amino acid sequences of p75 as deduced from cloning experiments for a number of species, are shown in Figure 3 (Johnson et al., 1986; Radeke et al., 1987; Large et al., 1989). The molecule has a calculated molecular weight of less than 50,000 for all species identified to date, although bound carbohydrate raises the molecular mass, probably by as much as 10- to 15-kD. (On SDS-PAGE, it migrates as a band of ~75 kD, but this value is probably anomalously high.)

Topographically, p75 is organized as a typical type 1 receptor (with a single transmembrane segment). The extracellular domain has four repeating motifs of about 40 residues that are characterized by three disulfide bonds. Several homologous receptors and transmembrane proteins, for example, tumor necrosis factor (TNF) receptors (Banner et al., 1993), have similar domains and by analogy predict a half-cystine pairing of I–II, III–V, and IV–VI. The crystallographic data upon which this is based also predicts that the extracellular domains will exist as elongated (linear) structures. The intracellular domain is poorly conserved relative to other members of the family and offers little information regarding potential signal transducing capabilities. The transmembrane and juxtamembrane regions, in contrast, show a high percentage identity and may be important in interactions with other membrane-bound proteins (such as the Trk family; see following).

Evidence that p75 can induce intracellular signals (in the absence of a Trk receptor) is sparse. p75 has been implicated in both apoptosis (Rabiadeh et al., 1993) and migratory responses in Schwann cells (Anton et al., 1994) and it has been demonstrated to be involved in the sphingomyelin pathway (Dobrowsky et al., 1994). It also appears to bind intracellular kinases, but the physiological significance of this observation is unknown as they are apparently not activated by this interaction (Volonte et al., 1993).

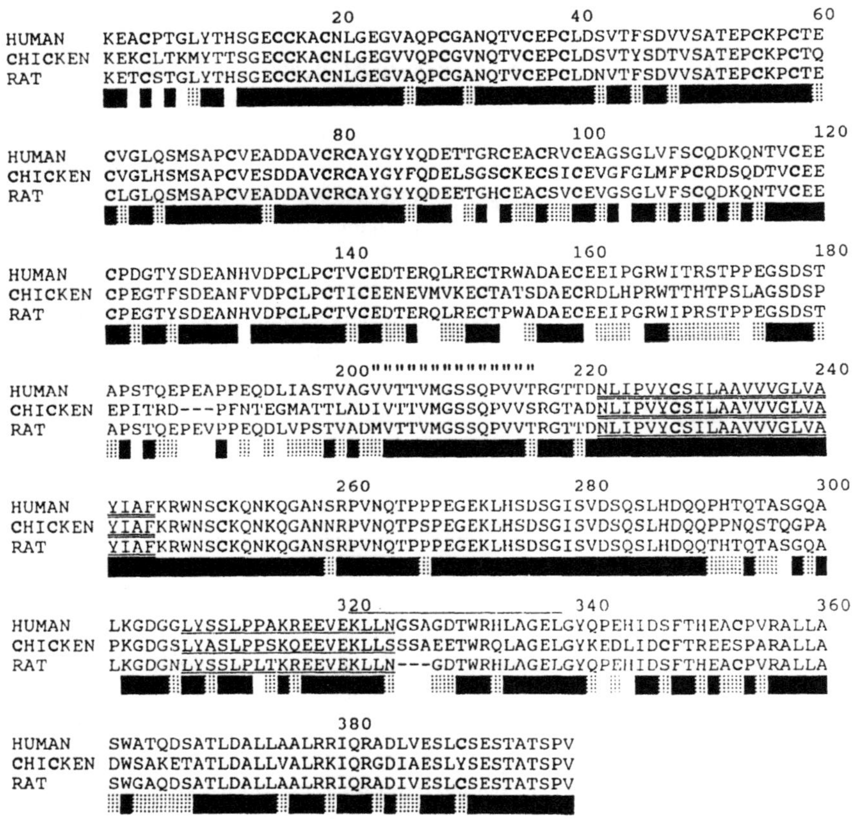

Figure 3. p75 nerve growth factor receptors. Completely conserved residues are indicated by solid square below the sequences, and conservative substitutions are indicated by checkered squares. Cys residues appear in bold type. The probable membrane-spanning region is double underlined, and the hydrophobic region with probable β-strand–turn–β-strand structure is denoted by "" above the sequences. The cytoplasmic region with homology to the proposed iron-binding region of ferritin has a single underline below each of the three sequences, and the region with homology to enzymes that act on phosphorylated substrates is denoted by a single line above the sequences. Sequence from Johnson et al. (1986), Radeke et al. (1987), and Large, et al. (1989). (Reproduced, with permission, from the *Annual Review of Biochemistry*, Vol. 62, © 1993, by Annual Reviews Inc. Raffioni et al., 1993.)

B. Trk Family of Receptors

The second family of plasma membrane-bound receptors, that bind either NGF or other neurotrophins, is the Trk family. There are at present five proposed members of this family. Three of them labeled A[1], B, and C specifically bind NGF or one of the other members of the neurotrophin family. Thus, TrkB binds to and is activated by BDNF and NT4/5 and TrkC rather specifically binds to and is activated by NT3 (Bothwell, 1995). NGF only interacts significantly with TrkA. Thus, the distribution of these receptors, which has been studied extensively in both neuronal and non-neuronal tissue, is a reasonable indicator of responsiveness of individual members of the neurotrophin family with respect to both developing and mature tissues (Meakin and Shooter, 1992). Many of the expectations advanced from *in situ* hybridization and immununological localization experiments have also been confirmed using transgenic ("knockout") animals (Snider, 1994).

The other two proposed members of the Trk family, DTrk and TrkE, are less well defined in terms of both function and ligand specificity. The former has been identified in *Drosophilia* and has been classed as a Trk entity based primarily on sequential similarities to the other members of the Trk family (Pulido et al., 1992). The ligand for this entity has not yet been identified and it is not known whether it represents a homolog to the neurotrophins. Neurotrophin homologs have not yet been identified in such species. TrkE had been described as a putative NGF receptor in human keratinocytes (Di Marco et al., 1993a). It has been suggested to be the entity that is responsible for an autocrine dependence of keratinocytes on NGF during certain stages of their development (Di Marco et al., 1993b) and as such, could function in other non-neuronal cells of epithelial origin, many of which have been suggested to show various NGF responsiveness. However, as with Dtrk, the extracellular domain of this receptor is entirely dissimilar to the other members of the Trk family, being rather characterized by a discoidin-like domain, and it has not yet been rigorously demonstrated to specifically bind NGF with high affinity. Its inclusion in this group is also based on sequence similarity and organization of the intracellular domain. If TrkE is eventually established to be a specific NGF receptor like TrkA and p75, it will mean that three entirely different NGF-binding domains will have evolved apparently independently. It may be rationalized that an independent Trk entity for neuronal vs. non-neuronal cells could have important implications, particularly during development, that might explain this seeming redundancy. However, such hypotheses are best held until further detailed analyses with regard to ligand specificity have been carried out.

All of the Trk family members (A, B, and C) are composed of extracellular binding domains that are, in the usual fashion, characterized by various recognizable motifs (based on structural features) and include ample glycosylation. The intracellular domain, separated from the ligand-binding domain by a single transmembrane segment, contains, in each case, a tyrosine kinase which when stimulated

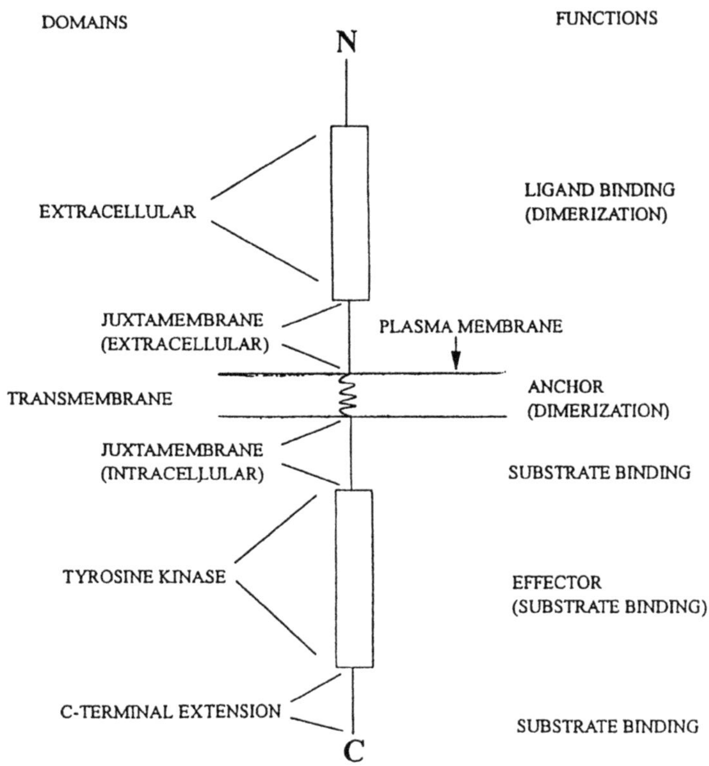

Figure 4. Tyrosine kinase-containing receptors (RTK). (Taken from Bradshaw (1995) with permission from Springer-Verlag, Heidelberg, © 1995.)

following ligand binding and protomer dimerization leads to various autophosphorylation events (assumed to occur as the result of intermolecular catalysis between the two kinases of the dimer structure) (Ullrich and Schlessinger, 1990). The kinase entity is flanked by a juxta-membrane region and a short C-terminal extension. In TrkA, these provide sites of substrate interaction (via phosphorylated tyrosine residues in the TrkA sequence) that are clearly of major importance to its signal transduction capacity (Figure 4). There is also a short insert in the kinase domain, similar to that found in other RTKs, but its significance in the Trk family is unknown. Overall, there is almost 70% sequence identity among these members of the Trk family.

The three principal members of the Trk family are also synthesized in isomeric forms, arising from alternate mRNA splicing events, although apparently not as extensively as in the fibroblast growth factor receptor family (Jaye et al., 1992). For TrkA, only two forms, differing only by a short insert in the juxtamembrane region on the extracellular side, are known. However, with Trk B and C, more extensive alterations, resulting from truncations in the intracellular region, have been observed that are clearly compromised with respect to signal transduction capability (Bothwell, 1995). These forms have been suggested to be dominant–negative regulators (by forming heterodimers with full-length receptors and preventing cross-phosphorylation and hence, activation) or "reservoirs" of ligand that would help to sustain high concentrations of neurotrophin at the cell surface.

C. Trk–p75 Interactions

Although p75 was the first NGF receptor to be defined in molecular terms (Johnson et al., 1986; Radeke et al., 1987), it was appreciated from other studies, such as equilibrium binding, cross-linking, and direct isolation, that the receptor situation on responsive cells was complex and not readily interpreted with only a single molecular species (Bothwell, 1995; Raffioni et al., 1993). Indeed, clear evidence suggested that a higher molecular weight form existed (Kouchalakos and Bradshaw, 1986; Massagué et al., 1981); nonetheless, the discovery of TrkA was still viewed with some surprise since it was generally assumed that the higher molecular weight form was generated by association of p75 with another intracellular (or possibly transmembrane) effector. The most unexpected aspect was the realization that TrkA could "function" by itself, thus apparently eliminating a role for p75, at least in the direct signaling pathways induced. This enigma has still not been fully resolved. Clearly, transfection experiments, placing TrkA in a non-neuronal context as well as normal cells expressing TrkA in the absence of p75, have demonstrated its ability to induce a variety of tyrosine kinase-dependent pathways and responses (Barbacid, 1994; Bothwell, 1995; Meakin and Shooter, 1992; Raffioni et al., 1993) and chimeric receptors, containing the transmembrane and intracellular domains of TrkA that presumably do not interact with p75 (because no NGF ligand is present), can induce normal responses in an NGF-responsive paradigm (PC12 cells; Obermeier et al., 1993, 1994). In addition, antibodies that block binding of NGF to p75 do not interfere with NGF responses in PC12 cells (Weskamp and Reichardt, 1991) and responsive neurons and NGF-derivatives that do not bind to p75, but still stimulate TrkA, produce normal NGF responses (Ibanez et al., 1992). Finally, experiments to detect direct interactions between p75 and TrkA, such as cross-linking or co-immunoprecipitation, have proven unsuccessful (Bothwell, 1995).

What is, then, the role of p75 and does it exert any effect on TrkA and TrkA-induced signals in the presence of NGF? A number of possibilities exist, other than the proposed functions described above, that p75 may perform in the absence of a Trk receptor, but these depend almost entirely on indirect evidence. In the main, they suggest a modulating activity for p75 that can *affect* TrkA-binding specificity and responses without necessarily inducing signals directly (Bothwell, 1995; Raffioni et al., 1993). This view is supported by transgenic animals that do not express p75, which show both sympathetic and sensory deficits and losses in nocioceptive function (Lee et al., 1992).

It is accepted that the influence of p75 on NGF responses results from p75–TrkA interactions, no matter how transient they may be; the most likely model would be a p75–TrkA heterodimer formed by the mutual binding of an NGF dimer (the normal structure for the ligand) by one molecule of each receptor. Homodimers of p75 and of TrkA, each with bound ligand, might also form, but only the latter would presumably be functional as a signaling entity. It has been suggested that a p75–TrkA heterodimer may be an intermediate (although obviously not an obligatory one) in the formation of NGF–TrkA homodimer complexes (Bothwell, 1995). From the present knowledge of the structure of NGF (McDonald et al., 1991) and the proposed sites of interaction of NGF with TrkA and p75 (Ibanez et al., 1993), it does not seem likely that a hetero-tetrameric complex, composed of p75 and TrkA dimers, can form bound to a single NGF homodimer.

In a recent review, Bothwell (1995) has suggested an intriguing alternative model that postulates interactions between the homodimers of p75–NGF and TrkA–NGF. Thus, the interactions of ligand would be with each receptor-type (either with preformed dimers or ones resulting from the NGF binding) and only *occupied* receptors (of each type) would bind. In this regard, it has been observed that phosphorylation of the intracellular domain prevents dimerization and this may be of regulatory significance (Grob et al., 1985). The principal features of this model are that NGF (as well as other neurotrophins) binds p75 in a positive cooperative manner (NGF is less potent than BDNF and NT3 in this regard) and that the conformation induced (for the p75–NGF bound state) in turn induces the dimerization of Trk, thus facilitating the binding of NGF to the main signal generating moiety. In this model, NGF can also bind directly to and induce the dimerization of TrkA independently of p75. Among other things, this model provides an explanation for the long-debated controversy concerning the molecular nature of the "high" and "low" affinity receptors for NGF (Hempstead et al., 1991; Klein et al., 1991) and suggests that p75 will be more important in cells with lower numbers of TrkA receptors, such as sensory neurons, where it will increase the effective affinity of the TrkA present. Although, this model has yet to be rigorously proven, it is attractive from the point of view that it potentially resolves many existing conflicts in the NGF receptor field.

III. SIGNAL TRANSDUCTION

Most information on NGF-induced signal transduction comes from studies on PC12 cells (Greene and Tischler, 1982). In these cells, NGF has been found to induce rapid tyrosine phosphorylation by activation of the tyrosine kinase activity of TrkA (Kaplan et al., 1991a, b; Klein et al., 1991). Consequently, TrkA phosphorylates several substrate signaling proteins including phospholipase Cγ (PLCγ), phosphatidylinositol-3 kinase (PI-3K), and Shc (Raffioni and Bradshaw, 1992; Kim et al., 1991; Vetter et al., 1991; Rozakis-Adcock et al., 1992; Soltoff et al., 1992; Borrello et al., 1994). These modifications are accomplished through the specific interaction of the candidate substrates with the phosphotyrosines of TrkA formed by autophosphorylation. The formation of these interaction sites presumably occurs by intermolecular catalysis (between the two protomers of the dimer). However, it has not been ruled out that cytoplasmic kinases, such as $pp60^{src}$, do not participate in at least some of these reactions. This is particularly germane in the case of NGF responses in PC12 cells where an obligatory requirement for the Src protein has been demonstrated (Kremer et al., 1991).

The interaction of signaling substrates occurs primarily through SH2 domains, protein structures that recognize and bind to Tyr (PO_4)–X–X–X sequences (the nature of the X moieties provides the individual specificities; Pawson, 1994). These structures are named for the prototype which occurs as a subdomain of $pp60^{src}$ and, hence the name, Src homology (SH2). Another domain of $pp60^{src}$, SH3, also participates in interactions between signaling molecules, but does not directly involve phosphotyrosine residues.

The binding and subsequent phosphorylation of the signaling substrates leads either to direct activation or to the initiation of further interactions that result eventually in the activation of one or more downstream kinases. The more complex pathways provide a greater opportunity for amplification and regulation.

A. TrkA Pathways

The association of TrkA with PLCγ is mediated through interaction with a tyrosine residue (Tyr 785) in the carboxyl-terminal extension (Kim et al., 1991; Vetter et al., 1991; Obermeier et al., 1993). This enzyme cleaves phosphinositides (PI) to the corresponding diacylglycerols (DAG) and tri-inositol phosphates. The DAG formed apparently acts primarily to stimulate certain isozymes of protein kinase C (PKC) while the inositol phosphates act on receptors on the endoplasmic reticulum to release intracellular calcium ions. However, DAG assays strongly suggest that the activation of PLCγ induced by NGF is transient (Pessin et al., 1991) and down-regulation of PKC by phorbol esters does not inhibit the neurotrophic effect of NGF (Reinhold and Neet, 1989) suggesting that PLCγ is not essential for neurite outgrowth. Indeed, chimeric receptors with a Tyr→Phe mutation in this position readily promote neurite outgrowth on PC12 cells (Obermeier et al., 1993, 1994).

PI-3K is tyrosine phosphorylated on its 85 kD subunit after association with TrkA (Raffioni and Bradshaw, 1992; Soltoff et al., 1992). Association of PI-3K has been reported to occur by interaction with Tyr751 of TrkA and conversion of this residue to phenylalanine is also without effect on neurite outgrowth in PC12 cells (Obermeier et al., 1994). This is in keeping with the pattern of stimulation of this enzyme by non-neurotrophic agents, such as EGF (Raffioni et al., 1993). PI-3K is one of two isozymes that catalyze phosphorylations of the inositol ring in PIs. The

3′-substituted derivatives are not cleared by PLCγ and apparently act directly as second messengers although their role in signal transduction is not well understood in either trophic or mitotic processes.

The third major pathway induced by NGF–TrkA complexes involves the activation of $p21^{Ras}$ (Hagag et al., 1986; Altin et al., 1991b), a GTP-binding protein known to play a central role in signal transduction in a wide variety of cells (Lowy and Willumsen, 1993). This entity acts as a molecular switch and controls a downstream cascade of kinases that eventually activate a group of pre-existing transcription factors leading ultimately to the modulations in phenotypic profile that are characteristic of the response induced. In PC12 cells, this pathway is activated by the binding and tyrosine phosphorylation of Shc, an effector protein that, in turn, binds grb2 and SOS, two additional linking proteins. The latter specifically binds and stabilizes the activated form of Ras (with GTP bound; Qiu and Green, 1991; Muroya et al., 1992). All of these proteins apparently assemble in a single complex on the receptor intracellular domain via Tyr-490, the Shc binding site (Obermeier et al., 1994). The activated Ras protein then initiates the kinase cascade by binding a Raf kinase which, in turn, activates mitogen-activated protein kinase kinase (MAPKK) which activates mitogen-activated protein kinase (MAPK) (Boulton et al., 1991; Oshima et al., 1991; Dent et al., 1992) (Figure 5). These involve phosphorylations of the serine/threonine type (and, in the last case, tyrosine phosphorylation as well.) The phosphorylation of MAPK not only leads to its activation, but also to its translocation to the nucleus where additional phosphorylation/activation events occur. An example is the activation of S6 kinase II (Scimeca et al., 1992). This rather complex and yet incompletely defined pathway of protein phosphorylation ultimately must trigger the information in the nucleus that induces specific gene expression and protein synthesis. This is manifested initially in immediate early gene expression following NGF stimulation, as exemplified by increases of Myc, Fos, Jun, Src, as well as other TIS gene transcripts (Greenberg et al., 1985; Kujubu et al., 1987; Wu et al., 1989; Altin et al., 1991a). However, these early changes in gene expressions are not specific to NGF since they also occur with other neurotrophic, as well as mitogenic factors, such as FGF or EGF (Altin et al., 1991a). Using high-resolution 2D electrophoresis of cells labeled with [^{35}S]-methionine, it has been shown that NGF and FGF induce early changes in protein synthesis in PC12 cells that are quite similar (Hondermarck et al., 1994). By comparison with the modifications induced by EGF, several neurotrophic-specific proteins have been detected, providing evidence for molecular specificity of the differentiative versus proliferative pathways.

B. Proliferative versus Differentiative Responses

One of the more intriguing questions remaining to be resolved is the molecular bases for signaling events that produce mitotic responses as opposed to trophic (often differentiative) ones (Fantl et al., 1993). PC12 cells provide an excellent

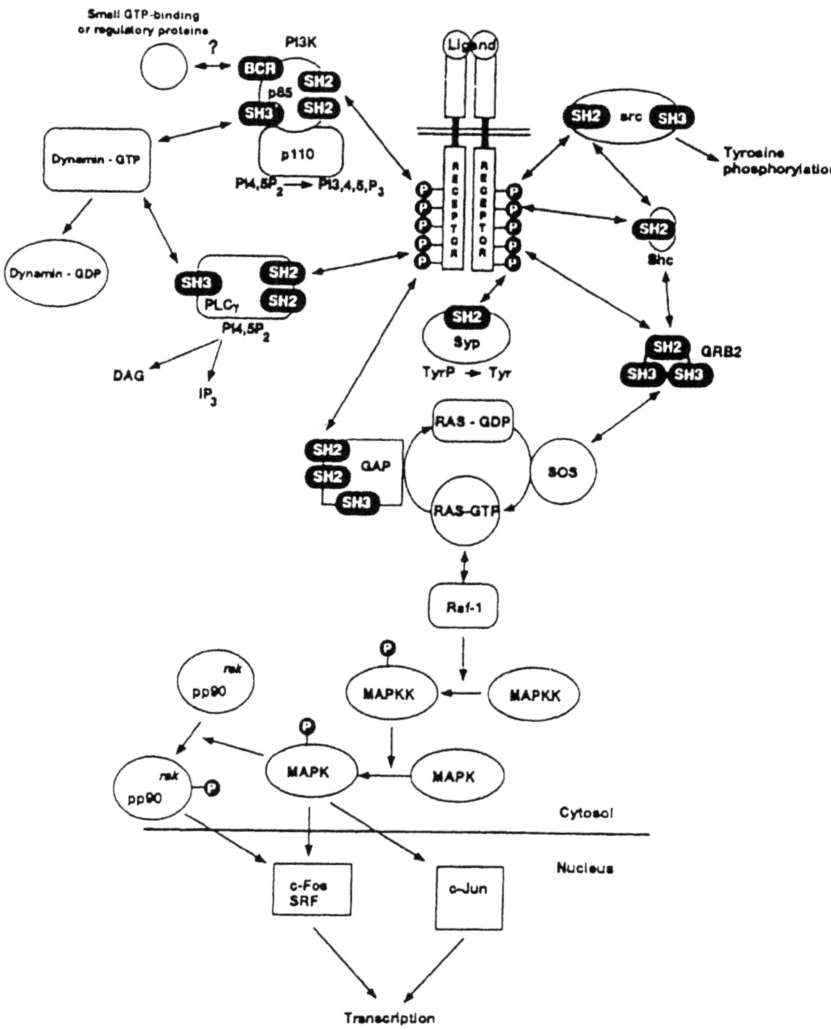
Small GTP-binding
or regulatory proteins
?
PI3K
BCR
SH2
p85
SH2
SH3
p110
PI4,5P2 → PI3,4,5,P3
Dynamin - GTP
Dynamin - GDP
SH3
SH2
SH2
PLCγ
PI4,5P2
DAG
IP3
Ligand
RECEPTOR
RECEPTOR
P
SH2
src
SH3
Tyrosine
phosphorylation
SH2
Shc
SH2
Syp
TyrP → Tyr
SH2
GRB2
SH3
SH3
RAS - GDP
SH2
SH2
GAP
SH3
SOS
RAS-GTP
Raf-1
MAPKK
MAPKK
rsk
pp90
MAPK
MAPK
rsk
pp90
Cytosol
Nucleus
c-Fos
SRF
c-Jun
Transcription

opportunity to dissect this question since NGF (as well as FGF) are trophic while EGF is normally only weakly hyperplastic. All utilize tyrosine kinase-containing receptors. More importantly, all appear to have the capacity to induce the same signaling pathways.

Recent results have provided important clues to explain the phenotypic differences. Most importantly, they relate the phenotypic differences to the strength and duration of the initial signal produced by the ligand–receptor complex. A variety of experiments have shown that EGF lacks neurotrophic activity not because of qualitative differences, but because it fails to sustain the activation of MAPK (and its translocation to the nucleus; Traverse et al., 1992). This apparently results from the rapid down-regulation (via ligand-induced endocytosis) of the occupied receptor, an event that is greatly stimulated by the phosphorylation of two serine residues in the C-terminal extension (Theroux et al., 1992). Alterations or conditions that retain high concentrations of EGF receptor on the cell surface invariably lead to the induction of neurite outgrowth (by EGF; Raffioni and Bradshaw, 1995). These observations strongly support the view that the "state of differentiation" of the target cell will govern its response to external signals and that the same pathways can be utilized to different physiological ends depending on the context in which they are induced.

IV. CONCLUSIONS

NGF recognition by specific cell surface receptors is provided by at least two independent entities. These differ in both structure and function. TrkA clearly induces tyrosine phosphorylations that are important in many responses, including differentiation in PC12 cells, and is likely the principal effector in neuronally-responsive cells. TrkE may serve a similar function in non-neuronal cells, albeit

Figure 5. The signal transduction pathways, mediated by SH2 and SH3. The many signal outputs that can potentially be achieved by a single class of activated PTK receptor. The initial receptor signaling molecule interactions are mediated by SH2-domain phosphorylated interactions, and the subsequent steps are mediated either by further protein–protein interactions involving SH2 and SH3 domain interactions, or by second messenger molecules generated by the primary interacting SH2 domain-containing proteins, e.g., phospholipase Cγ (PLCγ) and phosphatidylinositol (PI)-3 kinase (PI3K). The recently described pathway mediating the flow of signaling information from an activated RTK receptor via Ras to the nucleus is included. Two-headed arrows indicate direct protein–protein interactions, whereas single-headed arrows indicate enzymatic transformations. GTP, guanosine triphosphate; GDP, guanosine diphosphate; SOS, son of sevenless; MAPK, MAPK kinase; DAG, diacylglycerol; IP_3, inositol triphosphate; PI4,5P_2, phosphoinositide-4, 5-biphosphate; PI3,4,5P_3, phosphoinoside-3, 4, 5-triphosphate; SRF, somatotrophin-releasing factor. (Taken from Fry et al., 1993 with permission from Oxford University Press.)

that its rigorous identity as a NGF-specific receptor remains to be satisfied. p75, although more abundant and widely distributed, is less well understood in physiological terms. However, recent evidence supports the view that it can induce some intracellular signals and it likely plays a role in modulating TrkA activity (Bothwell, 1995). However, such effects seem more related to overall cellular responses than they are to changes in molecular events, i.e., modification of signal transduction cascades. More general, non-signaling roles, have also been suggested. In the sense that the same signaling pathway may be utilized in different cells to different ends, it is also likely that other pathways, not directly induced by NGF via TrkA, may produce NGF-like responses. A PC12 cell variant (PC12-E2), responds to IL-6 in a manner quite similar to NGF (native PC12 cells are generally refractive to IL-6), but it does not activate the Ras–MAPK pathway. It is likely that a pathway involving STAT proteins, transcription factors that are activated in the cytoplasm by receptor-mediated phosphorylations and translocated to the nucleus, are involved (Wu and Bradshaw, 1995). This emphasizes the multiplicity of mechanisms available to accomplish similar biological phenomena. It is in keeping with the view that NGF (as well as the other related neurotrophins) participates in many biological responses and has generated at least two principal receptor entities to accomplish these.

NOTE

1. TrkA, the first member of the family to be identified is often referred to, primarily in the older literature, simply as Trk. However, more recent use has generally adopted the designation A to avoid confusion with the other family members.

REFERENCES

Altin, J. G., & Bradshaw, R. A. (1993). Nerve growth factor and related substances: structure and mechanism of action. In: Neurotrophic Factors (Loughlin, S. E. & Fallon, J. H., Eds.) pp. 129–180. Academic Press, San Diego.

Altin, J. G., Kujubu, D. A., Raffioni, S., Eveleth, D. D., Herschman, H. R., & Bradshaw, R. A. (1991a). Differential induction of primary-response (TIS) genes in PC12 pheochromocytoma cells and the unresponsive variant PC12nnr5. J. Biol. Chem. 266, 5401–5406.

Altin, J. G., Wetts, R., & Bradshaw, R. A. (1991b). Microinjection of a p21ras antibody into PC12 cells inhibits neurite outgrowth induced by nerve growth factor and basic fibroblast growth factor. Growth Factors 4, 145–155.

Angeletti, R. H., Bradshaw, R. A., & Wade, R. D. (1971). Subunit structure and amino acid composition of mouse submaxillary gland nerve growth factor. Biochemistry 10, 463–469.

Anton, E. S., Weskamp, G., Reichardt, L. F., & Matthew, W. D. (1994). Nerve growth factor and its low-affinity receptor promote Schwann cell migration. Proc. Natl. Acad. Sci. USA 91, 2795–2799.

Banner, D. W., D'Arey, A., Janes, W., Gentz, R., Schoenfeld, H.-J., Broger, C., Loetscher, H., & Lesslauer, W. (1993). Crystal structure of the soluble human 55 kD TNF receptor–human TNFβ complex: Implications for TNF receptor activation. Cell 73, 431–445.

Barbacid, M. (1994). The Trk family of neurotrophin receptors. J. Neurobiol. 25, 1386–1403.

Bleasby, A. J., & Wotten, J. C. (1990). Construction of validated, nonredundant composite protein-sequence databases. Protein Eng. 3, 153–159.

Borrello, M. G., Pelicci, G., Arighi, E., De Filippis, L., Greco, A., Bongarzone, I., Rizzetti, M. G., Pelicci, P. G., & Pierotti, M. A. (1994). The oncogenic versions of the Ret and Trk tyrosine kinases bind Shc and Grb2 adaptor proteins. Oncogene 9, 1661–1668.

Bothwell, M. (1995). Functional interactions of neurotrophins and neurotrophin receptors. Ann. Rev. Neurosci. 18, 223–253.

Boulton, T. G., Nye, S. H., Robbins, D. J., Ip, N. Y., Radziejewska, E., Morgenbesser, S. D., DePinho, R. A., Panayotatos, N., Cobb, M. H., & Yancopoulos, G. D. (1991). ERKs: A family of protein-serine/threonine kinases that are activated and tyrosine phosphorylated in response to insulin and NGF. Cell 65, 663–675.

Bourne, H. R., Sanders, D. A., & McCormick, F. (1991). The GTPase superfamily: Conserved structure and molecular mechanism. Nature 349, 117–127.

Bradshaw, R. A. (1996). Growth factors. In: *Comprehensive Human Physiology*: Vol. 1 (Greger, R., Windhorst, U., Mommaerts, W., & Köpchen, H.-P., Eds.) Springer-Verlag, Heidelberg, pp. 431–450.

Bradshaw, R. A., & Rubin, J. S. (1980). Polypeptide growth factors: Some structural and mechanistic considerations. J. Supramolecular Struct. 14, 183–199.

Bradshaw, R. A., Blundell, T. L., Lapatto, R., McDonald, N. Q., & Murray-Rust, J. (1993). Nerve growth factor revisited. Trends in Biochem. Sci. 18, 48–52.

Bradshaw, R. A., Murray-Rust, J., Ibanez, C. F., McDonald, N. Q., Lapatto, R., & Blundell, T. L. (1994). Nerve growth factor: Structure/function relationships. Prot. Sci. 3, 1901–1913.

Deckwerth, T. L., & Johnson, Jr., E. M. (1993). Neurotrophic factor deprivation-induced death. Annals New York Acad. Sci. 679, 121–131.

Dent, P., Haser, W., Haystead, T. A., Vincent, L. A., Roberts, T. M., & Sturgill, T. W. (1992). Activation of mitogen-activated protein kinase kinase by v-Raf in NIH 3T3 cells and *in vitro*. Science 257, 1404–1407.

Di Marco, E., Cutuli, N., Guerra, L., Cancedda, R., & De Luca, M. (1993a). Molecular cloning of trkE, a novel trk-related putative tyrosine kinase receptor isolated from normal human keratinocytes and widely expressed by normal human tissues. J. Biol. Chem. 268, 24290–24295.

Di Marco, E., Mathor, M., Bondanza, S., Cutuli, N., Marchisco, P. C., Cancedda, R., & DeLuca, M. (1993b). Nerve growth factor binds to normal human keratinocytes through high and low affinity receptors and stimulates their growth by a novel autocrine loop. J. Biol. Chem. 268, 22838–22846.

Dobrowsky, R. T., Werner, M. H., Castellino, A. M., Chao, M. V., & Hennan, Y. A. (1994). Activation of the sphingomyclin cycle through the low-affinity neurotrophin receptors. Science 265, 1596–1599.

Fantl, W. J., Johnson, D. E., & Williams, L. T. (1993). Signaling by receptor tyrosine kinases. Ann. Rev. Biochem. 62, 453–481.

Feinstein, D. L., & Larhammar, D. (1990). Identification of a conserved protein motif in a group of growth factor receptors. FEBS Letters, 272, 7–11.

Fry, M. J., Panayotou, G., Booker, G. W., & Waterfield, M. D. (1993). New insights into protein-tyrosine kinase receptor signaling complexes. Protein Sci. 2, 1785–1797.

Greenberg, M. E., Greene, L. A., & Ziff, E. B. (1985). Nerve growth factor and epidermal growth factor induce rapid transient changes in proto-oncogene transcription in PC12 cells. J. Biol. Chem. 260, 14101–14110.

Greene, L. A., & Tischler, A. S. (1982). PC12 pheochromocytoma cultures in neurobiological research. Adv. Cell. Neurobiol. 3, 373–414.

Grob, P. M., Ross, A. H., Koprowski, H., & Bothwell, M. (1985). Characterization of the human melanoma nerve growth factor receptor. J. Biol. Chem. 260, 8044–8049.

Hagag, N., Halegoua, S., & Viola, M. (1986). Inhibition of growth factor-induced differentiation of PC12 cells by microinjection of antibody to ras p21. Nature 319, 680–682.

Hempstead, B. L., Martin-Zanca, D., Kaplan, D. R., Parada, L. F., & Chao, M. V. (1991). High-affinity NGF binding requires coexpression of the Trk proto-oncogene and the low-affinity NGF receptor. Nature 350, 678–683.

Hondermarck, H., McLaughlin, C. S., Patterson, S. D., & Bradshaw, R. A. (1994). Early changes in protein synthesis induced by basic fibroblast growth factor, nerve growth factor, and epidermal growth factor in PC12 pheochromocytoma cells. Proc. Natl. Acad. Sci. USA 91, 9377–9381.

Ibáñez, C. F., Ebendal, T., Barbany, G., Murray-Rust, J., Blundell, T. L., & Persson, H. (1992). Disruption of the low affinity receptor-binding site in NGF allows neuronal survival and differentiation by binding to the trk gene product. Cell 69, 329–341.

Ibáñez, C. F., Ilag, L. L., Murray-Rust, J., & Persson, H. (1993). An extended surface of binding of Trk tyrosine kinase receptors in NGF and BDNF allows the engineering of a multifunctional pan-neurotrophin. EMBO J. 12, 2281–2293.

Jaye, M., Schlessinger, J., & Dionne, C. A. (1992). Fibroblast growth factor receptor tyrosine kinases: Molecular analysis and signal transduction. Biochim. Biophys. Acta. 1135, 185–199.

Jing, S., Tapley, P., & Barbacid, M. (1992). Nerve growth factor mediates signal transduction through trk homodimer receptors. Neuron 9, 1067–1079.

Johnson, D., Lanahan, A., Buck, C. R., Sehgal, A., Morgan, C., Mercer E., Bothwell M., & Chao, M. (1986). Expression and structure of the human NGF receptor. Cell 47, 545–554.

Kaplan, D. R., Hempstead, B. L., Martin-Zanca, D., Chao, M. V., & Parada, L. F. (1991). The trk proto-oncogene product: A signal transducing receptor for nerve growth factor. Science 252, 554–558.

Kaplan, D. R., Martin-Zanca, D., & Parada, L. F. (1991b). Tyrosine phosphorylation and tyrosine kinase activity of the trk proto-oncogene product induced by NGF. Nature 350, 158–160.

Kim, U. H., Fink, Jr., O., Kim, H. S., Park, D. J., Contreras, M. L., Guroff, G., & Rhee, S. G. (1991). Nerve growth factor stimulates phosphorylation of phospholipase C-γ in PC12 cells. J. Biol. Chem. 266, 1359–1362.

Klein, R., Jing, S. Q., Nanduri, V., O'Rourke, E., & Barbacid, M. (1991). The trk proto-oncogene encodes a receptor for nerve growth factor. Cell 65, 189–197.

Kouchalakos, R. N., & Bradshaw, R. A. (1986). Nerve growth factor receptor from rabbit sympathetic ganglia membranes: Relationship between subforms. J. Biol. Chem. 261, 16054–16059.

Kremer, N. E., D'Arcangelo, G., Thomas, S. M., DeMarco, M., Brugge, J. S., & Halegoua, S. (1991). Signal transduction by nerve growth factor and fibroblast growth factor in PC12 cells requires a sequence of src and ras actions. J. Cell Biol. 115, 809–819.

Kujubu, D. A., Lim, R. W., Varnum, B. C., & Herschman, H. R. (1987). Induction of transiently expressed genes in PC12 pheochromocytoma cells. Oncogene, 1, 257–262.

Lapthorn, A. J., Harris, D. C., Littlejohn, A., Lustbader, J. W., Canfield, R. E., Machin, K. J., Morgan, F. J., & Isaacs, N. W. (1994). Crystal structure of human chorionic gonadotropin. Nature 369, 455–461.

Large, T. H., Weskamp, G., Helder, J. C., Radeke, M. J., Misko, T. P., Shooter, E. M., & Reichardt, L. F. (1989). Structure and developmental expression of the nerve growth factor receptor in the chicken central nervous system. Neuron 2, 1123–1134.

Lee, K. F., Li, E., Huber, L. J., Landis, S. C., Sharpe, A. H., Chao, M. V., & Jaenisch, R. (1992). Targeted mutation of the gene encoding the low affinity NGF receptor p75 leads to deficits in the peripheral sensory neuron system. Cell 69, 737–749.

Levi-Montalcini, R., & Angeletti, P. U. (1968). Nerve growth factor. Physiol. Revs. 48, 534–569.

Longo, F. M., Holtzman, D. M., Grimes, M. L., & Mobley, W. C. (1993). Nerve growth factor: Actions in the peripheral and central nervous system. In: Neurotrophic Factors (Loughlin, S. E., & Fallon, J. H., Eds.) pp. 209–256. Academic Press, San Diego.

Lowy, D. R., & Willumsen, B. M. (1993). Function and regulation of RAS. Ann. Rev. Biochem. 62, 851–891.

Luporini, P., Vallesi, A., Miceli, C., & Bradshaw, R. A. (1994). Ciliate pheromones as early growth factors and cytokines, In: *Primordial Immunity: Foundations for Vertebrate Immune Systems* (Beck, G., Cooper, E. L., Habict, G. S., & Marchalonis, J. J., Eds.) pp. 195–206. Ann. N.Y. Acad. Sci.

Massagué, J., Guillete, B. J., Czech, M. P., Morgan, C. J., & Bradshaw, R. A. (1981). Identification of a nerve growth factor receptor protein in sympathetic ganglia membranes by affinity labeling. J. Biol. Chem. 256, 9419–9424.

McDonald, N. Q., Lapatto, R., Murray-Rust, J., Gunning, J., Wlodawer, A., & Blundell, T. L. (1991). New protein fold revealed by a 2.3-Å resolution crystal structure of nerve growth factor. Nature 354, 411–414.

Meakin, S. O., & Shooter, E. M. (1992). The nerve growth factor family of receptors. Trends in Neurosci. 15, 323–331.

Muroya, K., Hattori, S., & Nakamura, S. (1992). Nerve growth factor induces rapid accumulation of the GTP-bound form of p21ras in rat pheochromocytoma PC12 cells. Oncogene 7, 277–281.

Murray-Rust, J., McDonald, N. Q., Blundell, T. L., Hosang, M., Oefner, C., Winkler, F., & Bradshaw, R. A. (1993). Topological similarities in TGF-β 2, PDGF-BB, and NGF define a superfamily of polypeptide growth factors. Structure 1, 153–159.

Obermeier, A., Halfter, H., Weismuller, K.-H., Jung, G., Schlessinger, J., & Ullrich, A. (1993). Tyrosine 785 is a major determinant of Trk-substrate interaction. EMBO J. 12, 933–941.

Obermeier, A., Bradshaw, R. A., Seedorf, K., Choidas, A., Schlessinger, J., & Ullrich, A. (1994). Neuronal differentiation signals are controlled by nerve growth factor receptor/Trk binding sites for SHC and PLCγ. EMBO J. 13, 1585–1590.

Oshima, M., Sithanandam, G., Rapp, U. R., & Guroff, G. (1991). The phosphorylation and activation of B-raf in PC12 cells stimulated by nerve growth factor. J. Biol. Chem. 266, 23753–23760.

Pawson, T. (1994). SH2 and SH3 domains in signal transduction. Adv. Cancer Res. 64, 87–110.

Pessin, M. S., Altin, J. G., Jarpe, M., Tansley, F., Bradshaw, R. A., & Raben, D. M. (1991). Carbachol stimulates a different phospholipid metabolism than nerve growth factor and basic fibroblast growth factor in PC12 cells. Cell Regulation 2, 382–390.

Pimental, E. (1994). Handbook of Growth Factors, Vols. 1–3. CRC Press, Boca Raton.

Pulido, D., Campuzano, S., Koda, T., Modolell, J., & Barbacid, M. (1992). Dtrk, a *Drosophilia* gene related to the Trk family of neurotrophin receptors, encodes a novel class of neural cell adhesion molecules. EMBO J. 11, 391–404.

Qiu, M. S., & Green, S. H. (1991). NGF and EGF rapidly activate p21ras in PC12 cells by distinct, convergent pathways involving tyrosine phosphorylation. Neuron 7, 937–946.

Rabiadeh, S., Oh, J., Zhong, L., Yang, J., Bitler, C. M., Butcher, L. L., & Bredesen, D. E. (1993). Induction of apoptosis by the low-affinity NGF receptor. Science 261, 345–348.

Radeke, M. J., Misko, T. P., Hsu, C., Herzenberg, L. A., & Shooter, E. M. (1987). Gene transfer and molecular cloning of the rat nerve growth factor receptor. Nature 325, 593–597.

Raffioni, S., & Bradshaw, R. A. (1992). Activation of phosphatidylinositol 3-kinase by epidermal growth factor, basic fibroblast growth factor and nerve growth factor in PC12 pheochromocytoma cells. Proc. Natl. Acad. Sci. USA 89, 9121–9125.

Raffioni, S., & Bradshaw, R. A. (1995). Staurosporine causes epidermal growth factor to induce differentiation in PC12 cells via receptor up-regulation. J. Biol. Chem. 270, 7568–7572.

Raffioni, S., Bradshaw, R. A., & Buxser, S. E. (1993). The receptors for nerve growth factor and other neurotrophins. Ann. Rev. Biochem. 62, 823–850.

Reinhold, D. S., & Neet, K. E. (1989). The lack of a role for protein kinase C in neurite extension and in the induction of ornithine decarboxylase by nerve growth factor in PC12 cells. J. Biol. Chem. 264, 3538–3544.

Rozakis-Adcock, M., McGlade, J., Mbamalu, G., Pelicci, G., Daly, R., Li, W., Batzer, A., Thomas, S., Brugge, J., & Pelicci, P. G., Schlessinger, J., & Pawson, T. (1992). Association of the Shc and

Grb2/Sem5 SH2-containing proteins is implicated in activation of the Ras pathway by tyrosine kinases. Nature 360, 689–692.

Scimeca, J. C., Nguyen, T. T., Filloux, C., & Van Obberghen, E. (1992). Nerve growth factor-induced phosphorylation cascade in PC12 pheochromocytoma cells. Association of S6 kinase II with the microtubule-associated protein kinase, ERK1. J. Biol. Chem. 267, 17369–17374.

Snider, W. D. (1994). Functions of the neurotrophins during nervous system development: What the knockouts are teaching us. Cell 77, 627–638.

Soltoff, S. P., Rabin, S. L., Cantley, L. C., & Kaplan, D. R. (1992). Nerve growth factor promotes the activation of phosphatidylinositol 3-kinase and its association with the trk tyrosine kinase. J. Biol. Chem. 267, 17472–17477.

Sporn, M. B., & Roberts, A. B. (Eds.). (1990). Peptide Growth Factors and Their Receptors, Vols. 1 & 2. Springer-Verlag, Berlin.

Stahl, N., Farruggella, T. J., Boulton, T. G., Zhong, Z., Darnell, J. E., Jr. & Yancopoulos, G. (1995). Choice of STAT and other substrates specified by modular tyrosine-based motifs in cytokine receptors. Science 267, 1349–1353.

Theroux, S. J., Stanley, K., Campbell, D. A., & Davis, R. J. (1992). Mutational removal of the major site of serine phosphorylation of the epidermal growth factor receptor causes potentiation of signal transduction: Role of receptor down-regulation. Mol. Endo. 6, 1849–1857.

Traverse, S., Gomez, N., Patterson, H., Marshall. C. J., & Cohen, P. (1992). Sustained activation of the mitogen-activated protein (MAP) kinase cascade may be required for differentiation of PC12 cells. Comparison of the effects of nerve growth factor and epidermal growth factor. Biochem. J. 288, 351–355.

Ullrich, A., & Schlessinger, J. (1990). Signal transduction by receptors with tyrosine kinase activity. Cell 61, 203–212.

Vetter, M. L., Martin-Zanca, D., Parada, L. F., Bishop, J. M., & Kaplan, D. R. (1991). Nerve growth factor rapidly stimulates tyrosine phosphorylation of phospholipase C-γ 1 by a kinase activity associated with the product of the trk protooncogene. Proc. Natl. Acad. Sci. USA 88, 5650–5654.

Volonte, C., Angelastro, J. M., & Greene, L. A. (1993). Association of protein kinases ERK1 and ERK2 with p75 nerve growth factor receptors. J. Biol. Chem. 268, 21240–21245.

Weskamp, G., & Reichardt, L. F. (1991). Evidence that biological activity of NGF is mediated through a novel subclass of high affinity receptors. Neuron 6, 649–63.

Wu, B. Y., Fodor, E. J., Edwards, R. H., & Rutter, W. J. (1989). Nerve growth factor induces the proto-oncogene c-jun in PC12 cells. J. Biol. Chem. 264, 9000–9003.

Wu, Y. Y., & Bradshaw, R. A. (1995). PC12-E-2 cells: A stable variant with altered responses to growth factor stimulation. J. Cell. Physiol. 164, 522–532.

PART II

CHANNELS

VOLTAGE-GATED POTASSIUM CHANNELS

Olaf Pongs

I. INTRODUCTION

It is likely that in most, if not all, excitable as well as inexcitable cells the membrane resting potential is under the control of potassium (K) channels (Hille, 1993). These channels are highly selective for K^+ over Na^+ ions. Since the extracellular K^+ concentration is generally one order of magnitude lower than the intracellular one, K channels mediate outward currents which may be either inwardly or outwardly rectifying. K channels may be distinctly activated by intracellular or extracellular

Biomembranes
Volume 6, pages 199–220.

ISBN: 1-55938-663-0.

ligands, for example, cAMP, ATP, Ca^{2+}, O_2, G-proteins, by mechanical perturbations (stretch, pressure), or by changes in membrane potential. This chapter concentrates on the molecular biology of voltage-activated K (K_v) channels. They are ubiquitous and, at the same time, highly diverse (Rudy, 1988). The biological significance of K_v channel diversity may be related to their widely varying functions in controlling cellular signal transduction, secretion, and (or) cell proliferation. In the nervous system, K_v channels are involved in the regulation and control of action potential wave forms and frequencies, thresholds of excitation, setting the resting potential, and attenuation of cell excitability. Thus, the expression of different K_v channels may aid the complex nervous system in encoding distinct pre- and post-synaptic signals, which is required in the specific spatial and temporal integration of synaptic inputs and in the firing of correspondent output signals. In general, K_v channel diversity may be manifest in variations of activation and inactivation behaviors, of single channel open durations and conductancies, and of gating properties (Pongs, 1992a, 1993; Jan and Jan, 1992, 1994; Bezanilla and Stefani, 1994; Chandy and Gutman, 1994).

The past years have witnessed rapid progress in the cloning and functional characterization of many K_v channels from both the animal and the plant kingdom. From these studies, it emerged that K_v channels are formed by a hetero-oligomeric assembly of α- and β-subunits. (Rehm and Lazdunski, 1988; Parcej and Dolly, 1989; Parcej et al., 1992; Rettig et al., 1994; Scott et al., 1994a). α-subunits appear to be the K channel forming subunits belonging to a superfamily of ion channel forming proteins (Jan and Jan, 1994). β-subunits are auxiliary subunits which do not form K channels by themselves. Whether they also belong to a protein superfamily is not established yet. Since α-subunits are the principal K_v channel-forming subunits, variations in their primary sequences may account for much of the K_v channel diversity which is observed in excitable cells. Therefore, the molecular biology of α-subunits will be discussed first. Note the cloning of another type of protein (minK) which is a relatively small polypeptide of ~200 amino acids that mediates, in *in vitro* expression systems, very slowly activating K^+ outward currents (Swanson et al., 1993). minK proteins are unrelated to the K_v protein superfamily. Whether they represent K channel forming proteins or only some kind of auxiliary subunits (Attali et al., 1993) is still a matter of conjecture. Thus, they will not be considered in this article.

II. BASIC DESIGN OF K_v CHANNEL α-SUBUNITS

The characterization of K_v channel genes in *Drosophila* has laid the ground work for what we presently know about the structure and function of K_v channels. Five independent K_v channel α-subunit encoding genes were cloned (*Shaker*, *Shab*, *Shal*, *Shaw*, *ether-à-go-go* (*eag*)) and the gene products functionally characterized in *in vitro* expression systems (Figure 1) (Salkoff et al., 1992; Brüggemann et al., 1993). Subsequently, the discovery of these genes has lead to an explosion in

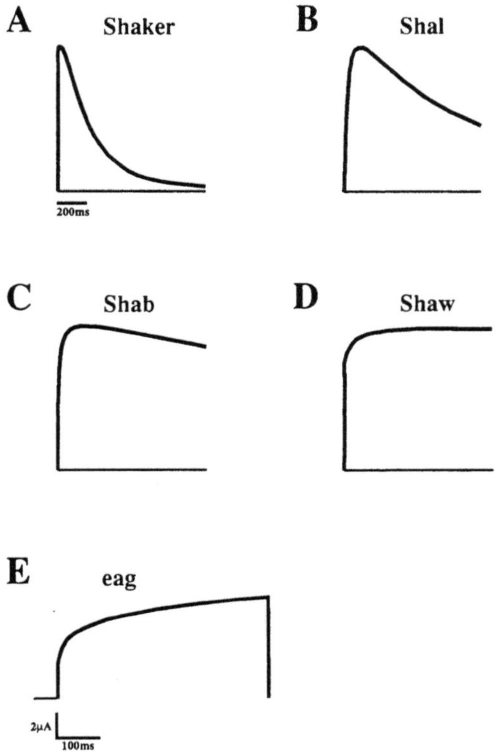

Figure 1. Typical outward current traces which are elicited upon depolarization in *Xenopus* oocytes injected with mRNAs encoding *Shaker* (**A**), *Shal* (**B**), *Shab* (**C**), *Shaw* (**D**), and *eag* (**E**) K_v channel α-subunits. Outward currents were recorded in the two-electrode voltage clamp configuration. Data are adapted from Salkoff et al. (1992) and Brüggemann et al. (1993).

cloning and expressing related genes ranging from *Aplysia* to man (Chandy and Gutman, 1994). The presently available data suggest that K_v channels belong to a gene superfamily, which may be divided into subfamilies as indicated in Table 1. A general nomenclature is now widely used, in which each K_v family member is given a number indicating its subfamily kinship (Chandy et al., 1991). In the case of human K_v genes, a different nomenclature has been adopted. "K_v" is replaced by "KCN" and subfamily kinship is indicated by a letter of the alphabet (see Table 1). Several, but by far not all, K_v channel genes have been mapped to human chromosomal loci (Pongs, 1992a; Chandy and Gutman, 1994). So far, in only one case has a correlation been established between a mutation in a K_v channel gene and a heritable disease (Browne et al., 1994).

Table 1. Subfamilies of Voltage-gated Potassium Channel α-subunits

Drosophila gene to which subfamily is related	*Subfamily*	*Human genome nomenclature*
Shaker	$K_v1.1$ to $K_v1.n$	KCNA1 to KCNAn
Shab	$K_v2.1$ to $K_v2.n$	KCNB1 to KCNBn
Shaw	$K_v3.1$ to $K_v3.n$	KCNC1 to KCNCn
Shal	$K_v4.1$ to $K_v4.n$	KCND1 to KCNDn
Unknown	$K_v5.1$ to $K_v5.n$	KCNE1 to KCNEn
Unknown	$K_v6.1$ to $K_v6.n$	KCNF1 to KCNFn
eag	$K_v7.1$ to $K_v7.n$	KCNG1 to KCNGn

Analysis of the protein sequences for the various K_v channel α-subunits, which were derived from many different cDNAs, has revealed a basic structural design (Figure 2). It consists of relatively hydrophilic amino- and carboxy-terminal ends flanking a core region that is highly conserved in both structure and sequence. The hydrophilic termini are facing the cytoplasm. The core domain constitutes the membrane spanning part of the K_v channel subunits. The core region always comprises six hydrophobic segments (S1–S3, S5, H5, and S6) and a positively charged amphipathic segment (S4; see Figure 3). This has lead to the proposal of a detailed membrane topology for the core domain (Durrell and Guy, 1992). According to this hypothetical model, segments S1 to S6 traverse the membrane and segment H5 is tucked into the lipid bilayer from the extracellular side. This topology implies that major activities of K_v channels like voltage-sensing, gating, and ion conduction are controlled by domains and amino acid residues, respectively, located in the core domain. Indeed, many structure–function studies, in which *in vitro* mutagenesis was combined with biophysical experiments have provided a large body of evidence supporting the general topology shown in Figure 2.

The α-subunits of voltage-activated Na^+ and Ca^{2+} channels correspond to single large polypeptides which contain four homologous domains (Catterall, 1988). The polypeptides express *in vitro* functional channels. It has been proposed that the four domains form an active channel by folding in a symmetrical fashion around a central pore. The sequence and topology of each of these domains is similar to that of K_v α-subunits, suggesting that K_v channels form tetramers. This notion is supported by experiments in which the assembly of subunits having differing drug sensitivities was investigated (MacKinnon, 1991b; Liman et al., 1992). Accordingly, K_v channel α-subunits may assemble in homo- or hetero-tetrameric complexes. Co-expression of different K_v channel α-subunits having distinct kinetic and pharmacological properties leads to the formation of K_v channels with novel properties (Christie et al., 1990; Isacoff et al., 1990; Ruppersberg et al., 1990). Hetero-multimeric assembly may also occur *in vivo* in excitable cells (Sheng et al., 1993; Scott et al., 1994b). The existence of hetero-multimeric K_v channels may contribute to functional diversity of K_v channels. Assembly of K_v channel α-subunits in hetero-multimeric

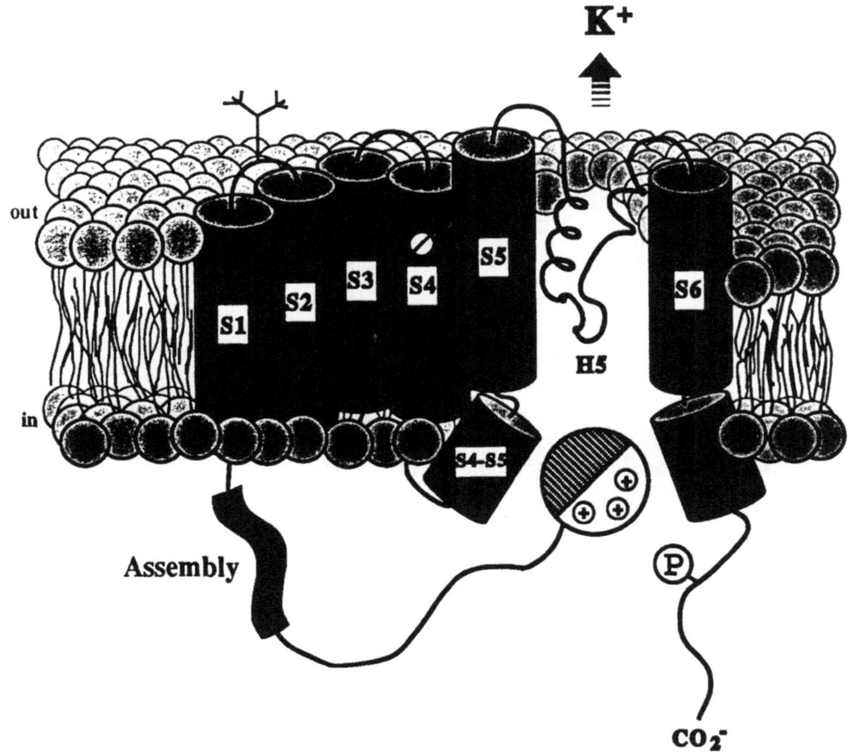

Figure 2. Proposed membrane-spanning orientation of one protein subunit of a voltage-activated K_v channel. Putative membrane spanning segments are designated S1–S6. H5 describes a hydrophobic region that is highly conserved among K_v channel α-subunits. It is thought that the H5 sequence enters and exits the lipid bilayer from the extracellular side. Amino acid side chains and/or peptide backbone residues of the H5 sequence may contribute to the K_v channel pore and outer mouth. Amino acid residues and/or peptide backbone residues of the S4/S5 linker region and the carboxy-terminal end of S6 may also contribute to the K_v channel pore and the inner mouth of the channel. The sequence between S1 and S2 is frequently N-glycosylated as shown by small branches. The amino- and carboxy-terminal sequences are located intracellularly. The amino-terminal domain may contain sequences involved in subunit interaction and assembly. Four subunits may be assembled in making a functional K_v channel. Amino-terminal sequences of some K_v channel α-subunits contain an inactivating domain drawn as a circle. This domain may occlude after activation the K_v channel pore and, thereby inactivate the channel. Positive charges within the circle indicate that the inactivating domain requires positively charged amino acid residues for function. Encircled P indicates putative consensus motif for phosphorylation by protein kinases. Arrow indicates direction of K^+ ion flow though the pore from inside (in) to outside (out). See text for references.

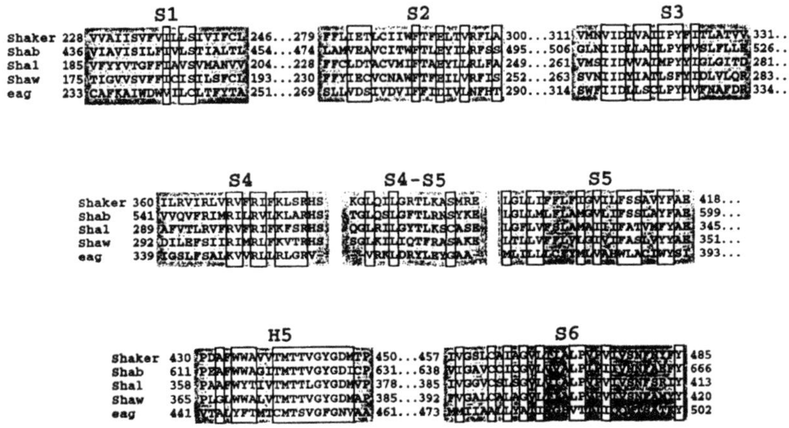

Figure 3. Amino acid sequence alignment of K_V channel α-subunit hydrophobic core sequences derived from the *Drosophila* genes *Shaker, Shab, Shal, Shaw,* and *eag*. The alignment was edited manually. Amino acids conserved in the five K_V α-subunit sequences have not been shaded. Gaps (–) have been introduced for optimal alignment. Numbers on the left- and right-hand side of each column refer to the first- and last-amino acid in this column. Sequence numbers are from Pongs et al. (1988) (*Shaker*), from Butler et al. (1990) (*Shab, Shal, Shaw*), and from Warmke et al. (1991) (*eag*).

K_V channels seems to be restricted, however, to members of the same K_V channel subfamily. Thus, a given cell may express distinct K_V channels by expressing different K_V subfamily genes. On the other hand, how assembly into heteromultimers is controlled, is presently not known. Several reports have implicated a distinct domain in the amino-terminus as well as segment S1 as being responsible for the specific assembly of K_V α-subunits (Li et al., 1992; Shen et al., 1993). However, the available data are not completely consistent with each other. It has been reported that a deletion of 114 amino acids in the amino terminus of *Shaker* channels determines subunit assembly (Li et al., 1992). Deletion of the corresponding amino acids in the amino-terminus of the *Shaker*-related mouse $K_V1.3$ channel does not, however, impair $K_V1.3$ channel function (Aiyar et al., 1993). Presumably, assembly is not affected in this case. It is quite possible that distinct amino-terminal domains in each subfamily are implicated in assembly of α-subunits.

III. CORRELATION BETWEEN K_V CHANNEL FUNCTION AND α-SUBUNIT DOMAINS

Hodgkin and Huxley (1952) developed an empirical kinetic model to describe the voltage-dependent activation and inactivation of ion channels. Channel opening was proposed to be controlled by several independent "gating particles" bearing an electrical charge that functions as voltage sensor. Note that, in principal, the charge of the gating particles may be negative or positive. The charges make the distribution of the gating particles within the membrane dependent on membrane voltage; they move in the membrane between a permissive- and nonpermissive-position. The gating particles have been shown to be integral components of the channel protein. Most importantly, movement of gating particles in the electric field across the membrane can be measured directly as a gating current (Bezanilla and Stefani, 1994). Mutations in segment S4 affect voltage-dependent activation and gating currents in K_v channels (Liman et al., 1991; Lopez et al., 1991; Papazian et al., 1991; Logothetis et al., 1992, 1993). Therefore, residues in S4 may contribute to the voltage sensor. This implies that S4 should move within the electric field upon a change in membrane potential and that this movement is somehow linked to channel gating (see following). The hallmark of S4 segments in voltage-gated ion channels is the occurrence of several positively charged amino acid residues (lysines and arginines) which are regularly spaced along the sequence in the general pattern $(\text{Arg/Lys–X–X})_n$, where X may be any amino acid (mostly hydrophobic) and *n* may vary between 4 and 7 (see Figure 3). The voltage-dependent activation and the gating currents have been investigated in mutant *Shaker* channels in which positively charged amino acids of segment S4 had been neutralized (Papazian et al., 1991). This affected the mid-points of voltage-dependent activation (voltage-sensing) and/or the slope of the gating current–voltage relationships indicating a reduction in the charge carried by the channel's gating particles across the membrane. These results suggested that voltage-sensing is linked to the movement of positive charges within the S4 segment of K_v channels. However, the presence or absence of positively charged amino acid side chains might not be correlated naively with the number of gating charges making up the gating current (Bezanilla and Stefani, 1994). Some substitutions do not shift the voltage-dependence of activation and alter the slope of the gating current–voltage relationship in a comparable way. The effect of some of these mutations may be explained by the existence of several components of gating charge movement, one of which has its voltage-dependent equilibrium shifted by the mutation(s). Moreover, gating charges may not move independently of each other during voltage-dependent activation (for a detailed discussion, see the review by Bezanilla and Stefani, 1994). Also, conservative substitutions of hydrophobic residues of the S4 sequence of *Shaker* channels caused comparable or even larger shifts of the activation along the voltage axis than the neutralization of basic amino acid residues (Lopez et al., 1991). The activity of the voltage sensor critically depends not only on the number and kind of basic amino acid residues, but also on the nature of the neighboring amino acid chains which may affect the interaction with other parts of the channel molecule or the lipid bilayer. The available results suggest that charge movements and the accompanying transitions during channel activation may occur in several separate steps, which have different voltage dependencies. In the absence of a high resolution structure of ion channels, one can only speculate about the conformational changes that accompany voltage-sensor movements and gating. Most likely, the conformational changes involve different parts of the channel molecule which probably alter their relative location within the electric field in a cooperative manner. The results on S4 mutant *Shaker* channels strongly argue that segment S4 plays a key role. Obviously, it contributes directly to the gating current and provides the side-chain interactions to other parts of the channel molecule that are involved in the voltage-dependent conformational transitions required to transform a closed channel in the resting state into an active, ion conducting channel.

A key player in the coupling of voltage-dependent transitions to channel gating might be the domain which links segments S4 and S5 (McCormack et al., 1991; Schopa et al., 1992). The linker region contains a leucine-heptad repeat region that has been discovered in many, but not all, cloned K_v channel α-subunits (McCormack et al., 1989). Substitution of a highly conserved leucine in this repeat by valine produces a large effect on the observed voltage-dependence of channel activation (McCormack et al., 1991; Schopa et al., 1992). The effect is comparable to the ones observed with S4 mutants, that is, the conductance is shifted along the voltage axis and the slope of the gating current–voltage relationship is reduced. A detailed analysis of this type of mutant came to the conclusion that K_v channels may follow two separate pathways during activation with the implication that activated channels may equilibrate between a relaxed- and a tight-conformation (McCormack et al., 1994). The former allows rapid gating and channel opening, the latter does not (Figure 4). The mutation in the S4/S5 linker region may affect the equilibrium between the two pathways and thereby alter the voltage-dependence of channel activation. This situation is reminiscent of the conformational changes which have

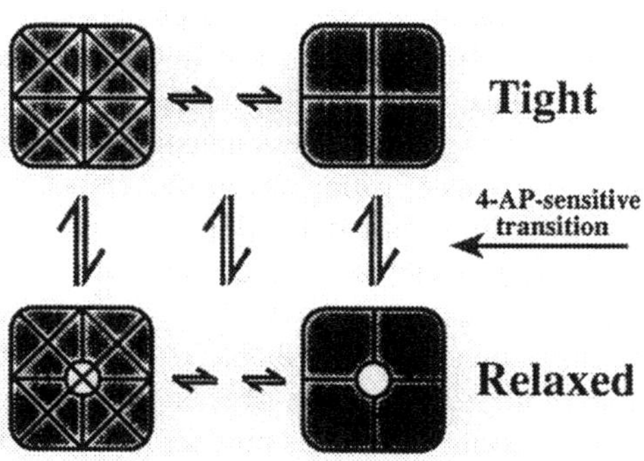

Figure 4. Allosteric mechanism of K_v channel gating (McCormack et al., 1994). K_v channels may exist in two conformations, tight (T) or relaxed (R). Upon depolarization of the membrane, either conformation becomes activated in a voltage-dependent manner. This may involve several independent steps abbreviated by arrows. The exact number of steps is uncertain. Transitions between the T and R conformations may occur during activation. The transitions are sensitive to intracellularly applied 4-aminopyridine (4-AP).

been observed for binding O_2 to hemoglobin which can exist in two conformational states—one having a high activity with respect to binding O_2, the other a low one.

Extensive *in vitro* mutagenesis experiments were carried out to define domains and single amino acid residues which may be involved in pore formation and lining the wall of the pore. Before discussing the results of these experiments, it may be useful to repeat some of the general concepts (and prejudices) which have been elaborated for K_v channel pores and which have been very influential in interpreting the molecular biology data. The K_v channel pore resembles a water-filled hole in the membrane, having a funnel like shape (Hille, 1993). On the cytoplasmic side there is a large vestibule. It is closed by a gate such that intracellular blockers may become trapped in the inner entrance to the pore when the channel closes. This does not occur with extracellular open channel blockers. K_v channel pores may be described as multi-ion pores through which ions pass in single file. Consequently, the pore has several sites which contact passing ions (and water molecules). Changes in any one of these sites may alter single channel conductance and possibly also single channel behavior. Furthermore, it is assumed that the pore has one selectivity filter which is responsible for the channel's selectivity for K^+ over both smaller ions like Na^+ and Li^+ and larger ions like Rb^+ and Cs^+. This hypothesis implies that changes in the permeability ratios between K^+ and any one of these ions may be attributed to an altered selectivity filter. These general considerations also argue that many more mutations might lead to alterations in single channel behavior than to changes in selectivity.

A crucial, but most difficult question concerns the structure of the selectivity filter and how it binds K^+ ions selectively. Ions (Rb^+, Cs^+) larger than K^+ may be simply filtered away in some kind of meshlike mechanism, but smaller ions (Na^+, Li^+) have to be actively discriminated against. In simple terms, this may mean that the binding (coordination) of K^+ is favorable, that of Na^+ and Li^+ is not. Regardless, whether one considers a situation in which the K^+ ion might shed only its outer aqueous envelope or its entire hydration shell during the interaction with the selectivity filter, two opposing views may be realized. K^+ ion coordination is accomplished by amino acid side chains; alternatively, K^+ ion coordination is brought about by a distinct backbone fold within the aqueous pore. Clearly, the *in vitro* mutagenesis experiments cannot provide a definite answer to these problems, but they have been instrumental in delineating the K_v channel domains which contribute decisively to pore formation and the outer- and inner-mouth of the channel.

Extensive structure–function studies in several laboratories, analyzing the selectivity and conductance properties of mutagenized K_v channels in *in vitro* expression systems, has demonstrated that three domains, the S4/S5 linker region (Kirsch et al., 1993; Slesinger et al., 1993), the H5 segment (referred to as P(ore)-region) (Yool and Schwarz, 1991; Hartmann et al., 1991; Kirsch et al., 1992a, b), and the carboxy-terminal end of S6 (Lopez et al., 1993) are determinants of ion conduction through the K_v channel pore. Among these domains, the ion selectivity filter may be formed by a small subset of amino acids, the so-called K_v channel signature

```
Shaker   DAFWWAVVTMTTVGYGDMT
Shab     E-----G-----------Y
Shaw     LGL---L----------IY
Shal     A---YTI----L------V
eag      T-LYFTMTC--S--F-NVA
```

Figure 5. Comparison of H5 sequences of *Shaker, Shab, Shaw, Shal,* and *eag* sequences. Filled dots indicate amino acid residues which contribute in *Shaker* K_v channels to ion selectivity (Heginbotham et al., 1994). Striped box marks a single amino acid residue in the H5 region, the nature of which is very important for TEA sensitivity of the respective *Shaker* related K_v channel.

sequence (Heginbotham et al., 1994), located within the P-region. K channels are highly homologous over a stretch of eight amino acids corresponding to the signature sequence TMTTVGYG in the *Shaker* K_v channel P-region (Figure 5). This sequence has been systematically mutated in the case of *Shaker* channels and to some degree in the case of the *Shaker*-related K_v 2.1 channel construct as well as in the case of *eag* channels. However, only the latter two studies also included, in addition to point mutations, double- and multiple-mutations in the signature sequence. The combined results indicate that many signature-sequence mutations may distinctly affect K_v channel selectivity (Heginbotham et al., 1994). Amino acid substitutions at positions 1 to 5, 7, and 9 in the signature sequence reveal that their effects on ion selectivity depend both on the kind of mutation that was introduced and on the sequence context in which the substitution was made. For example, a substitution of tyrosine by phenylalanine at position 7 of the *Shaker* signature sequence does not affect the ion selectivity of *Shaker* channels (Heginbotham et al., 1994), whereas the substitution of phenylalanine by tyrosine at position 7 of *eag* channels renders the mutant *eag* channel unselective for K^+ over Na^+ (Soto, Pardo, and Pongs, unpublished observation). Similar, seemingly contradictory results have also been obtained at other positions of the signature sequence. Most likely, the amino acid residues at positions 1 to 5, 7, and 9 do not directly interact with permeant ions in the K_v channel pore in the sense that they do not directly contribute to the selectivity filter. This leaves the two glycines at positions 6 and 8. Substitution or deletion of either glycine in *Shaker* channels always renders their pore unselective for cations (Heginbotham et al., 1992, 1994). Unfortunately, comparable *in vitro* mutagenesis studies have not been carried out with other types of K channels in order to find out how general the effects of glycine mutations in the signature sequence are. The implication of the results on mutating the signature sequence may, in fact, be that no side-chain is essential for ion selection in K_v channels contrary to some earlier interpretations and proposals. It is quite possible that the flexibility of the backbone around the two glycine residues may allow backbone-carbonyls in the vicinity to make up a selective K^+ ion-binding site in accordance with an earlier proposal that a coordination of the K^+ ion through interactions with oxygen atoms may provide the molecular basis of K channel selectivity (Hille, 1973; Bezanilla and Armstrong, 1972). Thus, K^+ selectivity may be due to interactions with oxygen atoms of backbone-carbonyl groups in the highly conserved signature sequence, similar to the K^+ selective-binding sites found in the enzyme, dialkylglycine decarboxylase, and in compounds such as valinomycin, nonactin, and crown ethers (Toney et al., 1993; Ovchinnikov et al., 1974). However, without any detailed structural knowledge provided by high-resolution techniques, any structural hypothesis on the K^+ channel selectivity filter must remain speculative.

IV. K_v CHANNEL INACTIVATION

Three types of K_v channel inactivation have been described in molecular terms: N-type, P-type, and C-type inactivation. N-type inactivation is linked to the presence of an amino-terminal inactivating domain (Hoshi et al., 1990; Zagotta et al., 1990; Ruppersberg, et al., 1991a), P-type inactivation to the presence of certain amino acid side-chains within the pore (DeBiasi et al., 1993), and C-type inactivation depends on residues in S6 (Choi et al., 1991, Lopez-Barneo et al., 1993). N-type inactivation refers to a rapid inactivation of K_v channels via an amino (N)-terminal inactivating domain. It has been proposed that N-type inactivation may operate in a ball-and-chain type mechanism (Zagotta et al., 1990). The amino-terminus would behave like a tethered ball which swings upon depolarization of the membrane into the K_v channel pore and occludes it from the inside. It has been shown that deletion of amino-terminal inactivating domains leads to a loss of N-type inactivation. Furthermore, the inactivating domain can be added in the form of a peptide back to inside-out patches containing K_v channels. This leads to a rapid inactivation of otherwise non-inactivating K_v channels. Without going into detailed biophysical considerations, this simple mechanistic view of N-type inactivation has the following important implications: (1) the inactivating domain does not bind to K_v channels in the closed resting state; (2) the inactivating domain has a receptor site close to the inner entrance of the K_v channel pore; (3) this site must be in the electrical field; (4) the inactivating domain occludes the open pore and thereby locks also the voltage sensor; and (5) recovery from inactivation requires removal of the inactivating domain from the occluded, but still open K_v channel pore. The results of structure–function studies on N-type inactivation in various channels are consistent with this general view of the mechanism of N-type inactivation.

Apparently, the inactivating domains, which have been identified (Figure 6), have two subdomains. One subdomain contains uncharged amino acids like the serine/cysteine motif in vertebrate A-type K_v channels, the other is characterized by a high density of the positively charged amino acids, lysine and arginine. They play an important role in determining the on-rate of binding inactivating domains to the

```
KVβ1                MQVSIACTEHNLKSRNGEDRLLSKQS
KV1.4      MEVAMVSAESSGCNSHM··RARERERLAHSRAA
KV3.4               MISSVCVSSY-RGKKSGNKPPSKTC
Shaker               MAAVAGLYGL----GEDRQHRKKQ
```

Figure 6. Alignment of the N-terminal inactivating domains of K_v α- and β-subunits of K_v channels. Gaps were introduced for optimal alignment and are indicated by dashes. Serines and cysteines in the proposed cysteine/serine motif of the inactivating domain are shaded, positively charged amino acids accumulated in the inactivating domains are boxed. Inactivating domains were defined for K_v1 β-subunits in Rettig et al. (1994), $K_v1.4$, and $K_v3.4$ in Ruppersberg et al. (1991a), and for *Shaker* in Zagotta et al., (1990).

receptor site (Murrell-Lagnado and Aldrich, 1993). The positive charges within the inactivating domain may sense the electrical field and thus direct the inactivating domain towards its receptor site. This is consistent with the observation that the on-rate for inactivating domain binding is slightly voltage-dependent. The uncharged inactivating subdomain seems to be important for the stability of inactivating domain–receptor site interaction. The subdomain contains, in vertebrate A-type K_v channels, a cysteine residue. It renders the activity of the inactivating domain in vertebrate A-type K_v channels very sensitive to oxidation (Ruppersberg et al., 1991b). Oxidation of the cysteine causes loss of inactivating domain activity. It has been shown that amino acid substitutions in the S4/S5 linker region of *Shaker* channels influence the interaction of the inactivating domain with the receptor site (Isacoff et al., 1991). Therefore, part of the S4/S5 linker may contribute to the receptor site. Phosphorylation of the cytoplasmic carboxy-terminus may also be involved (Drain et al., 1994).

The S4/S5 linker may also contain residues which participate in forming the gate of K_v channels as well part of the inner vestibule and, respectively, entrance to the pore. Mutations which affect N-type inactivation may at the same time influence single channel conductance (Isacoff et al., 1991) and binding of intracellular open channel blockers like tetraethylammonium (TEA) (Yellen et al., 1991). Under the assumption that the mutational alterations are due to direct effects on gating, conductance, and N-type inactivation, these results can be rationalized by a simple model (see also Figure 2). Upon depolarization the voltage-sensor dislocation leads to a conformational change in the S4/S5 linker region. This allows the channel's gate to open. The open gate allows K^+ ions to enter and, at the same time, creates the receptor site for the inactivating domain. Then the inactivating domain swings in, binds to the receptor site, and thereby occludes the open pore. Concomitantly, the opened gate is blocked in the open position. This simple model predicts that N-type inactivation freezes the open channel in an open state and that the blocked gate may also lock in the gating particles in the activated channel. Consistent with this view is that raising external K^+ concentration accelerates the off-rate of inactivating domain binding indicated by an accelerated rate of recovery from inactivation (Choi et al., 1991; Pardo et al., 1992). Furthermore, removal of the inactivating domain leaves the channel in an open state or in a state which allows the K_v channel to open at resting potential in an apparently voltage-independent manner (Ruppersberg et al., 1991a). Also, the frozen gating charges are relieved giving rise to an off-gating current (Bezanilla and Stefani, 1994).

Several residues in the P-region may affect slow inactivation of K_v channels (P-type inactivation). Replacing at position 7 tyrosine by valine in the signature sequence of *Shaker* channels accelerated inactivation ~50-fold (Heginbotham et al., 1994). Replacing isoleucine by leucine at position 4 in rat $K_v2.1$ channels also lead to a marked increase in inactivation (Kirsch et al., 1992b). Furthermore, two amino acid residues downstream of the signature sequence (position "10") are also found to regulate P-type inactivation (Kirsch et al., 1992b; Pardo et al., 1992). Rapid P-type inactivation is observed when *Shaker* (Lopez-Barneo et al., 1993) or $K_v1.4$ channels (Pardo et al., 1992) contain either lysine, glutamic acid or glutamine at this position. Possibly, the occurrence of P-type inactivation in mutant K_v channels is correlated with a requirement that there is some K^+ present in the external medium to produce outward currents. At very low external K^+ concentrations, these channels do not conduct. It remains to be seen whether this may be one feature by which P-type inactivation may be discerned from other types of K_v channel inactivation. Also, P-type inactivation is distinguished by C-type inactivation, which is slowed by external TEA, whereas P-type inactivation is not.

C-type inactivation is correlated with the presence of certain amino acid residues in the segment S6 of *Shaker* channels. Replacement of valine in one type of *Shaker* S6 segment with an alanine produces a slow C-type inactivation. When these residues are exchanged between different *Shaker* isoforms, C-type inactivation properties are transferred with the alanine (Hoshi et al., 1991; Wittka et al., 1991).

V. BASIC STRUCTURE OF K_v β-SUBUNITS

The mamba snake venom polypeptide α-dendrotoxin (α-DTX) has been of great importance in biochemically analyzing the structure of *Shaker*-related vertebrate K_v channels (Rehm and Lazdunski, 1988, Parcej and Dolly, 1989). The existence of high-affinity binding sites on K_v α-subunits has greatly aided the purification of K_v channels from rat and bovine synaptic plasma membranes. The purified preparations revealed a broad glycosylated α-subunit band (~70–80 kD) together with a smaller β-subunit (~40 kD) in SDS-PAGE. Sequencing of an N-terminal peptide showed that it corresponded to $K_v1.2$, a *Shaker*-related K_v channel α-subunit previously cloned from rat and bovine brain (Reid et al., 1992). The β-subunit was also analyzed by limited proteolysis. This revealed sequences which were not related to previously cloned ion channel subunits (Scott et al., 1994b).

```
rat Kv β1   MQVSIACTEHNLKSRNGEDRLLSKQSSTAPNVVN-------AARAKFRTVAIIARSLGTF    60
rat Kv β2                                            MYPESTTGSPARLSLRQ-G    19
rat Kv β3   ---------Q--R--SS----CGPRPGPGGGNGGPVGGGHGNPPGGGGLGSKSRTAVVPR    60

rat Kv β1   TPQHHISLKESTAKQTGMKYRNLGKSGLRVSCLGLGTWVTFGGQISDEVAERLMTIAYES   120
rat Kv β2   S-GMIY-TRYGSP-RQLQF--------------------------T--M--H---L--DN    79
rat Kv β3   P-APAGA-R---GRG---------------------------S-----T--D-L-V---H   120

rat Kv β1   GVNLFDTAEVYAAGKAEVILGSIIKKKGWRRSSLVITTKLYWGGKAETERGLSRKHIIEG   180
rat Kv β2   -I----------------V--N-----------------IF-------------------   139
rat Kv β3   -----------------RT--N-L-S-------Y-----IF---Q---------------   180

rat Kv β1   LKGSLQRLQLEYVDVVFANRPDSNTPMEEIVRAMTHVINQGMAMYWGTSRWSAMEIMEAY   240
rat Kv β2   --A--E----------------P------T---------------------S--------   199
rat Kv β3   -Q---D--------I-----S-PSS----------Y----L-L----------A------   240

rat Kv β1   SVARQFNMIPPVCEQAEYHLFQREKVEVQLPELYHKIGVGAMTWSPLACGIISGKYGNGV   300
rat Kv β2   -------L---I-------M-------------F-----------------V----DS-I   259
rat Kv β3   -M-----L---------NHF-----M--------------SV-------SL-TS--DGQ-   300

rat Kv β1   PESSRASLKCYQWLKERIVSEEGRKQQNKLKDLSPIAERLGCTLPQLAVAWCLRNEGVSS   360
rat Kv β2   -PY------G-----DK-L-----R--A---E-QA-------------I-----------   319
rat Kv β3   -DACK-TV-G------KVQ--D-K---ARVT--L---HQ----VA---I-----S-----   360

rat Kv β1   VLLGSSTPEQLIENLGAIQVLPKMTSHVVNEIDNILRNKPYSKKDYRS               408
rat Kv β2   ----A-NA---M--I--------LS-SI-H---S--G-----------               367
rat Kv β3   ----V-SA---M-H--SL---GQL-PQT-M---AL-G--SH--                    403
```

Figure 7. Deduced protein sequences of cloned rat brain cDNAs encoding voltage-gated potassium channel β-subunits. Amino acids are given in single letter code and are numbered on the right. Identical amino acids are indicated by dashes. Potential sites for phosphorylation by cAMP/cGMP-dependent protein kinase(s), and by protein kinase C or casein kinase II are marked by filled and open circles, respectively. K_v1 and K_v2 β-subunit sequences are from Rettig et al. (1994). K_v3 β-subunit sequence is from J. Rettig and O. Pongs (unpublished).

Figure 8. Alignment of K_v1 β-subunit with sequences derived from *Oryza (O.) sativa (sat.), Aradopsis (A.) thaliana (thal.),* and *Nicotiana (N.) tabacum (tab.)* cDNAs and from *Pseudomonas sp.* igrA gene. The *O. sat.* sequences are derived from two different cDNAs (D24673) and (D24756) cloned by expressed sequence tagging. The alignment suggests that both sequences are derived from exons belonging to the same transcription unit. Similarly, the *A. thal.* sequences (Z188389, Z30863) may also have been derived from exons belonging to the same transcription unit. *A. thal.* sequences aligned with K_v1 β-subunit residues 121–183 corresponds to the entire open reading frame of Atts07 cDNA (Z188389). *A. thal.* sequences Atts26 (Z30863) aligned with K_v1 β-subunit residues 198–232 and, respectively, 233–294 may have each 20 residues longer open reading frames. *N. tab.* cDNA (X56267) encodes an auxin-regulated protein. The *Pseudomonas sp.* gene corresponds to the PSEIGRA locus (M37389). The derived open reading frame may be longer than the first in frame-stop codon (residue 179) since the alignment with K_v1 β-subunit can be extended until the next in frame-stop codon in the sequence (residue 216).

Accession numbers refer to entries in the EMBL data bank. Sequences were aligned using Genetics Computer Group Inc. software. Gaps (–) are introduced for optimal alignment. Numbers on the right hand side correspond to K_v1 β-subunit sequence. Identical amino acids (in reference to the K_v1 β-subunit sequence) are darkly, conservative replacements lightly shaded.

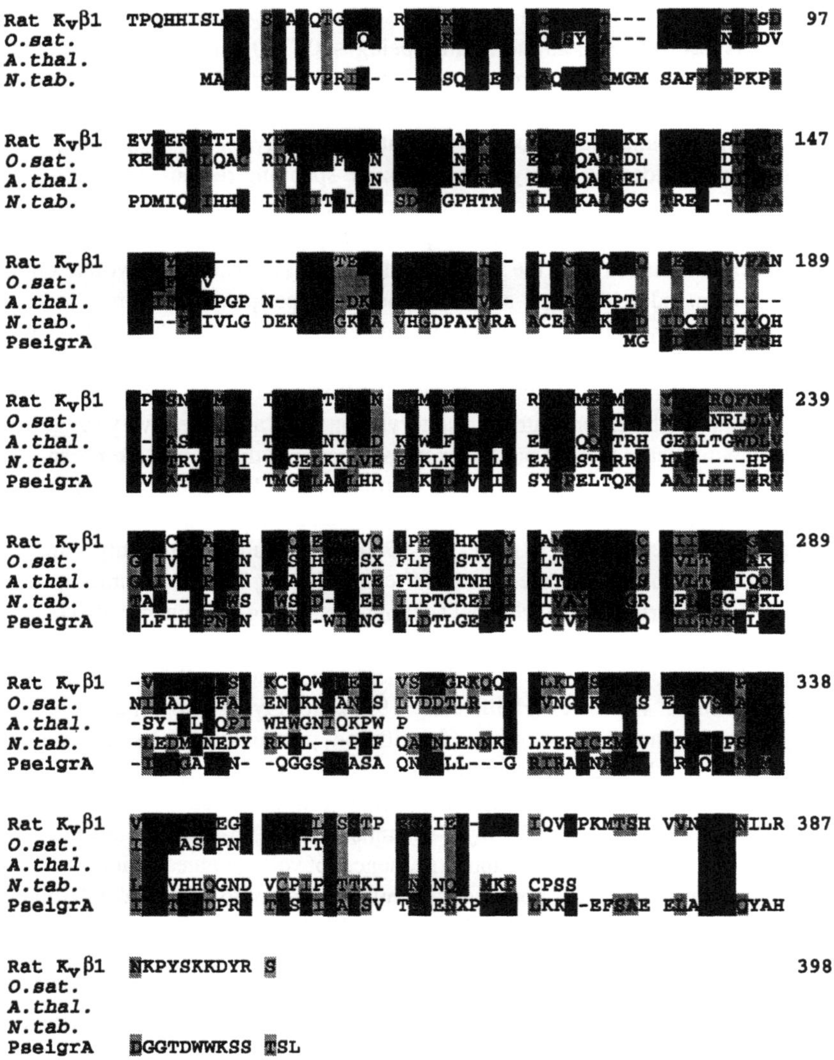
Rat KVβ1
O.sat.
A.thal.
N.tab.
PseigrA
97
147
189
239
289
338
387
398

The primary structure of several K_v β-subunits has been determined (Figure 7; Rettig et al., 1994). The molecular weights vary between 39- and 45-kD in accordance with biochemical data. The sequence alignment suggests that K_v β-subunits have variable amino-termini being ~40–100 amino acids long, and a highly conserved carboxy-terminus of ~325 amino acids. This may indicate that major functional differences between distinct β-subunits are due to variations in their amino-terminal sequences. Hydropathicity analyses of K_v β-subunit sequences did not identify any typical membrane-spanning regions (segments of at least 19 residues with an average hydropathy index > 1.6) or the presence of a hydrophobic amino-terminal signal sequence. Potential N-glycosylation sites are also absent, consistent with the lack of evidence for attached carbohydrate on the native protein. Indeed, the K_v β-subunits appear to be peripheral membrane proteins based on an analysis by the method of Klein et al. (1985), which correlates well with its dominant hydrophilic character. The pronounced hydrophilicity and absence of typical membrane-spanning regions, leader sequence, and N-glycosylation, together with demonstrated phosphorylation sites, are collectively suggestive of this subunit being located on the cytoplasmic side of the plasma membrane. As judged by the probabilistic method of Garnier et al. (1978), the β-subunit may contain four major α-helical domains being ~20–30 amino acids long and regularly spaced along the sequence (Ruth et al., 1989). K_v β-subunits are only remotely related to β-subunits of voltage-gated Ca^{2+} channels (Ca_v). Ca_v β-subunits are, like K_v β-subunits, peripheral ion channel subunits containing presumably also α-helical segments regularly spaced along the primary sequence. Unlike K_v β-subunits Ca_v β-subunits appear to have not only variable amino-terminal, but also variable carboxy-terminal sequences (Hullin et al., 1992). Both types of β-subunits contain many potential phosphorylation sites for different protein kinases including PKA, PKC, and casein kinase II. It could be demonstrated for both types of β-subunits that they are phosphorylated in their native state. But, so far, the kind and function of the phosphorylation sites in K_v β-subunits and Ca_v β-subunits are not well understood. In contrast to the α-subunit sequences of voltage-gated cation channels, β-subunit sequences appear not obviously related and not to belong to a β-subunit protein superfamily (Pongs, 1995). If at all, K_v β-subunits and Ca_v β-subunits may have diverged very early in evolution. On the other hand, vertebrate K_v β-subunit sequences have an interesting similarity with primary sequences derived from several plant cDNAs originating from *Oryza sativa*, *Nicotiana tabacum*, *Arabidopsis thaliana* and, respectively, from a bacterial (*Pseudomonas sp.*) gene (Figure 8). The *Nicotiana tabacum* cDNA encodes an auxine-regulated protein of unknown function. The *Oryza sativa* and *Arabidopsis thaliana* sequences were cloned by expressed sequence tagging. The *Pseudomonas sp.* gene sequence, also of unknown function, belongs to the igrA locus. The extensive sequence identity of these sequences to the K_v β-subunits (Figure 8) suggests that the encoded proteins are related and belong to a gene family having a common ancestor. However, the function of the plant and bacterial proteins is not established. Therefore, it is still speculative whether they might have a β-subunit related function in plants and in *Pseudomonas sp.* If so, primordial K_v channels existed already in the form of α,β-heterooligomeric complexes.

VI. K_v β-SUBUNIT AND INACTIVATION OF K_v CHANNELS

The K_v β-subunits cloned so far appear to bind specifically to members of the K_v1 subfamily, but not to members of other K_v subfamilies, for example, K_v2, K_v3, K_v4. It is, however, unlikely that only K_v1 α-subunits assemble with β-subunits in hetero-oligomeric complexes. Possibly, many more K_v β-subunits are encoded in the genome than have been cloned to date. Presumably, each K_v α-subunit subfamily has its own distinct K_v β-subunit(s). K_v1 α-subunits and K_v1 β-subunits assemble in K_v channels with novel properties in expression systems *in vitro* (Rettig et al., 1994). In particular, this is due to the presence of an inactivating domain in the K_v1 β-subunit amino-terminus (see Figure 6). The inactivating domain has a similar sequence to the amino-terminal K_v α-subunit ones. The structural similarity has its functional correlate. K_v1 channels, which do not possess an α-subunit inactivating domain may acquire this through assembly with K_v1 β-subunits. Accordingly, K_v α-subunits, which by themselves have delayed rectifier type properties, are converted to rapidly inactivating K_v channels by N-type inactivation via the β-subunit inactivating domain. The α/β co-expression studies indicate that N-type inactivation behavior of K_v channels may be due to the presence of an inactivating domain either on the α- or on the β-subunit. Note that not all K_v β-subunits contain an amino-terminal inactivating domain. Therefore, not each possible α/β hetero-oligomeric assembly yields rapidly inactivating K_v channels. *In situ* hybridization experiments with rat brain sections have shown (Rettig et al., 1994) that different K_v β-isoforms are differentially expressed in different areas of the brain. The distinct K_v β-isoform expression patterns are not comparable to the ones obtained for the different K_v1 α-subunits (Kues and Wunder, 1992). The implications of the results are that the assembly of α- and β-subunits in K_v channels is neuron specific and cannot be subunit specific. As a consequence, one K_v1 α-subunit type may assemble with different β-subunit isoforms in different neurons and *vice versa*. Potentially, this situation may give rise to a bewildering number of possible α/β-subunit combinations in excitable cells. It remains to be elucidated which combinations are, in fact, realized in the nervous system.

VII. K_v CHANNEL PHARMACOLOGY

Domain swapping experiments and detailed *in vitro* mutagenesis experiments, mainly with *Shaker*-related channels, indicate good agreement between the results obtained on the binding sites, which are important for binding K_v channel blockers to the outer and inner mouths of the pore, and the results obtained on domains involved in determining kinetic and conductance properties of the pore (Pongs,

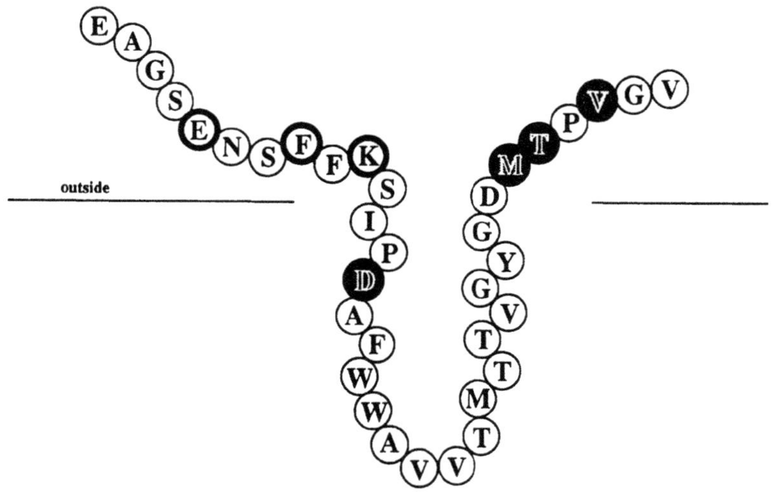

Figure 9. Diagram in single-letter code for amino acid sequence linking the S5–H5 region of *Shaker* K_v channels with the conserved H5 region inserted into the membrane. The diagram does not imply a certain secondary- or tertiary-structure of the sequence shown nor the relative position of each amino acid to the lipid bilayer. Amino acid side-chains which contribute to toxin-binding sites are encircled black. Shading marks amino acids which may contribute to sensitivity towards K_v channel block by external TEA.

1992b). In fact, the results of both kinds of studies are exceedingly complementary. Only a few amino acids on the outside of the K_v channel seem to determine whether or not it is sensitive to certain open channel blockers like dendrotoxin, charybdotoxin, angiotoxin, and TEA (Figure 9; Pongs, 1992b; Gross et al., 1994). Thus, it may only take the presence of three or four amino acids in and near to the H5 region to make a K_v channel resistant or sensitive to an open channel blocker. Note, however, this does not apply to every open channel blocker which has been studied. For example, the mast cell degranulating peptide (MDCP) binding site on *Shaker*-related channels is not confined to the S5/S6 linker region (Stocker et al., 1991). In this case, as yet unidentified amino acids outside of this region also contribute to MDCP binding. Binding sites for K_v channel blockers near or at the inner mouth of K_v channels are not as well-defined as the binding sites on the outside. This may support the view that several domains of the channel protein contribute to the inner mouth since mutations in the S4/S5 linker, segment S5, the H5 region, and in S6 may influence the binding of K_v blockers from the cytoplasmic side (Yellen et al., 1991, Lopez et al., 1993; Browne et al., 1994).

The available data strongly argue that small sequence variations in the extracellular K_v channel domain(s) may suffice for determining K_v channel sensitivities/insensitivities to antagonists. Nevertheless, it may be possible to develop specific drugs which selectively block K_v channel isoforms (Garcia et al., 1994).

REFERENCES

Aiyar, J., Grissmer, S., & Chandy K. G. (1993). Full-length and truncated $K_v1.3$ potassium channels remodulated by $5HT_{1C}$ receptor activation, and independently, by protein kinase C. Amer. J. Physiol.

Attali, B., Guillemare, E., Lesage, F., Honoré, E., Romey, G., Lazdunski, M., & Barhanin, J. (1993). The protein IsK is a dual activator of K^+ and Cl^- channels. Nature 365, 850–852.

Bezanilla, F., & Armstrong, C. M. (1972). Negative conductance caused by entry of sodium and cesium ions into the potassium channels of squid axons. J. Gen. Physiol. 60, 588–608.

Bezanilla, F., & Stefani, E. (1994). Voltage-dependent gating of ionic channels. Ann. Rev. Biophys. Biomol. Structure 23, 819–846.

Browne, D. L., Gancher, S. T., Nutt, J. G., Brunt, E. R. P., Smith, E. A., Kramer, P., & Litt, M. (1994). Episodic ataxia/myokymia syndrome is associated with point mutations in the human potassium channel gene, KCNA1. Nature Genetics 8, 136–140.

Brüggemann, A., Pardo, L. A., Stühmer, W., & Pongs, O. (1993). *Ether-à-go-go* encodes a voltage-gated channel permeable to K^+ and Ca^{2+} and modulated by cAMP. Nature 365: 445–448.

Butler, A., Wei, A., & Salkoff, L. (1990). *Shal, Shab,* and *Shaw*: Three genes encoding potassium channels in *Drosophila.* Nucl. Acid. Res. 18, 2173–2174.

Catterall, W. A. (1988). Structure and function of voltage-sensitive ion channels. Science, 242, 50–61.

Chandy, K. G., & Gutman, G. A. (1994). Voltage-gated potassium channels. In: CRC Handbook of Receptors and Channels. (North, P.A., ed.) CRC, Boca Raton, FL.

Chandy, K. G., Douglass, J., Gutman, G. A., Jan, L., Joho, R., Kaczmarek, L., Mc Kinnon, R., North, R. A., Numa, S., Phlipson, L., Ribeira, A. B., Rudy, B., Salkoff, L., Choi, K. L., Aldrich, R. W., & Yellen, G. (1991). Tetraethylammonium blockade distinguishes two inactivation mechanisms in voltage-activated K^+ channels. Proc. Natl. Acad. Sci. USA. 88, 5092–5095.

Choi, K. L., Aldrich, R. W., & Yellen, G. (1991). Tetraethylammonium blockade distinguishes two inactivation mechanisms in voltage-activated K^+ channels. Proc. Natl. Acad. Sci. USA 88, 5092–5095.

Christie, M. J., North, R. A., Osborne, P. B., Douglass, J., & Adelman, J. P. (1990). Heteropolymeric potassium channels expressed in *Xenopus* oocytes from cloned subunits. Neuron 4, 405.

DeBiasi, M., Drewe, J. A., Kirsch, G. E., & Brown, A. M. (1993). Histidine substitution identifies a surface position and confers Cs^+ selectivity on a K^+ pore. Biophys. J. 65, 1235–1242.

Drain, P., Dubin, A. E., & Aldrich R. W. (1994). Regulation of *Shaker* K^+ channel inactivation gating by the cAMP-dependent protein kinase. Neuron 12, 1097–1109.

Durell, S. R., & Guy, H. R. (1992). Atomic scale structure and functional models of voltage-gated potassium channels. Biophys. J. 632, 238–247.

Garcia, M. L., Garcia-Calvo, M., Hidalgo, P., Lee, A., & MacKinnon, R. (1994). Purification and characterization of three inhibitors of voltage-dependent K^+ channels from *Leiurus quinquestriatus* var hebraeus venom. Biochemistry 33, 6834–6839.

Garnier, J., Osguthorpe, D. J., & Robson, B. (1978). Analysis of the accuracy and implications of simple methods for predicting the secondary structure of globular proteins. J. Mol. Biol. 120, 97–120.

Gross, A., Abramson, T. S., & McKinnon, R. (1994). Transfer of the scorpion toxin receptor to an insensitive potassium channel. Neuron 13, 961–966.

Hartmann, H. A., Kirsch, G. E., Drewe, J. A., Tagliatela, M., Joho, R. H., & Brown, A. M. (1991). Exchange of conduction pathways between two related K^+ channels. Science 251, 942–944.

Heginbotham, L., Abramson, T., & MacKinnon, R. (1992). A functional connection between the pores of distantly related ion channel as revealed by mutant K^+ channels. Science 258, 1152–1155.

Heginbotham, L., Lu, Z., Abramson, T., & MacKinnon, R. (1994). Mutations in the K^+ channel signature sequence. Biophys. J. 66, 1061–1067.

Hille, B. (1973). Potassium channels in myelinated nerve. Selective permeability to small cations. J. Gen. Physiol. 61, 669–686.

Hille, B. (1993). Ionic Channels of Excitable Membranes 2nd ed. Sinauer, Sunderland, MA.

Hodgkin, A. L., & Huxley, A. F. (1952). A quantitative description of membrane current and its application to conduction and excitation in nerve. J. Physiol. Lond. 117, 500–544.

Hoshi, T., Zagotta, W. N., & Aldrich, R. W. (1990). Biophysical and molecular mechanisms of *Shaker* potassium channel inactivation. Science 250, 533–538.

Hoshi, T., Zagotta, W. N., & Aldrich, R. W. (1991). Two types of inactivation in *Shaker* K^+ channels: Effects of alternations in the carboxy terminal region. Neuron 7, 547–556.

Hullin, R., Singer-Lahat, D., Freichel, M., Biel, M., Dascal, N., Hofmann, F., & Flockerzi, V. (1992). Calcium channel β subunit heterogeneity: functional expression of cloned cDNA from heart, aorta and brain. EMBO J. 11, 885–890.

Isacoff, E. Y., Jan, Y. N., & Jan. L. Y. (1990). Evidence for formation of heteromultimeric potassium channels in *Xenopus* oocytes. Nature 345, 530–534.

Isacoff, E. Y., Jan, N. J., & Jan, L. Y. (1991). Putative receptor for the cytoplasmic inactivation gate in the *Shaker* K^+ channel. Nature, 353, 86–90.

Jan, L. Y., & Jan, Y. N. (1992). Structural elements involved in specific K^+ channel functions. Ann. Rev. Physiol. 54, 537–555.

Jan, L. Y., & Jan, Y. N. (1994). Potassium channels and their evolving gates. Nature 371, 119–122.

Kirsch, G. E., Drewe, J. A., Tagliatela, M., Joho, R. H., DeBiasi, M., Hartmann, H. A., & Brown, A. M. (1992a). A single nonpolar residue in the deep pore of altered K^+ channels acts as a K^+:Rb^+ conductance switch. Biophys. J. 62, 136–143.

Kirsch, G. E., Drewe, J. A., Hartmann, H. A., Tagliatela, M., DeBiasi, M., Brown, A. M., & Joho, R. H. (1992b). Differences between the deep pores of K^+ channels determined by an interacting pair of nonpolar amino acids. Neuron, 8, 499–505.

Kirsch, G. E., Shieh, C. C., Drewe, J. A., Vener, D. F., & Brown, A. M. (1993). Segmental exchanges define 4-aminopyridine binding and the inner mouth of K^+ pores. Neuron, 11, 1–20.

Klein, P., Kanehisa, M., & Delisi, C. (1985). The detection and classification of membrane-spanning proteins. Biochim. Biophys. Acta 815, 468–476.

Kues, W. A., & Wunder, F. (1992). Heterogeneous expression patterns of mammalian potassium channel genes in developing and adult rat brain. Eur. J. Neurosci. 4, 1296–1308.

Li, M., Jan, Y. N., & Jan. L. Y. (1992). Specification of subunit assembly by the hydrophilic amino-terminal domain of the *Shaker* potassium channel. Science 257, 1225–12340.

Liman, E. R., Hess, P., Weaver, F., & Koren, G. (1991). Voltage-sensing residues in the S4 region of a mammalian K^+ channel. Nature 353, 752–7456.

Liman, E. R., Tytgat, J., & Hess, P. (1992). Subunit stochiometry of a mammalian K^+ channel determined by construction of multimeric cDNAs. Neuron 9, 861–871.

Logothetis, D. E., Movehedi, S., Satler, C., Lind-Paintner, K., & Nadal-Ginard, B. (1992). Incremental reductions of positive charge within the S4 region of voltage-gated K^+ channel result in corresponding decreases in gating charge. Neuron 8, 531–540.

Logothetis, D. E., Kamen, B. F., Lindpaintner, K., Bisbas, D., & Nadal-Ginard, B. (1993). Gating charge differences between two voltage-gated K^+ channels are due to the specific charge content of their respective S4 regions. Neuron, 10, 1121–1129.

Lopez, G. A., Jan, Y. N., & Jan. L. Y. (1991). Hydrophobic substitution mutations in the S4 sequence after voltage-dependent gating in *Shaker* K^+ channels. Neuron 7, 327–336.

Lopez, G. A., Jan, Y. N., & Jan., L. Y. (1993). Mutations in the S6 domain affecting conductance, TEA, and barium blockade in the permeation pathway of the *Shaker* potassium channel. Biophys. J. 64, A113.

Lopez-Barneo, J., Hoshi, T., Heinemann, S. H., & Aldrich, R. W. (1993). Effects of external cations and mutations in the pore region on C-type inactivation of *Shaker* potassium channels. Receptors & Channels 1, 66–71.

MacKinnon, R. (1991a). New insights into the structure and function of potassium channels. Curr. Opinion in Neurobiol. 1, 14–19.

MacKinnon, R. (1991b). Determination of the subunit stochiometry of a voltage-activated potassium channel. Nature 350, 232–235.

McCormack, K., Campanelli, J. T., Ramaswami, M., Mathew, M. K., & Tanouye, M. A. (1989). Leucine zipper motif update. Nature 340, 103.

McCormack, K., Tanouye, M. A., Iverson, L. E., Lin, J. W., Ramaswami, M., McCormack, T., Pampanelle, J. T., Mathew, M. K., & Rudy, B. (1991). A role for hydrophobic residues in the voltage-dependent gating of *Shaker* K^+ channels. Proc. Natl. Acad. Sci. USA 88, 2931–2935.

McCormack, K., Joiner, W. J., & Heinemann, S. H. (1994). A characterization of the activating structural rearrangements in voltage-dependent *Shaker* K^+ channels. Neuron 12, 301–315.

Murrell-Lagnado, R. D., & Aldrich, R. W. (1993). Interactions of amino terminal domains of *Shaker* K^+ channel with a pore blocking site studied with synthetic peptides. J. Gen. Physiol. 102, 949–975.

Ovchinnikov, Y. A., Ivanov, V. T., & Shkrob, A. M. (1974). Membrane-Active Complexones. Elsevier Scientific Publishing Co., New York.

Papazian, D. M., Timpe, L. C., Jan, Y. N., & Jan, L. Y. (1991). Alteration of voltage-dependence of *Shaker* potassium channel by mutations in the S4 sequence. Nature 349, 305–310.

Parcej, D. N., & Dolly, J. O. (1989). Dendrotoxin receptor from bovine synaptic plasma membranes. Binding properties, purification and subunit composition of a putative constituent of certain voltage-activated K^+ channels. Biochem. J. 257, 899–903.

Parcej, D. N., Scott, V. E. S., & Dolly, J. O. (1992). Oligomeric properties of alpha-dendrotoxin-sensitive potassium ion channels purified from bovine brain. Biochemistry 31, 11084–11088.

Pardo, L. A., Heinemann, S. H., Terlau, H., Ludewig, U., Lorra, C., Pongs, O., & Stühmer, W. (1992). Extracellular K^+ specifically modulates a rat brain potassium channel. Proc. Natl. Acad. Sci. USA 89, 2466–2470.

Pongs, O. (1992a). Molecular biology of voltage-dependent potassium channels. Physiol. Rev. 72., S69–S88.

Pongs, O. (1992b). Structural basis of voltage-gated K^+ channel pharmacology. Trends in Pharmacol. Sci. 13, 359–365.

Pongs, O. (1993). *Shaker*-related K^+ channels. Sem. Neurosci. 5, 93–100.

Pongs, O. (1995). Regulation of the activity of voltage-gated potassium channels by β subunits. Sem. in Neurosci. 7, 137–146.

Pongs, O., Kecskemethy, N., Müller, R., Krah-Jentgens, I., Baumann, A., Kiltz, H. H., Canal, I., Llamazares, S., & Ferrus, A. (1988). *Shaker* encodes a family of putative potassium channel proteins in the nervous system of *Drosophila*. EMBO J. 7, 1087–1096.

Rehm, H., & Lazdunski, M. (1988). Purification and subunit structure of a putative K^+ channel protein identified by its binding properties for dendrotoxin I. Proc. Natl. Acad. Sci. USA 85, 4919–4923.

Reid. P. F., Pongs, O., & Dolly, J. O. (1992). Cloning of a bovine voltage-gated K^+ channel gene utilizing partial amino acid sequence of a dendrotoxin-binding protein from brain cortex. FEBS Lett. 302, 31–34.

Rettig, J., Heinemann, S. H., Wunder, F., Lorra, C., Parcej, D. N., Dolly, J. O., & Pongs, O. (1994). Inactivation properties of voltage-gated K^+ channels altered by presence of β-subunit. Nature 369, 289–294.

Rudy, B. (1988). Diversity and ubiquity of K^+ channels. Neurosci. 25, 729–749.

Ruppersberg, J. P., Schröter, K. H., Sakmann, B., Stocker, M., Sewing, S., & Pongs, O. (1990). Heteromultimeric channels formed by rat brain potassium-channel proteins. Nature 345, 535–537.

Ruppersberg, J. P., Frank, R., Pongs, O., & Stocker, M. (1991a). Cloned neuronal Ik(a) channels reopen during recovery from inactivation. Nature 353, 657–660.

Ruppersberg, J. P., Stocker, M., Pongs, O., Heinemann, S. H., Frank, R., & Koenen, M. (1991b). Regulation of fast inactivation of cloned mammalian Ik(a) channels by cysteine oxidation. Nature 352, 711–714.

Ruth, P., Röhrkasten, A., Biel, M., Bosse, E., Regulla, S., Meyer, H. E., Flockerzi, V., & Hofmann, F. (1989). Primary structure of the β subunit of the DHP-sensitive calcium channel from skeletal muscle. Science 245, 1115–1118.

Salkoff, L., Baker, K., Butler, A., Covarrubias, M., Pak, M. D., & Wei, A. (1992). An essential set of K^+ channels conserved in flies, mice and humans. Trends in Neurosci. 15, 161–166.

Schopa, N., McCormack, K., Tanouye, M. A., & Sigworth, F. J. (1992). The size of gating charge in wild-type and mutant *Shaker* potassium channels. Science 255, 1712–1715.

Scott, V. E. S., Rettig, J., Parcej, D. N., Keen, J. N., Findlay J. B. C., Pongs, O., & Dolly, J. O. (1994a). Primary structure of a β subunit of α-dendrotoxin-sensitive K^+ channels from bovine brain. Proc. Natl. Acad. Sci. USA 91, 1637–1641.

Scott, V. E. S., Muniz, Z. M., Sewing, S., Lichtinghagen, R., Parcej, D. N., Pongs, O., & Dolly, O. (1994b). Antibodies specific for distinct subunits unveil a hetero-oligomeric basis for subtypes of α-dendrotoxin-sensitive K^+ channels in bovine brain. Biochemistry 33, 1617–1623.

Shen, N. V., Chen, X., Boyer, M. M., & Pfaffinger, P. J. (1993). Deletion analysis of K^+ channel assembly. Neuron 11, 67–76.

Sheng, M., Liao, Y. J., Jan, Y. N., & Jan, L. Y. (1993). Presynaptic A-current based on heteromultimeric K^+ channels detected *in vivo*. Nature 365, 72–75.

Slesinger, P. A., Jan. Y. N., & Jan, L. Y. (1993). The S4–S5 loop contributes to the ion-selective pore of potassium channels. Neuron 11, 739–749.

Stocker, M., Pongs, O., Hoth, M., Heinemann, S. H., Stühmer, W., Schröter, K.-H., & Ruppersberg, J. P. (1991). Swapping of functional domains in voltage-gated K^+ channels. Proc. Roy. Soc. Lond. 245, 101–107.

Swanson, R., Hice, R. E., Folander, K., & Sanguinetti, M. C. (1993). The I_{sK} protein, a slowly inactivating voltage-dependent K^+ channel. Sem. Neurosci. 5, 117–124.

Toney, M. D., Hohenester, E., Cowan, S. W., & Jansonius, J. N. (1993). Dialkylglycine decarboxylase structure: Bifunctional active site and alkali metal sites. Science 261, 756–759.

Warmke, J., Drysdale, R., & Ganetzky, B. (1991). A distinct potassium channel polypeptide encoded by the *Drosophila* eag locus. Science 252, 1560–1562.

Wittka, R., Stocker, M., Boheim, G., & Pongs, O. (1991). Molecular basis for different rates of recovery from inactivation in the *Shaker* potassium channel family. FEBS Lett. 286, 193–200.

Yellen, G., Jurman, M. E., Abramson, T., & MacKinnon, R. (1991). Mutations affecting internal TEA blockade identify the probable pore-forming region of a K^+ channel. Science 251, 939–942.

Yool, A. J., & Schwarz, T. L. (1991). Alteration of ionic selectivity of a K^+ channel by mutation of the H5 region. Nature 349, 700–704.

Zagotta, W. N., Hoshi, T., & Aldrich, R. W. (1990). Restoration of inactivation in mutants of *Shaker* potassium channels by a peptide derived from *Sh*B. Science 250, 568–571.

VOLTAGE-GATED CALCIUM CHANNELS

Gabor Mikala, John L. Mershon, and Arnold Schwartz

Biomembranes
Volume 6, pages 221–248.

ISBN: 1-55938-663-0.

I. INTRODUCTION

It is well established that calcium ions (Ca^{2+}) act as intracellular messengers. Several important cellular events including muscle contraction, vesicular exocytosis, synaptic transmission, and even the expression of certain genes, are initiated or modulated when the concentration of free cytoplasmic Ca^{2+} rises into the micromolar range. The concentration of free Ca^{2+} ions is most critical for cellular function. Ca^{2+} is chelated by a variety of metabolites (e.g., citrate) and proteins inside the cell, and is also sequestered by organelles such as mitochondria and the endoplasmic reticulum. In addition, cells possess several pumps and exchangers which can quickly remove free Ca^{2+} ions from the cytoplasm when the Ca^{2+} concentration rises. These components or "sinks" are essential for maintaining the resting concentration of Ca^{2+} in the cytoplasm at very low levels—in the range of 0.1 μM. Since the Ca^{2+} concentration in the extracellular compartment is in the millimolar range, there is a steep concentration gradient, supported by the electric potential gradient, across the cell membrane. It is notable that the magnitude of this electrochemical gradient is much greater than that for Na^+ or K^+ ions.

Transient increases in the free cytoplasmic Ca^{2+} concentration act as important and, in some cases, vital intracellular signals. Changes in free Ca^{2+} are initiated via two major pathways: a release from intracellular storage sites or an influx from the extracellular space. The latter is due to an increase in the Ca^{2+} permeability of the plasma membrane. The transient increase in permeability is achieved by the opening of ion channels through which Ca^{2+} can pass down its electrochemical gradient. With respect to the signals responsible for channel-opening and -closing (gating), there are two major classes of channels that allow Ca^{2+} to move into the cytoplasm. One class, which can be directly opened subsequent to the binding of an agonist, is called ligand-gated channels (e.g., N-methyl-D-aspartate receptor or nicotinic-acetylcholine receptor). The other class, voltage-gated calcium channels (VGCC), open in response to depolarization of the plasma membrane. These voltage-dependent channel openings, together with a high selectivity for Ca^{2+}, are the critical hallmarks of a VGCC.

Voltage-gated calcium channels have been extensively investigated by conventional electrophysiological and biochemical techniques. However, the advances of cloning and functional expression techniques have opened a new, broader view of these macromolecular structures. This chapter first summarizes the information obtained using conventional techniques. We will then focus on more recent studies addressing the molecular and functional diversity of VGCCs which have provided an insight into the functional subunit structure of these channels.

II. THE FIVE CLASSICAL TYPES OF VGCCs AND THEIR PROPERTIES

Excitable cells can contain multiple types of VGCCs. Different criteria are used to classify these channels. When defined by the pattern of channel activation, two major classes appear: low voltage-activated (LVA) and high voltage-activated (HVA) channels. It is customary to term LVA channels as T-type ("T" stands for *transient* or *tiny*), while HVA channels have been further divided into four subclasses: L-type (where "L" stands for *long*-lasting), N-type (where "N" stands for *neither* "L" nor "T," or more recently, *neuronal*), P-type (first described in cerebellar *Purkinje* cells), and a fifth channel type (undesignated) which cannot be placed into any of these subclasses.

A. T-type Channels

The existence of a low-threshold calcium current has been identified in numerous tissues including cardiac myocytes (Bonvallet, 1987), the specialized conducting system of the heart (Hagiwara et al., 1988), and smooth muscle cells of the vascular and visceral system as well as skeletal muscle myotubes. Several regions of the central and peripheral nervous system, e.g., the inferior olivary nucleus (Llinás and Yarom, 1981), the substantia nigra (Bertolino and Llinás, 1992), dorsal root ganglia (Carbone and Lux, 1984, 1987) and so forth, are also rich sources of these channels.

T-type calcium channels require only a slight depolarization for activation and carry a transient current at negative membrane potentials that inactivates rapidly during prolonged depolarizations (Hirano et al., 1989). The T-type current of rat dorsal root ganglion neurons (Carbone and Lux, 1987) activates at approximately –50 mV and reaches its peak between –40 and –10 mV. Channel inactivation is

Table 1. Classification of VGCCs Based on Their Physiological Properties

Channel Class	*Properties*	*Conductance*	*Blockers*
T-type	LVA, transient currents	8 pS	Ni^{2+}, octanol, amiloride
L-type	HVA, long lasting responsive to DHP agonists and antagonists	25 pS (heart) 12 pS (skeletal muscle)	Dihydropyridines Penylalkylamines, benzothiazepines
N-type	HVA, moderate rate of inactivation	13–16 pS	-Conotoxin-GVIA -MVIIA
P-type	HVA, non-inactivating	14 pS (also 9 pS and 19 pS)	AGA-IVA, FTx, (S)FTX
Non-classified	HVA, fast inactivation	12–15 pS	Ni^{2+}

Source: The table has been compiled from references cited in the text. The conductance value for skeletal muscle L-channels is from Yatani et al. (1988).

very rapid; at holding potentials more positive than -60 mV, the channel is completely inactivated; however, inactivation is gradually eliminated as the holding potential decreases from −50 mV to −100 mV. Interestingly, no Ca^{2+}-dependent inactivation component has been identified in the behavior of these channels (Dupont et al., 1986). These channels have similar maximal conductances to Ca^{2+} and Ba^{2+}; in the presence of 110 mM Ba^{2+} as the charge carrier, the conductance value is in the order of 8 pS (Nowycky et al., 1985). It is also of considerable interest that T-type channels show a high resistance to β-adrenergic stimulation and to "washout/run down" during intracellular perfusion or isolated patch studies (Llinás et al., 1989; Pelzer et al., 1992). These features are quite different from those of L-type channels, which often coexist in the same cell membranes.

Selective inhibition by different pharmacological agents is a convenient and widely used tool to identify different channel subtypes. T-type calcium channels are uniquely sensitive to Ni^{2+} ions (100 μM) (Carbone and Lux, 1984; Nowycky et al., 1985). Lower-chain aliphatic alcohols (octanol, nonanol, or decanol) also exert a selective blocking action on these channels (Llinás and Yarom, 1986). Menthol, a cyclo-aliphatic alcohol (0.1–1 mM) reduces the amplitude of the low-threshold Ca^{2+} current in a dose-dependent manner without affecting inactivation kinetics (when applied to dorsal root ganglion neurons) (Swandulla et al., 1987). The pyrazine-class diuretic, amiloride, causes a partial block of the T-type calcium channel (Tang et al., 1988), but also affects Na^+/H^+ and Na^+/Ca^{2+} exchangers (Siegl et al., 1984), glutamate receptors (Manew et al., 1990), and epithelial Na^+ channels (Bertolino and Llinás, 1992).

The precise role of T-type Ca^{2+} channels in cardiac cells has not yet been established (Hirano et al., 1989), but their cellular distribution and activation at negative membrane potentials suggest a possible function in pacemaker activity (Hagiwara et al., 1988). T-type channels may also be involved in the release of trigger Ca^{2+} from the sarcoplasmic reticulum (Morad and Clyman, 1987), although this activity is probably very minor in view of the low channel density and very small currents at plateau potentials in cardiac myocytes (Droogmans and Nilius, 1989). In neural cells, where T-type currents are activated at negative membrane potentials close to the resting potential, the T-type channel is thought to be responsible for neuronal oscillatory activity (Llinás, 1988; White et al., 1989). This phenomenon represents spontaneous membrane potential fluctuations not mediated by synaptic activity. It has also been suggested that T-type channels can mediate Ca^{2+} entry triggering the Ca^{2+}-dependent release of dopamine from the dendrites of *pars compacta* neurons of the substantia nigra (Bertolino and Llinás, 1992).

B. L-Type Channels

L-type calcium channels are widely distributed throughout the cells of the body; they are found in most excitable cells (muscle, nerve, or gland cells), but, curiously, nonexcitable cells like fibroblasts and macrophages harbor them as well. These channels constitute an essential link between transient changes in membrane potential and a variety of cellular responses including: initiation of contraction in skeletal, cardiac, and smooth muscles, hormone secretion; and activation of second-messenger systems in many cell types (Catterall, 1988).

The kinetic and conduction properties of L-type Ca^{2+} channels clearly place this channel type as an essential link between membrane depolarization and an intracellular Ca^{2+} signal; they generate currents that are activated by large depolarizations, particularly from depolarized holding potentials, and inactivate with a slow time course (Nowycky et al., 1985). Using physiological extracellular divalent ion concentrations, the threshold for activation of the L-type current in cardiac myocytes is around –40 mV (Beeler and Reuter, 1970; Cavalié et al., 1983). The current begins to inactivate at holding potentials more positive than –60 mV and it reaches its maximal amplitude in the range of +10 mV. In cardiac cell-attached patches, the current–voltage relationship of the L-type Ca^{2+} channel is linear over most of the voltage range, whether the charge carrier is Ca^{2+} or Ba^{2+}. However, for external divalent cation concentrations in the range of 20–110 mM, unitary Ba^{2+} currents are much larger than Ca^{2+} currents; the limiting conductances are about 20- to 25-pS and 8- to 10-pS, respectively (Cavalié et al., 1983). Inactivation of the L-type channel appears to be governed by both voltage-dependent and Ca^{2+}-dependent mechanisms, each supporting the other. The relative importance of each mechanism varies under different conditions. A physiologically important point of regulation, β-adrenergic stimulation, increases Ca^{2+} current through L-type channels (Armstrong and Eckert, 1987; Hosey and Lazdunski, 1988).

Interestingly, while most of the physiological data regarding L-type channels have been obtained from cardiac muscle cells, the biochemical structure of the L-type channels has been most widely studied in skeletal muscle T-tubules because of the high concentration of these channels in this tissue. This Ca^{2+} channel is composed of five different polypeptide subunits (Catterall, 1988): the α_1-subunit (175 kD) which contains the pore, the gating mechanism, and the organic Ca^{2+}-antagonist binding sites; the α_2-subunit (143 kD), which is associated with the α_1-subunit and is highly glycosylated; the intracellular β-subunit (54 kD), which can be phosphorylated (together with the α_1-subunit) by protein kinases; the transmembrane γ-subunit (30 kD) and δ-subunit (27 kD; Figure 1).

L-type calcium channels are inhibited by transition metal ions, small organic molecules, and certain polypeptide toxins. These channels are effectively blocked by the "classical" divalent cation blockers such as Cd^{2+}, Co^{2+}, La^{3+} (Nathan et al., 1988) and, to a lesser extent, by Ni^{2+}. The sensitivities of the different L-channel subtypes show significant differences (e.g., the cardiac L-type channel responds to the multivalent blockers with the sequence $La^{3+} > Cd^{2+} > Co^{2+} > Ca^{2+}$, while the skeletal muscle channel responds: $Ca^{2+} > Cd^{2+} \sim Co^{2+}$; Tsien et al., 1987). The cardiac channel is virtually impermeable to Mg^{2+}, while skeletal channels show measurable Mg^{2+} currents (Lansman et al., 1986). However, the most characteristic blockers of L-type Ca^{2+} channels are the "organic Ca^{2+} channel blockers." These

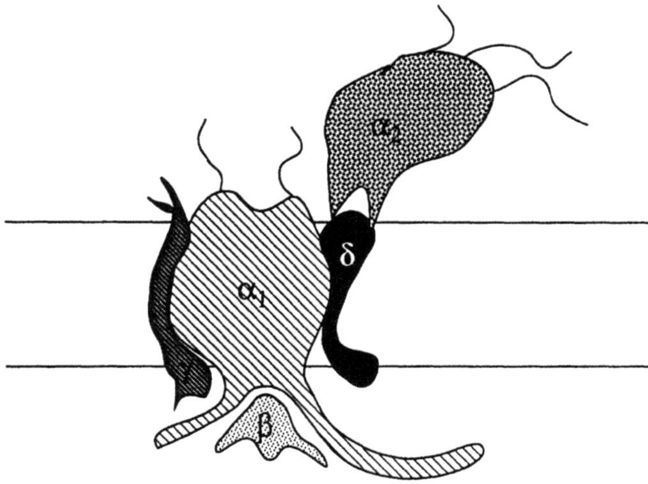

Figure 1. Subunit structure of VGCCs. The α_1-subunit is a highly hydrophobic protein, with cytoplasmic N- and C-termini. The α_2-subunit is a highly glycosylated extracellular protein covalently attached to the membrane-spanning γ-subunit. The small β-subunit is cytoplasmic and tightly bound to the α_1-subunit. The γ-subunit is a transmembrane protein that may be specific to the skeletal muscle L-channel.

include the 1,4- dihydropyridines (DHP), such as nifedipine; phenylalkylamines (PAA), such as verapamil; benzothiazepines (BTZ), such as diltiazem; and diphenylmethyl-piperazines, such as flunarizine. There are many other mostly inhibitory organic molecules that affect L-type Ca^{2+} channel activity that are not mentioned here because of their lack of specificity (for review see Porzig, 1990). The DHPs are the most characteristic organic modulators of the L-type calcium channel (Kongsamut et al., 1985) because of their high specificity and affinity. Thus, sensitivity to DHPs is considered a defining characteristic for L-type Ca^{2+} channels. Interestingly, optical enantiomers of DHPs often have opposites effect on Ca^{2+} channel function (Franckowiak et al., 1985). For example, (–)Bay K 8644 potentiates Ca^{2+} currents (Brown et al., 1984) while (+)Bay K 8644 is inhibitory. A further example of the complexity of DHP modulation is that even a single enantiomer can exert a dual effect (Kass, 1987; Sanguietti et al., 1986): (+)Bay K 8644 has a minor stimulatory effect when ventricular myocytes are pulsed from –80 mV and an inhibitory effect when pulsing is at –30 mV; (–)Bay K 8644 is a potent stimulator from –80 mV and a moderate blocker from –30 mV. Clearly, the three dimensional conformations of the drug-sensitive regions are critical in determining the final effect. Recently, polypeptide toxins have been discovered that block Ca^{2+} channels, including agatoxins (Mintz et al., 1991), conotoxins (Olivera et al., 1990), and certain toxins from snake venoms, such as calciseptine which blocks cardiac and neuronal L-channels, but does not affect skeletal muscle L-channels (Weille et al., 1991). These toxins may prove to be valuable experimental tools for the evaluation of the L-type channel.

L-type Ca^{2+} channels have multiple functions. The most understood is the skeletal muscle L-type channel, which plays a unique role in mediating (skeletal muscle type) excitation–contraction coupling. The latter process in this tissue is observed even in Ca^{2+}-free external medium, providing evidence, though still much debated, that Ca^{2+} entry through L-channels is not essential nor significant in the initiation of contraction. However, excitation–contraction coupling is, under certain conditions, sensitive to DHPs, suggesting that the skeletal muscle Ca^{2+} channel does play an important role in it (Miller, 1992). It is hypothesized that the Ca^{2+} channel protein (i.e., the α_1-subunit) responds to depolarization with a conformational change that reaches the Ca^{2+}-release channel of the sarcoplasmic reticulum. The Ca^{2+}-release channel (otherwise known as the "ryanodine receptor") opens and allows Ca^{2+} to flow into the cytoplasm, triggering muscle contraction. In contrast to skeletal muscle, the cardiac L-type calcium channel provides excitation–contraction coupling in a more "balanced" way. Although a conformational change probably does reach the Ca^{2+}-release channel, Ca^{2+} influx during the "plateau phase" of the action potential is indispensable. In the case of neuroendocrine L-channels, the role of the L-type Ca^{2+} channel in excitation–secretion coupling is probably similar to that in cardiac channels (in the excitation–contraction coupling). Neuronal L-channels are located on the cell bodies and proximal dendrites of the neurons, especially clustered at high density at the base of major dendrites (Westenbroek et al., 1990). It is hypothesized that neural L-channels are mainly involved in postsynaptic processes; nevertheless, the wide distribution of L-channels probably reflects multiple functions for these proteins (Bertolino and Llinás, 1992).

C. N-type Channels

N-type channels are HVA Ca^{2+} channels that are widely distributed in the nervous system, but appear to be absent from muscle tissues. They are also represented in some endocrine cells, such as those of the anterior pituitary, adrenal gland, and pancreas. The N-type Ca^{2+} channel is insensitive to DHPs and is defined by the characteristic irreversible blockade by ω-conotoxin (ω-CgTx; Nowycky et al., 1987).

N-type Ca^{2+} channels seem to represent a subclass of Ca^{2+} channels with considerably heterogeneous physiological properties. Even within one type of neural tissue, there is still controversy concerning the detailed properties and identification of these channels. In dorsal root ganglion neurons, the N-type channel is distinguished by having a conductance of 13 pS, a range of inactivation between −120 mV and −30 mV, and a decay-time constant between 50 ms and 80 ms (Nowycky et al., 1985). Single-channel analysis of rat sympathetic neurons has demonstrated the existence of N-type Ca^{2+} channels with a unitary conductance of 20 pS and a subconductance state of 13 pS (Plummer et al., 1989). This Ca^{2+} current seems to have a slowly inactivating component and a long-lasting sustained component (Thayer et al., 1987), and the same N-type channel may be responsible for both of these two components (Plummer and Hess, 1991). Until recently, the subunit structure of N-type channels was not clearly understood. ω-CgTx binding sites were identified in marine ray electric organs having a molecular mass of ~200 kD under reducing conditions (Miller, 1992). This labeled protein represents the α_1-subunit. Antibodies raised against skeletal muscle α_2/δ-subunits which co-precipitate brain DHP receptors, co-precipitate only a small percentage of ω-CgTx binding sites (Miller, 1992). On the contrary, antibodies against the skeletal muscle β-subunit co-precipitate the majority of ω-CgTx binding sites (Sakamoto and Campbell, 1991). Recently, a functional N-type Ca^{2+} channel complex was purified from rabbit brain (Witcher et al., 1993). It was shown that the channel consists of a 230 kD α_1-subunit, a 160 kD $\alpha_2\delta$-subunit, a 57 kD β-subunit (a class β_3-subunit), and a unique 95 kD glycoprotein subunit (Witcher et al., 1993).

N-type channels seem to be quite similar to cardiac L-type channels with regard to ion permeation and response to multivalent transition metal blockade: high sensitivity to Cd^{2+} ($K_i \sim 1\ \mu M$) and lower sensitivity to Ni^{2+}, Co^{2+}, and Mn^{2+} ($K_i \geq 50\ \mu M$; Tsien et al., 1987). N-type channels are characteristically blocked by a peptide toxin from the piscivorous marine snail *Conus geographus* (Olivera et al., 1985), ω-conotoxin GVIA (Nowycky et al., 1987). The different subtypes of N-channels show a slightly different sensitivity to ωCgTx and can also be distinguished by using a more narrow specificity conotoxin from *Conus magus*, ω-conotoxin MVIIA (Hillyard et al., 1992).

The exact physiological roles for N-type channels are not fully understood. It is clear, however, that these channels play a critical role in some forms of neurotransmitter release that are ω-CgTx sensitive and DHP-resistant (Miller, 1992). Direct measurements have shown that a predominant Ca^{2+} channel of the presynaptic terminals has N-type pharmacology, that is, DHP-resistant and ω-CgTx sensitive. Furthermore, ω-CgTx binding sites have been localized to "active zones" of neuron terminals (Westenbroek et al., 1992) that are known to represent Ca^{2+} channels supplying Ca^{2+} for neurotransmitter release (Miller, 1992). Interestingly, injection of ω-CgTx into mammalian cerebrospinal fluid produced only minor effects (Olivera et al., 1984), suggesting that the Ca^{2+} entry through N-type channels plays only a minor role in the transmitter release in the mammalian central nervous system. Nevertheless, it is possible that N-type channels exert their role in excitation–secretion coupling primarily through protein–protein interactions, analogously to the role of the skeletal muscle L-type channel in the skeletal muscle type of excitation–contraction coupling.

D. P-type Channels

P-type Ca^{2+} channels were only recently recognized as DHP- and ω-CgTx-resistant channels with unique properties (Regan et al., 1991; Llinás et al., 1992).

These channels were initially discovered in the Purkinje cells of the cerebellar cortex, exhibiting little inactivation and an intermediate voltage-dependence of activation (between those for the N-type and T-type channels). Now it is known that although these channels are highly abundant in Purkinje cells (Usowicz et al., 1992), they are not restricted to them, but are widely present in the brain stem, olfactory bulb and neurocortex, and even in the pituitary (Llinás et al., 1992).

P-type channels activate in the range of –45 mV to –35 mV, and show essentially no inactivation. To investigate P-type channels, it is essential to use specific blockers of these channels. Two neurotoxins of the funnel web spider *Agelenopsis aperta* are used for this purpose: the polyamine, FTX (Llinás et al., 1989) and the polypeptide ω-Aga-IVA (Mintz et al., 1992). The FTX-sensitive P-currents have a conductance of ~ 10- to 15-pS using 80 mM Ba^{2+} and 5- to 8-pS using 100 mM Ca^{2+} as the charge carrier (Llinás et al., 1989). The open probability of the P-type channel is voltage-dependent and the channel is blocked by Cd^{2+} and Co^{2+} ions. These channels are insensitive to DHPs or ω-conotoxin-GVIA and have a unique monovalent ion selectivity in the absence of divalent cations. The permeability sequence is: $Rb^+ > Na^+ > K^+ > Li^+ > Cs^+$ (Bertolino and Llinás, 1992).

The observations that P-type Ca^{2+} channels are widely present in the central nervous system and that ω-Aga-IVA blocks more than 70% of the voltage-dependent Ca^{2+} entry in synaptosomes prepared from whole rat brain indicate that these channels may play a very basic role in synaptic function throughout the brain (Mintz et al., 1992; Uchitel et al., 1992). P-channels also appear to be important in neural integration. Interestingly, these channels co-localize with a Ca^{2+}-binding protein designated calbindin (Mintz et al., 1992). During the process of aging, in those cells which undergo degeneration, the ratio of P-channels to calbindin is increased (Llinás et al., 1992). These changes may be an early marker of impending neural cell death.

E. Non-classified VGCCs

There are HVA Ca^{2+} channel subtypes in certain neurons which fall outside the categories of L-, N- and P-type. Under some conditions, these channels may substantially contribute to Ca^{2+} entry into the cell. However, since a thorough investigation of these channels is only beginning, very limited information is presently available about these channels.

Recently, a clear example of non-L, non-N, non-P, non-T Ca^{2+} channels was identified and characterized in rat cerebellar granular cells (Ellinor et al., 1993). These are rapidly decaying, Ni^{2+}- and Cd^{2+}-sensitive, high-voltage activated currents. In the presence of ωCgTx-GVIA, ω-Aga-IVA, and nimodipine (blockers for L-, N-, and P-currents), a transient current appears in these neurons; channel activation starts at –60 mV and peak currents can be obtained around 0 mV. The rapid inactivation is illustrated by a small τ_h (22.2 ± 1.3 ms at 0 mV) and by the fact that steady-state inactivation occurs at relatively negative potentials ($V_{1/2}$ = 62 mV in 5 mM Ba^{2+}).

The fast inactivation of these VGCCs may be very important for certain neural functions; the total Ca^{2+} entry (especially during repetitive depolarizations) is significantly smaller than in the case of the opening of other Ca^{2+} channels (except T-type). This reduced Ca^{2+} entry might result in less K^+ channel activation and, therefore, preserve the excitability of the cell (Ellinor et al., 1993).

III. MOLECULAR PROPERTIES OF VGCCs

The cloning and functional-expression of VGCC cDNAs has opened a new window into the investigation of Ca^{2+} channel function. In this chapter, we will summarize the structural knowledge concerning the different VGCC subunits and their interactions obtained through molecular cloning and expression studies.

A. The α_1-Subunits

The molecular architecture of the Ca^{2+} channel was first described for skeletal muscle L-type VGCCs. These initial studies were founded on the previous biochemical characterization of this channel, the abundance of which is very high in skeletal muscle relative to other tissues. Indeed, each of the five subunits of this channel was cloned based on the detailed biochemical knowledge of the skeletal muscle "DHP-receptor" (amino acid sequences, antibodies) and opened the way to the cloning of corresponding subunits from other tissues.

Table 2 summarizes the current molecular biological knowledge of the cloned full-length α_1-subunits. In addition, this table broadly classifies the different α_1-subunits into six major groups; at this time, classification of cloned Ca^{2+} channel subtypes still exists without a widely accepted, clear and convenient nomenclature. Currently, molecular cloning has identified six primary types of α_1-subunits which are encoded by (at least) six distinct, but related genes. There are three different α_1-subunit genes encoding L-type Ca^{2+} channel subunits, and three "non-L" type genes. There has not been any report of the cloning of a T-type VGCC as of the time of preparation of this chapter.

All the α_1-subunit sequences show striking similarity to each other and (to a lesser extent) to the α-subunit of Na^+ channels. There are conserved basic structural units, that is, in each α_1-subunit there are four homologous repeating units (designated motifs I through IV). Each of these motifs contains six putative transmembrane α-helices (S1 to S6; Catterall, 1988). The fourth segment (S4) of each motif is exceedingly highly conserved and has a positively-charged amino acid (arginine or lysine) at every third (or fourth) position. This region is postulated to serve as the voltage sensor. There is another highly conserved region between S5 and S6 membrane spanning α-helices. It has been suggested that this region (divided into short segments SS1 and SS2) folds back into the membrane and forms the pore

Table 2. Properties of Cloned Full-length VGCC α_1-Subunits

Classes	*Alternative Classification*	*cDNA Clones*	*Amino Acids*	*Deduced MW(kD)*	*Distribution*	*Chromosomal Localization of its Gene*
Cardiac-type	CACHL1A1	pCARD3 (Mikami et al., 1989)	2,171	243	Ubiquitous:	
	CaCh2	pSCal (Biel et al., 1990)	2,166	242	heart, brain,	2p13.3
	type C/classC	rbC-I (Snutch et al., 1991)	2,140	240	smooth muscles,	
	CACN2	rbC-II (Snutch et al., 1991)	2,143	240	kidney,	
	CaCNL1A1	VSMα_1 (Koch et al., 1990)	2,169	244	fibroblasts	
		hHt-1 (Schultz et al., 1993)	2,180	244		
Skeletal-type	CaChl	pCaC6 (Tanabe et al., 1987)			skeletal muscle	1q32.1
	CACN1	DHPRα_1 (Perez-Reyes et al., 1989)	1,873	212		
	CChl1a3					
	CACNL1A3					
Neuroendocrine L-type	CaCh3	α_1D (Williams et al., 1992a)	2,161	245	brain, endocrine	
	type D/class D	CaCN4 (Seino et al., 1992b)	2,181	248	organs, kidney	3p14.3
	CACNL1A2	RBα_1 (Hui et al., 1991)	1,634	187		
		HCa3a (Yaney et al., 1992)	1,610	182		

(continued)

Table 2. (Continued)

Classes	Alternative Classification	cDNA Clones	Amino Acids	Deduced MW(kD)	Distribution	Chromosomal Localization of its Gene
Brain "P-like"	CaCh4	BI-1 (Mori et al., 1991)	2,273	257	brain, especially	
	BI	BI-2 (Mori et al., 1991)	2,424	273	cerebellum, pituitary	?
	type A/Class A CACN3	rBA-1 (Starr et al., 1991)	2,212	250		
Brain N-type	CaCh5	BIII (Fujita et al., 1993)	2,339	261		
	BIII	rbB-1 (Dubel et al., 1992)	2,336	262	brain, pituitary	?
	type B/Class B	doe-4 (Horne et al., 1993)	2,326	264		
		α_1B-1 (Williams et al., 1992b)	2,339	262		
		α_1B-2 (Williams et al., 1992b)	2,237	252		
Brain non-L	CaCh6	BII-1 (Niidome et al., 1992)	2,259	254	brain	
non-D	BII					
non-N	type E/Class E	BII-2 (Niidome et al., 1992)	2,178	245		
		doe-1 (Ellinor et al., 1993)	2,223	251		?
		rbE-II (Soong et al., 1993)	2,222	252		

Source: The table has been compiled from references cited in the text. In addition to these, chromosomal localization data are obtained from Schultz et al. (1993) Iles et al. (1993), and Seino et al. (1992b).

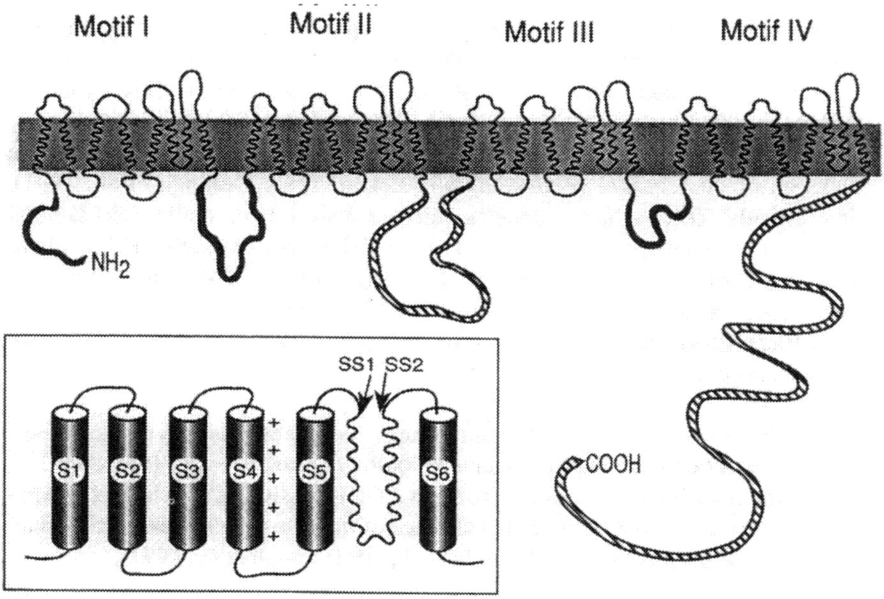

Figure 2. Transmembrane architecture of the VGCCα_1 subunit. The α_1-subunits have four motifs, each with six putative transmembrane α-helices. In each motif, there is a region suggested to fold into the membrane as a β-barrel structure (SS1–SS2 segments). The insert shows a single motif in detail, with α-helices (S1–S6) represented by cylinders. The putative pore-forming β-barrel structure is indicated by the arrows (SS1, SS2). Moderately and highly divergent intracellular regions are shown shaded or stippled, respectively.

lining of the channel (Guy and Conti, 1990; Kim et al., 1993; Mikala et al., 1993; Tang et al., 1993). Although these homologous regions show striking homology, there are also domains that have a high degree of variability among the different subunit sequences. The major divergence between the α_1-subunit clones of different classes lies in the amino- and carboxy-terminal tails, as well as in the intracellular linker regions between each of the motifs. The highest level of divergence is observed in the loop between motifs II and III; this is also the longest loop protruding into the cytoplasm. Interestingly, this domain has been shown to be critical in determining the mechanism of excitation–contraction coupling in muscle cells (Tanabe et al., 1990a).

The Skeletal Muscle α_1-Subunit

The α_1-subunit of the skeletal muscle L-channel was the first cloned Ca^{2+} channel subunit (Tanabe et al., 1987; Ellis et al., 1988). The cDNA harbors an open reading frame of 1873 amino acids, encoding a polypeptide of 212 kD. The α_1-subunit, isolated from skeletal muscle T-tubules, consists predominantly of a 165 kD isoform (Hosey and Lazdunski, 1988; DeJongh et al., 1991). This raises the possibility that post-translational proteolytic processing of this subunit, possibly occurring at or near the C-terminal tail, may be important in the regulation of the function of this channel (Armstrong and Eckert, 1987; DeJongh et al., 1991). Interestingly, *Xenopus laevis* oocytes, injected with *in vitro* synthesized skeletal muscle α_1-subunit-specific message, do not exhibit any expressed DHP-sensitive (barium) current (Lory et al., 1991). Expression of this cDNA was, however, successfully achieved in dysgenic myotubes (Tanabe et al., 1988) and mouse L-cells (Perez-Reyes et al., 1989; Lacerda et al., 1991; Varadi et al., 1991). These expression studies clearly demonstrated that:

1. The expression of the α_1-subunit restores the skeletal muscle type excitation–contraction coupling in dysgenic myotubules (Tanabe et al., 1988); and
2. The α_1-subunit alone is able to form a DHP-sensitive Ca^{2+} channel (Perez-Reyes et al., 1989), but in the absence of the accessory subunits, a channel with abnormal kinetics (Lacerda et al., 1991; Varadi et al., 1991).

Cardiac-type α_1-Subunits

α_1-subunit clones encoding the members of the "cardiac" Ca^{2+} channel family (type-C channels) have been isolated from various tissues including rabbit (Mikami et al., 1989) and human heart (Schultz et al., 1993), rat brain (Snutch et al., 1991), rabbit lung (Biel et al., 1990), rat aorta (Koch et al., 1990), and human fibroblasts (Soldatov, 1992). Surprisingly, these cloning studies have revealed not only a high level of interspecies homology, but have demonstrated a previously unanticipated degree of structural diversity within the same family of Ca^{2+} channels (Snutch and Reiner, 1992). All of these channels exhibit the same predicted structure with the four-times-six membrane-spanning segments as well as longer amino- and carboxy-terminal intracellular regions (compared to the skeletal muscle α_1-subunit). However, divergent regions have been detected in various domains throughout the Ca^{2+} channel cDNAs. It is likely (and in certain cases proven) that these Ca^{2+} channel isoforms are generated by alternative splicing mechanisms from a single cardiac-type gene. The regions demonstrated to be involved in these processes are: the 5′ untranslated region (Biel et al., 1991; Schultz et al., 1993), the amino-terminus (Biel et al., 1991; Snutch et al., 1991; Schultz et al., 1993), motif IS6 (Biel et al., 1991; Soldatov, 1992), the loop between motifs I and II (Biel et al., 1991), motif IIIS2 (Soldatov, 1992), motif IVS3 and the linker between IVS3 and IVS4 (Perez-Reyes et al., 1990; Snutch et al., 1991; Diebold et al., 1992), as well as the carboxy-terminal tail (Figure 3). Functional expression studies to date have been unable to demonstrate significant differences in the behavior of these different isoforms. Nevertheless, it is intriguing that the expression of some isoforms is

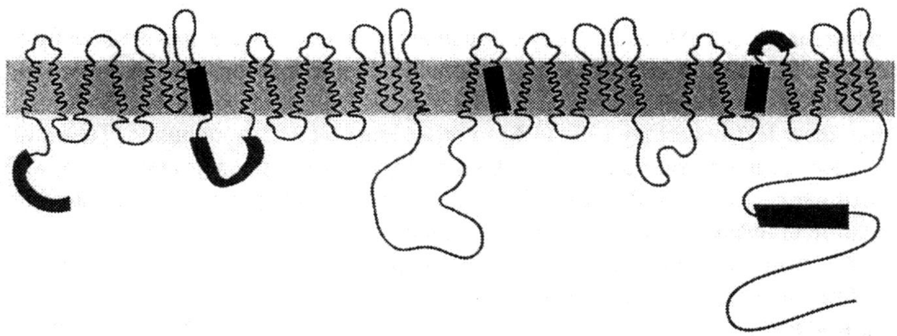

Figure 3. Regions involved in alternative splicing in cardiac-type L-channel α_1 subunits. Divergent domains suggested or proven to be generated by alternative splicing of a common cardiac L-channel α_1-subunit pre-mRNA, are shown in bold.

regulated both in a developmental (Diebold et al., 1992) and in a tissue-specific way (Snutch et al., 1991; Schultz et al., 1993), supporting functional significance.

Functional expression of the cardiac-type α_1-subunit cDNAs was first achieved in *Xenopus* oocytes (Mikami et al., 1989). Microinjection of RNA complementary to the cloned α_1-subunit was capable of producing voltage-dependent Ca^{2+} channel activity with native-like pharmacology (Mikami et al., 1989; Schultz et al., 1993). It was demonstrated that these α_1-subunits were also capable of forming a VGCC with the correct unitary conductance in the absence of the other subunits (Bosse et al., 1992; Schultz et al., 1993). However, the currents had slower than normal activation and inactivation kinetics, indicating a requirement for the other subunits. Upon injection of the α_1-subunit cDNA into the nuclei of dysgenic muscle myotubules, cardiac-type excitation–contraction coupling (i.e., exhibiting a strict dependence on extracellular Ca^{2+} and high sensitivity to Cd^{2+}) was produced as well as VGCC with kinetics resembling that of native cardiac channels (the myotubes do express the α_2-, β-, and γ-subunits of the VGCC; Tanabe et al., 1990). Mammalian cell expression of these α_1-subunits was achieved in CHO (Bosse et al., 1992) and L cells (Lory et al., 1993) as stable transfectants as well as in COS cells by transient expression (Perez-Reyes et al., 1992). These studies also showed VGCC expression with the expected pharmacology (Bosse et al., 1992; Perez-Reyes, 1992) and unitary conductance (Bosse et al., 1992), as well as the presence of a fully functional receptor for calcium-channel blockers in the cell membranes (Bosse et al., 1992). The slow kinetics of the currents were also noted in these studies.

Neuroendocrine α_1-Subunits

α_1-subunit cDNAs encoding the neuroendocrine L-type VGCC (type-D channels) have been isolated from rat (Hui et al., 1991) and human brain (Williams et al., 1992a), and from human (Seino et al., 1992a) and hamster (Yaney et al., 1992) pancreas. Partial cDNA clones of α_1-subunits of this class were also isolated from kidney (Yu et al., 1992). These α_1-subunits share a significant homology (~65%) with the other two members (skeletal muscle, cardiac) of the L-type VGCC class, but show less homology (33–44%) to other brain VGCC α_1-subunits. The most unique feature of the channel α_1-subunits of this class is the appearance of seven methionine codons at the beginning of the putative coding sequence, an unusual feature of unknown, but potential significance.

Alternative splicing, as a source of molecular diversity, has also been identified in this calcium channel family. Divergent regions that are likely to be alternatively spliced are very similar to those in cardiac-type channels: IS6 (Seino et al., 1992a; Williams et al., 1992a), the loop between IS6 and IIS1 (Hui et al., 1992), IVS3 (Perez-Reyes et al., 1990), the linker between IVS3 and IVS4 (Perez-Reyes et al., 1990), as well as the carboxy-terminal tail (Hui et al., 1992; Yaney et al., 1992). The divergence involving the length of the carboxy-terminal tail is a major point of interest. The clones isolated from rat brain and a hamster pancreatic clone exhibit a "truncated" version of this region (Hui et al., 1992; Yaney et al., 1992). Again, the functional significance of this truncation is unknown.

Expression of these α_1-subunits alone or in combination with α_2-subunits in *Xenopus* oocytes did not result in functional Ca^{2+} channel activity (Williams et al., 1992b). However, co-expression with the β-subunit did provide a high level of DHP-sensitive VGCC expression. These currents were different from those recorded by expressing cardiac-type VGCCs (Williams et al., 1992b). They exhibited different current–voltage relationships (the curve was shifted by approximately −20 mV to depolarized potentials); their tail currents were markedly prolonged after application of Bay K 8644; and they showed a partial and reversible block with 10–15 μM ω-CgTX. This weak ω-CgTx block may very well account for the observation in native cells that ω-CgTx blocked not only N-type but also L-type neuronal VGCCs (Williams et al., 1992b). It is also of particular importance that cardiac-type and neuroendocrine-type L-channels, which are often co-expressed in the same cells, may make unique contributions to the physiological or pathological behavior of these cells (Snutch and Reiner, 1992).

"Brain" α_1-Subunits

Three classes of VGCC α_1-subunits, predominantly or exclusively expressed in the brain, have been identified to date by molecular cloning: BI (Mori et al., 1991; Starr et al., 1991), BII (Niidome et al., 1992; Ellinor et al., 1993; Soong et al., 1993), and BIII (Dubel et al., 1992; Williams et al., 1992b; Fujita et al., 1993; refer to Table 2 for alternative designations). These channel α_1-subunits exhibit a significant sequence homology (58–82%) to each other and a lower homology (40–51%) to members of the DHP-sensitive L-type VGCC family. This sequence homology is the basis for classifying them as members of the "non-DHP-sensitive" VGCC subfamily (Fujita et al., 1993). The phylogenetic tree constructed on the sequence identities also indicates that the relationship between BI and BIII is closer than that between BI and BII or BII and BIII (Fujita et al., 1993). The most significant divergences are found in the loop between motifs II and III as well as in the carboxy-terminal region. It has been suggested that the loop between motifs II and III may interact with unique intracellular neural proteins in the same way the analogous region of the skeletal muscle α_1-subunit mediates skeletal-type excitation–contraction coupling.

Complementary DNA clones representing members of the BI brain Ca^{2+} channel α_1-subunit family have been isolated from rat (Starr et al., 1991) and rabbit (Mori et al., 1991) brain tissues. These cDNAs encode ~250–270 kD proteins that are significantly different from those encoded by the different L-type VGCC α_1-subunit cDNAs. The differences are most pronounced in two cytoplasmic regions: the linker between motifs II and III and the carboxy-terminal tail. It is notable that potential alternative splicing was also detected in the C-terminal region, indicated by the isolation of unique cDNAs that differ solely in this region (Mori et al., 1991; Starr et al., 1991). Northern blot analysis demonstrated high levels of expression of these α_1-subunits in the cerebellum. Nevertheless, variable amounts of their messages were present in every investigated region of the central nervous system with a lower level of expression detected in the heart and pituitary gland (Starr et al., 1991).

Functional expression of at least two isoforms of the α_1-subunit of this channel was achieved in *Xenopus* oocytes (Mori et al., 1991). The α_1-subunit alone was capable of inducing inward Ba^{2+} current. However, these currents were small. On the other hand, co-injection of α_1- and skeletal α_2-, and especially α_1- and skeletal β-subunit specific cRNAs resulted in a dramatic increase in peak Ba^{2+} currents. The currents obtained by co-expression of α_1-, α_2-, and β-subunits were highly sensitive to Cd^{2+} (IC_{50} ~ 0.5 μM), and insensitive to Ni^{2+} (100 μM), nifedipine (10 μM), or ω-CgTx (10 μM). Interestingly, the crude venom of *A. aperta* reduced the peak Ba^{2+} current by 40% (Mori et al., 1991). These data, taken together with the high cerebellar expression, suggest that the current expressed by the members of the BI channel family may represent P-type current; however, there are several conflicting points. The BI-expressed current cannot be completely blocked by *A. aperta* venom, these currents show significant inactivation, and the single-channel slope conductance does not match those obtained from native P-channels.

Complementary DNAs encoding members of the BII VGCC α_1-subunit family have been isolated from rabbit (Niidome et al., 1992) and rat (Soong et al., 1993) brain, as well as from the electric organ of the marine ray, *Discopygne omnata* (Ellinor et al., 1993). The deduced amino acid sequence of these clones revealed VGCC α_1-subunits distantly related to other known VGCC α_1-clones. Functional expression of the *D. omnata* clone of this class in *Xenopus* oocytes revealed unique channel properties (Ellinor et al., 1993). The currents obtained by the co-expression of doe-1-, α_2- and β_3-cRNAs were HVA, rapidly inactivating, and insensitive to DHPs and ω-Aga-IVA. However, these currents were sensitive to Cd^{2+} and Ni^{2+} (IC_{50} ~ 33 μM), and were blocked reversibly by ω-CgTX-GVIA. These features correspond well to those of the unclassified cerebellar granular cell Ba^{2+} current. Functional expression of the rat rbE-II clone resulted in Ba^{2+} currents with similar properties, except that these channels were LVA (but not sensitive to octanol and amiloride) and insensitive to ω-CgTX-GVIA (Soong et al., 1993). These features are closest to the LVA currents described in neurons isolated from the reticular nucleus of the thalamus.

BIII-type α_1-clones were obtained from human (Williams et al., 1992a), rat (Dubel et al., 1992), and rabbit (Fujita et al., 1993) brain. These clones were expressed in mammalian HEK293 cells (Williams et al. 1992a) and in dysgenic muscle myotubes (Fujita et al., 1993); antibodies to peptides taken from the deduced sequences precipitated ω-CgTx-binding sites (Dubel et al., 1992). Functional expression (in the presence of α_2- and β-subunits) revealed HVA Ba^{2+} currents which were irreversibly blocked by ω-CgTX-GVIA (5 μM; Williams et al., 1992a; Fujita et al., 1993). The expressed currents were Cd^{2+}-sensitive and DHP-insensitive and had a unitary slope conductance of 14.3 pS (Fujita et al., 1993). These data clearly indicate the successful cloning and expression of N-type Ca^{2+} channel cDNAs. It is also likely that the doe-4 clone of *D. omnata* represents a class BIII channel cDNA (Horne et al., 1993). Functional expression of the clones also revealed a single class of high-affinity ω-CgTx binding sites (K_D ~ 55 pM; Williams et al., 1992b), providing further evidence for the isolation of a presynaptic Ca^{2+} channel.

B. The α_2/δ Subunits

In addition to the pore-forming α_1-subunits, Ca^{2+} channels also contain the accessory α_2-, β-, γ-, and δ-subunits. It was shown that the α_2- and δ-subunits are synthesized as a single pre-protein which is proteolytically processed into the α_2-subunit (the N-terminal portion) and to a heterogeneous population of δ-peptides, all of which have a common N-terminus at Ala^{935} (DeJongh et al., 1990). The α_2- and δ-subunits remain bound to each other via disulfide bridges.

Cloning of the α_2/δ-subunit cDNAs has been achieved from rabbit skeletal muscle (Ellis et al., 1988), rat brain (Kim et al., 1992), and from human brain (Williams et al., 1992a). These cDNAs are highly homologous; the skeletal muscle clone encodes a 125 kD protein while the brain sequences yield a calculated molecular weight of the protein of ~ 123.5 kD. The major difference between these clones is a small divergent region where the brain isoforms (probably splice variants of the skeletal muscle subunit) have an insertion of seven amino acid residues and a deletion of a 19 amino acid segment (Kim et al., 1992; Williams et al., 1992b). It was also noted that the (human) aorta might express a third isoform of the α_2-subunit (Williams et al., 1992b). All of these α_2-subunit clones encode proteins with 18 conserved N-glycosylation sites, however, the functional role of the extensive glycosylation of these subunits is not known. There is also controversy concerning the physical arrangement of these subunits in relation to the membrane and also to the other subunits. It is, however, generally accepted that the δ-portion of the protein acts as a membrane anchor for the largely extracellular α_2-portion.

Several expression studies have suggested that the α_2/δ-subunits do have important functional interactions with the α_1-subunits of different classes. First, it was noted that the magnitude of the current in *Xenopus* oocytes induced by the expression of a cardiac α_1-subunit was enhanced by co-expression of these subunits (Mikami et al., 1989). In some studies, a similar co-expression also altered the kinetic properties of the expressed cardiac-type current (Singer et al., 1991). In the oocyte expression of the neuroendocrine L-channel α_1-subunit (Williams et al., 1992a), the BI channel (Mori et al., 1991), and also other "brain" channels, co-expression of α_2/δ-subunits enhanced the current density without a significant change in channel kinetics, suggesting a "stabilizing" role for these subunits. Co-expression of the skeletal muscle α_2/δ-subunit with the skeletal muscle α_1-subunit in mouse L-cells increased DHP-binding density (~ 2-fold increase), whereas the current kinetics remained unchanged (Varadi et al., 1991). Based on these observations, it is possible that the α_2/δ-subunits play a "membrane-anchor" role for the α_1-subunits, increasing their half-life (Lory et al., 1992).

C. The β-Subunits

The β-subunit of a Ca^{2+} channel was first identified as a small molecular weight (~ 55 kD) phosphorylated protein that consistently copurified with the skeletal muscle α_1-subunit (Catterall, 1988). Using partial amino acid sequences obtained from fragments of this protein, the skeletal muscle version of the β-subunit became the first cloned low molecular weight Ca^{2+} channel subunit (Ruth et al., 1989). Analysis of the deduced amino acid sequence of this clone indicated a hydrophilic protein that lacks any membrane-spanning regions (or signal sequence). The secondary structure predicted for this protein suggests that the β-subunit contains four major α-helical domains (I–IV). The hydrophobic residues within domains II, III, and IV were homologously arranged in a heptad repeat that is thought to interact with cytoskeletal proteins (Ruth et al., 1989). These properties support the previously obtained biochemical data indicating that the β-subunit is tightly associated with intracellular regions of the α_1-subunit. Several potential phosphorylation sites were also noted in the β-subunit sequence. This is consistent with its known biochemical properties and raises the possibility that phosphorylation of these sites has a functional role in the regulation of Ca^{2+} channel function.

Using the sequence information provided by the skeletal muscle β-subunit, an explosive growth has occurred recently in the field of β-subunit cloning and expression. The classification of these many β-subunit clones remains controversial. Table 3 summarizes the properties of the β-subunit clones published to date using a classification based on the suggestions of Castellano et al. (1993a). There appears to be at least four distinct β-subunit genes termed β_1 through β_4, each of which gives rise to alternatively spliced isoforms. The β_1 gene, which codes for the

Table 3. Properties of Cloned Full-length VGCC β-subunits

Classes	*Synonymous Names*	*Message Sizes*	*cDNA Clones*	*Amino Acids*	*Deduced MW (kD)*	*Distribution*
β1	Skeletal β(B_{sk})	3.0 kb	CaB1 (β_{1a})	524	58	Ubiquitous:
		1.9 kb	Brain β(β_{1b})	597	66	skeletal muscle,
	CaB1	1.6 kb	β_2	478	53	brain, heart,
			β_a	597	66	smooth muscles
			β_b	477	53	
			β_c	522	58	
β2		3.5 kb	rtβ_2	604	68	aorta, brain
	CaB2	~4 kb	CaB2a	606	68	lungs, trachea
		~6 kb	CaB2b	632	71	
β3	CaB3	~3 kb	CaB3	477	54	brain, heart
		2.7 kb	Bk201(β_3)	484	55	kidney, lungs, skeletal muscle
β4	~7 kb		βk213(β_4)	519	58	brain, kidney

Source: The table has been compiled from references cited in the text.

skeletal muscle β-subunit, encodes two more splice variants. The isoforms (β_a) that were cloned from rat brain (Pragnell et al., 1991; Perez-Reyes et al., 1992), rat and rabbit heart (Hullin et al., 1992; Perez-Reyes et al., 1992), and human heart (Collin et al., 1993) lack 45 amino acids of the α-helical domain II, but have a uniquely long C-terminus. The β_b isoforms that were cloned from human brain (Powers et al., 1992; Williams et al., 1992a) and human heart (Collin et al., 1993) also have the same 45 residue-long deletion, but share identical C-termini with the skeletal muscle version (β_c). Two splice variants of the β_2 gene have been isolated from rat brain (Perez-Reyes et al., 1992) and from rabbit heart (Hullin et al, 1992). Both isoforms contain all four helical domains in their structure, but differ in their amino-termini. cDNAs for the β_3 gene have been isolated from rat brain (Castellano et al., 1993a) and rabbit heart (Hullin et al., 1992). These clones have short amino-termini and several small "deletions" in the coding region. In the β_3 isoforms, only three of the four α-helical domains are retained and none of the protein kinase A consensus recognition sites are present. Recently, a cDNA clone was isolated from rat brain (Castellano et al., 1993b) encoding a product of the β_4 gene. This clone is similar to the other β-subunit sequences in the central core region, especially the β_3 sequences, but has unique N- and C-termini. β_4, like the β_3 clones, does not contain a consensus sequence for cAMP-dependent phosphorylation. Interestingly, a unique feature of the β_4 clone is a potential ATP/GTP-binding site between domains I and II.

The classification controversy in the field of β-subunits in part originates from the fact that they do not show strict tissue-specific expression. This is especially true for the β_1-gene products, they can be found in nearly every tissue expressing VGCCs. β_2- and β_3-gene products appear to be more restricted, but still can be found in a large number of tissues. Only β_4, at the time of this writing, appears to be a brain-specific β-subunit. The apparent lack of tissue specificity, however, does not mean that there is no cell-type specificity. In co-expression studies, the different α_1- and β-subunits widely cross-react, exhibiting almost identical or very similar functional consequences. The significance of subtle differences found in these interactions could not be attributed to any physiological role. Moreover, the wealth of different β-subunits in any tissue makes it nearly impossible to conclude that any particular subunit type specifically determines a certain physiological function.

β-subunits of different classes have been co-expressed with representatives of all six cloned classes of α_1-subunits. When skeletal muscle β-subunits were co-expressed with the skeletal muscle α_1-subunit in L cells, a dramatic increase was seen in the number of DHP-binding sites (Lacerda et al., 1991; Varadi et al., 1991). Interestingly, the current density was not significantly affected; however, the activation and inactivation kinetics were accelerated and the current–voltage relationship was shifted towards hyperpolarized potentials (Varadi et al., 1991). Similar effects have been noted with quantitative, but not qualitative differences in other co-expression experiments using different α_1- or β-subunits. In general, the effects of the β-subunit are:

1. Increase in the number of functional channel proteins;
2. Changes in Ca^{2+} channel activation and inactivation; and
3. Leftward shift of the current–voltage relationship.

In the case of cardiac calcium channels, it was suggested, based on single channel analyses, that the presence of β-subunits may also increase the open-state probability of the Ca^{2+} channel (Wakamori et al., 1993). It is also of considerable interest that in several co-expression studies a synergistic effect of α_2/δ- and β-subunits on the Ca^{2+} channel density has been noted (Mori et al., 1991; Williams et al., 1992a).

D. The γ-Subunits

The γ-subunit of the skeletal muscle "DHP receptor" is a 32 kD transmembrane protein that also co-purifies with the skeletal muscle α_1-subunit. Its cDNA has been cloned from rabbit skeletal muscle (Jay et al., 1990). The sequence of this cDNA reveals it to be highly hydrophobic with many consensus N-glycosylation sites. Northern blots, probed with the labeled skeletal muscle γ-subunit cDNA, have failed to detect significant levels of expression in tissues other than skeletal muscle. Therefore, it is possible that other Ca^{2+} channels do not possess this subunit.

Co-expression of the γ-subunit with skeletal muscle α_1-subunit (Varadi et al., 1991) did not produce a significant effect on the α_1-subunit activity (a slight decrease of current density and DHP binding was noted). Similarly unremarkable effects were observed with the BI channel α_1-subunit cDNAs. However, in one series of experiments (Singer et al., 1991), the skeletal muscle γ-subunit enhanced the level of cardiac γ-subunit expression in *Xenopus* oocytes, accelerated the Ba^{2+}-current inactivation and shifted the voltage-dependence of inactivation towards hyperpolarized potentials. Further studies are needed to clarify the role of the γ-subunit.

IV. CONCLUDING REMARKS

Voltage-gated Ca^{2+} channels are multi-subunit protein complexes which form a highly selective ion pore in the membrane. The α_1-subunit forms the channel pore, voltage sensor, and harbors the clinically important drug-binding sites. Molecular technology is beginning to elucidate roles for the other subunits. These channels represent an important group of macromolecular structures that contribute to the regulation of intracellular Ca^{2+} concentration, a critical factor in numerous cellular processes. Recent progress utilizing molecular biological techniques in conjunction with biochemical and biophysical analyses has revealed a diversity which is much greater than ever predicted using conventional techniques. However, it is likely that this period of extensive growth in structural information is ending. We predict that this era will be followed by a new growth of information concerning the *functional* properties of these cloned isoforms, utilizing structure–function studies which probe the critical regions of the different subunits involved in their physiological, regulatory, and pharmacological properties.

ACKNOWLEDGMENTS

The authors thank Liz Wendelmoot for typing the manuscript and Gwen Kraft for help with the figures. We also thank Dr. Gyula Varadi and Mark Strobeck for useful suggestions on the manuscript.

REFERENCES

Armstrong, D., & Eckert, R. (1987). Voltage-activated calcium channels that must be phosphorylated to respond to membrane depolarization. Proc. Natl. Acad. Sci. USA 84, 2518–2522.

Beeler, G. W., & Reuter, H. (1970). Membrane calcium current in ventricular myocardial fibers. J. Physiol. 207, 191–209.

Bertolino, M., & Llinás, R. R. (1992). The central role of voltage-activated and receptor-operated calcium channels in neuronal cells. Annu. Rev. Pharmacol. Toxicol. 32, 399–421.

Biel, M., Ruth, P., Hullin, R., Stühmer, W., Flockerzi, V., & Hofmann, F. (1990). Primary structure and functional expression of a high voltage activated calcium channel from rabbit lung. FEBS Letts. 269, 409–412.

Biel, M., Hullin, R., Freundner, S., Singer, D., Dascal, N., Flockerzi, V., & Hofmann, F. (1991). Tissue-specific expression of high-voltage-activated dihydropyridine-sensitive L-type calcium channels. Eur. J. Biochem. 300, 81–88.

Bosse, E., Bottlender, R., Kleppisch, T., Hescheler, J., Welling, A., Hofmann, F., & Flockerzi, V. (1992). Stable and functional expression of the calcium channel α_1-subunit from smooth muscle in somatic cell lines. EMBO J. 11, 2033–2038.

Bonvallet, R. (1987). A low threshold calcium current recorded at physiological Ca^{2+} concentrations in single frog atrial cells. Pflügers Arch. 408, 540–542.

Brown, A. M., Kunze, D. L., & Yatani, A. (1984). The agonist effect of dihydropyridines on Ca^{2+} channels. Nature 311, 570–572.

Carbone, E., & Lux, H. D. (1984). A low voltage-activated fully inactivating Ca^{2+} channel in vertebrate sensory neurons. Nature 310, 501–502.

Carbone, E., & Lux, H. D. (1987). Kinetics and selectivity of a low voltage-activated calcium current in chick and rat sensory neurons. J. Physiol. 386, 547–570.

Castellano, A., Wei, X., Birnbaumer, L., & Perez-Reyes, E. (1993a). Cloning and expression of a neuronal calcium channel β subunit. J. Biol. Chem. 12359–12366.

Castellano, A., Wei, X., Birnbaumer, L., & Perez-Reyes, E. (1993b). Cloning and expression of a third calcium channel β subunit. J. Biol. Chem. 268, 3450–3455.

Catterall, W. A. (1988). Structure and function of voltage-sensitive ion channels. Science 242, 50–61.

Cavalié, A., Ochi, R., Pelzer, D., & Trautwein, W. (1983). Elementary currents through Ca^{2+} channels in guinea pig myocytes. Pflügers Arch. 398, 284–297.

Collin, T., Wang, J-J., Nargeot, J., & Schwartz, A. (1993). Molecular cloning of three isoforms of the L-type voltage-dependent calcium channel β subunit from normal human heart. Circ. Res. 72, 1337–1344.

DeJongh, K. S., Warner, C., & Catterall, W. A. (1990). Subunits of purified calcium channels α_2 and δ are encoded by the same gene. J. Biol. Chem. 265, 14738–14741.

DeJongh, K. S., Warner, C., Colvin, A. A., & Catterall, W. A. (1991). Characterization of the two size forms of the α_1 subunit of skeletal muscle L-type calcium channels. Proc. Natl. Acad. Sci. USA 88, 10778–10782.

Diebold, R. J., Koch, W. J., Ellinor, P. T., Wang, J-J., Muthuchami, M., Wieczorek, D., & Schwartz, A. (1992). Mutually exclusive exon splicing of the cardiac calcium channel α_1 subunit gene generates developmentally regulated isoforms in the rat heart. Proc. Natl. Acad. Sci. USA 89, 1497–1501.

Droogmans, G., & Nilius, B. (1989). Kinetic properties of the cardiac T-type calcium channel in the guinea-pig. J. Physiol. 419, 627–650.

Dubel, S. F., Starr, T. V. B., Hell, J., Ahlijamian, M. K., Enyeart, J. J., Catterall, W. A., & Snutch, T. P. (1992). Molecular cloning of the α_1 subunit of an ω-conotoxin-sensitive calcium channel. Proc. Natl. Acad. Sci. USA 89, 5058–5062.

Dupont, J.-L., Bossu, J.-L., & Feltz, A. (1986). Effect of internal calcium concentration on calcium currents in rat sensory neurones. Pflügers Arch. 406, 433–435.

Ellinor, P. T., Zhang, F., Randall, A. D., Zhou, M., Schwarz, T. L., Tsien, R. W., & Horne, W. A. (1993). Functional expression of a rapidly inactivating neuronal calcium channel. Nature 363, 455–458.

Ellis, S. B., Williams, M. E., Ways, R., Brenner, R., Sharp, A. H., Leung, A. T., Campbell, K. P., McKenna, E., Koch, W. J., Hui, A., Schwartz, A., & Harpold, M. M. (1988). Sequence and expression of mRNAs encoding the α_1 and α_2 subunits of a DHP-sensitive calcium channel. Science 241, 1661–1664.

Franckowiak, G., Bechem, M., Schramm, M., & Thomas, G. (1985). The optical isomers of the 1,4-dihydropyridine Bay K 8644 show opposite effects on Ca^{2+} channels. Eur. J. Pharmacol. 114, 223–226.

Fujita, Y., Mynlieff, M., Dirksen, R. T., Kim, M. S., Niidome, T., Nakai, J., Friedrich, T., Iwanabe, N., Miyata, T., Furuichi, T., Furutama, D., Mikoshiba, K., Mori, Y., & Beam, K. G. (1993). Primary structure and functional expression of the ω-conotoxin-sensitive N-type calcium channel from rabbit brain. Neuron 10, 585–598.

Guy, H. R., & Conti, F. (1990). Pursuing the structure and function of voltage-gated channels. Trends. Neurosci. 13, 201–206.

Hagiwara, N., Irisawa, H., & Kameyawa, M. (1988). Contribution of two types of calcium currents to the pacemaker potentials of rabbit sinuatreal node cells. J. Physiol. 395, 233–253.

Hillyard, D. R., Monje, V. D., Mintz, I. M., Bean, B. P., Nadasdi, L., Ramachandran, F., Miljanich, G., Azimi-Zoonoz, A., McIntosh, J. M., & Cruz, F. (1992). A new *Conus* peptide ligand for mammalian presynaptic Ca^{2+} channels. Neuron 9, 69–77.

Hirano, Y., Fozzard, H. A., & January, C. T. (1989). Characteristics of L- and T-type Ca^{2+} currents in canine cardiac Purkinje cells. Am. J. Physiol. 256, H1478–H1492.

Horne, W. A., Ellinor, P. T., Inman, I., Zhou, M., Tsien, R. W., & Schwarz, T. L. (1993). Molecular diversity of Ca^{2+} channel α_1 subunits from the marine ray *Discopygne omnata*. Proc. Natl. Acad. Sci. USA 90, 3787–3791.

Hosey, M. M., & Lazdunski, M. (1988). Calcium channels: Molecular pharmacology, structure and regulation. J. Membrane Biol. 104, 81–105.

Hui, A., Ellinor, P. T., Krizanova, O., Wang, J.-J., Diebold, R. J., & Schwartz, A. (1991). Molecular cloning of multiple subtypes of a novel rat brain isoform of the α_1 subunit of the voltage-dependent calcium channel. Neuron 7, 35–44.

Hullin, R., Singer-Lahat, D., Freichel, M., Biel, M., Dascal, N., Hofmann, F., & Flockerzi, V. (1992). Calcium channel β subunit heterogeneity: Functional expression of cloned cDNA from heart, aorta and brain. EMBO J. 11, 885–890.

Iles, D. E., Segers, B., OldeWeghuis, D., Suijkerbuijk, R., Mikala, G., Schwartz, A., & Wieringa, B. (1994). Refined localization of the α_1 subunit of the skeletal muscle L-type voltage-dependent calcium channel to human chromosome 1q32.1 by *in situ* hybridization. Genomics 19, 561–563.

Jay, S. D., Ellis, S. B., McCue, A. F., Williams, M. E., Vedvick, T. S., Harpold, M. M., & Campbell, K. D. (1990). Primary structure of the γ subunit of the DHP-sensitive calcium channel from skeletal muscle. Science 248, 490–492.

Kass, R. S. (1987). Voltage-dependent modulation of cardiac calcium channel current by optical isomers of Bay K 8644: Implications for channel gating. Circ. Res. 61, 11–15.

Kim, H-L., Kim, H., Lee, P., King, R. G., & Chin, H. (1992). Rat brain expresses an alternatively spliced form of the dihydropyridine-sensitive L-type calcium channel α_2 subunit. Proc. Natl. Acad. Sci. USA 89, 3251–3255.

Kim, M., Mori, T., Sun, L., Imoto, K., & Mori, Y. (1993). Structural determinants of ion selectivity in brain calcium channel. FEBS Letts. 318, 145–148.

Koch, W. J., Ellinor, P. T., & Schwartz, A. (1990). cDNA cloning of a dihydropyridine-sensitive calcium channel from rat aorta: Evidence for the existence of alternatively spliced forms. J. Biol. Chem. 265, 17786–17791.

Kongsamut, S., Kamp, T. J., & Miller, R. J. (1985). Calcium channel agonist and antagonist effects of the stereoisomers of the dihydropyridine 202–791. Biochem. Biophys. Res. Commun. 130, 141–148.

Lacerda, A. E., Kim, H. S., Ruth, P., Perez-Reyes, E., Flockerzi, V., Hofmann, F., Birnbaumer, L., & Brown, A.M. (1991). Normalization of current kinetics by interation between the α_1 and β subunits of the skeletal muscle dihydropyridine-sensitive Ca^{2+} channel. Nature 352, 527–530.

Lansman, F. B., Hess, P., & Tsien, R. W. (1986). Blockade of current through single calcium channels by Cd^{2+}, Mg^{2+} and Ca^{2+}. Voltage and concentration dependence of calcium entry into the pore. J. Gen. Physiol. 88, 321–347.

Llinás, R. R. (1988). The intrinsic electrophysiological properties of mammalian neurons: Insight into central venous system function. Science 242, 1654–1664.

Llinás, R., & Yarom, Y. (1981). Electrophysiology of mammalian inferior olivary neurons *in vitro*. Different types of voltage-dependent ionic conductances. J. Physiol. 315, 549–567.

Llinás, R., & Yarom, T. (1986). Specific blockage of the low-threshold calcium channel by high molecular weight alcohols. Soc. Neurosci. Abstr. 12, 174.

Llinás, R., Sugimori, M., Lin, J. W., & Cherkey, B. (1989). Blocking and isolation of a calcium channel from neurons in mammals and cephalopods utilizing a toxin fraction (FTX) from funnel-web spider poison. Proc. Natl. Acad. Sci. USA 86, 1689–1693.

Llinás, R., Sugimori, M., Hillman, D. E., & Cherksey, B. (1992). Distribution and functional significance of the P-type, voltage-dependent Ca^{2+} channels in the mammalian central nervous system. Trends Neurosci. 15, 351–355.

Lory, P., Varadi, G., & Schwartz, A. (1991). Molecular insights into regulation of L-type Ca^{2+} channel function. New. Physiol. Sci. 6, 277–281.

Lory, P., Varadi, G., Slish, D. F., Varadi, M., & Schwartz, A. (1993). Characterization of β subunit modulation of a rabbit cardiac L-type Ca^{2+} channel α_1 subunit as expressed in mouse L cells. FEBS Letts. 315, 167–172.

Manew, H., Bertolino, M., & De-Erausquin, G. (1990). Amiloride blocks glutamate operated cationic channels and protects neurons in culture from glutamate-induced death. Neuropharmacology 29, 1103–1110.

Mikala, G., Bahinski, A., Yatani, A., Tang, S., & Schwartz, A. (1993). Differential contribution by conserved glutamate residues to an ion-selectivity site in the L-type Ca^{2+} channel pore. FEBS Letts. 335, 265–269.

Mikami, A., Imoto, K., Tanabe, T., Niidome, T., Mori, Y., Takeshima, H., Narumiya, H., & Numa, S. (1989). Primary structure and functional expression of the cardiac dihydropyridine sensitive calcium channel. Nature 340, 230–233.

Miller, R. J. (1992). Voltage-sensitive Ca^{2+} channels. J. Biol. Chem. 267, 1403–1406.

Mintz, I. M., Venema, V. J., Adams, M. E., & Bean, B. P. (1991). Inhibition of N- and L-type Ca^{2+} channels by the spider venom ω-Aga-IIIA. Proc. Natl. Acad. Sci. USA 88, 6628–6631.

Mintz, I. M., Venema, V. J., Swiederek, K. M., Lee, T. D., Bean, B. P., & Adams, M. E. (1992). P-type calcium channels blocked by the spider toxin ω-Aga-IVA. Nature 355, 827–829.

Morad, M., & Cleeman, L. (1987). Role of Ca^{2+} channel in development of tension in heart muscle. J. Mol. Cell. Cardiol. 19, 527–553.

Mori, Y., Friedrich, T., Kim, M. S., Mikami, A., Nakai, J., Ruth, P., Bosse, E., Hofmann, F., Flockerzi, V., Furuichi, T., Mokoshiba, K., Imoto, K., Tanabe, T., & Numa, S. (1991). Primary structure and functional expression from complementary DNA of a brain calcium channel. Nature 350, 398–402.

Nathan, R. D., Karai, K., Clark, R. B., & Giles, W. (1988). Selective block of calcium current by lanthanum in single bullfrog atrial cells. J. Gen. Physiol. 91, 549–572.

Niidome, T., Kim, M. S., Friedrich, T., & Mori, Y. (1992). Molecular cloning and characterization of a novel calcium channel from rabbit brain. FEBS Letts. 308, 7–13.

Nowycky, M. C., Fox, A. P., & Tsien, R. W. (1985). Three types of neuronal calcium channel with different calcium agonist sensitivity. Nature 316, 440–443.

Olivera, B. M., McIntosh, J. M., Cruz, L. J., Lugu, F. A., & Gray, W. R. (1984). Purification and sequence of a presynaptic peptide toxin from *Conus geographus* venom. Biochemistry 23, 5087–5090.

Olivera, B. M., Gray, W. R., Zeikus, R., Melntosh, J. M., Varga, J., Rivier, J., DeSantos, V., & Cruz, L. J. (1985). Peptide neurotoxins from fish-hunting cone snails. Science 230, 1338–1343.

Olivera, B. M., Rivier, F., Clark, C., Ramilo, C. A., Corpuz, G. P., Abogadie, F. C., Mena, E. E., Woodward, S. R., Hillyard, D. R., & Cruz, L. J. (1990). Diversity of *Conus* neuropeptides. Science 249, 257–263.

Perez-Reyes, E., Kim, H. S., Lacerda, A. E., Horne, W., Wei, W., Rampe, D., Campbell, K. P., Brown, A. M., & Brinbaumer, L. (1989). Induction of calcium currents by the expression of the α_1-subunit of dihydropyridine receptor from skeletal muscle. Nature 340, 233–236.

Perez-Reyes, E., Wei, X., Castellano, A., & Birnbaumer, L. (1990). Molecular diversity of L-type calcium channels. J. Biol. Chem. 265, 20430–20436.

Pelzer, D., Pelzer, S., & McDonald, T. F. (1992). Calcium channels in heart. In: *The Heart and Cardiovascular System*, 2nd ed. (Fozzard, H. A., Haber, E., Jennings, R. B., Katz, A. M., & Morgan, H. E., eds.) pp. 1049–1089. Raven Press, New York.

Perez-Reyes, E., Castellano, A., Kim, H. S., Bertrand, P., Baggstrom, E., Lacerda, A. E., Wei, X., & Birnbaumer, L. (1992). Cloning and expression of a cardiac/brain β subunit of the L-type calcium channel. J. Biol. Chem. 267, 1792–1797.

Plummer M. R., & Hess, P. (1991). Reversible uncoupling of inactivation in N-type calcium channels. Nature 351, 657–659.

Plummer, M. R., Logothetis, D. E., & Hess, P. (1989). Elementary properties and pharmacological sensitivities of calcium channels in mammalian peripheral neurons. Neuron 2, 1453–1463.

Porzig, H. (1990). Pharmacological modulation of voltage-dependent Ca^{2+} channels in intact cells. Proc. Physiol. Biochem. Pharmacol. 114, 209–262.

Powers, P. A., Liu, S. Hogan, K., & Gregg, R. G. (1992). Skeletal muscle and brain isoforms of a β-subunit of human voltage-dependent calcium channels are encoded by a single gene. J. Biol. Chem. 267, 22967–22972.

Pragnell, M., Sakamoto, J., Jay, S. D., & Campbell, K. P. (1991). Cloning and tissue specific expression of the brain calcium channel β subunit. FEBS Letts. 291, 253–258.

Regan, L. J., Sak, D. W. Y., & Bean, B. P. (1991). Ca^{2+} channels in rat central and peripheral neurons: High threshold current resistant to dihydropyridine blockers and ω-conotoxin. Neuron 6, 269–280.

Ruth, P., Röhrkastern, Biel, M., Bosse, E., Regulla, S., Meyer, H. E., Flockerzi, V., & Hofmann, F. (1989). Primary structure of the β subunit of the DHP-sensitive calcium channel from skeletal muscle. Science 245, 1115–1118.

Sakamoto, Y., & Campbell, K. P. (1991). A monoclonal antibody to the β subunit of the skeletal muscle dihydropyridine receptor immunoprecipitates the brain ω-conotoxin GVIA receptor. J. Biol. Chem. 266, 18914–18919.

Sanguinetti, M. C., Krafte, D. S., & Kass, R. S. (1986). Voltage-dependent modulation of Ca^{2+} channel current in heart cells by Bay K 8644. J. Gen. Physiol. 88, 369–392.

Schultz, D., Mikala, G., Yatani, A., Engle, D. B., Iles, D. E., Segers, B., Sinke, E. J., Weghuis, D. O., Klöcner, U. Wakamori, M., Wang, J-J., Molvin, D., Varadi, G., & Schwartz, A. (1993). Cloning, chromosomal localization and functional expression of the α_1 subunit of the L-type voltage-dependent calcium channel from normal human heart. Proc. Natl. Acad. Sci. USA 90, 6228–6232.

Seino, S., Chen, L., Seino, M., Blondel, O., Takeda, J., Johnson, J. H., & Bell, G. I. (1992a). Cloning of the α_1 subunit of a voltage-dependent calcium channel expressed in pancreatic β cells. Proc. Natl. Acad. Sci. USA 89, 584–588.

Seino, S., Yamada, Y., Espinosa, R., LeBeau, M. M., & Bell, G. I. (1992b). Assignment of the gene encoding the α_1 subunit of the neuroendocrine/brain-type calcium channel (CACNL1A2) to human chromosome 3, band p14.3. Genomics 13, 1375–1377.

Siegl, P. K. S., Cragse, Jr. E. J., Trumble, M. J., & Kaczorowski, G. J. (1984). Inhibition of Na^+/Ca^{2+} exchange in membrane vesicle and papillary muscle preparations from guinea pig heart by analogs of amiloride. Proc. Natl. Acad. Sci. USA 81, 3238–3242.

Singer, D., Biel, M., Latan, I., Flockerzi, V., Hofmann, F., & Dascal, N. (1991). The roles of the subunits in the function of the calcium channel. Science 253, 1553–1557.

Snutch, T. P., & Reiner, P. B. (1992). Ca^{2+} channels: Diversity of form and function. Curr. Opin. Neurobiol. 2, 247–253.

Snutch, T. P., Tomlinson, W. F., Leonard, F. P., & Gilbert, M. M. (1991). Distinct calcium channels are generated by alternative splicing and are differentially expressed in the mammalian CNS. Neuron 7, 45–57.

Soldatov, N. M. (1992). Molecular diversity of L-type Ca^{2+} channel transcripts in human fibroblasts. Proc. Natl. Acad. Sci. USA 89, 4628–4632.

Soong, T. W., Stea, A., Hodson, C. D., Dubel, S. J., Vincent, S. R., & Snutch, T. P. (1993). Structure and functional expression of a member of the low voltage-activated calcium channel family. Science 260, 1133–1136.

Starr, T. V. B., Prystay, W., & Snutch, T. P. (1991). Primary structure of a calcium channel that is highly expressed in the rat cerebellum. Proc. Natl. Acad. Sci. USA 88, 5621–5625.

Swamdulla, D., Carbone, E., Schafer, K., & Lux, H. D. (1987). Effect of menthol on two types of Ca^{2+} currents in cultured sensory neurons of vertebrates. Pflügers Arch. 402, 52–59.

Tanabe, T., Takeshima, H., Mikami, A., Flockerzi, V., Takahashi, H., Shangawa, K., Kojima, M., Matsus H., Hirose, T., & Numa, S. (1987). Primary structure of the receptor for calcium channel blockers from skeletal muscle. Nature 328, 313–318.

Tanabe, T., Beam, K. G., Powell, J. A., & Numa, S. (1988). Restoration of excitation–contraction coupling and slow calcium current in dysgenic muscle by dihydropyridine receptor complementary DNA. Nature 336, 134–139.

Tanabe, T., Beam, K. G., Adams, B. A., Niidome, T., & Numa, S. (1990a). Regions of the skeletal muscle dihydropyridine receptor critical for excitation–contraction coupling. Nature 346, 567–569.

Tanabe, T., Mikami, A., Numa, S., & Beam, K. G. (1990b). Cardiac-type excitation–contraction coupling in dysgenic skeletal muscle injected with cardiac dihydropyridine receptor cDNA. Nature 344, 451–453.

Tang, C. M., Presser, F., & Morad, M. (1988). Amiloride selectively blocks the low threshold (T) calcium channel. Science 20, 313–215.

Tang, S., Mikala, G., Bahinski, A., Yatani, A., Varadi, G., & Schwartz, A. (1993). Molecular localization of ion selectivity sites within the pore of a human L-type cardiac calcium channel. J. Biol. Chem. 268, 13026–13029.

Thayer, S. A., Hirning, L. D., & Miller, R. J. (1987). Distribution of multiple types of Ca^{2+} channels in rat sympathetic neurons *in vitro*. Mol. Pharmacol. 32, 579–586.

Tsien, R. W., Hess, P., McCluskey, E. W., & Rosenberg, R. L. (1987). Calcium channels: Mechanisms of selectivity, permeation, and block. Annu. Rev. Biophys. Biophys. Chem. 16, 265–290.

Uchitel, O. D., Protti, D. A., Sanchez, V., Cherksey, B. D., Sugimori, M., & Llinás, R. (1992). P-type voltage-dependent calcium channel mediates presynaptic calcium influx and transmitter release in mammalian synapses. Proc. Natl. Acad. Sci, USA 89, 3330–3333.

Usowicz, M. M., Sugimori, M., Cherksey, B., & Llinás, R. (1992). P-type calcium channels in the somata and dendrites of adult cerebeller Purkinje cells. Neuron 9, 185–199.

Varadi, G., Lory, P., Schultz, D., Varadi, M., & Schwartz, A. (1991). Acceleration of activation and inactivation by the β subunit of the skeletal muscle calcium channel. Nature 352, 159–162.

Wakamori, M., Mikala, G., Schwartz, A., & Yatani, A. (1993). Single-channel analysis of a cloned human heart L-type Ca^{2+} channel α_1 subunit: Effects of a cardiac β subunit. Biochem. Biophys. Res. Commun. 196, 1170–1176.

Wei, X., Perez-Reyes, E., Lacerda, A. E., Schuster, G., Brown, A. M., & Birnbaumer, L. (1991). Heterologous regulation of the cardiac Ca^{2+} channel α_1 subunit by skeletal muscle β and δ subunits. J. Biol. Chem. 266, 21943–21947.

Weille, F. R., Schweitz, H., Maes, P., Tartar, A., & Lazdunski, M. (1991). Calciseptine, a peptide isolated from black mamba venom, is a specific blocker of the L-type calcium channel. Proc. Natl. Acad. Sci. USA 88, 2437–2440.

Westenbroek, R. E., Ahlijanian, M. K., & Catterall, W. A. (1990). Clustering of L-type Ca^{2+} channels at the base of major dendrites in hippocampal pyramidal neurons. Nature 347, 281–284.

Westenbroek, R. E., Hell, F. W., Warner, C., Dubel, S. J., Snutch, T. P., & Catterall, W. A. (1992). Biochemical properties and subcellular distribution of an N-type calcium channel α_1 subunit. Neuron 9, 1099–1115.

White, G., Lovinger, D. M., & Weight, F. F. (1989). Transient low threshold Ca^{2+} current triggers burst firing through an after depolarization potential in an adult mammalian neuron. Proc. Natl. Acad. Sci. USA 86, 6802–6806.

Williams, M. E., Brust, P. F., Feldman, D. H., Patthi, S., Simerson, S., Maroufi, A., McCue, A. F., Veijcelebi, G., Ellis, S. B., & Harpold, M. M. (1992a). Structure and functional expression of an ω-conotoxin-sensitive N-type calcium channel. Science 257, 389–395.

Williams, M. E., Feldman, D. H., McCue, A. F., Brenner, R., Veijcelebi, G., Ellis, S. B., & Harpold, M. M. (1992b). Structure and functional expression of α_1, α_2 and β subunits of a novel human neuronal calcium channel subtype. Neuron 8, 71–84.

Witcher, D. R., Waard, M., Sakamoto, J., Franzini-Armstrong, C., Pragnell, M., Kahl, S. D., & Campbell, K. P. (1993). Subunit identification and reconstitution of the N-type Ca^{2+} channel complex purified from brain. Science 261, 486–689.

Yaney, G. C., Wheeler, M. B., Wei, X., Perez-Reyes, E., Birnbaumer, L., Boyd, A. E., & Moss, L. G. (1992). Cloning of a novel α_1-subunit of the voltage-dependent calcium channel from the β-cell. Mol. Endocrinol. 6, 2143–2152.

Yatani, A., Imoto, Y., Codina, F., Hamilton, S. L., Brown, A. M., & Birnbaumer, L. (1988). The stimulatory G protein of adenylyl cyclase, G_s, also stimulates dihydropyridine-sensitive Ca^{2+} channels. J. Biol. Chem. 263, 9887–9895.

Yu, A. S. L., Hebert, S. C., Brenner, B. M., & Lytton, J. (1992). Molecular characterization and nephron distribution of a family of transcripts encoding the pore-forming subunit of Ca^{2+} channels in the kidney. Proc. Natl. Acad. Sci. USA 89, 10494–10498.

CYCLIC NUCLEOTIDE-GATED CHANNELS

Paulus W. Wohlfart and Neil J. Cook

Biomembranes
Volume 6, pages 249–272.

ISBN: 1-55938-663-0.

I. INTRODUCTION

Since the discovery of the 3′,5′-cyclic nucleotides, cAMP and cGMP, more than three decades ago, the role of these second messengers in mediating cellular functions has been the focus of intensive biochemical and biological research. Besides characterizing the mechanisms by which the intracellular metabolism of these nucleotides is regulated, a major aspect of this research has been the identification and characterization of cyclic nucleotide receptors. It is now known that cyclic nucleotide-dependent kinases are not the only receptor proteins mediated by the binding of cAMP or cGMP in eucaroytic systems. Other cyclic nucleotide-binding proteins, including cyclic nucleotide-stimulated or -inhibited phosphodiesterases and cyclic nucleotide-gated channels, are now becoming the subjects of intensive investigation.

Along-side the voltage-gated and the ligand-gated channel families, cyclic nucleotide-gated channels comprise a new steadily growing class of proteins, initially thought to be concentrated solely in sensory tissues. There is however increasing evidence, stemming primarily from cDNA cloning experiments, that this channel type is also present in nonsensory tissues essentially wherever intracellular cyclic-nucleotide metabolism might be linked directly with the conductive properties of the plasma membrane. This review is written (1) to give a short survey about our knowledge of cyclic nucleotide-gated channel function in different tissues, (2) to summarize electrophysiological and pharmacological behavior of the different members of this channel class, and (3) to review the sparse biochemical knowledge of the rod photoreceptor channel and our knowledge regarding the structure and topology of this channel class.

II. FUNCTION OF CYCLIC NUCLEOTIDE-GATED CHANNELS IN VARIOUS TISSUES

Cyclic nucleotide-gated channels were first discovered in sensory organs specialized for visual or olfactory transduction, where an extracellular activating stimulus, light or odor, is coupled to changes in the intracellular concentrations of cGMP or cAMP. In 1985, it was demonstrated by Fesenko et al. (1985), for the first time, that a cyclic nucleotide (cGMP) is able to open cation channels by directly binding to a membrane receptor (the channel protein) when patch–clamp studies on excised plasma membrane patches of the outer segment plasma membrane of vertebrate rod photoreceptor cells were performed. These measurements revealed that the application of cGMP to the cytosolic side of the patch in the absence of ATP or GTP (necessary for protein phosphorylation) induced a cationic conductance identical to that known to be regulated by light in these cells. Cyclic GMP-gated channels were subsequently also demonstrated to exist in vertebrate cone photoreceptors (Haynes and Yau, 1985; Cobbs et al., 1985). The light-induced hyperpolarization of these rod photoreceptors could thus be explained by the following mechanism (reviewed in Stryer, 1986): after illumination, the light receptor, rhodopsin, undergoes a conformational change and is able to activate a GTP-binding protein (transducin). Transducin, in turn, activates a phosphodiesterase that rapidly degrades cGMP in the photoreceptor cytosol. This leads to cGMP-gated cation channel closure and hyperpolarization, an electrical event that is subsequently transferred to the brain culminating in the process known as vision.

In olfactory transduction, a process similar to visual transduction, cyclic nucleotide-gated channels have been identified by patch–clamp recordings of excised patches from olfactory cilia (Nakamura and Gold, 1987). In contrast to photoreceptors, the likely internal messenger involved in olfactory transduction is cAMP and not cGMP (Pace et al., 1985; Sklar et al., 1986; Lowe et al., 1989; Breer et al., 1990). This correlates well with the fact that the native olfactory channel is equally sensitive to cGMP and cAMP (see following). The present view of olfactory signal transduction involves stimulation of odorant receptors coupled via G-proteins to the stimulation of adenylyl cyclase, leading to an increase in intracellular cAMP, which directly opens the cGMP/cAMP-gated channels in the plasma membrane, thereby depolarizing the entire odorant-sensitive cell (Levy et al., 1991).

Completing the picture of sensory transduction, two other cyclic nucleotide-gated channels have been found in tissues sensitive to extra-corporal stimuli: a cAMP-dependent channel has been described in cochlea hair cells from guinea pig (Kolesnikov et al., 1991). These cells are the mechano-receptors of the inner ear, raising the possibility that the mechano-transduction might be mediated by cyclic nucleotides similar to olfaction and photoreception. Another channel has been found in the cells of the chick pineal gland (Dryer and Henderson, 1991). These nonretinal photosensitive cells express a variety of proteins which are normally found in retinal photoreceptors, and they also display light-dependent responses. At least one of them, the light-induced inhibition of melatonin secretion, might be mediated by the cGMP-dependent channels.

The existence of cGMP-gated channels in retinal bipolar cells has also been reported (Nawy and Jahr, 1990, 1991; Schiells and Falk, 1992), although it has been demonstrated that they must be immunologically distinct from the rod photoreceptor channel (Wässle et al., 1992).

Besides cells specialized for sensory transduction, cyclic nucleotide-gated channels are now detected in a variety of other nonsensory tissues (Distler et al., 1994). For example, specific probes against the mRNA of the cGMP-gated channel from the rod photoreceptor have been used to screen by Northern blotting and PCR amplification a number of tissues for the expression of this channel (Ahmad et al., 1990). Messenger RNA for this channel was also detected in heart and kidney, but not in cerebellum, cerebral cortex, liver, muscle, olfactory bulb, spleen, testes, or thymus. The renal mRNA hybridizing with the rod photoreceptor cGMP-gated channel might not be attributable to a cGMP-dependent conductance previously described electrophysiologically in cortical collecting duct cells (Light et al., 1989; Ahmad et al., 1992) since, in contrast to the photoreceptor or the olfactory channel, the opening probability of this conductance has been described to be *reduced* by the direct binding of cGMP. One explanation might be that these renal cells contain at least two structurally and functionally different channels, one of which is closed by cGMP and is detectable electrophysiologically, while the other one is opened by cGMP, but is expressed in low amounts and, therefore, only detectable by Northern blotting in conjunction with PCR amplification.

Using a pair of degenerate oligodeoxynucleotide primers that flank the cGMP-binding region of the bovine rod photoreceptor channel, Biel et al. (1993) amplified by PCR cDNA derived from a variety of tissues and found amplification products of the predicted size in rabbit heart, aorta, sino-atrial node, bovine cerebellum, bovine kidney, and to a lesser extent in human thyroid C cells. A full-length channel clone was identified by screening of a rabbit aorta cDNA library with the radio-labeled PCR-fragment. The protein encoded by the open reading frame of this aortic channel clone seems to be more homologous to the bovine olfactory channel than to the bovine photoreceptor channel. Transcripts of this aortic clone could not be detected in conventional northern blotting experiments with mRNA from total rabbit aorta, however this channel might be present only in small amounts in endothelial or smooth muscle vascular cells. Endothelial cells are known to be a rich source of an enzyme called constitutive nitric oxide synthase, releasing the diffusible second messenger nitric oxide after stimulation with a plethora of agents (for a review see Nathans, 1992). Both endothelial and smooth muscle cells contain soluble guanylate cyclases which are activated by nitric oxide, thereby increasing intracellular cGMP-concentration. It is possible that one of the intracellular molecules responding to cGMP is a cGMP-dependent cation conductance in the plasma membrane of one of these cell types.

Another new member of the cyclic nucleotide-gated channel family, different in its primary sequence from both the photoreceptor and olfactory channel, was described by Biel et al. (1994) independently by Weyand et al. (1994). Using Northern blot experiments, this channel type was found in testis, kidney cortex, and kidney medulla, and to a lesser extent in cardiac atrium and ventricle, whereas no signal could be detected with mRNA derived from ileum, tongue, aorta, cerebellum, and total brain. A positive signal in the kidney would mean that at least two different types of the channel family could be expressed in the same tissue. The Northern blot experiments described by Biel et al. (1994) and Weyand et al. (1994) for this new member of the channel family, and those by Ahmad et al. (1990) for the photoreceptor channel, could be further investigated by immunohistochemistry using selected sequence-specific anti-peptide antibodies in order to determine the exact tissue-distribution of the different members of the cyclic nucleotide-gated channel family. In a first attempt, Weyand et al. (1994) found, using immuncyto-chemistry, high expression of the testis gene in bovine retinal cone photoreceptors but not in bovine testis. Further studies will give an impression of the real channel distribution in different cells and provide a greater insight into the functional roles of these channels in nonsensory tissues.

It is questionable whether one should include in the family of cyclic nucleotide-gated channels newly reported channels which are controlled by voltage and cyclic nucleotides dually. For example, a pacemaker cationic conductance is present in the heart that is directly activated by intracellular cAMP-binding by shifting its voltage-activation curve to more positive voltages (DiFrancesco and Tortora, 1991). Although nothing is known about the structure of this cardiac channel, it might resemble the recently cloned, voltage-dependent *eag* channel in *Drosophila*, where a cyclic nucleotide-binding domain is present in the sequence (Warmke et al., 1991). It is interesting to speculate about the existence of a new channel family, on one hand, gated through voltage and on the other hand, modulated by cyclic nucleotides.

III. ELECTROPHYSIOLOGICAL PROPERTIES OF CYCLIC NUCLEOTIDE-GATED CHANNELS

The native cyclic nucleotide-gated channels of which the electrophysiological properties have been the most extensively characterized, are the photoreceptor channel followed by the olfactory channel. Since its discovery by Fesenko et al. (1985), there has been a plethora of electrophysiological studies on the photoreceptor channel (see Yau and Baylor, 1989 for a review). Before the discovery and identification of the cGMP-gated channel, it was referred to as the "light-sensitive channel." Since there are two types of light-transducing cells in the retina, the rod cells responsible for contrast vision at low illumination levels and the cone cells responsible for normal color vision at daylight illumination levels, it is necessary to distinguish between the cone- and the rod-photoreceptor channels when discussing their electrophysiological properties.

The rod photoreceptor channel is known to strongly outwardly rectify (Bodoia and Detwiler, 1985; Baylor and Nunn, 1986) and displays virtually no voltage sensitivity (Yau and Baylor, 1989). The channel responds within milliseconds to changes of cGMP-concentration (Karpen et al., 1988) and is permeable to divalent as well as monovalent cations (Yau and Nakatani, 1984; Hodgkin et al., 1985). Under physiological conditions, it mediates a cation conductance carried by Na^+, K^+, Ca^{2+}, and Mg^{2+} at a ratio of 0.7: 0.15: 0.15: 0.05, respectively (Nakatani and Yau, 1988a). The permeability to Ca^{2+} has some important consequences for adaption of rod cells to steady illumination, which is a fundamental process in visual signal transduction (Matthews et al., 1988; Nakatani and Yau, 1988b). The effective unit conductance under physiological conditions is very low, due to the block of Na^+-influx by Ca^{2+} and Mg^{2+} (Haynes et al., 1986; Zimmermann and Baylor 1986). In the absence of divalent cations, the unit conductance is of the order of 26 pS which correlates well with that of the purified channel protein after reconstitution into artificial lipid bilayers (Hanke et al., 1988). Both divalent cations act as a permeable blocker from the extracellular side (Stern et al., 1986; Nakatani and Yau, 1988a), independent of the cGMP concentration, indicating that Ca^{2+} and Mg^{2+} bind with similar affinity to open and closed states of the channel (Karpen et al.

1993). In some rare cases, these electrophysiological data have been supplemented with cation selectivity data obtained by other biophysical methods (Schnetkamp, 1990; Wohlfart et al., 1990).

The cone photoreceptor channel seems to be similar in some, but not all of its electrophysiological properties compared to the rod photoreceptor channel. The most important difference seems to be the monovalent cation conductivity and selectivity (see Picones and Korenbrot, 1992).

Like the photoreceptor channel, the cyclic nucleotide-gated channels from vertebrate olfactory epithelial cells are also nonselective cation channels with a permeability ratio for the most important monovalent cations Na^+ and K^+ of 0.8: 1.1 (Nakamura and Gold, 1987; Suzuki, 1989; Bruch and Teeter, 1990; Dhallan et al., 1990; Kolesnikov et al., 1990; Kurahashi, 1990; Frings and Lindemann, 1991). In similarity to the rod photoreceptor channel, the high permeability of the olfactory channel to Ca^{2+} might be important for the sensory adaption in olfaction (Kurahashi and Shibuya, 1990). Handling dendritic membranes as a source with lower channel density, single-channel activity could be resolved, indicating a single-channel conductance of 40 pS (Firestein et al., 1991). The channel is blocked by intracellular Ca^{2+} (Zufall et al., 1991a), and does not inactivate after opening (Zufall et al., 1991b).

IV. PHARMACOLOGY

Different channel blockers have been investigated in order to allow a pharmacological comparison of cyclic nucleotide-gated channels with other classes of channels. Tetrodotoxin, a blocker of voltage-dependent sodium channels, had no effect on the photoreceptor channel (Nicol et al., 1987). Amiloride has been described to act as a blocker of the olfactory channel, but only in a concentration range where blocking of other channels and ion transporters might occur (Frings et al., 1992). The photoreceptor channel is not blocked by a derivative of amiloride, 5-(-N-ethyl-N-isopropyl)-amiloride (EIPA), whereas another analog of amiloride, 3′,4′-dichlorobenzamil seems to be an effective inhibitor for both of the above cyclic nucleotide-gated channels (Nicol et al., 1987; Kolesnikov et al., 1990), suggesting possible structural homologies between cyclic nucleotide-gated channels and the T-Type Ca^{2+}-channel.

Another class of channel blockers, the benzothiazepines, has been used to look for similarities between cyclic nucleotide-gated channels and other Ca^{2+}-channels. Diltiazem inhibits, in a voltage-dependent manner, the photoreceptor (Stern et al., 1986; Schnetkamp, 1990) and the olfactory channel (Frings et al., 1992). However, in both cases, the channel is inhibited stereo-selectively by the L-*cis*- and not the D-*cis*-isomer. Blockage of voltage-dependent calcium channels exhibits the reverse stereo-selectivity (Reynolds and Snyder, 1988). Blocking the photoreceptor channel by L-*cis*-diltiazem has been shown to be voltage-dependent and cGMP-independent (Quandt et al., 1991; Haynes, 1992; McLatchie and Matthews, 1992).

There are two single observations regarding photoreceptor channel blocking compounds which merit further investigation. (1) Marinov et al. (1992) described inhibition of the channel in excised patch measurements by μM concentrations of the native plant pigment, pelargonidin. This inhibitory action could be eliminated by rotenone, an insecticidal, anti-protozoal constituent of derris root. (2) Donner et al. (1990) described the effect of sulfhydryl-binding reagents on the channel. N-ethyl-maleimide (NEM) and iodoacetamide (IAA), transiently increased rod conductance. However these SH-reagents may act not only by increasing the photoreceptor current, but also by inhibition of the cGMP-hydrolysis that occurs in rod outer segments after illumination. Recently it was suggested, that the rather unspecific guanylyl cyclase inhibitor LY83583 is a potent blocker of the cyclic nucleotide-gated channel in isolated olfactory receptor neurons (Leinders et al., 1995). However, these results have to be confirmed on a purified channel preparation to rule out influence of the soluble guanylyl cyclase present in this whole cell voltage-clamp recordings.

V. CYCLIC NUCLEOTIDE SPECIFICITY

While conductive properties of cyclic nucleotide-gated channels revealed by the electrophysiology appear to be similar, the sensitivity of the various channels towards cyclic nucleotides is dependent on the tissue and the type of channel (see Table 1).

K_M- or EC_{50}-values for the activation of the retinal rod photoreceptor channels by cGMP between 17–130 μM have been reported; cAMP is much less effective (K_M 1.85 mM; see Altenhofer et al., 1991). Although the K_M for cAMP is in the mM range, a role of physiological concentrations of cAMP (<100 μM) on the activation of the photoreceptor channel has been proposed, because cAMP in concentrations below its K_M, enhances the total current of subsaturating concentrations of cGMP (Furman and Tanaka, 1989; Ildefonse et al., 1992). The cGMP-dependence of the rod photoreceptor channel shows sigmoidicity, indicating cooperative binding of more than one cGMP-molecule per active channel complex. Hill coefficients between 1.7–3.1 were obtained (Koch and Kaupp, 1985; Yau and Nakatani, 1985, 1988; Puckett and Goldin, 1986; Fesenko et al, 1985; Zimmerman et al., 1985); therefore, it could be concluded that at minimum two, but probably three or more cGMP-molecules are necessary for the opening of one channel complex.

Modifications to the cyclic phosphate, ribose, and most positions on the guanine-ring system generally produced compounds that were less effective than cGMP itself (Fesenko et al., 1985; Zimmermann et al., 1985; Karpen et al., 1988; Tanaka et al., 1989). In contrast some modifications at the C_8-position of the guanine ring, for example 8-Br-cGMP, result in analogs that were more effective compared to cGMP (Koch and Kaupp, 1985; Zimmermann et al., 1985; Tanaka et al., 1989). It has been shown that the affinity of the photoreceptor channel towards the cyclic nucleotide can be modulated by different phosphatases (Gordon et al., 1992), indicating that phosphorylation of the cGMP-binding domain might be important for different cyclic nucleotide affinities.

Table 1. Cyclic Nucleotide Specificity of Channels from Various Tissues

Tissue / Cell Type	*Selectivity*	*K_M (cGMP)*	*K_M (cAMP)*	*Species*	*References*
Aorta	cGMP >> cAMP	0.45 μM	32 μM	rabbit (cloned)	Biel et al., 1994
Cochlea (hair cells)	cGMP >> cAMP	>>1mM	20–30 μM	guinea pig (native)	Kolesnikov et al., 1991
Olfactory mucosal	cGMP ~ cAMP	1.0–2.4 μM	2.5–4 μM	rat, frog, toad (native)	Nakamura and Gold, 1987; Frings et al., 1992
Epithelium	cGMP > cAMP	1.5–2.4 μM	64–68 μM	rat, bovine (cloned)	Dhallan et al., 1990; Altenhofen et al., 1991
Pineal gland	cGMP >> cAMP	n.d.[a]	n.d.	chicken	Dryer and Henderson, 1991
Retinal rod outer segment	cGMP >> cAMP	17–130 μM	>>1 mM	mammalian, amphibian (native)	Fesenko et al., 1985; Koch and Kaupp 1985; Zimmermann and Baylor, 1986
		52–86 μM	n.d.	bovine, human (cloned)	Kaupp et al., 1989; Dhallan et al., 1992
Testis, kidney, and heart	cGMP >> cAMP	18 μM	1.9 mM	bovine (cloned)	Biel et al., 1994

Note: [a] n.d.: not determined.

In contrast to the photoreceptor channel, cyclic nucleotide-gated channels in native membranes of olfactory cells were activated with similar affinities by cAMP and cGMP with activation constants of 2.5 μM (rat) to 4.0 μM (frog); Hill coefficients were 1.4–1.8 (Kolesnikov et al., 1990; Frings et al., 1992). Differences in cGMP/cAMP-specificity were observed when membranes containing expressed cloned olfactory channels were used; the cloned channels possess a higher affinity toward cGMP (Dhallan et al., 1990; Altenhofen et al., 1991). For example, the rat cloned channel displays a higher K_M for cAMP (68 μM) compared to the native form, while cGMP-affinity is similar (2.4 μM; Dhallan et al., 1990). These differences might be due to different post-translational modifications and/or regulatory mechanisms in the expression system compared to the native systems. In native membranes, it has been shown that the channel becomes less sensitive to cAMP when the intracellular Ca^{+2} concentration is increased (Zufall et al., 1991a; Kramer and Siegelbaum, 1992). This inhibition is not due to a direct interaction between Ca^{2+} and the channel protein and it could be washed out in excised patches, indicating an indirect effect via a regulatory Ca^{2+}-binding protein. An attractive explanation was given by Liman and Buck (1994) and independently by Bradley et al. (1994), when a second subunit of the olfactory cyclic nucleotide-gated channel was cloned and characterized. It was demonstrated that after coexpression of this new subunit with the first olfactory subunit, high sensitivity to cAMP was gained that resembles more closely the native conductance. These data indicate that native olfactory cyclic nucleotide-gated channel is likely to be a heterooligomer of two subunits.

VI. BIOCHEMICAL PROPERTIES

In order to investigate the structural properties of a particular protein, it is necessary to at least partially purify the protein of interest. Until now, the rod photoreceptor outer segment channel is the only member of this channel class that could be successfully purified from a native source to near homogeneity. This is due to two major reasons: (1) bovine rod outer segments can be rapidly prepared in large quantities, and (2) the density of photoreceptor channels in such preparations is relatively high, so that purification ends in amounts of channel protein visible as Coomassie-blue stained bands after gel electrophoresis.

A. Purification of the Photoreceptor Rod Channel

The purification of the cGMP-gated channel from rod outer segments required a means of detection of the channel protein after chromatographic separation, for example channel activity or ligand binding. However, at the beginning, no ligand with a sufficiently high affinity was available for the cGMP-gated channel (but see also Hurwitz and Hocombe, 1991). Therefore, a functional assay for the channel protein after solubilization and reconstitution into Ca^{2+}-containing liposomes was developed, utilizing the fact that the cGMP-gated channel is appreciably permeable to Ca^{2+} and that this cation can be readily measured using spectroscopic techniques (Cook et al., 1986). With this assay, a rapid procedure could be established, involving anion-exchange chromatography on DEAE-Fractogel TSK and dye-affinity chromatography on AF-Red-Fractogel (Cook et al., 1987). SDS-polyacrylamide gel electrophoresis revealed that the purified channel extract contained two polypeptides of M_r 63- and 240-kD. The higher molecular weight band was found to resemble spectrin (see following); thus the 63 kD or a multimer of it, is sufficient to form the cGMP-gated photoreceptor channel. The availability of the purified channel protein allowed the production and characterization of specific antibodies which were used to demonstrate that the channel is localized exclusively in the plasma membrane of the rod photoreceptor outer segment (Cook et al., 1989).

The cGMP-dependence of purified reconstituted channel was found to saturate at about 50 μM cGMP, and to be highly cooperative (n_H = 3.1). After fusing liposomes containing the purified protein with an artificial lipid bilayer, it was possible to investigate the electrophysiological behavior of the purified channel (Hanke et al., 1988). The electrophysiological data so obtained and the biochemical results demonstrate that the purified channel exhibits properties virtually identical to the channel *in situ*. More recently, an affinity-purification procedure based on an immobilized cGMP-affinity matrix has been described for the purification of the rod photoreceptor cGMP-gated channel (Hurwitz and Hocombe, 1991).

B. Association of the Photoreceptor Rod Channel with Cytoskeletal and Cytosolic Proteins

The molecular identity of the high molecular weight band (240 kD) in the purified channel extract is not yet known. This protein was found to be very susceptible to proteolysis and in some experiments, was absent without loss of channel activity. Molday and coworkers were able to demonstrate some immuno-crossreactivity between the 240 kD protein and the α-subunits of erythrocyte spectrin and brain fodrin, cytoskeletal proteins of identical molecular weight (Wong and Molday, 1986; Molday et al., 1990). A monoclonal antibody, PMs 4B2, which cross-reacts strongly with erythrocyte spectrin, specifically binds to the 240 kD component in purified extracts. The addition of erythrocyte spectrin was found to block this binding to the purified 240 kD channel component. Immunoprecipitation experiments revealed that the PMs 4B2 antibody could coprecipitate the 240- and the 63-kD components from the same extract. From these results, it was concluded that the 240 kD protein is a spectrin-like cytoskeletal protein, not directly essential for channel activity, but directly associated with the cGMP-gated channel protein.

Recent biochemical results indicate however, that the 240 kD protein might have an important indirect function by linking cytosolic proteins in a transient way to the channel complex inside the rod outer segment plasma membrane (Hsu and Molday, 1993). It was shown that the Ca^{2+}-binding protein, calmodulin, which is also present in the rod outer segment, increases the apparent K_M of the channel for cGMP without affecting the cooperativity and the V_{max}. Using an overlay technique, [^{125}I] calmodulin bound in the presence of Ca^{2+} to immunopurified and electro-blotted channel extract, but only to the 240 kD protein and not to the 63 kD protein, indicating that the 240 kD protein is important for the calmodulin-dependent channel modulation. The exact nature and role of the 240 kD protein, as well as the way in which the 240 kD protein interacts with the 63 kD protein, is still unclear. A first important step would involve the identification of the primary sequence of the 240 kD protein, to demonstrate whether or not this protein is really a peripheral, cytoskeletal protein.

C. Molecular Properties (I): Primary Structures of Cyclic Nucleotide-gated Channels

Once one member of the cyclic nucleotide channel family had been purified, its primary structure could be obtained be subjecting the purified channel protein to limited trypsin cleavage in order to obtain amino acid sequence information, followed by probing a cDNA-library with specific oligonucleotides, and cloning and sequencing the positive clones (Kaupp et al., 1989). The sequence information obtained for the photoreceptor cGMP-gated channel was then used by others to detect further non-retinal members of this channel family (Table 2).

In Figure 1, five selected sequences are aligned. The rod photoreceptor channel bears ~60% amino acid homology to the olfactory channel. In contrast, the aortic- and the olfactory-channel type seems to be homologous. The recently cloned kidney channel constitutes a third type. The highest degree of variation between these

Table 2. Primary Sequences of Cyclic Nucleotide-gated Channels

Channel Type	*Number of Amino Acids*	*Suggested Monomer Molecular Weight*	*Reference(s)*
Rod photoreceptor (bovine)	690	79.6 kD	Kaupp et al., 1989
Olfactory channel (rat)	664	n.g.[a]	Dhallan et al., 1990
Olfactory channel (bovine)	663	76.0 kD	Ludwig et al., 1990
Rod photoreceptor (human)	686	n.g.	Dhallan et al., 1992
Rod photoreceptor (human)	690	n.g.	Pittler et al., 1992
Rod photoreceptor (mouse)	683	n.g.	Pittler et al., 1992
Olfactory channel (catfish)	682	n.g.	Goulding et al., 1992
Rod photoreceptor, new	623	70.8 kD	Chen et al., 1993
Subunit (human)	909	102.3 kD	
Aortic channel (rabbit)	732	84.0 kD	Biel et al., 1993
Testis, kidney, and heart channel (bovine)	706	81.0 kD	Biel et al., 1994

Note: [a] n.g.: not given.

```
Rod photoreceptor channel (bov)      .......... .......... .......... .......... .......... .......... .......... ..........
Rod photoreceptor channel (hum)      .......... .......... .......... .......... .......... .......... .......... ..........
Rod channel subunit 2 (hum)          MPRELSRIEE EKEDEEEEEE EEEEEEEEEV TEVLLDSCVV SQVGVGQSEE DGTRPQSTSD QKLWEEVGEE AKKEAEEKAK 80
Olfactory channel (bov)              .......... .......... .......... .......... .......... .......... .......... ..........
Aortic channel (rabbit)              .......... .......... .......... ........MS SWRSCCARAP LSGSAWRRSA ATRRSRRCLK TKRKRWSSGK 42
Testis, kidney & heart (bov)         .......... .......... .......... .......... .......... .......... .......... ..........

Rod photoreceptor channel (bov)      .......... .......... .......MKK VIINTWHSFV NIPNVVGPDV EKEITRM..E NGACSS..FS GDDDDSASMF 49
Rod photoreceptor channel (hum)      .......... .......... ...MKLSMKN NIINTQQSFV TMPNVIVPDI EKEIRRM..E NGACSS..FS EDDDSAYTS. 53
Rod channel subunit 2 (hum)          EEAEEVAEEE AEKEPQDWAE TKEEPEAEAE AASSGVPATK QHPEVQVEDT DADSCPLMAE ENPPSTVLPP PSPAKSDTLI 160
Olfactory channel (bov)              .......... .......... .......MTE KANGVKSSPA NNHNHHAPPA IKASGKD..D HRASSRPQSA AADDTSSELQ 51
Aortic channel (rabbit)              GTPMQSTQCE TRRRAQTPCE STGHTWRMTE KSNGVKSSPA NNHNNHVPAT IKANGKD..E SRTRSRPQSA ADDDTSSELQ 120
Testis, kidney & heart (bov)         .......... .......... .........M AKISTQYSHP TRTHPSVRTM DRDLDCI..E NGLSRT..HL PCEETSSELQ 47

                                                                                               ROS-N-Terminus
                                                                                                    ▼
Rod photoreceptor channel (bov)      EESETENPHA RDSFRSNTHG SG........ .......... ..QPSQREQY LPGAIALFNV NNSSNKEQ.. ....EPKEKK 103
Rod photoreceptor channel (hum)      EESENENPHA RGSFSYKSLR KG                    ..GPSQREQY LPGAIAIFNV NNSSNKDQ.. ....EPEEKK 106
Rod channel subunit 2 (hum)          VPSSASGTHR KKLPSEDDEA EELKALSPAE SPVVAWSDPT TPKDTDGQD. ..RAASTAST NSAIINDRLQ ELVKLFKERT 237
Olfactory channel (bov)              QLAEMDAPQQ R........R GGFRRIARLV GVLREWAYRN FREEEPRPDS FLERFRGPEL HTVTT..... .......... 108
Aortic channel (rabbit)              RLAEMDAPQQ R........R GGFRRIVRLV GVIRQWANRN FREEEARPDS FLERFRGPEL QTVTT..... .......... 177
Testis, kidney & heart (bov)         EGIAMETRGL AESRQSSFTS QGPTRLSRLI ISLRAWSARH LHQEDQRPDS FLERFRGAEL QEVSSRESHV QFNVGSQEPP 127

                                                                                                          *****
Rod photoreceptor channel (bov)      KKKKEKKSKP DDKNENKKDP EKKKKKEKDK DKKKKEEKGK DKKEEEKKEV VV........ ........ID PSGNTYYN.W 166
Rod photoreceptor channel (hum)      KKKKEKKSKS DDKNENKNDP EKKKKK.KDK EKKKKEEKSK DKKEHHKKEV VV........ ........ID PSGNTYYN.W 168
Rod channel subunit 2 (hum)          EKVKEKLIDP DVTSDEESPK PSPAKKAPEP APDTKPAEAE PVEEEHYCDM LCCKFKHRPW KKYQFPQSID PLTNLMYVLW 317
Olfactory channel (bov)              .......... .......... ........QQ GDGKGDKDGE GKGTKKKFEL FV........ ........LD PAGDWYYR.W 143
Aortic channel (rabbit)              .......... .......... ........QQ GDGKGDKDGD GKGTKKKFEL FV........ ........LD PAGDWYYR.W 214
Testis,kidney & heart (bov)          DRGRSAWPLA RNNTNTCNNS EKDDKAKKEE KEKKEEKKEN PKKEEKKKDS VV........ ........MD PSSNMYYH.W 192
                                                                                                          *****
```

```
                                          GS1
                                           ▼
                                 ****** S1 ********                        *********** S2 ******                                                ****
Rod photoreceptor channel (bov)  LFCITLPVMY NWTMIIARAC FDELQSDYLE YWLAFDYLSD VVYLLDMFV. RTRTGYLEQG LLVKEERKLI DKYKSTFQFK 245
Rod photoreceptor channel (hum)  LFCITLPVMY NWTMVIARAC FDELQSDYLE YWLILDYVSD IVYLIDMFV. RTRTGYLEQG LLVKEELKLI NKYKSNLQFK 247
Rod channel subunit 2 (hum)      LFFVVMAWNW NCWLIPVRWA FPYQTPDNIH HWLLMDYLCD LIYFLDITVF QTRLQFVRGG DIITDKKDMR NNYLKSRRFK 397
Olfactory channel (bov)          LFLIALPVLY NWCLLVARAC FSDLQKGYYI VWLVLDYVSD VVYIADLFI. RLRTGFLEQG LLVKDTKKLR DNYIHTMQFK 222
Aortic channel (rabbit)          LFVIAMPVLY NWCLLVARAC FSDLQRGYFL VWLVLDYSSD VVYIADLFI. RLRTGFLEQG LLVKDPKKLR DNYIHTLQFK 291
Testis, kidney & heart (bov)     LTVIAVPVFY NWCLLVCRAC FDELQSEHLM LWLVLDYSAD ILYGMDMLV. RARTGFLEQG LMVMDASRLW KHYTQTLHFK 269
                                 ******************               ********************                                                ****

                                 ******* S3.********    *********.S4.**********            *********** S5.**********
Rod photoreceptor channel (bov)  LDVLSVIPTD LLYIKFGWNY PEIRLNRLLR ISRMFEFFQR TETRTNYPNI FRISNLVMYI IIIHWNACV YFSISKAIGF 325
Rod photoreceptor channel (hum)  LDVLSLIPTD LLYFKLGWNY PEIRLNRLLR FSRMFEFFQR TETRTNYPNI FRISNLVMYI VIIIHWNACV FYSISKAIGF 327
Rod channel subunit 2 (hum)      MDLLSLLPLD FLYLKVGVN. PLLRLPRCLK YMAFFEFNSR LESILSKAYV YRVIRTTAYL LYSLHLNSCL YYWASAYQGL 476
Olfactory channel (bov)          LDVASIIPTD LIYFAVGIHN PEVRFNRLLH FARMFEFFDR TETRTSYPNI FRISNLILYI LIIIHWNACI YYAISKSIGF 302
Aortic channel (rabbit)          LDVASIIPTD LIYFAVGIHN PELRFNRLLH FARMFEFFDR TETRTSYPNI FRISNLVLYI LVIIHWNACI YYAISKSIGF 371
Testis, kidney & heart (bov)     LDVLSLVPID LAYFKLGMNY PELRFNRLLK LARLFEFFDR TETRTNYPNM FRIGNLVLYI LIIIHWNACI YFAISKFIGF 349
                                 *******************    ***********************            *************************

                                 GS2      GS                                                                                 GS
                                  ▼        ▼                ═══════ pore ═══════           *********** S6 *********          ▼
Rod photoreceptor channel (bov)  GNDTWVYPDV NDPDFGRLAR KYVYSLYWST LTLTTIGETP PPVRDSEYFF VVADFLIGVL IFATIVGNIG SMISNMNAAR 405
Rod photoreceptor channel (hum)  GNDTWVYPDI NDPEFGRLAR KYVYSLYWST LTLTTIGETP PPVRDSEYVF VVVDFLIGVL IFATIVGNIG SMISNMNAAR 407
Rod channel subunit 2 (hum)      GSTHWVYDGV GNS....... .YIRCYYFAV KTLITIGGLP DPKTLFEIVF QLLNYFTGVF AFSVMIGQMR DVVGAATAGQ 548
Olfactory channel (bov)          GVDTWVYPNI TDPEYGYLSR EYIYCLYWST LTLTTIGETP PPVKDEEYLF VIFDFLIGVL IFATIVGNVG SMISNMNATR 382
Aortic channel (rabbit)          GVDTWVYPNI TDPEYGYLAR EYIYCLYWST LTLTTIGETP PPVKDEEYLF VIFDFLIGVL IFATIVGNVG SMISNMNATR 451
Testis, kidney & heart (bov)     GTDSWVYPNV SNPEYGRLSR KYIYSLYWST LTLTTIGETP PPVKDEEYLF VVIDFLVGVL IFATIVGNVG SMISNMNASR 429
                                                       ═════(SS1-SS2)═════          *************************

                                              GS3
                                               ▼
Rod photoreceptor channel (bov)  AEFQARIDAI KQYMHFRNVS KDMEKRVIKW FDYLWTNKKT VDEREVLKYL PDKLRAEIAI NVHLDTLKKV RIFADCEAGL 485
Rod photoreceptor channel (hum)  AEFQARIDAI KQYMHFRNVS KDMEKRVIKW FDYLWTNKKT VDEKEVLKYL PDKLRAEIAI NVHLDTLKKV RIFADCEAGL 487
Rod channel subunit 2 (hum)      TYYRSCMDST VKYMNFYKIP KSVQNRVKTW YEYTWHSQGM LDESELMVQL PDKMRLDLAI DVNYNIVSKV ALFQGCDRQM 628
Olfactory channel (bov)          AEFQAKIDAV KHYMQFRKVS KEMEAKVIRW FDYLWTNKKS VDEREVLKNL PAKLRAEIAI NVHLSTLKKV RIFQDCEAGL 462
Aortic channel (rabbit)          AEFQAKIDAV KHYMQFRKVS KEMEAKVIKW FDYLWTNKKT VDEREVLKNL PAKLRAEIAI NVHLSTLKKV RIFQDCEAGL 531
Testis, kidney & heart (bov)     AEFQAKIDSI KQYMQFRKVT KDLETRVIRW FDYLWANKKT VDEKEVLKSL PDKLKAEIAI NVHLDTLRKV RIFQDCEAGL 509
```

Figure 1. (Continued)

```
                                        ==================cyclic nucleotide binding domain=============================
Rod photoreceptor channel (bov)    LVELVLKLQP QVYSPGDYIC KKGDIGREMY IIKEGKLAVV AD.DGITQFV VLSDGSYFGE ISILNIKGSK AGNRRTANIK 564
Rod photoreceptor channel (hum)    LVELVLKLQP QVYSPGDYIC KKGDIGREMY IIKEGKLAVV AD.DGVTQFV VLSDGSTFGE ISILNIKGSK AGNRRTANIK 566
Rod channel subunit 2 (hum)        IFDMLKRLRS VVYLPNDYVC KKGEIGREMY IIQAGQVQVL GGPDGKSVLV TLKAGSVFGE ISLLAVGG.. .GNRRTANVV 705
Olfactory channel (bov)            LVELVLKLRP QVFSPGDYIC RKGDIGKEMY IIKEGKLAVV AD.DGVTQYA LLSAGSCFGE ISILNIKGSK MGNRRTANIR 541
Aortic channel (rabbit)            LVELVLKLRP QVFSPGDYIC RKGDIGKEMY IIKEGKLAVV AD.DGVTQYA LLSAGSCFGE ISILNIKGSK MGNRRTANIR 610
Testis, kidney & heart (bov)       LVELVLKLRP AVFSPGDYIC KKGDIGREMY IIKEGKLAVV AE.DGITQFV VLGDGSYFGE ISILNIKGSK SGNRRTANIR 588
                                        ===============================================================================

                                   =============
Rod photoreceptor channel (bov)    SIGYSDLFCL SKDDLMEALT EYPDAKGMLE EKGKQILMKD GLLDINIANA GSDPKDLEEK V.......TR MESSVDLLQT 637
Rod photoreceptor channel (hum)    SIGYSDLFCL SKDDLMEALT EYPDAKTMLE EKGKQILMKD GLLDLNIANA GSDPKDLEEK V.......TR MEGSVDLLQT 639
Rod channel subunit 2 (hum)        AHGFTNLFIL DKKDLNEILV HYPESQKLIR KKARRMLRSN NKPKEEKSVL ILPPRAGTPK LFNAALAMTG KMGGKGAKGG 785
Olfactory channel (bov)            SLGYSDLFCL SKDDLMEAVT EYPDAKRVLE ERGREILMKE GLLDENEVAA SMEV.DVQEK L.......EQ LETNMDTLYT 613
Aortic channel (rabbit)            SLGYSDLFCL SKDDLMEAVT EYPDAKKVLE ERGREILMKE GLLDENEVAA SMEV.DVQEK L.......KQ LETNMETLYT 682
Testis, kidney & heart (bov)       SIGYSDLFCL SKDDLMEALT EYPEAKKALE EKGRQILMKD NLIDEELAKA GADPKDIEEK V.......EH LETSLDSLQT 661
                                   =============

                                              GS
                                              ▼
Rod photoreceptor channel (bov)    RFARI...LA EYESMQQKLK QRLTKVEKFL KPLIDTEFSA IEGSGTESGP TDSTQD.... .......... .......... 690
Rod photoreceptor channel (hum)    RFARI...LA EYESMQQKLK QRLTKVEKFL KPLIDTEFSS IEGPWSESGP IDST...... .......... .......... 690
Rod channel subunit 2 (hum)        KLAHLRARLK ELAALEAAAK QQELVEQAKS SQDVKGEEGS AAPDQHTHPK EAATDPPAPR TPPEPPGSPP SSPPPASLGR 865
Olfactory channel (bov)            RFARL...LA EYTGAQQKLK QRITVLETKM KQNNEDDSLS DGMNSPEPPA EKP....... .......... .......... 660
Aortic channel (rabbit)            RFGRL...LA EYTGAQQKLK QRITVLEVKM KQNTEDDYLS DGMNSPEPAA AEQP...... .......... .......... 727
Testis, kidney & heart (bov)       RFARL...LA EYNATQMKVK QRLSQLESQV KMGLPPDGDA PQTEASQP.. .......... .......... .......... 706

Rod channel subunit 2 (hum)        PEGEEEGPAE PEEHSVRICM SPGPEPGEQI LSVKMPEERE EKAE 909
```

Figure 1. Alignment of five amino acid sequences of the cyclic nucleotide-gated channel family (GS: potential N-glycosylation site; GS1–GS3: potential N-glycosylation sites of the rod photoreceptor channel; S1–S6: transmembrane helices).

selected references occurs at the N-terminal region. This correlates well with the fact that the bovine rod photoreceptor channel's N-terminus is post-translationally processed by a specific protease without loss of channel activity (see following). This suggests that variation in this sequence region does not necessarily result in inactive or differently active channel proteins.

Common to each of the sequences is a typical pattern of hydrophobic sequences, called S1–S6. In the first sequence published (Kaupp et al., 1989), these hydrophobic putative transmembrane segments were named in a different way (S1→H2, S2→H2, S3→H3, S4 not assigned, S5→H5). This mode of naming was changed when it was recognized that the cyclic nucleotide-gated channels structurally resemble voltage-gated channels in their transmembrane topology. The cyclic nucleotide-gated channel sequences also contain a voltage-gating motif (S4) with a typical pattern of positively charged amino acids (reviewed by Hille, 1991), even though the opening of each cyclic-nucleotide gated channel is only slightly voltage dependent (Yau and Baylor, 1989). This might be explained by some important mutations, leading to the introduction of negative residues, inside the S4-segment compared to sequences of the voltage-gated channel family.

The relationship between cyclic nucleotide-gated and voltage-gated channels was further strengthened when similarities inside the pore region between the two channels families could be observed (Guy et al., 1991; Goulding et al., 1992; Heginbotham et al., 1992). The pore region of the cyclic-nucleotide-gated channels seems to be located between the S5 and the S6 transmembrane helix (Figure 1). Its sequence is very similar to that of voltage-gated K^+ channels (for an alignment, see Heginbotham et al., 1992). In contrast to the cyclic nucleotide-gated channels, the voltage-gated K^+ channels display two major electrophysiological differences: (1) they are highly selective for K^+, and (2) they are not efficiently blocked by the divalent cations Ca^{2+} and Mg^{2+}. It is possible to speculate that a gap of two amino acids inside the pore region of cyclic nucleotide-gated channels (i.e., between Gly_{362} and Glu_{363} of the rod photoreceptor sequence), compared to the K^+ channel pore region sequences, could confer a different ionic selectivity on these channels (Heginbotham et al., 1992). Further evidence for the relationship between cyclic nucleotide-gated channels and voltage-gated channels was given by Kramer et al. (1994). A 20 amino acid peptide called "ball peptide" derived from the Shaker-voltage dependent potassium channel and responsible for its rapid inactivation also blocks the olfactory cyclic nucleotide-gated channel.

Unlike voltage-gated channels, the cyclic nucleotide-gated channels contain a cyclic-nucleotide binding sequence near their C-terminal, which is very homologous to the cyclic-nucleotide binding site of cGMP-dependent protein kinase, cAMP-dependent protein kinase, and the *Escherichia coli* catabolite gene activator protein, CAP (Takio et al., 1984; Weber et al., 1987; Kaupp et al., 1989). In analogy to these other cGMP-binding proteins, it was concluded that each polypeptide chain accounts for the binding of one cGMP-molecule. The cyclic nucleotide-binding domain is highly conserved between the bovine olfactory channel and the bovine rod photoreceptor channel (see for a further discussion, see Ludwig et al., 1990). It will therefore be interesting to see whether the small sequence differences among the two cyclic nucleotide-binding domains can really account for the observed difference in the cGMP/cAMP-selectivity between these channels.

It was initially thought that the active-channel moiety is formed by a homo-oligomer, each monomer containing between four and six integral membrane subunits. This view was questioned when Chen et al. (1993) cloned and sequenced another protein from human retina, which has only low amino acid homology to the previously described cGMP-gated channel subunits (Figure 1, "human subunit 2") and did not form functional channels by itself. Co-expression of this new subunit with the human rod photoreceptor channel induces two properties different to expression of the photoreceptor channel alone: (1) rapid flickering of the channel openings and (2) higher sensitivity toward L-*cis*-diltiazem. Because both properties are also present in native channels (Koch et al., 1987), it was concluded that the active-channel moiety is a hetero-oligomer of the new cloned subunit and the previously known subunit. However, if this thesis is correct, several questions remain to be answered. Purification of the bovine rod photoreceptor channel reveals only two distinct bands: the 240 kD band with spectrin like immunoreactivity, and the 63 kD band, thought to build the homo-oligomeric transmembrane pore. When the 63 kD subunit was subjected to limited proteolysis, several peptide fragments were obtained that could be sequenced (Kaupp et al., 1989). Each sequence obtained in this way could be refound in the amino acid sequence corresponding to the protein's cDNA clone. Furthermore, N-terminal sequencing of the 63 kD subunit revealed the presence of only one sequence (Molday et al., 1991). It would, therefore, appear that the 63 kD band after the purification is not a mixture of two different transmembrane subunits with only minor sequence homology. One way to confirm the existence of a new transmembrane subunit would involve the cloning and sequencing of the bovine counterpart of the new human subunit cloned by Chen et al. (1993), followed by generation of anti-peptide antibodies specific for this clone. These sequence- specific antibodies, when used to test the fractions of the purification protocol, would allow us to answer the question as to whether or not this new transmembrane subunit is really present in rod outer segment membranes and, if so, whether or not it is lost during the purification procedure.

D. Molecular Properties (II): Post-translational Modifications

The purified rod photoreceptor channel has an apparent molecular weight of 63 kD, which is considerably less than the molecular weight 79.6 kD calculated from the DNA-sequence (see Table 2). It could be shown that the lower molecular weight of the purified channel is in part due to the absence of the first 92 amino acids as revealed by N-terminal sequencing (Molday et al., 1991). The lower molecular weight was also observed when specific anti-channel antibodies were probed against other mammalian (non-bovine) rod outer segment membranes, whereas channel DNA expressed in COS-1 cells and *Xenopus* oocytes displayed the higher molecular weight predicted from the cDNA sequence. Therefore, this proteolytic cleavage seems to be specific for rod outer segments only. Using cDNAs corresponding to the processed and nonprocessed channel forms in eucaryotic expression systems, in combination with electrophysiological methodology might provide further insights into whether or not this processing is of functional relevance.

Glycosylation at specific amino acids is another important post-translational modification of membrane proteins. The consensus sequence for N-glycosylation (NXS or NXT) could be find in the channel sequences in different numbers and at different sites. Only in the case of the rod photoreceptor channel has this glycosylation been studied in detail at the protein level (Wohlfart et al., 1989). Using lectin-binding and enzymatic deglycosylation, the rod channel could be shown to be N-glycosylated with a high-mannose oligosaccharide at one site only. Reconstitution of purified and enzymatically deglycosylated channel protein revealed no difference in channel function compared to the normal glycosylated channel form so it could be concluded that the N-glycosylated sugar does not interact with the cGMP-binding site or the pore region of the channel. Although of no apparent importance for the function of the channel protein, it was important to determine at which site the glycosylation occurs in order to arrive at a topological model (see following). Polyclonal antibodies against peptides containing the potential N-glycosylation sites of the bovine rod photoreceptor channel sequence were used in immunochemistry, localizing the amino acid sequence 321–339 to the extracellular side of rod outer segments (Wohlfart et al., 1992). Antibodies against a shortened peptide of this glycosylation site displayed in Western blotting experiments a higher affinity toward deglycosylated than glycosylated channel protein. Therefore, it could be concluded that the rod photoreceptor channel is glycosylated at its second consensus site, Asn_{327}.

Other post-translational modifications of the sequences are possible, but until now have not been investigated. For example, it would be very interesting to know whether or not these channel proteins are acylated in native membranes; this would provide additional membrane anchors besides the described hydrophobic transmembrane regions. Nearly all of the sequences contain at least one consensus sequence for this modification (not shown in Figure 1). Because it could be shown that at least the rod photoreceptor channel is modulated by phosphatases (Gordon et al., 1992), phosphorylation might be important for the regulation of this channel protein class. Searching in the sequences of the channel proteins, a lot of stronger and weaker consensus sequences could be found for phosphorylation by different protein kinases (not shown in Figure 1). By using protein kinase-specific inhibitors or by working with purified kinases and channel proteins, we will be able in the future, to determine if tissue-specific phosphorylation of this channel class takes place.

VII. TOPOLOGY OF CYCLIC NUCLEOTIDE-GATED CHANNELS

Although biochemical evidence is available at the moment only for the rod photoreceptor channel protein, one can propose a topological model for the folding of the polypeptide chain through the plasma membrane that is valid for the whole class of cyclic nucleotide-gated channels, because all members of this channel class are very homologous in their structural elements.

Using peptide-specific antibodies, it was possible to show that the N-terminus of the rod photorecptor channel is located on the cytoplasmic side of the rod outer segment plasma membrane (Molday et al., 1991). This, combined with the fact that the cGMP-binding site which contains the C-terminus, must also be on the cytosolic side means that the channel polypeptide must traverse the membrane an even number of times. Given the resemblance of these channels with voltage-gated cation channels (Hegonbotham, 1992; Jan and Jan, 1992), it is reasonable to assume that cyclic nucleotide-gated channels exist as tetramers of six-transmembrane helix-containing subunits. When one also takes into consideration that Asn_{327} of the rod photoreceptor channel protein is N-glycosylated and must be present extracellularly, it is possible to propose the topological model shown in Figure 2.

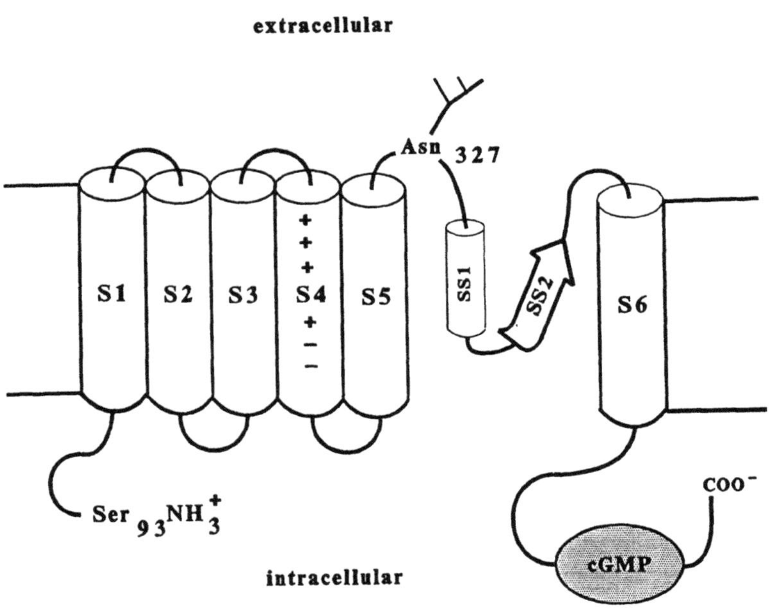

Figure 2. Transmembrane topology of the bovine rod photoreceptor channel subunit.

Fulfilling the step from the transmembrane topology of one polypeptide chain to a three-dimensional model of the whole channel complex, various assumptions have to be made. Exact experimental evidence is missing about how many transmembrane subunits are necessary to form one active channel complex. Realizing the Hill-coefficients for the cooperativity of the cyclic nucleotide-dependent channel opening, and taking into account that one cGMP molecule binds to one channel polypeptide chain, one can assume that the channel exists as a tetra- or pentamer. Because of their resemblance to voltage-gated cation channels, a tetramer, as presented in Figure 3, is most likely for cyclic nucleotide-gated channels.

Whether this channel complex is formed as a homotetramer of one single polypeptide chain or as a heterotetramer of two different polypeptide chains, as proposed by Chen et al. (1993), Liman and Buck (1994), and Bradely et al. (1994), can not be exactly answered at the moment. Examining the biochemical evidence gained by the purification and cloning of the bovine rod photoreceptor channel, a homo-oligomer of one single polypeptide seems to be the most probable population of active channels inside the rod outer segment plasma membrane, although the presence of a small subpopulation of hetero-oligomeric channels in the native plasma membranes can not be excluded.

In reference to the three-dimensional rearrangement of the transmembrane segments, it can be assumed that the region between S5 and S6 (referred to as SS1–SS2 in analogy to voltage dependent K^+ channels) apparently lines the channel

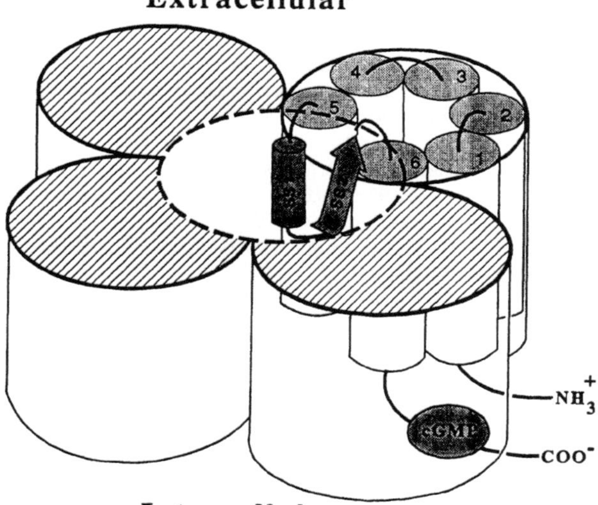

Figure 3. Three-dimensional model of the cyclic nucleotide-gated channel.

pore providing an internal hydrophilic environment area for the transmembrane flux of positive charges. The more hydrophobic transmembrane segments of the polypeptide form the connection between the pore region of the channel and the lipophilic lipid bilayer. In contrast to voltage-gated cation channels, the S4-transmembrane segment inside the cyclic nucleotide-gated channels is not a voltage sensor element, perhaps due to fact that this segment is arrested in the membrane plane by negatively charged amino acids which are not present in S4-segments found in voltage-gated channels. Although the gating function of this S4-segment seems to be lost in cyclic-nucleotide gated channels, it still might be important for maintaining the architecture of the transmembrane segment rearrangement.

In conclusion, cyclic nucleotide-gated channels comprise a new, interesting class of channel proteins. Detailed information about their function in sensory tissues is available, but this knowledge is still lacking for nonsensory tissues. Structurally, the cyclic nucleotide-gated channels seem to possess a homologous architecture compared to voltage-dependent cation channels, where some structural elements have been inactivated and others have been built in.

ACKNOWLEDGMENT

The authors wish to thank Dr. S. Stengelin for his help in sequence searching and editing.

REFERENCES

Ahmad, I., Redmond, L. J., & Barnstable, C. J. (1990). Developmental and tissue-specific expression of the rod photoreceptor cGMP-gated ion channel gene. Biochem. Biophys. Res. Com. 173, 463–470.

Ahmad, I., Korbmacher. C. H., Segal, A. S., Cheung, P., Boulpaep, E. L., & Barnstable, C. J. (1992). Mouse cortical collecting duct cells show nonselective cation channel activity and express a gene related to the cGMP-gated rod photoreceptor channel. Proc. Natl. Acad. Sci. USA 89, 10262–10266.

Altenhofen, W., Ludwig, J., Eismann, E., Kraus, W., Bonigk, W., & Kaupp, U. B. (1991). Control of ligand specificity in cyclic nucleotide-gated channels from rod photoreceptors and olfactory epithelium. Proc. Natl. Acad. Sci. USA 88, 9868–9872.

Baylor, D. A., & Nunn, B. J. (1986). Electrical properties of the light-sensitive dark noise in retinal rods from the lizard and frog. J. Physiol. 371, 115–145.

Biel, M., Altenhofen, W., Hullin, R., Ludwig, J., Freichel, M., Flockerzi, V., Dascal, N., Kaupp, U. B., & Hofmann, F. (1993). Primary structure and functional expression of a cyclic nucleotide-gated channel from rabbit aorta. FEBS Letts. 329, 134–138.

Biel, M., Zong, X., Distler, M., Bosse, E., Klugbauer, N., Murakami, M., Flockerzi, V., & Hofmann, F. (1994). Another member of the cyclic-nucleotide-gated channel family, expressed in testis, kidney, and heart. Proc. Natl. Acad. Sci. USA 91, 3505–3509.

Bodoia, R. D., & Detwiler, P. B. (1985). Patch–clamp recordings of the light-sensitive dark noise in retinal rods from the lizard and frog. J. Physiol. 367, 183–216.

Bradley, J., Li, J., Davidson, N., Lester, H. A., & Zinn, K. (1994). Heteromeric olfactory cyclic nucleotide-gated channels: A subunit that confers increased sensitivity to cAMP. Proc. Natl. Acad. Sci. USA 91, 8890–8894.

Breer, H., Boekhoff, I., & Tareilus, E. (1990). Rapid kinetics of second messenger formation in olfactory transduction. Nature 345, 65–68.

Bruch, R. C., & Teeter, J. H. (1990). Cyclic AMP links amino acid chemoreceptors to ion channels in olfactory cilia. Chemical Senses 15, 419–430.

Chen. T.-Y., Peng Y.-W., Dhallan, R. S., Ahamed, B., Reed, R. R., & Yau, K.-W. (1993). A new subunit of the cyclic nucleotide-gated cation channel in retinal rods. Nature 362, 764–767.

Cobbs, W. H., Barkdoll, A. E., III, & Pugh, E. N., Jr. (1985). Cyclic GMP increases photocurrent and light sensitivity of retinal cones. Nature 317, 64–66.

Cook, N. J., Zeilinger, C., Koch, K.-W., & Kaupp, U. B. (1986). Solubilization and functional reconstitution of the cGMP-dependent cation channel of bovine rod outer segments. J. Biol. Chem. 261, 17033–17039.

Cook, N. J., Hanke, W., & Kaupp, U. B. (1987). Identification, purification, and functional reconstitution of the cyclic GMP-dependent channel from rod photoreceptors. Proc. Natl. Acad. Sci. USA 84, 585–589.

Cook, N. J., Molday, L. L, Reid, D., Kaupp, U. B., & Molday R. S. (1989). The cGMP-gated channel of bovine rod photoreceptors is localized exclusively in the plasma membrane. J. Biol. Chem. 264, 6996–6999.

Dhallan, R. S., Yau, K.-W., Schrader, K. A., & Reed, R. R. (1990). Primary structure and functional expression of a cyclic nucleotide-activated channel from olfactory neurons. Nature 347, 184–187.

Dhallan, R. S., Macke, J. P., Eddy, R. L., Shows, T. B., Reed, R. R., Yau, K.-W., & Nathans, J. (1992). Human rod photoreceptor cGMP-gated channel: Amino acid sequence, gene structure, and functional expression. J. Neurosci. 12, 3248–3256.

DiFrancesco, D., & Tortora, P. (1991). Direct activation of cardiac pacemaker channels by intracellular cyclic AMP. Nature 351, 145–147.

Distler, M., Biel, M., Flockerzi, V., & Hoffmann, F. (1994). Expression of cyclic nucleotide-gated cation channels in non-sensory tissues and cells. Neuropharmacology 33, 1275–1282.

Donner, K., Hemila, S., Kalamkarov, G., Koskelainen A., Pogozheva, I., & Rebrik, T. (1990). Sulfhydryl binding reagents increase the conductivity of the light-sensitive channel and inhibit phototransduction in retinal rods. Exp. Eye Res. 51, 97–105.

Dryer, S. E., & Henderson, D. (1991). A cyclic GMP-activated channel in dissociated cells of the chick pineal gland. Nature 353, 756–758.

Fesenko, E. E., Kolesnikov, S. S., & Lyubarsky, A. L. (1985). Induction by cyclic GMP of cationic conductance in plasma membrane of retinal rod outer segment. Nature 313, 310–313.

Firestein, S., Zufall. F., & Shepherd, G. M. (1991). Single odor-sensitive channels in olfactory receptor neurons are also gated by cyclic nucleotides, J. Neurosci. 11, 3565–3572.

Frings, S., Lynch, J. W., & Lindemann, B. (1992). Properties of cyclic nucleotide-gated channels mediating olfactory transduction—activation, selectivity, and blockage, J. Gen. Physiol. 100, 45–67.

Furman, R. E., & Tanaka, J. C. (1989). Photoreceptor channel activation: Interaction between cAMP and cGMP. Biochemistry 28, 2785–2788.

Gordon, S. E., Brautigan, D. L., & Zimmermann, A. L. (1992). Protein phosphatases modulate the apparent agonist affinity of the light-regulated channel in retinal rods. Neuron 9, 739–748.

Goulding, E. H., Ngai, J., Kramer, R. H., Colicos, S., Axel, R., Siegelbaum, S. A., & Chess, A. (1992). Molecular cloning and single-channel properties of the cyclic nucleotide-gated channel from catfish olfactory neurons. Neuron 8, 45–56.

Guy, H. R., Durell, S. R., Warmke, J., Drysadale, R., & Ganetzky, B. (1991). Similarities in amino acid sequences of *Drosophila eag* and cyclic nucleotide-gated channels Science 254, 730.

Hanke, W., Cook, N. J., & Kaupp, U. B. (1988). cGMP-dependent channel protein from photoreceptor membranes: Single-channel activity of the purified and reconstituted protein. Proc. Natl. Acad. Sci. USA 85, 94–98.

Haynes, L. W. (1992). Block of the cyclic GMP-gated channel of vertebrate rod and cone photoreceptors by L-*cis*-diltiazem. J. Gen. Physiol. 100, 783–801.

Haynes, L. W., & Yau, K.-W. (1985). Cyclic GMP-sensitive conductance in outer segment membrane of catfish cones. Nature 317, 61–64

Haynes, L. W., Kay, A. R., & Yau, K.-W. (1986). Single cyclic GMP-activated channel activity in excised patches of rod outer segment membrane. Nature 321, 66–70.

Heginbotham, L., Abramson, T., MacKinnon, R. (1992). A functional connection between the pores of distantly related ion channels as revealed by mutant K^+-channels. Science 258, 1152–1155.

Hille, B. (1991). Ionic Channels of Excitable Membranes. Sinauer Associates, Sunderland, Massachusetts.

Hodgkin, A. L., McNaughton, P. A., Nunn, B. J., & Yau, K.-W. (1985). Effect of ions on retinal rods from *Bufo marinus*. J. Physiol. 350, 649–680.

Hsu, Y.-T., & Molday, R. S. (1993). Modulation of the cGMP-gated channel of rod photoreceptor cells by calmodulin. Nature 361, 76–79.

Hurwitz, R., & Hocombe, V. (1991). Affinity purification of the photoreceptor cGMP-gated cation channel. J. Biol. Chem. 266, 7975–7977.

Iledfonse, M., Crouzy, S., & Bennett, N. (1992). Gating of retinal rod cation channel by different nucleotides: Comparative study of unitary currents. J. Membr. Biol. 130, 91–104.

Jan, L. Y., & Jan, Y. L. (1992). Tracing the roots of ion channels. Cell 69, 715–718.

Karpen, J. W., Zimmerman, A. L., Stryer, L., & Baylor, D. A. (1988). Gating kinetics of the cyclic-GMP-activated channel of retinal rods: Flash photolysis and voltage-jump-studies. Proc. Natl. Acad. Sci. USA 85, 1287–1291.

Karpen, J. W., Brown, L., Stryer, L., & Baylor, D. A. (1993). Interactions between divalent cations and the gating machinery of cyclic GMP-activated channels in salamander retinal rods. J. Gen. Physiol. 101, 1–25.

Kaupp, U. B., Niidome, T., Tanabe, T., Terada, S., Bönigk, W., Stühmer, W., Cook, N. J., Kangawa, K., Matsuo, H., Hirose, T., Miyata, T., Numa, S. (1989). Primary structure and functional expression from complementary DNA of the rod photoreceptor cyclic GMP-gated channel. Nature 342, 762–766.

Koch, K.-W., & Kaupp, U. B. (1985). Cyclic GMP directly regulates a cation conductance in membranes of bovine rods by a cooperative mechanism. J. Biol. Chem. 260, 6788–6800.

Koch, K.-W., Cook, N. J., & Kaupp, U. B. (1987). The cGMP-dependent channel of vertebrate rod photoreceptors exists in two forms of different cGMP-sensitivity and pharmacological behaviour. J. Biol. Chem. 262, 14415–14421.

Kolesnikov. S., Zhainazarov, A. B., & Kosolapov, A. V. (1990). Cyclic nucleotide-activated channels in the frog olfactory receptor plasma membrane. FEBS Letts. 266, 96–98.

Kolesnikov, S., Rebrik, T. I., Zhainazarov, A. B., Tavartkiladze, G. A., & Kalamkarov, G. R. (1991). A cyclic-AMP-gated conductance in cochlear hair cells. FEBS Letts. 290, 167–170.

Kurahashi, T. (1990). The response induced by intracellular cyclic AMP in isolated olfactory receptor cells of the newt. J. Physiol. 430, 355–371.

Kramer, R. H., & Siegelbaum, S. A. (1992). Intracellular Ca^{2+} regulates the sensitivity of cyclic nucleotide-gated channels in olfactory receptor neurons. Neuron 9, 897–906.

Kramer, R. H., Goulding, E., & Siegelbaum, S. A. (1994). Potassium channel inactivation peptide blocks cyclic nucleotide-gated channels by binding to the conserved pore domain. Neuron 12, 655–662.

Kurahashi, T., & Shibuya, T. (1990). Ca^{2+}-dependent adaptive properties in the solitary olfactory receptor cell of the newt. Brain. Res. 515, 1–2.

Leinders-Zufall, T., & Zufall, F. (1995). Block of cyclic nucleotide-gated channels in salamander olfactory receptor neurons by the guanylyl cyclase inhibitor LY83583. J. Neurophysiol. 74, 2759–2762.

Levy, N. S., Bakalyar, H. A., & Reed, R. R. (1991). Signal transduction in olfactory neurons. J. Steroid Biochem. Mol. Biol. 39, 633B–637B.

Light, D. B., Corbin, J. D., & Stanton, B. A. (1989). Dual ion-channel regulation by cyclic GMP and cyclic GMP-dependent protein kinase. Nature 344, 336–339.

Liman, E. R., & Buck, L. B. (1994). A second subunit of the olfactory cyclic nucleotide-gated channel confers high sensitivity to cAMP. Neuron 13, 611–621.

Lowe, G., Nakamura, T., & Gold, G. H. (1989). Adenylate cyclase mediates olfactory transduction for a wide variety of odorants. Proc. Natl. Acad. Sci. USA 86, 5641–5645.

Ludwig, J., Margalit, T., Eismann, E., Lancet, D., & Kaupp, U. B. (1990). Primary structure of cAMP-gated channel from bovine olfactory epithelium. FEBS Letts. 270, 24–29.

Marinov, B. S., Filatov, G. N., Nikonov, S. S., Ruzhieva, R. Ch., Lubarsky, A. L., & Fesenko E. E. (1992). Regulation of conductance of retinal rod outer segment membrane by redox agents. Biol. Membrane 9, 166–173.

Matthews, H. R., Murphy, R. L. W., Fain, G. L., & Lamb, T. D. (1988). Photoreceptor light adaptation is mediated by cytoplasmic calcium concentration. Nature 334, 67–69.

McLatchie, L. M., & Matthews, H. R. (1992). Voltage-dependent block by L-*cis*-diltiazem for the cyclic GMP-activated conductance of salamander rods. Proc. R. Soc. London, B. Biol. Sci. 247, 113–119.

Molday, L. L., Cook, N. J., Kaupp, U. B., & Molday, R. S. (1990). The cGMP-gated cation channel of bovine rod photoreceptor cells is associated with a 240 kDa protein exhibiting immunochemical crossreactivity with spectrin. J. Biol. Chem. 265, 18690–18695.

Molday, R. S., Molday, L. L., Dose, A., Lewis, I.-C., Illing, M., Cook, N. J., Eismann, E., & Kaupp, U. B. (1991). The cGMP-gated channel of the rod photoreceptor cell—characterization and orientation of the N-terminus J. Biol. Chem. 266, 21917–21922.

Nakamura, T., & Gold, G. H. (1987). A cyclic nucleotide-gated conductance in olfactory receptor cilia. Nature 325, 442–444.

Nakatani, K., & Yau, K.-W. (1988a). Calcium and magnesium fluxes across the plasma membrane of the toad rod outer segment. J. Physiol. 395, 695–729.

Nakatani, K., & Yau, K.-W. (1988b). Calcium and light adaptation in retinal rods and cones. Nature 334, 69–71.

Nathans, C. (1992). Nitric oxide as a secretory product of mammalian cells. FASEB J. 6, 3051–3064.

Nawy, S., & Jahr, C. E. (1990). Suppression by glutamate of cGMP-activated conductance in retinal bipolar cells. Nature 346, 269–271.

Nawy, S., & Jahr, C. E. (1991). cGMP-gated conductance in retinal bipolar cells is suppressed by the photoreceptor transmitter. Neuron 7, 677–683.

Nicol, G. D., Schnetkamp, P. P. M., Saimi, Y., Gragoe, E. J., Jr., & Bownds, M. D. (1987). A derivative of amiloride blocks both the light-regulated and cyclic GMP-regulated conductances in rod photoreceptors. J. Gen. Physiol. 90, 651–669.

Pace, U., Hanski, E., Salomon, Y., & Lancet, D. (1985). Odorant-sensitive adenylate cyclase may mediate olfactory reception. Nature 316, 255–258.

Picones, A., & Korenbrot, J. I. (1992). Permeation and interaction of monovalent cations with the cGMP-gated channel of cone photoreceptors. J. Gen. Physiol. 100, 647–673.

Pittler, S. J., Lee, A. K., Altherr, M. R., Howard, T. A., Seldin, M. F., Hurwitz, R. L., Wasmuth, J. J., & Baehr, W. (1992). Primary structure and chromosomal localization for human and mouse rod photoreceptor cGMP-gated cation channel. J. Biol. Chem. 267, 6257–6262.

Puckett, K. L., & Goldin, S. M. (1986). Guanosine 3′, 5′-cyclic monophosphate stimulates release of actively accumulated calcium in purified disks from rod outer segments of bovine retina. Biochemistry 25, 1739–1746.

Quandt, F. N., Nicol, G. D., & Schnetkamp, P. P. M. (1991). Voltage-dependent gating and block of the cyclic-GMP-dependent current in bovine rod outer segments. Neuroscience 42, 629–638.

Reynolds, I. J., & Snyder, S. H. (1988). Calcium antagonist receptors. In: Ion Channels (Narahashi, T., Ed.) pp. 213–249. Plenum Publishing Corp., New York.

Schnetkamp, P. P. M. (1990). Cation selectivity of and cation binding to the cGMP-dependent channel in bovine rod outer segments. J. Gen. Physiol. 96, 517–534.

Shiells, R. A., & Falk, G. (1992). Properties of the cGMP-activated channel of retinal on-bipolar cells. Proc. R. Soc. London B. Biol. Sci. 247, 21–35.

Sklar, P. B., Anholt, R. R. H. M., & Snyder, S. H. (1986). The odorant sensitive adenylate cyclase of olfactory receptor cells. J. Biol. Chem. 261, 15538–15543.

Stern, J. H., Kaupp, U. B., & MacLeish, P. R. (1986). Control of the light-regulated current in rod photoreceptors by cyclic GMP, calcium, and *L-cis*-diltiazem. Proc. Natl. Acad. Sci. USA 83, 1163–1167.

Stryer, L. (1986). Cyclic GMP cascade of vision. Annu. Rev. Neurosci. 9, 87–119.

Suzuki, N. (1989). Voltage and cyclic nucleotide-gated currents in isolated olfactory receptor cells. In: Chemical Senses, Vol. 1 (Brand, J. G., Teeter, J. H., Cagan, R. H., & Kare, M. R., Eds.) pp. 469–493. Marcel Dekker, Inc. New York.

Takio, K., Wade, R. D., Smith, S. B., Krebs, E. G., Walsh, K. A., & Titani, K. (1984). Guanosine cyclic 3′, 5′-phosphate dependent protein: A chimeric protein homologous with two separate protein families. Biochemistry 23, 4207–4218.

Tanaka, J. C., Eccelston, J. F., & Furman, R. E. (1989). Photoreceptor channel activation by nucleotide derivatives. Biochemistry 28, 2776–2784.

Warmke, J., Drysdale, R., & Ganetzky, B. (1991). A distinct potassium channel polypeptide encoded by the *Drosophila eag* locus. Science 252, 1560–1562.

Wässle, H., Grünert, U., Cook, N. J., & Molday, R. S. (1992). The cGMP-gated channel of rod outer segments is not localized in bipolar cells of the mammalian retina. Neurosci. Lett. 134, 199–202.

Weber, I. T., Steitz, T. A., Bubis, J., & Taylor, S. S. (1987). Predicted structures of cAMP binding domains of type I and II regulatory subunits of cAMP-dependent protein kinase. Biochemistry 26, 343–351.

Weyand, I., Godde, M., Frings, S., Weiner, J., Muller, F., Altenhofen, W., Hatt, H., & Kaupp, U. B. (1994). Cloning and functional expression of acyclic-nucleotide-gated channel from mammalian sperm. Nature 368, 859–863.

Wohlfart, P., Müller, H., & Cook, N. J. (1989). Lectin binding and enzymatic deglycosylation of the cGMP-gated channel from bovine rod photoreceptors. J. Biol. Chem. 264, 20934–20939.

Wohlfart, P., Vienhues, R., & Cook, N. J. (1990). Spectrophotometric determination of cGMP-gated channel Mg^{2+}-fluxes using dichlorophosphonazo III. Biochim. Biophys. Acta 1022, 283–290.

Wohlfart, P., Haase, W., Molday, R. S., & Cook, N. J. (1992). Antibodies against synthetic peptides used to determine the topology and site of glycosylation ot the cGMP-gated channel from bovine rod photoreceptors. J. Biol. Chem. 267, 644–648.

Wong, S., & Molday, R. S. (1986). A spectrin like protein in retinal rod outer segments. Biochemistry 25, 6294–6300.

Yau, K.-W., & Baylor, D.A. (1989). Cyclic-GMP-activated conductance of retinal photoreceptor cells. Annu. Rev. Neurosci. 12, 289–327.

Yau, K.-W., & Nakatani, K. (1984). Cation selectivity of light-sensitive conductance in retinal rod outer segments. Nature 309, 352–354.

Yau, K.-W., & Nakatani, K. (1985). Light-suppressible, cyclic GMP-sensitive conductance in the plasma membrane of a truncated rod outer segment. Nature 317, 252–255.

Zimmermann, A. L., Yamanaka, G., Eckstein, F., Baylor, D. A., & Stryer, L. (1985). Interaction of hydrolysis-resistant analogs of cyclic GMP with the phosphodiesterase and light-sensitive channel of retinal rod outer segments. Proc. Natl. Acad. Sci. USA 82, 8813–8817.

Zimmermann, A. L., & Baylor, D. A. (1986). Cyclic GMP-sensitive conductance of retinal rods consists of aqueous pores. Nature 321, 70–72.

Zufall, F., Shepherd, G. M., & Firestein, S. (1991a). Inhibition of the olfactory cyclic nucleotide-gated ion channel by intracellular calcium. Proc. R. Soc. Lond. B. 246, 225–230.

Zufall, F., Firestein, S., & Shepherd, G. M. (1991b). Analysis of single cyclic nucleotide-gated channels in olfactory receptor cells. J. Neurosci. 11, 3573–3580.

IP3-SENSITIVE CALCIUM CHANNEL

Katsuhiko Mikoshiba, Teiichi Furuichi, and Atsushi Miyawaki

Biomembranes
Volume 6, pages 273–289.

ISBN: 1-55938-663-0.

I. INTRODUCTION

Various extracellular signals utilize the phosphoinositide signaling pathway to regulate cellular function. Many of the signals are recognized by receptors which are coupled to G-proteins. G-proteins then activates phospholipase C (PLC) to hydrolyze phosphatidylinositol 4,5-bisphosphate, generating inositol 1,4,5-trisphosphate (IP3) and 1,2-diacylglycerol (DAG). DAG activates protein kinase C, which phosphorylates various kinds of proteins to exert physiological functions (Nishizuka, 1988). IP3 binds to the IP3 receptor (IP3R) that releases Ca^{2+} from intracellular storage sites. It is now known that IP3R itself is an IP3-sensitive Ca^{2+} channel. Ca^{2+} regulates various physiological functions by associating with calmodulin and Ca^{2+}/calmodulin-dependent protein kinase II (CaM kinase II), and also other Ca^{2+} binding proteins (Berridge and Irvine 1989; Berridge, 1993; Figure 1).

IP3R was originally characterized as a protein, P400, enriched in the normal cerebellum, but absent in Purkinje-cell-deficient cerebellar mutant mice (Mallet et al., 1976; Mikoshiba and Changeux, 1978; Mikoshiba et al., 1979, 1985; Maeda et al., 1988, 1989, 1990). IP3R was also characterized as a cerebellar phosphoprotein PCPP-260 (Walaas et al., 1986) or GP-A (Groswald and Kelly, 1984). These

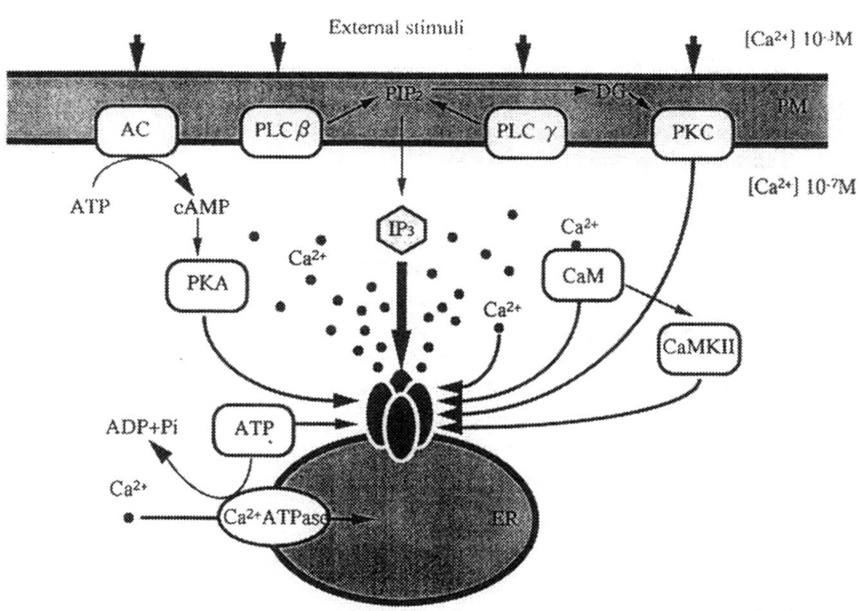

Figure 1. Cross-talks with other signaling systems. IP3, inositol 1,4,5 triphosphate; PIP2, phosphatidylinositol 4,5-bisphosphate; DG, diacylglycerol, CaM, calmodulin; CaMKII, calmodulin-dependent protein kinase II; PKA, cAMP dependent protein kinase.

characterizations were started before the importance of IP3 was recognized as a second messenger to release Ca^{2+}. IP3-binding protein and P400 protein were found to be identical by using specific monoclonal antibodies (Maeda et al., 1990).

The primary structure of IP3R has been determined through cDNA cloning from many species and the information of the primary structure facilitated an understanding of the molecular properties of IP3R.

Here we describe the structure and function of IP3R, IP3-sensitive Ca^{2+} channel, and also their role in Ca^{2+} signaling.

II. STRUCTURE–FUNCTION RELATIONSHIP OF IP3 RECEPTOR

A. Biochemical Properties of the IP3 Receptor

IP3Rs have been shown to interact with various lectins, such as concanavalin A (ConA; Mikoshiba et al., 1979; Maeda et al., 1988; Supattapone et al., 1988), lentil lectin (Maeda et al., 1990), wheat germ agglutinin (Mikoshiba et al., 1979; Chadwick et al., 1990; Khan et al. 1992), and *Limulus* polyphemus agglutinin (Khan et al., 1992). Endo-β-N-acetylglucosaminidase F digestion of IP3R revealed that IP3R has asparagine-linked oligosaccharide chains. Although the apparent molecular weight estimated by SDS/PAGE is 250- or 260-kD, the molecular weight deduced from the amino acid sequence is 313 kD. The IP3R was found to be phosphorylated (Mikoshiba et al., 1985) by protein kinase A (see Section IIIB), Ca^{2+}/calmodulin dependent protein kinase II (Yamamoto et al., 1989; Ferris et al., 1991), and protein kinase C (Ferris et al., 1991). Cross-linking experiments showed that it forms a homotetramer (Maeda et al., 1991).

B. Molecular Properties of the IP3-binding Domain

Cloning of the cDNA for the mouse IP3 receptor has provided much information about the structure of the receptor (Figure 2). From the hydropathy profile, the receptor was predicted to have membrane-spanning domains that cluster around the C-terminus (Furuichi et al., 1989). It has long N-terminal and short C-terminal ends faced into the cytoplasm. IP3R can be divided into three functional domains: the IP3-binding domain, the regulatory domain (or coupling domain), and the channel domain. About 650 N-terminal amino acid residues within the large cytoplasmic portion of the IP3R are highly conserved among different species (Mignery and Sudhof, 1990; Miyawaki et al., 1990). Deletion of a small fragment within the region abolished IP3-binding activity, suggesting the critical sequences for IP3 binding: probably the three-dimensionally restricted binding site (Miyawaki et al., 1991). The binding region is enriched in positively charged Arg and Lys residues (Miyawaki et al., 1991). Polymeric anions, such as heparin, strongly interact with the IP3R, thereby preventing IP3 from binding to it (Supattapone et al., 1988; Gill

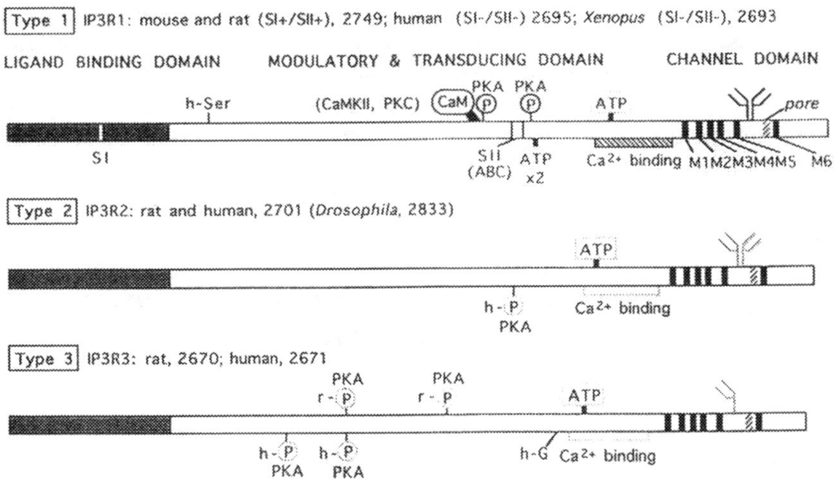

Figure 2. Structure of IP_3Rs. **Type 1 IP_3R** (IP_3R1), 2749 amino acids (SI^+/SII^+ splicing subtype) from the mouse (Furuichi et al., 1989) and rat (Mignery et al., 1990), 2695 amino acids (SI^-/SII^-) from human (Yamamoto-Hino et al., 1994) and 2693 amino acids (SI^-/SII^-) from *Xenopus* (Kume et al., 1993). In contrast to the rodent SI^-/SII^- splicing subtype, *h-Ser* is a human-specific insertion (Ser-666 in human) between residues 681 and 682 in the rodent SI^+ subtype (Yamamoto-Hino et al., 1994). The following sites and regions are revealed: splicing segments SI (residues 318–332) and SII (residues 1692–1731, subsegments A, 1692–1714; B, 1715; C, 1716–1731) (Nakagawa et al., 1991a, 1991b); ligand-binding domain with a dotted box (IP_3-binding site, N-terminal 650 amino acids) (Miyawaki et al., 1991); modulatory and transducing domain including CaM (CaM-binding site (Yamada et al., 1994), PKA (Ser residues, 1588 and 1755, for PKA phosphorylation (Ferris et al., 1992), ATP (potential ATP-binding sites, 1773–1778, 1775–1780 and 2016–2021) (Furuichi et al., 1989), and another potential site, 1768–1773, detected by completely splicing out the SII segment (Ferris et al., 1992), and Ca^{2+}-binding with a shaded horizontal bar (Ca^{2+}-binding site, residues 1961–2219) (Mignery et al., 1992; Channel domain (six solid vertical bars, six putative MSDs M1–M6, between residues 2276–2589 in the mouse) (Furuichi et al., 1993) including two N-glycosylation sites (two branched bars, Asn residues 2475 and 2503) (Jan and Jan, 1992) and one putative 'pore' forming sequence (a shaded vertical bar, residues 2530–2552) between M5 and M6 (Jan and Jan, 1992). The CaMKII and PKC phosphorylation sites have not been identified yet. **Type 2 IP_3 R** (IP_3R2), 2701 amino acids from the rat (Sudhof et al., 1991) and man (Yamanoto-Hino et al., 1994). *Drosophila* IP_3R (2833 amino acids) (Yoshikawa et al., 1992) is, if anything, similar to mammalian IP_3R2. The putative ligand-binding domain (N-terminal 649 amino acids in human) homologous to that of IP_3R1 is indicated by a dotted box. Putative binding sites for ATP (ATP, residues 1968–1973) and Ca^{2+} (Ca^{2+}, residues 1914–2173) are conserved. The following sites and regions are indicated: h-P-PKA, potential PKA phosphorylation site in human IP_3R2 (Ser residue 1687); the putative channel domain, M1–M6 (between residues 2230–2541) in man; *(continued)*

et al., 1989). A C-terminal truncated IP3R, which has only the IP3-binding domain, is soluble since the transmembrane domain is missing. Cross-linking experiments show that the deletion mutant stays as a monomer, suggesting that the IP3R monomer itself can bind IP3 (Miyawaki et al., 1991). The regulatory domain contains various sites for regulation which couple with other signaling systems (see Section II D).

C. Transmembrane Topology of the IP3 Receptor

There were two models of the membrane topology of the IP3 receptor: one is the 8-transmembrane model (Mignery and Sudhof, 1990; Mignery et al., 1990; Sudhof et al., 1991) which has no putative sugar-containing region in the receptor. We have proposed a six-transmembrane model (Yoshikawa et al., 1992; Kume et al., 1993; Mikoshiba et al., 1993; Michikawa et al. 1994; Yamamoto-Hino et al., 1994) (Figure 3).

The transmembrane topology of the IP3R was analyzed by two approaches using immunocytochemical and molecular biological techniques. An immunocytochemical study using the site-specific antibody showed that the residues 2504–2523 are located on the luminal side of the ER. Among the 20 putative glycosylation sites in the IP3R, only two were the candidates of N-glycosylation according to the 6-transmembrane model. Indeed, by ConA column chromatography of the site-directed mutant receptors, both the two Asn residues, Asn-2475 and Asn-2503, were glycosylated, indicating that these two Asn residues are located on the ER luminal side. These results led us to conclude that IP3R traverses the membrane six times (Michikawa et al., 1994).

D. Alternative Splicing Results in Receptor Heterogeneity

Type 1 IP3R has various isoforms resulting from RNA-splicing which includes two splicing sites, SI and SII. SI contains 45 nucleotides in the IP3-binding domain

Figure 2. (Continued) two putative N-glycosylation sites (Asn residues 2430 and 2456 in man). **Type 3 IP_3R** (IP_3R3), 2670 amino acids from the rat (Blondel et al., 1993) and 2671 amino acids from man (Maranto et al., 1994; Yamamoto-Hino et al., 1994). h-G is a human specific insertion (Gly-1837). The putative ligand-binding domain (N-terminal 650 amino acids) homologous to that of IP_3R1 is indicated by a dotted box. The following sites and regions are indicated: putative binding sites for ATP (residues 1921–1926 in the rat) and Ca^{2+} (residues 1865–2147 of rat receptor that include a large insertion of 29 amino acids, however); two potential PKA phosphorylation sites in the rat (r-P-PKA, Ser residues 1130 and 1457) and human (h-P-PKA, residues 934 and 1133); putative channel domain, M1–M6 (between residues 2205–2517 or 2203–2520) in man; one putative N-glycosylation site (Asn residue 2405) in man. Note, functional sites which have not yet been experimentally determined are encircled by dotted lines. See the text for details.

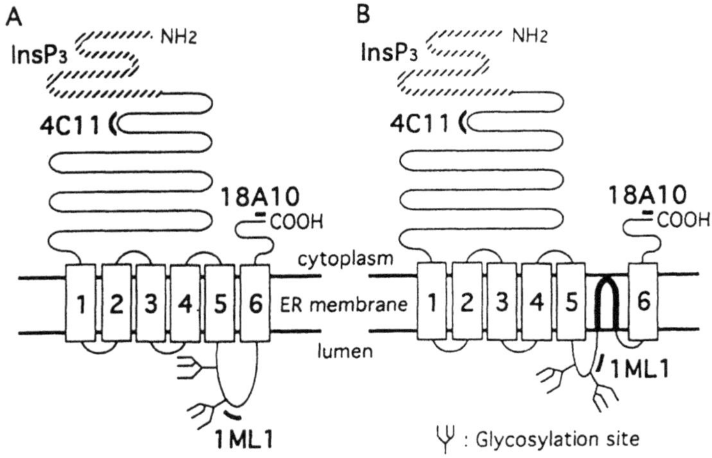

Figure 3. Transmembrane topology models for the mouse InsP3R1 within the ER membrane. (**A**) The putative membrane-spanning segments are indicated as the open boxes. The epitopes of 1ML1 (residues 2504–2523), 4C11 (within the residues 679–727) (Furuichi et al., 1989), and 18A10 (within the residues 2736–2747; Nakade et al., 1991) are indicated. The InsP3 binding site (within the residues 1–650; Miyawaki et al., 1991) is indicated as the shaded line. (**B**) The putative pore-forming region (boldface line) is inserted into the ER membrane from the luminal side.

(SI+ denotes its presence and SI- denotes its absence) (Danoff et al., 1991; Nakagawa et al., 1991a,b). SII resides between the two phosphorylation sites in the regulatory domain. SII contains 120 nucleotides divided into three further subsegments, A, B, and C (Nakagawa et al., 1991a, 1991b). Isoforms SII+, SIIB-, and SII(BC)- which contain any segment of A, B, or C are only expressed in the central nervous system. When analyzed in the brain at the adult stage, the ratio of the subtypes differ from region to region. In the mouse CNS, the SIIB- subtype is predominant (50–54%) and the SIIABC-subtype is present at lower levels. On the other hand, the SIIABC-subtype is a predominant splicing subtype in spinal cord (54%). In the peripheral tissues tested, only the SIIABC- subtype mRNA was detected (Nakagawa et al., 1991a,b). Thus, the SII, SIIB-, and SIIBC- subtypes are brain-specific. However, SII(ABC)-subtypes are expressed in both central nervous system and peripheral tissues (Nakagawa et al., 1991b). PKA phosphorylation enhances the Ca^{2+} releasing activity in type 1 receptor (Nakade et al., 1994), and the SII segment is located just between the two phosphorylation sites. Therefore, splicing at the SII segment may play an important role in the regulation of the function of IP3R.

During the development of the cerebellum, SI shows a peak at P12 and SI- is expressed at lower level than SI within first two weeks of age, but gradually increased to the adult stage. The alternative splicing pattern, therefore, changes in a regional-specific and developmental-specific manner, and this may produce a great diversity by forming tetramer structures in IP3-induced Ca^{2+} release.

III. FUNCTIONAL PROPERTIES OF IP3 RECEPTORS

A. IP3 Receptor as a Ca^{2+} Channel

Purified mouse IP3R incorporated into (Maeda et al., 1991) or not microsomes fused with (Watras et al., 1991) a planar lipid bilayer showed IP3-induced cation-selective channel activity. Incorporation of purified receptor into proteoliposome also showed IP3-induced Ca^{2+} release channel activity (Ferris et al., 1989; Nakade et al., 1994). The neuroblastoma/glioma hybrid cell line, NG108-15, and fibroblast L cells transfected with IP3R cDNA showed enhanced IP3-binding activity (Miyawaki et al., 1990). The membrane fraction from the stable transformant L-fibroblast cell that expresses the type I IP3R showed an about doubled V_{max}, and the EC50 value for IP3 induced Ca^{2+} release was about 10-fold lower than that in the nontransfected cells (Miyawaki et al., 1990). From these results, it is concluded that IP3R is a Ca^{2+} releasing channel.

B. Cross-talks with Other Signaling Systems

IP3-induced Ca^{2+} release (IICR) is modulated by various endogenous mediators such as ATP (Smith et al., 1985; Suematsu et al., 1985; Maeda et al., 1991), GTP (Ghosh et al., 1989), and fatty acids (Chow and Jondal, 1990). IP3R binds ATP with molar stoichiometry. ATP enhances IICR in reconstituted membranes as well as in microsomal membrane fractions (Forsberg et al., 1987; Ferris et al., 1990, Maeda et al., 1991). The binding is selective for adenine nucleotide and the affinity is in the order of ATP>ADP>>AMP. A nucleotide-binding consensus sequence, Gly–X–Gly–XX–Gly, was found in the N-terminal cytoplasmic domain of the mouse type I IP3R (amino acid residues 2016–2021) (Furuichi et al., 1989). Calmodulin (CaM) was found to bind to IP3R. Addition of CaM did not affect IP3 binding to the IP3R or IICR activity. IICR is found to be regulated by Ca^{2+} (Joseph et al., 1989; Bezprozvanny et al., 1991). The maximum probability of opening occurred at 0.2 μM free Ca^{2+}, with sharp decreases on either side of the maximum. The ryanodine receptor works as a Ca^{2+} induced Ca^{2+} release (CICR) channel, but the maximum activity of the ryanodine receptor was maintained between 1μM and 100 μM Ca^{2+}. Therefore, the IP3-gated channel itself, within the physiological range of cytoplasmic Ca^{2+}, is positively regulated at the resting Ca^{2+} levels and then negatively for Ca^{2+} release, whereas the ryanodine receptor behaves solely as a Ca^{2+} activated channel. The presence of different types of channels with different Ca^{2+} sensitivities may provide a basis for the complex pattern of intracellular Ca^{2+} regulation.

Cyclic AMP-dependent protein kinase (PKA) predominantly phosphorylates the IP3R (Walaas et al., 1986; Supattapone et al., 1988; Weeks et al., 1988; Yamamoto et al., 1989; Ferris et al., 1991; Nakade et al., 1994). Type 1 IP3R has two consensus amino acid sequences that fulfill the criteria to be phosphorylated by PKA. There is evidence that the phosphorylation of the IP3R by PKA increases IICR in platelets (Enouf et al., 1987), hepatocytes (Burgess et al., 1991), and cerebellar membranes (Volpe and Alderson-Lang, 1990). On the contrary, Supattapone et al. (1988) and Quinton and Dean (1992) reported that the phosphorylation by PKA of cerebellar membranes and platelet membranes reduces the IICR. The explanation for this discrepancy might be the crude preparations used. Phosphorylation by PKA might alter the function of other molecules of the Ca^{2+} regulation system such as a Ca^{2+} pump or IP3 formation by phospholipase C. In addition, the heterogeneity of IP3R may result in different functional regulation by PKA. Nakade et al. (1994) recently purified a single type of IP3R, type I, by affinity purification using a subtype-specific antibody and reconstituted it into lipid vesicles. Using this novel method, IP3R I was found to increase in IICR activity by PKA phosphorylation (Nakade et al., 1994).

IV. MULTIPLE MOLECULAR FORMS OF THE IP3 RECEPTOR

Many pieces of information about the structure and function of IP3R have been obtained through the cloning and sequencing of the cDNAs of various types of IP3Rs. Now, full-length cDNAs for three types of IP3Rs from different species have been reported: Type 1 receptor from mouse (Furuichi et al., 1989), rat (Mignery et al., 1992), human (Yamada et al., 1994), and *Xenopus laevis* (Kume et al., 1993); Type 2 receptor from rat (Sudhof et al., 1991), human (Yamamoto-Hino et al., 1994), and *Drosophila melanogaster* (Yoshikawa et al., 1992); Type 3 receptor from rat (Blondel et al., 1993) and human (Maranto, 1994; Yamaoto-Hino et al., 1994). The overall homology of human type 2 IP3R to that of rat is 95%, however it is less with type 1 (mouse, rat, *Xenopus*; all 68%) and with type 3 (human, rat; all 64%). The homology of human type 3 to that of rat is 95%, but type 1 in mouse, rat, *Xenopus* are all 62%. Each type of the receptor is highly conserved in its sequence, and shows a homology higher than 60% to other types. It is clear that they compose one family.

The genes for three receptors were mapped to chromosomes by *in situ* hybridization. The genes for IP3R1 (*Insp3r1*; Furuichi et al., 1993) and IP3R2 (*Insp3r2*) were mapped to different loci of mouse chromosome 6. IP3R3 (*Insp3r3*) was mapped to mouse chromosome 17. The genes of the human receptors were mapped to loci with linkage conservation between human and mouse genomes; *Insp3r1* to human chromosome 3p25–26 (Yamada et al., 1994), *Insp3r2* to 12p11 (Yamamoto-Hino et al., 1994), and *Insp3r3* to 6p21 (Yamamoto-Hino et al., 1994).

The presence of several types of IP3R suggest different regulatory mechanisms specific for the tissues and cell types. Type I IP3R is expressed relatively weakly in the human hematopoietic and lymphoma cell lines except THR-1 monocytes as detected by Northern blot hybridization (Yamamoto-Hino et al., 1994). However, type 1 IP3R in HL-60 myeloid cells, when induced to differentiate toward the neutrophilic lineage by either retinoic acid or DMSO, exhibited a substantial increase in its level (Yamada et al., 1994). Type 2 receptor is expressed at certain levels in HL-60 myeloid cells, Namalwa cells, and Jurkat cells. Type 3 receptor is highly expressed in HL-60 cells, Jurkat T lymphoma, Namalwa, and Raji preB cells and monocytes (Yamamoto-Hino et al., 1994). It is clear that each IP3R is differentially expressed among various cells. There is a difference in the expression of type 1 receptor between the two monocyte cell lines. THP-1 expresses a large amount of the receptor; U937 exhibits no expression. The expression of type 1 IP3R in monocytes might be associated with their phagocytosis as THP-1 has phagocytotic activity (Hemmi and Breitman, 1985) and U937 does not.

V. DISTRIBUTION OF THE IP3 RECEPTOR IN VARIOUS TISSUES

Immunohistochemistry and *in situ* hybridization showed that the IP3R is predominantly enriched in cerebellar cortex. Purkinje cells are the most enriched sites (Maeda et al., 1988, 1989; Nakanishi et al., 1991). It is also distributed in the cerebral cortex, nucleus accumbens septi, anterior olfactory nucleus, caudate-putamen, and cerebellar nuclei. The content of IP3R is relatively low in the amygdaloid cortex, prepiriform cortex, dentate gyrus, olfactory tubercle, precommissural hippocampus, hypothalamus, substantia nigra, and pons. The distribution agrees well with the sites for [^{3}H]IP3 binding (Worley et al., 1989; Maeda et al., 1990). *In situ* hybridization showed that mRNAs are located in peripheral tissue such as thymus, heart, lung, liver, spleen, kidney, uterus, oviduct, and testis (Furuichi et al., 1990). Immunohistochemistry showed that bronchiles and arteries were labeled in lung. Smooth muscle cells of oviduct and uterus are strongly labeled: tunica muscularis of the oviduct, and myometrium of the uterus. IP3-induced Ca^{2+} release might play an important role in muscle contaction. The fact that the IP3R is enriched in smooth muscle may suggest a function in smooth muscle contraction.

The IP3R mRNA are also enriched in the secondary oocytes within the Graafian follicles. By electron microscopic observation, IP3R was clearly localized on ER (Maeda et al., 1989; Ross et al., 1989; Otsu et al., 1990; Satoh et al., 1990).

VI. THE IP3 RECEPTOR ON THE PLASMA MEMBRANE

IP3R was believed to be localized on the ER, but not on the plasma membrane. Several reports, however, suggest that IP3R is also located in the plasma membrane in human T cells (Kuno and Gardner, 1987), Jurkat cells (Khan et al., 1992). B cells (Brent et al., 1993), olfactory neurons (Fadool and Ache, 1992; Restrepo et al., 1990), and cerebellar Purkinje cells (Kuno et al., 1994).

Electron microscopic localization by immunogold technique showed that IP3R-immunoreactivities are present on the caveola structure of the plasma membrane in the endothelium, smooth muscle cell and keratinocyte (Fujimoto et al., 1992). Surface biotinylation indicated that the anti-IP3R monoclonal activity recognizes a protein of 240 kD. Immunogold localization on the olfactory cilia was also demonstrated (Cunningham et al., 1993). It should be tested whether the molecule immunoreactive with the antibody is really IP3R or not. These observations suggest that IP3 somehow regulate the Ca^{2+} entry across the plasma membrane.

VII. ROLE OF THE IP3 RECEPTOR IN CA^{2+} OSCILLATIONS AND WAVES.

Ca^{2+} oscillations and Ca^{2+} waves commonly occur in various cells in response to neurotransmitters, hormones, and growth factors. These ligands activate polyphosphinositide turnover leading to IP3-induced Ca^{2+} release (IICR). IICR is suggested to occur in fertilized eggs of hamster (Miyazaki, 1988), frog (Busa et al., 1985), and sea urchin (Galione et al., 1991). However, Ca^{2+} induced Ca^{2+} release (CICR) also occurs in these eggs. Ca^{2+} oscillations and Ca^{2+} waves were expected to occur by a combination of IICR and CICR from separate intracellular stores, but decisive evidence was lacking. Nakade et al. (1991) demonstrated that the monoclonal antibody (mAb 18A10) blocks IICR using cerebellar microsomal fraction. The mAb recognizes an epitope at the COOH-terminus of the IP3R, but does not inhibit IP3-binding activity of the receptor. The mAb introduced into the hamster egg inhibited both Ca^{2+} release upon injection of IP3 and Ca^{2+} (Miyazaki et al., 1992). The mAb completely blocked sperm-induced Ca^{2+} waves and Ca^{2+} oscillations. These results indicate that Ca^{2+} waves and oscillations in fertilized hamster eggs is mediated solely by the IP3R, and Ca^{2+}-sensitized IICR, but not CICR.

VIII. ACTIVITY-DEPENDENT MOVEMENT OF THE IP3 RECEPTOR

The *Xenopus laevis* IP3R cDNA was cloned and sequenced. The antibody against the fusion protein of the *Xenopus* IP3R stained the receptor in the ER localized in yolk-free corridors in the animal hemisphere and the perinuclear region of fully grown stage VI oocytes (Kume et al., 1993). The polarized localization with more concentrated IP3R in the animal hemisphere agrees well with the asymmetry of responsiveness of exogenously expressed phospholipase C-activating receptors and Ca^{2+} gated ion channels in *Xenopus* oocytes previously described (Oron et al., 1988; Peter et al., 1991). The strong immuno-staining signal with the anti-IP3R antibody disappeared coincidentally with germinal vesicle breakdown (GVBD) and with the appearance of immunopositive yolk-free patches. The localization of the receptor in both cytoplasm and cortex is consistent with spiral waves of intracellular Ca^{2+} following activation of introduced receptor (Lechleiter et al., 1991a,b) and injection of GTPγS or IP3 (Lechleiter and Clapham, 1992). A drastic redistribution of the IP3R occurs in fertilized eggs. The distinct immunopositive patches observed in the cortical region of the unfertilized eggs became fuzzy and rather continuous and stronger cortical signals were observed in the animal hemisphere than in the vegetal hemisphere. The role of IP3R upon egg activation was examined by antisense ablation of the receptor. Antisense phosphorothioate oligonucleotide complementary to the 5′ flanking and translation start site or the corresponding sense oligonucleotide was microinjected into fully grown stage VI oocytes. Eggs, after incubation with progesterone, were assayed for IP3-sensitive cortical contraction which represent egg activation. Antisense nucleotide clearly suppressed egg activation.

IX. THE IP3 RECEPTOR AS A FAMILY OF VOLTAGE-GATED AND SECOND MESSENGER-GATED CHANNELS

The IP3R is composed of four subunits; each has six membrane-spanning segments. This suggests that IP3R is a member of the superfamily including voltage-sensitive ion channels (Michikawa et al., 1994; Jan and Jan, 1992; Figure 4). Recently, the second messenger-gated ion channels on the plasma membrane, such as cyclic nucleotide-gated cation channels, putative Ca^{2+} activated K^+ channels, and plant K^+ channel for phosphoinoditide-mediated Ca^{2+} entry, and plant K^+ channels/transporters. It is clear from the membrane topology that the voltage- and second messenger-gated ion channels share a basic design that consists of a set of six putative membrane-spanning segments (SI–S6) and a pore-forming region between the S5 and S6 segment known as H5, or P (Miller, 1992). A hydrophobic region (residues 2529–2552) exists between the M5 and M6 segments of the mouse IP3R1 (Michikawa et al., 1994). There is no significant sequence homology with the pore-forming regions of the voltage- and other second messenger-gated channels. The acidic residue of the Gly–Glu pair in the H5 region was critical to the formation of a functional channel (Heiginbotham et al., 1992). The Gly–Asp pair is highly conserved among all types of IP3R. Therefore the Gly–Asp pair may act as part of the pore-forming region of the IP3R. From this, the IP3R is thought to share the basic design of the channel-forming domain with the voltage- and second messenger-gated ion channels on the plasma membrane. In spite of the similarity of the proposed basic design of the channel-forming domain, the gating machinery of the IP3R is different from that of the voltage- and second messenger-gated channels on the plasma membrane. The S4 segment is proposed to serve as a voltage sensor in the voltage-gated ion channels. IP3R has no homology with the S4 segment. It fits well with the report that the cerebellar IP3R reconstituted into a planar lipid bilayer exhibited a slight but not significant voltage dependency (Watras et al., 1991; Bezprozpany et al., 1991; Maeda et al., 1991).

The topology of the second messenger-gated ion channels and IP3R are similar since both the N- and C-termini of the channel protein are cytoplasmic (Figure 4). The IP3-binding site locates within the N-terminal fourth of the IP3R. However,

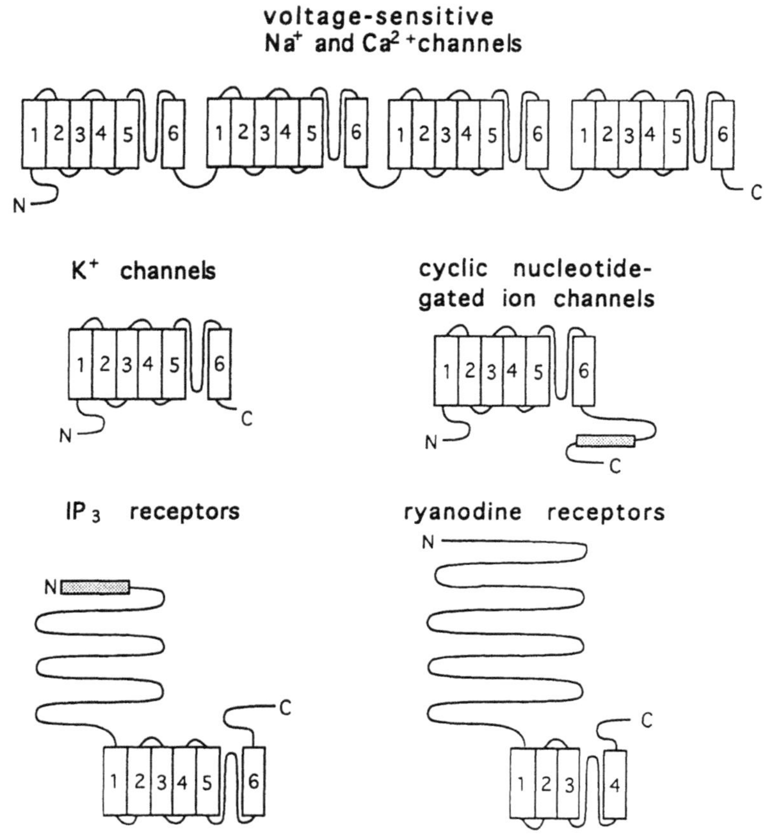

Figure 4. IP3 receptor, a family of voltage- and second messenger-gated channels. From the membrane topology that the voltage- and second messenger-gated ion channels share a basic design that consists of a set of six putative membrane-spanning segment (SI–S6) and a pore-forming region between the S5 and S6 segment known as H5, P, or SSsl membrane. A hydrophobic region (residues 2529–2552) exists between the M5 and M6 segments of the mouse IP3R1.

ligand-binding sites of the second messenger-gated channels are in the C-terminal region. The positional difference of the ligand-binding sites within the channel protein suggests a diversity in the gating mechanisms.

SUMMARY

IP3 is a second messenger that releases Ca^{2+} from the Ca^{2+} store site. cDNA cloning and sequencing of the IP3 receptor revealed the functional structure of the IP3R. IP3R has an IP3 binding domain of about 650 amino acids at the N-terminus. A regulatory (or coupling) domain lies at the middle part that links both the IP3-binding domain and the Ca^{2+} channel domain at the C-terminus. IP3-induced Ca^{2+} releasing activity is enhanced by ATP binding and by PKA phosphorylation. CaM kinase and PKC also phosphorylate IP3R. Ca^{2+} regulates the IP3-induced Ca^{2+} releasing activity in a bell-shape form. The channel domain has six membrane-spanning segments (M1–M6). A hydrophobic region between the M5 and M6 segments is considered to be a pore-forming region. Overall homology in the channel domains indicates that IP3R belongs to a family of voltage-sensitive and second messenger-gated channels. IP3R isoforms are produced from RNA splicing in the IP3-binding and regulatory domains. Each isoform is expressed in a region-specific and developmentally specific manner. Complementary DNA sequences of type 1, 2, and 3 IP3R are determined. These subtypes are expressed differentially: Type 1 IP3R is expressed at high level in the central nervous system, while type 2 and 3 expressions are observed in hematopoietic and lymphatic cell types.

Monoclonal antibody against IP3R blocked sperm-induced Ca^{2+} waves and oscillations in hamster egg. Antisense nucleotide injected in the *Xenopus* oocyte suppressed egg activation. These experiments show that IP3R is involved in Ca^{2+} waves and oscillations as well as egg activation.

REFERENCES

Berridge, M. J. (1993). Inositol triphosphate and calcium signalling. Nature 361, 315–325.

Berridge, M. J., & Irvine, R. F. (1989). Inositol phosphates and cell signalling. Nature 341, 197–205.

Bezprozvanny, I., Watras, J., & Ehrlich, B. E. (1991). Bell-shaped calcium response curves of Ins(1,4,5)P_3-and calcium-gates channels from endoplasmic reticulum of cerebellum. Nature 351, 751–754.

Blondel, O., Takeda, J., Janssen, H., Seino, S., & Bell, G. I. (1993). Sequence and functional characterization of a third inositol triphosphate recepter subtype, IP3R-3, expressed in pancreatic islets, kidney, gastrointestinal tract, and other tissues. J. Biol. Chem. 268, 11356–11363.

Brent, L. H., Gong, O., Ross, J. M., & Wieland, S. J. (1993). Mitogen-activated Ca^{2+} channels in human β lymphocytes. J. Cell. Physiol. 155, 520–529.

Burgess, G. M., Bird, G. St. J., Obie, J. F., & Putney, J. W. Jr. (1991). The mechanism for synergism between phospholipase C and adenylase-linked hormones in liver J. Biol. Chem. 266, 4772–4781.

Busa, W. B., Ferguson, J. E., Joseph, S. K., Williamson, J. R., & Nuccitelli, R. (1985). Activation of frog (*Xenopus laevis*) egg by inositol triphosphate. I. Characterization of Ca^{2+} release from intracellular stores. J. Cell Biol. 101, 677–682.

Chadwick, C. C., Sito, A., & Fleischer, S. (1990). Isolation and characterization of the inositol triphosphate receptor from smooth muscle. Proc. Natl. Acad. Sci. USA 87, 2132–2136.

Chow, S. C., & Jondal, M. (1990). Polyunsaturated free fatty acids stimulate an increase in cytosolic Ca^{2+} by mobilizing the inositol 1,4,5-triphosphate-sensitive Ca^{2+} pool in T cells through a mechanism independent of phosphoinositide turnover. J. Biol. Chem. 265, 902–907.

Cunningham, A. M., Rygo, D. K., Sharp, A. H., Reed, R. R., Snyder, S. H., & Ronnet, G. V. (1993). Neuronal inositol 1,4,5-triphosphate receptor localized to the plasma membrane of olfactory cilia. Neurosci. 57, 339–352.

Danoff, S. K., Ferris, C. D., Donath, C., Fischer, G., Munemitsu, S., Ullrich, A., Snyder, S. H., & Ross, C. A. (1991). Inositol 1,4,5-triphosphate receptors: Distinct neuronal and non-neuronal forms derived by alternative splicing differ in phosphorylation. Proc. Natl. Acad. Sci. USA 88, 2951–2955.

Enouf, J., Giraud, F., Bredoux, R., Bordeau, N., & Toledano-Levy, S. T. (1987). Possible role for cAMP-dependent phosphophate in the calcium release mediated by inositol 1,4,5-trisphophate in human platelet membrane vesicles. Biochim. Biophys. Acta 928, 76–82.

Fadool, D. A., & Ache, B. W. (1992). Plasma membrance inositol 1,4,5-triphosphate-activated channels mediate signal transduction in lobster olfactory receptor neuron. Neuron 9, 907–918.

Ferris, C. D., Huganir, R. L., Supattapone, S., & Snyder, S. H. (1989). Purified inositol 1,4,5-triphosphate receptor mediates calcium flux in reconstituted lipid vesicles. Nature 342, 87–89.

Ferris, C. D., Huganir, R. L., & Snyder, S. H. (1990). Calcium flux mediated by purified inositol 1,4,5-triphosphate receptor in reconstituted lipid vesicles is allosterically regulated by adenine nucleotides. Proc. Natl. Acad. Sci. USA 87, 2147–2151.

Ferris, C. D., Huganier, R. L., Bredt, D. S., Cameron, A. M., & Snyder, S. H. (1991). Inositol trisphsphate receptor: phosphorylation by protein kinase C and calcium-calmodulin dependent protein kinases in reconstituted lipid vesicles. Proc. Natl. Acad. Sci. USA 88, 2232–2235.

Forsberg, E. J., Feueerstein, G., Shohami, E., & Pollard, H. B. (1987). Adenosine triphosphate stimulates inositol phospholipid metabolism and prostacyclin formation in adrenal medullary endothelial cells by means of P2-purinergic receptors. Proc. Natl. Acad. Sci. USA 84, 5630–5634.

Fujimoto, T., Nakade, S., Miyawaki, A., Mikoshiba, K., & Ogawa, K. (1992). Localization of inositol 1,4,5-triphosphate receptor-like protein in plasmalemmal caveolae. J. Cell Biol. 119, 1507–1513.

Furuichi, T., Yoshikawa, S., Miyawaki, A., Miyawaki, K., Maeda, N., & Mikoshiba, K. (1989). Primary structure and functional expression of the inositol 1,4,5-triphosphate-binding protein P400. Nature 342, 32–38.

Furuichi, T., Shiota, C., & Mikoshiba, K. (1990). Distribution of inositol 1,4,5-triphosphate receptor mRNA in mouse tissues. FEBS Letts. 267, 85–88.

Furuichi, T., Simon-Chazottes, D., Fujino, I., Yamada, N., Hasegawa, M. A., Yoshikawa, S., Guenet, J. L., & Mikoshiba, K. (1993). Widespread expression of inositol 1,4,5-triphosphate receptor type 1 gene (Insp3rl) in the mouse central nervous system. Receptors & Channels 1, 11–24.

Galione, A., Lee, H. C., & Busa, W. (1991). Ca^{2+}-induced Ca^{2+} release in sea urchin egg homogenates: Modulation by cyclic ADP-ribose. Science 253, 1143–1146.

Ghosh, T. K., Mullaney, J. M., Tarazi, F. I., & Gill, L. D. (1989). GTP-activated communication between inositol 1,4,5-triphosphate sensitive and insensitive pools. Nature 340, 236–223.

Groswald, D. E., & Kelly, P. T. (1984). Evidence that a cerebellum-enriched, synaptic junction glycoprotein is related to fodrin and resists extraction with triton in a calcium-dependent manner. J. Neurochem. 42, 534–546.

Heiginbotham, L., Abramson, T., & Machinnon, R. (1992). A functional connection between the pores of distantly related ion channels as revealed by mutant K^+ channels. Science 258, 1152–115519.

Hemmi, H., & Breitman, T. R. (1985). Induction of functional differentiation of a human monocytic leukemia cell line (THP-1) by retinoid acid and cholera toxin. Jpn. J. Cancer Res. 76, 345–351.

Jan, L. Y., & Jan, Y. N. (1992). Tracing the roots of ion channels. Cell 69, 715–718.

Joseph, S. K., Rice, H. L., & Williamson, J. R. (1989). The effect of external calcium and pH on inositol triphosphate-mediated calcium release from cerebellum microsomal fractions. Biochem. J. 258, 261–265.

Khan, A. A., Steiner, J. P., & Snyder, S. H. (1992). Plasma memebrance inositol 1,4,5-triphosphate receptor of lymphocytes; Selective enrichment in sialic acid and unique binding specificity. Proc. Natl. Acad. Sci. USA 89, 2849–2853.

Kume, S., Muto, A., Aruga, J., Nakagawa, T., Michikawa, T., Furuichi, T., Nakade, S., Okano, H., & Mikoshiba, K. (1993). The *Xenopus* IP3 receptor: structure, function, and localization in oocytes and eggs. Cell 73, 555–570.

Kuno, M., & Gardner, P. (1987). Ion channels activated by inositol 1,4,5-triphosphate in plasma membrane of human T-lymphocytes. Nature 326, 301–304.

Kuno, M., Maeda, N., & Mikoshiba, K. (1994). IP3-activated calcium-permeable channels in the inside-out patches of cultured cerebellar Purkinje cells. Biochem. Biophys. Res. Comm. 199, 1128–1135.

Lechleiter, J. D., & Clapham, D. E. (1992). Molecular mechanisms of intracellular calcium excitability in *X. laevis* oocytes. Cell 69, 283–294.

Maeda, N., Niinobe, M., Nakahira, K., & Mikoshiba, K. (1988). Purification and characterization of P400 protein, a glycoprotein characteristic of Purkinje cell, from mouse cerebellum. J. Neurochem. 51, 1724–1730.

Maeda, N., Niinobe, M., Inoue, Y., & Mikoshiba, K. (1989). Developmental expression and intracellular location of P400 protein characteristic of Purkinje cells in the mouse cerebellum. Dev. Biol. 133, 67–76.

Maeda, N., Niinobe, M., & Mikoshiba, K. (1990). A cerebellar Purkinje cell marker P400 protein is an inositol 1,4,5-triphosphate ($InsP_3$) receptor protein. Purification and characterization of $InsP_3$ receptor complex. EMBO J. 9, 61–67.

Maeda, N., Kawasaki, T., Nakade, S., Yokota, N., Taguchi, T., Kasai, M., & Mikoshiba, K. (1991). Structural and functional characterization of inositol 1,4,5-triphosphate receptor channel from mouse cerebellum. J. Biol. Chem. 266, 1109–1116.

Mallet, J., Huchet, M., Pougeois, R., & Changeux, J.-P. (1976). Anatomical, physiological and biochemical studies on the cerebellum from mutant mice III. Protein differences associated with the Weaver, staggerer and nervous mutation. Brain Res. 103, 291–312.

Maranto, A. R. (1994). Primary structure, ligand binding and localization of the human type 3 inositol 1,4,5-triphosphate receptor expressed in intestine epithelium. J. Biol. Chem. 269, 1222–1230.

Michikawa, T., Hamanaka, H., Otsu, H., Yamamoto, A., Miyawaki, A., Furuichi, T., Tashiro, Y., & Mikoshiba, K. (1994). Transmembrance topology and sites of N-glycosylation of inositol 1,4,5-triphosphate receptor. J. Biol. Chem. (in press).

Mignery, G. A., & Sudhof, T. C. (1990). The ligand binding site and transduction mechanism in the inositol-1,4,5-triphosphate receptor. EMBO J. 9, 3893–3898.

Mignery, G. A., Newton, C. L., Archer, B. T., III, & Sudhof, T. C. (1990). Structure and expression of the rat inositol 1,4,5-triphosphate receptor. J. Biol. Chem. 265, 12679–12685.

Mignery, G., Johnston, P. A., & Sudhof, T. C. (1992). Mechanism of Ca^{2+} inhibition of inositol 1,4,5-triphosphate ($InsP_3$) binding to the cerebellar $InsP_3$ receptor. J. Biol. Chem. 267, 7450–7455.

Mikoshiba, K., & Changeux, J.-P. (1978). Morphological and biochemical studies on isolated molecular and granular layers from bovine cerebellum. Brain Res. 142, 487–504.

Mikoshiba, K., Huchet, M., & Changeux, J. P. (1979). Biochemical and immunological studies on the P400 protein, a protein characteristic of the Purkinje cell from mouse and rat cerebellum. Dev. Neurosci. 2, 254–275.

Mikoshiba, K., Okano, H., & Tsukada, Y. (1985). P400 protein characteristic to Purkinje cells and related proteins in cerebella from neuropathological mutant mice: autoradiographic study by ^{14}C-leucine and phosphorylation. Dev. Neurosci. 7, 179–187.

Miller, C. (1992). Hunting for the pore of voltage-gated channels. Curr. Biol. 2, 573–575.

Miyawaki, A., Furuichi, T., Maeda, N., & Mikoshiba, K. (1990). Expressed cerebellar-type inositol 1,4,5-triphosphate receptor, P400, has calcium release activity in a fibroblast L cell line. Neuron 5, 11–18.

Miyawaki, A., Furuichi, T., Ryou, Y., Yoshikawa, S., Nakagawa, T., Saitoh, T., & Mikoshiba, K. (1991). Structure–function relationships of the mouse inositol 1,4,5-triphosphate receptor. Proc. Natl. Acad. Sci. USA 88, 4911–4915.

Miyazaki, S. (1988). Inositol 1,4,5-triphosphate-induced calcium release and guanine nucleotide-binding protein-mediated periodic calcium releases in golden hamster eggs. J. Cell Biol. 106, 345–353.

Miyazaki, S., Yuzaki, M., Nakade, K., Shirakawa, H., Nakanishi, S., Nakade, S., & Mikoshiba, K. (1992). Block of Ca^{2+} wave and Ca^{2+} oscillation by antibody to the inositol 1,4,5-triphosphate receptor in fertilized hamster egg. Science 257, 251–255.

Nakade, S., Maeda, N., & Mikoshiba, K. (1991). Involvement of the C-terminus of the inositol 1,4,5-triphosphate receptor in Ca^{2+} release analysed using region-specific monoclonal antibodies. Biochem. J. 277, 125–131.

Nakade, S., Rhee, S. K., Hamanaka, H., & Mikoshiba, K. (1994). Cyclic AMP-dependent phosphorylation of an immunoaffinity-purified homotetrameric inositol 1,4,5-triphosphate receptor(type I) increases Ca^{2+} flux in reconstituted lipid vesicles, J. Biol. Chem.

Nakagawa, T., Okano, H., Furuichi, T., Aruga, J., & Mikoshiba, K. (1991a). The subtypes of the mouse inositol 1,4,5-triphosphate receptor are expressed in a tissue-specific and developmentally specific manner. Proc. Natl. Acad. Sci. USA 88, 6244–6248.

Nakagawa, T., Shiota, C., Okano, H., & Mikoshiba, K. (1991b). Differential localization of alternative spliced transcripts encoding inositol 1,4,5-triphosphate receptors in mouse cerebellum and hippocampus: *In situ* hybridization study. J. Neurochem. 57, 1807–1810.

Nakanishi, S., Maeda, N., & Mikoshiba, K. (1991). Immunohistochemical localization of an inositol 1,4,5-triphosphate receptor, P400, in neural tissue: studies in developing and adult mouse brain. J. Neurosci. 11, 2075–2086.

Nishizuka, Y. (1988). The molecular heterogeneity of protein kinase C and its implications for cellular regulation. Nature 334, 661–665.

Oron, Y., Gillo, B., & Gershengorn, M. (1988). Differences in receptor-evoked membrane electrical responses in native and mRNA-injected *Xenopus* oocytes. Proc. Natl. Acad. Sci. USA 85, 3820–3824.

Otsu, H., Yamamoto, A., Maeda, N., Mikoshiba, K., & Tashiro, Y. (1990). Immunogold localization of inositol 1,4,5-triphosphate ($InsP_3$) receptor in mouse cerebellar Purkinje cells using three monoclonal antibodies. Cell Struct. Funct. 15, 163–173.

Peter, A. B., Schittny, J. C., Niggli, V., Reuter, H., & Sigel, E. (1991). The polarized distribution of poly(A+)-mRNA-induced functional ion channels in the *Xenopus* oocyte plasma membrane is prevented by anticytoskeletal drugs. J. Cell Biol. 114, 455–464.

Quinton, T. M., & Dean, W. L. (1992). Cyclic AMP-dependent phosphorylation of the inositol-1,4,5-triphosphate receptor inhibits Ca^{2+} release from platelet membranes. Biochem. Biophys. Res. Commun. 184, 893–899.

Restrepo, D., Miyamoto, T., Bryant, B. P., & Teeter, J. H. (1990). Odor Stimuli trigger influx of calcium into olfactory neurons of the channel catfish. Science 249, 1166–1168.

Ross, C. A., Meldolesi, J., Milner, T. A., Satoh, T., Supattapone, S., & Snyder, S. H. (1989). Inositol 1,4,5-triphosphate receptor localized to endoplasmic reticulum in cerebellar Purkinje neurons. Nature 339, 468–470.

Satoh, T., Ross, C. A., Villa, A., Supattapone, S., Pozzan, T., Snyder, S. H., & Meldolesi, J. (1990). The inositol 1,4,5-triphosphate receptor in cerebellar Purkinje cells: Quantitative immunogold labeling reveals concentration in an ER subcompartment. J. Cell Biol. 111, 615–624.

Smith, J. B., Smith, L., & Higgins, B. L. (1985). Temperature and nucleotide dependence of calcium release by myo-inositol 1,4,5-triphosphate in cultured vascular smooth muscle cells. J. Biol. Chem. 260, 14413–14416.

Sudhof, T. C., Newton, C. L., Newton, B. T. III, Ushkaryov, Y. A., & Mignery, G. A. (1991). Structure of a novel $InsP_3$ receptor. EMBO J. 10, 3199–3206.

Suematsu, E., Hirata, M., Hashimoto, T., & Kuriyama, H. (1985). Inositol 1,4,5-triphosphate releases Ca^{2+} from intracellular store sites in skinned single cells of porcine coronary artery. Biochem. Biophys. Res. Commun. 120, 481–485.

Supattapone, S., Danoff, S. K., Theibert, A., Joseph, S. K., Steiner, J., & Snyder, S. H. (1988). Cyclic AMP-dependent phosphorylation of a brain inositol triphosphate receptor decreases its release of calcium. Proc. Natl. Acad. Sci. USA 85, 8747–8750.

Volpe, P., & Alderson-Lang, B. H. (1990). Regulation of inositol 1,4,5-triphosphate-induced Ca^{2+} release. II. Effect of cAMP-dependent protein kinase. Am. J. Physiol. 258, C1086–C1091.

Walaas, S. I., Nairn, A. C., & Greengard, P. (1986). PCPP-260, a Purkinje cell-specific cyclic AMP-regulated membrane phosphoprotein of Mr 260,000. J. Neurosci. 6, 954–961.

Walker, J. W., Somlyo, A. V., Goldman, Y. E., Somlyo, A. P., & Trentham, D. R. (1987). Kinetics of smooth and skeletal muscle activation by laser pulse photolysis of caged inositol 1,4,5-triphosphate. Nature 327, 249–252.

Watras, J., Beprovanny, I., & Ehrlich, B. (1991). Inositol 1,4,5-triphosphate-gated channels in cerebellum: presence of multiple conductance states. J. Neurosci. 11, 3239–3245.

Weeks, G., Picciotto, M., Nairn, A. C., Walaas, S. I., & Greengard, P. (1988). Purification and characterization of PCPP-260: A Purkinje cell-enriched cyclic AMP-regulated membrane phosphoprotein of Mr 260,000. Synapse 2, 89–96.

Wojcikiewicz, R. J. H., Nakade, K., Mikoshiba, K., & Nahorski, S. R. (1992). Inositol 1,4,5-trisphophate receptor immunoreactivity in SH-SY5Y human neuroblastoma cells is reduced by chromic muscarinic receptor activation. J. Neurochem. 59, 383–386.

Worley, P. F., Baraban, J. M., & Snyder, S. H. (1989). Inositol 1,4,5-triphosphate receptor binding: autoradiographic localization in rat brain. J. Neurosci. 9, 339–346.

Yamada, N., Makino, Y., Clark, R. A., Pearson, D. W., Mattei, M. G., Guenet, J. L., Ohama, Ohama, E., Fujino, I., Miyawaki, A. Miyawaki, Furuichi, T., & Mikoshiba, K. (1994). Human inositol 1,4,5-triphosphate type 1 receptor, $InsP_3R1$: Structure, function, regulation of expression and chromosomal localization. Biochem. J. (in press).

Yamamoto, A., Otsu, H., Yoshimori, T., Maeda, N., Mikoshiba, K., & Tashiro Y. (1991). Stacks of flattened smooth endoplasmic reticulum highly enriched in inositol 1,4,5-triphosphate (InsP3) receptor in mouse cerebellar Purkinje cells. Cell Structure and Function 16, 419–432.

Yamamoto, H., Maeda, N., Niinobe, M., Miyamoto, E., & Mikoshiba, K. (1989). Phosphorylation of P400 protein by cyclic dAMP-dependent protein kinase and Ca^{2+}/calmodulin-dependent protein kinase II. J. Neurochem. 53, 917–923.

Yamamoto-Hino, M., Sugiyama, T., Hikichi, K., Mattei, M. G., Hasegawa, K., Sekine, S., Sakurada, K., Miyawaki, A., Furuichi, T., Hasegwa, M., & Mikoshiba, K. (1994). Cloning and characterization of human type 2 and type 3 inositol 1,4,5-triphosphate recepters. Receptors and Channels.

Yoshikawa, S., Tanimura, T., Miyawaki, A., Nakamura, M., Yuzaki, M., Furuichi, T., & Mikoshiba, K. (1992). Molecular cloning and characterization of the inositol 1,4,5-triphosphate receptor in *Drosophila melanogaster*. J. Biol. Chem. 267, 16613–16619.

THE RYANODINE RECEPTOR

A. G. Lee

Biomembranes
Volume 6, pages 291–326.

ISBN: 1-55938-663-0.

I. EXCITATION–CONTRACTION COUPLING

The central question in excitation–contraction (E–C) coupling in muscle is how depolarization of the outer membrane (the sarcolemma) of the muscle cell leads to release of Ca^{2+} from the sarcoplasmic reticulum. In skeletal muscle, an action potential initiated at the neuromuscular junction spreads over the plasma membrane and passes into the muscle fiber through invaginations of the sarcolemma, the transverse tubules (or T-tubules), resulting in near-synchronous initiation of Ca^{2+} release from the sarcoplasmic reticulum (SR) throughout the fiber volume. The T-tubules form junctions with the terminal cisternae region of the SR, referred to

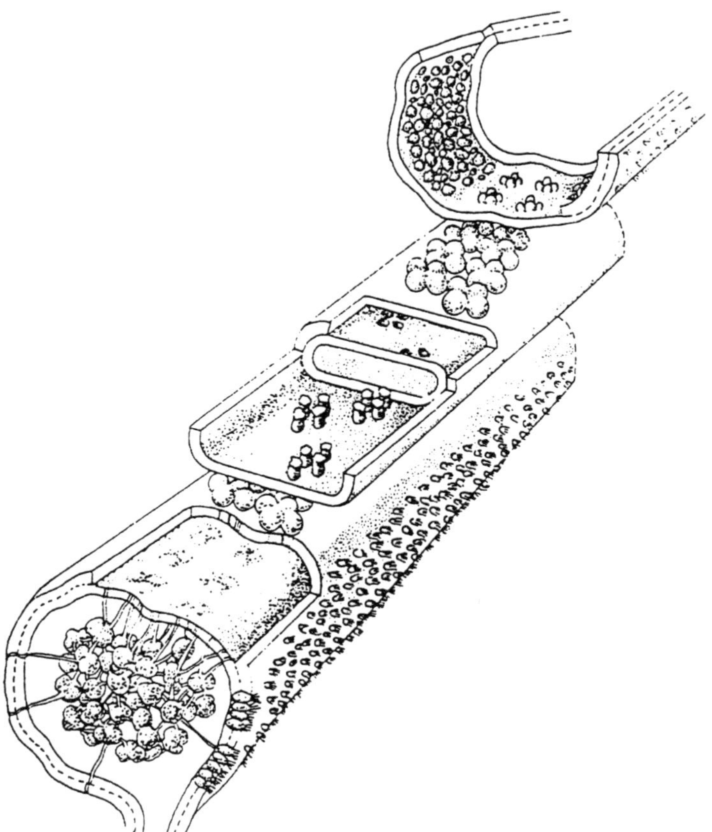

Figure 1. Three-dimensional image of a triad showing the relationship between calsequestrin, feet proteins, and tetrads. (From Block et al., 1988, with permission).

as triad junctions as each is composed of one T-tubule and two apposed terminal cisternae (Figure 1). The release of Ca^{2+} occurs at the triad. The gap between the T-tubule and the SR is ca. 10 nm and is bridged by "foot structures" or "feet" (Figure 2; Fleischer and Inui, 1989; Franzini-Armstrong and Jorgensen, 1994). These feet have been shown to be identical to a protein identified by its ability to bind the plant alkaloid ryanodine, and thus called the ryanodine receptor (RyR). The RyR has intrinsic Ca^{2+} channel activity and is the protein involved in Ca^{2+} release from SR.

The morphology of heart muscle cells (myocytes) is distinct from that of skeletal muscle fibers. The number of T-tubules varies between animal species, with frog heart myocytes having none. Where they exist, heart T-tubules are of greater

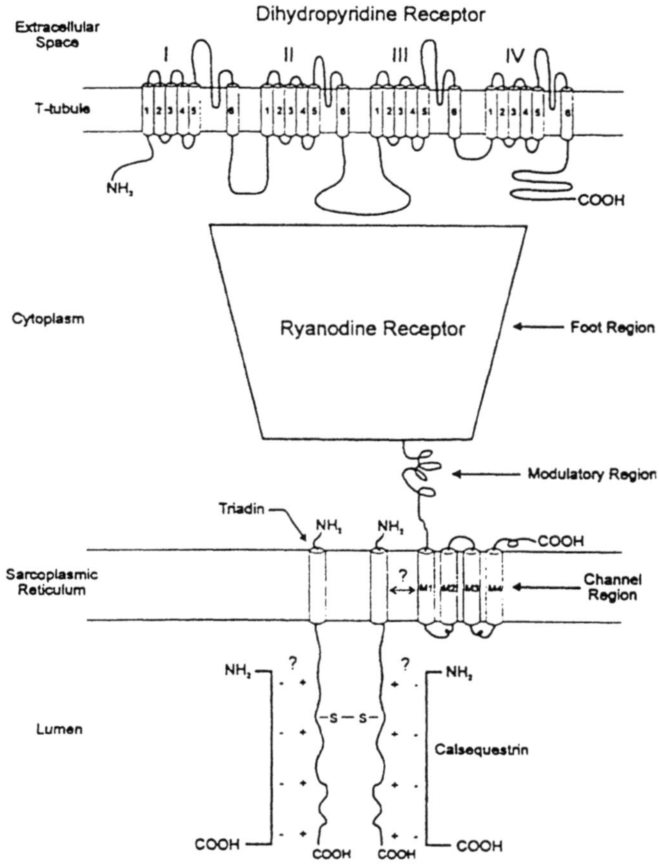

Figure 2. A model for the triad junction. Only a single subunit of the RyR and the α_1-subunit of the DHPR are shown. The structure of the ryanodine receptor is that of Takeshima et al. (Takeshima et al., 1989). A possible interaction of triadin and calsequestrin is shown. (From McPherson and Campbell, 1993b, with permission).

diameter than those in skeletal muscle. As well as triads, mammalian hearts contain dyads linking a single terminal cisternae with a T-tubule and peripheral couplings where terminal cisternae are connected directly to the plasmalemma. Smooth muscle contains little if any T-tubule, but does contain dyads connecting SR with the plasmalemma (Fleischer and Inui, 1989; Franzini-Armstrong and Jorgensen, 1994).

It is now clear that the release of Ca^{2+} involves two proteins, the dihydropyridine receptor (DHPR) in the T-tubule and the RyR in the SR membrane (Figure 2). The mechanism linking the DHPR and the RyR is different in cardiac and skeletal muscle. In cardiac muscle, depolarization of the sarcolemma triggers a Ca^{2+} influx from the external medium by opening the DHPRs. The Ca^{2+} ions entering the myocyte through the DHPRs bind to the RyR in the SR, opening the channels and leading to a larger, secondary release of Ca^{2+} into the cytoplasm (Figure 3). Binding of Ca^{2+} to troponin C then initiates muscle contraction. The process is referred to as Ca^{2+}-induced Ca^{2+} release. Suggestions that inositol-1,4,5- trisphosphate ($InsP_3$) could also be a diffusible messenger in E–C coupling have been shown to be incorrect, even though $InsP_3$ has been shown to activate Ca^{2+} release in smooth muscle (Hannon et al., 1992).

In skeletal muscle, the link between the DHPR and the RyR is by direct mechanical coupling, with the DHPR acting as a voltage sensor rather than as a Ca^{2+} channel (Figure 3). Schneider and Chandler (1973) first observed a voltage-dependent charge movement across the T-tubule membrane using voltage–clamp techniques under conditions where all ionic currents were blocked. They proposed

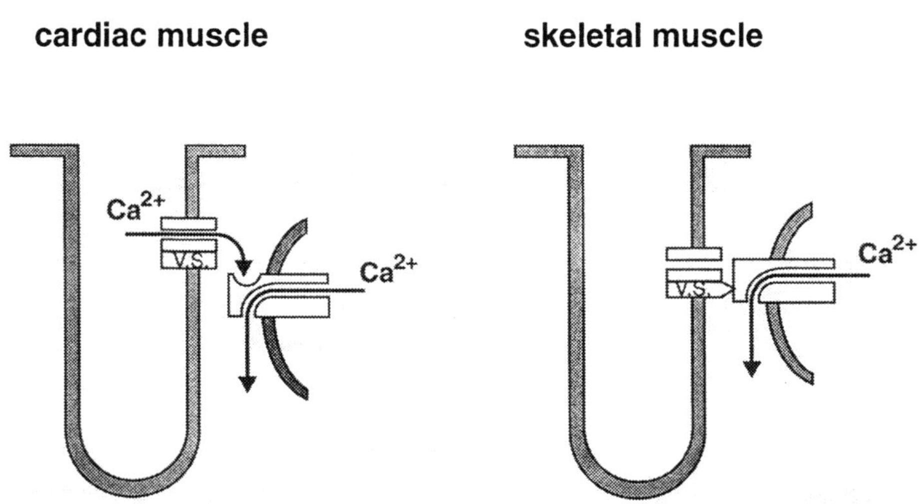

Figure 3. EC coupling in (**a**) cardiac and (**b**) skeletal muscle, showing the link between the DHPR in the T-tubule membrane and the RyR in the SR membrane.

that depolarization of the T-tubule membrane led to the movement of charged species in the membrane, these movements providing the link between depolarization of the T-tubule and Ca^{2+} release from the SR. It was suggested that in its resting conformation, the voltage sensor in the T-tubule membrane directly blocked the release channel in the SR membrane, and that this block was removed by a voltage-dependent movement of the voltage sensor on depolarization of the T-tubule.

A number of experiments argue against any major role for Ca^{2+}-induced Ca^{2+} release in E–C coupling in skeletal muscle. Thus, it has been observed that Ca^{2+} release occurs in intact fibers following membrane depolarization to beyond the reversal potential for Ca^{2+} (Brum et al., 1987, 1988). Ca^{2+} release has also been observed in the presence of high concentrations of Ca^{2+} buffers that lower the external free Ca^{2+} concentration to low values (Miledi et al., 1984). Surface-bound Ca^{2+} is also unlikely to be involved since injection of Ca^{2+} buffers into the myoplasm does not stop release of Ca^{2+} from the SR on depolarization of the muscle (Csernoch et al., 1993). In contrast, in cardiac muscle, E–C coupling is dependent on extracellular Ca^{2+} (Stern and Lakatta, 1992). The different mechanisms of Ca^{2+} release in heart and skeletal muscle is consistent with the finding of a significant fraction of the RyRs in regions of the cardiac SR without any contact with the plasmalemma in mammalian (corbular SR) and avian (extended junctional SR) heart (Sommer and Johnson, 1979; Jorgensen et al., 1993).

Although in skeletal muscle the primary release of Ca^{2+} from SR occurs by a Ca^{2+}-independent mechanism, a secondary role for Ca^{2+}-induced Ca^{2+} release is possible. Binding studies have suggested that in many muscle types there is an excess of RyRs over DHPRs, which would mean that some RyRs could not be under the direct control of DHPRs (Meissner, 1994).

II. THE DHPR AND ITS INTERACTION WITH THE RyR

Charge movement in the T-tubule was found to be blocked by drugs in the dihydropyridine (e.g., nifedipine, nitrendipine) and phenylalkylamine (verapamil, D600) and benzothiazepine (diltiazem) classes of Ca^{2+} channel antagonists, and since these drugs also blocked release of Ca^{2+} from SR, the charge movement was proposed to be due to gating of the DHPR (Rios et al., 1992). DHPRs have been located in the surface membranes of a variety of cells and shown to be the L-type Ca^{2+} channel. Ca^{2+} channel antagonists block Ca^{2+} currents by preferential binding to the inactivated state of the channel (Bean, 1984). Inactivation of the voltage sensors in the skeletal muscle T-tubule membrane has also been observed following depolarizations of seconds or tens of seconds, with repolarization leading to recovery of both charge movement and E–C coupling (Schneider, 1994). Although L-type Ca^{2+} channel activity is observed in skeletal muscle, the kinetics of channel opening are normally slow, and as described above, there is strong evidence against Ca^{2+} influx through these, or any other Ca^{2+} channels in the sarcolemma or T-tubule, being involved in E–C coupling (Hain et al., 1994). Nevertheless, conditions are known under which Ca^{2+} influx through the L-type Ca^{2+} channel is fast, and it appears that the channel can act both as a voltage sensor controlling release of Ca^{2+} from the SR and as a voltage-gated Ca^{2+} channel allowing Ca^{2+} influx from the external medium (Hain et al., 1994). In cardiac myocytes, Ca^{2+} influx through L-type Ca^{2+} channels is faster than in skeletal muscle and, as described, is essential for activating Ca^{2+} release from SR (Hain et al., 1994).

The DHPR has been purified from skeletal muscle and contains one each of the five subunits α_1, α_2, β, γ, and δ, all of which have been sequenced (Jay et al., 1991; Miller, 1992). The α_1-subunit has structural and sequence homology to the α-subunit of the plasma membrane sodium channel, which is thought to contain the voltage sensor and form the channel pore (Armstrong, 1992). Correspondingly, functional expression and labeling studies of the α_1-subunit of the DHPR have shown that this subunit is the channel forming and DHP-binding portion of the molecule (Nakayama et al., 1991; Perez-Reyes et al., 1992). Three α_1-isoforms have been detected in skeletal muscle and one in cardiac muscle (Franzini-Armstrong and Jorgensen, 1994). The functions of the other subunits are not yet clear.

The α_1-subunit contains four internal repeats each including five hydrophobic helices and one positively charged helix (helix 4), thought to act as a voltage sensor (Figure 4). The importance of the α_1-subunit of the DHPR has been confirmed in experiments using dysgenic myotubes lacking the α_1-subunit, which show neither charge movement following depolarization nor E–C coupling; both could be restored by transfection with α_1-cDNA (Numa et al., 1990).

In skeletal muscle, the α_1-subunit of the DHPR in the T-tubule and the RyR in the SR probably make physical contact (Block et al., 1988). A close relationship is also suggested by their subcellular distributions which are very similar at all stages of the development of T-tubules and triads (Yuan et al., 1991). The loop regions between domains I and II, and between domains II and III on the DHPR exhibit

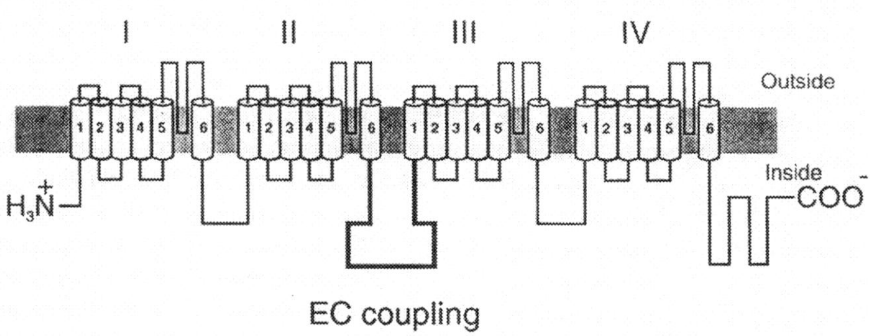

Figure 4. The α_1-subunit of the DHPR, showing the loop between internal repeats II and III which, in skeletal muscle, is involved in interaction with the RyR.

major differences in sequence between the skeletal and cardiac muscle forms (Tanabe et al., 1987; Mikami et al., 1989), and so could be responsible for tissue-specific differences in the interactions between DHPRs and RyRs. The importance of the loop between domains II and III has been confirmed using a dysgenic skeletal muscle culture incapable of E–C coupling because of the lack of the α_1-subunit of DHPR (Tanabe et al., 1990). Injection of cDNA for skeletal α_1-DHPR restored E–C coupling in the absence of extracellular Ca^{2+} whereas cardiac-type E–C coupling (dependent on extracellular Ca^{2+}) was observed following injection of cardiac α_1-DHPR (Tanabe et al., 1990). Studies with chimeric cDNAs identified the loop region between domains II and III as the major determinant of the type of coupling (Tanabe et al., 1990). Studies with these loop regions expressed as fusion proteins in *E. coli* show that the loop from both the skeletal and the cardiac DHPR can activate the skeletal RyR, increasing channel open probability and increasing ryanodine binding, but has no effect on the cardiac RyR (Lu et al., 1994). These experiments show that the II–III loop region of skeletal and the cardiac DHPR α_1-subunit interact in a functional manner with the skeletal, but not with the cardiac RyR (Lu et al., 1994). Thus, the nature of E–C coupling depends both on the DHPR and on the RyR receptor.

These experiments suggest a direct interaction between the DHPR and the RyR in skeletal muscle. This would also be consistent with the observation that immunoprecipitation of solubilized triads with antibodies directed against the RyR led to co-immunoprecipitation of the DHPR and conversely, immunoprecipitation of solubilized triads with antibodies directed against the DHPR led to co-immunoprecipitation of the RyR (Marty et al., 1994a).

It has been suggested that interaction between the DHPR and the RyR could be mediated by another protein, triadin, a 95 kD protein found in junctional SR (Caswell et al., 1991). However, the sequence of triadin suggests that it contains only a single transmembrane α-helix, with the bulk of the protein being located in the lumen of the SR with only 47 amino acids exposed on the cytoplasmic side of the SR membrane and available for binding to the DHPR (Figure 2; Knudson et al., 1993). A more likely possibility is that triadin links calsequestrin to the RyR. The luminal domain of triadin has been shown to bind to both the RyR and calsequestrin; binding to calsequestrin only occurs in the presence of Ca^{2+} (Guo and Campbell, 1995).

Simon and Hill (1992) observed that the rate of Ca^{2+} release was proportional to the fourth-power of the ryanodine channel-controlling component of the charge movement in the T-tubule membrane. This is consistent with a model requiring the movement of four independent and identical voltage sensors in the T-tubule membrane to activate one RyR. The observed kinetics would require that if any three of the four voltage sensors were in the activating state, then the movement of the fourth voltage sensor between the activating and inactivating states would lead to opening and closing, respectively, of the ryanodine receptor, with negligible delay (Simon and Hill, 1992). The observed charge movement in the T-tubule membrane would correspond to the movement of the fourth voltage sensor. The four voltage sensors could correspond to the four S4 segments in each DHPR. An alternative possibility is that a group of four neighboring DHPRs act together to gate one RyR—this would be consistent with the observation in toadfish swim-bladder skeletal muscle, of an arrangement of four particles (the DHPR) opposite each RyR (Block et al., 1988). More complex models have also been proposed. Thus, Rios et al. (1993) have suggested that each of the four voltage sensors could act as an allosteric ligand for one of the four subunits making up the RyR (see following). It was suggested that the RyR could either open on its own or, with increasing probability, as each of the four voltage sensors move.

III. STRUCTURE OF THE RYR

The RyR is an integral membrane protein of the SR. It has been purified and its sequence determined. It consists of a homotetramer of identical 565 kD subunits (skeletal muscle form). Electron microscopy shows the receptor as a complex with four-fold symmetry with overall dimensions and shape matching those of the 'feet' structures seen in skeletal muscle spanning the gap between the SR and the transverse tubules at the triad junctions. Up to four copies of the immunophilin, FK-506-binding protein ($M_r = 12$ kD) are tightly associated with the purified RyR and should be considered an integral component of the RyR (Section X; Jayaraman et al., 1992; Timerman et al., 1993, 1994; Brillantes et al., 1994).

A. 3D Structure

In electron micrographs, feet appear as four large subunits symmetrically located around a central less dense area, with adjacent feet being closely apposed in orderly arrays (Franzini-Armstrong and Jorgensen, 1994). Rotary shadowing of isolated vesicles shows a foot structure with a clear depression in the center of a tetrameric structure (quatrefoil) formed by four identical subunits (Loesser et al., 1992). It has now been established that the feet proteins and the RyR are the same protein (Loesser et al., 1992). The T-tubule membrane contains orderly groups of four proteins, referred to as tetrads. It is believed that each protein corresponds to a DHPR and that in skeletal muscle every other RyR interacts with a tetrad, with the four DHPRs being located almost immediately below the four lobes of the RyR with which they interact, suggesting interaction between the feet and tetrads (Franzini-Armstrong and Jorgensen, 1994).

A representation of the surface of the RyR obtained from electron microscopic studies at a resolution of ca. 3 nm is shown in Figure 5 (Radermacher et al., 1994). The RyR is composed of two main components, the larger with the shape of a square prism 29 × 29 × 12 nm, with, attached to one of its faces, a smaller protruding feature. The larger component is the foot structure and the smaller feature is believed to be the transmembrane region of the RyR. The transmembrane assembly

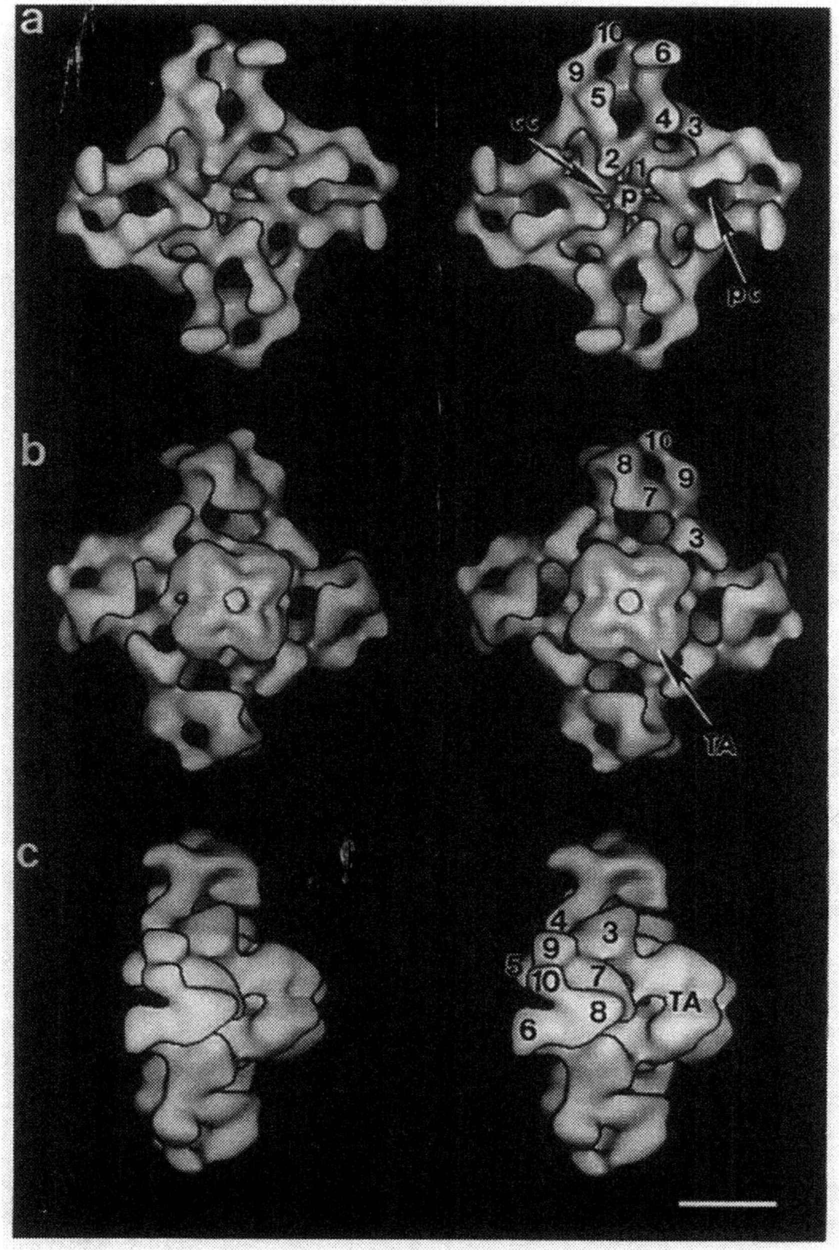

Figure 5. Surface representation of the RyR (stereo pairs). (**a**) View along the fourfold symmetry axis showing the surface that would face the cytoplasm and the apposing transverse tubule in a triad junction. (**b**) View onto the face that would interact with the SR. (**c**) Side view, normal to the fourfold axis. The putative structural domains are numbered 1–10. Other abbreviations: p, plug; pc, major peripheral cavity; cc, central cavity; TA, transmembrane assembly. Scale bar, 10 nm. (From Radermacher et al., 1994, with permission).

is square-shaped in cross-section and tapered along its length so that the edge-length is ca. 12 nm near the site of attachment to the cytoplasmic assembly and ca. 6 nm at its distal end. The total length of the transmembrane assembly is 7 nm, and thus presumably includes small regions on either side of the SR membrane as well as the region actually spanning the lipid bilayer region of the membrane. Down the center of the channel is a region of low electron density (cc in Figure 5), of diameter ca. 2–3 nm, which, it has been suggested, could represent the pathway for Ca^{2+} movement across the membrane (Radermacher et al., 1994). In the open state of the channel, an actual central opening can be seen in the transmembrane region (Serysheva et al., 1995; Orlova et al., 1996). At the top of the channel, nearest to the cytoplasmic assembly, is a globular mass of diameter ca. 3.0–3.5 nm, which is thought to act as a channel plug (p in Figure 5). To the sides of the plug are four small cavities leading to the exterior of the transmembrane assembly which could represent the exit site for Ca^{2+} ions into the myoplasm. Radermacher et al. (1994) suggested that channel gating could involve movements of this plug.

The cytoplasmic assembly is arranged as domains, of which 10, each repeated 4 times, have been identified in Figure 5 (Radermacher et al., 1994). Large cavities (pc in Figure 5) between the domains mean that greater than 50% of the total volume of the cytoplasmic assembly is occupied by solvent. The only visible connection between the cytoplasmic and transmembrane assemblies is provided by domain 1. Domains 4–10 form the large lobes at the corners of the cytoplasmic assembly. Radermacher et al. (1994) described the cytoplasmic assembly as forming a scaffolding whose main role is a mechanical one of holding together the transverse tubule and the SR membrane against the potentially disruptive forces generated during cycles of muscle contraction.

The flat structure of the top of the RyR (Figure 5) is consistent with its proposed interaction with the tetrads in the T-tubule. Domains 4, 5, and 6 lie at the top of the RyR and are thus the domains most likely to be involved in the interaction (Radermacher et al., 1994). The spacing between the centers of domains 4–6 on the top surface of the RyR is 13.5 nm, matching the spacing of 13–14 nm between the DHPRs in the tetrads with which they are proposed to interact (Franzini-Armstrong and Jorgensen, 1994; Radermacher et al., 1994).

The mechanism of the link between changes in the DHPR and the opening of the Ca^{2+} channel in the RyR is still obscure: it obviously involves very long-range interactions. Since the only visible link between the cytoplasmic and transmembrane regions of the RyR receptor appear to be domain 1, Radermacher et al. (1994) have suggested that interactions between domain 1 and the central plug could be involved.

B. Sequence

Three different RyR genes have now been identified in mammals (Section IV). The high degree of homology between the three RyRs suggest a very similar overall structure. Hydropathy plot analysis of the RyR suggests that it contains a short

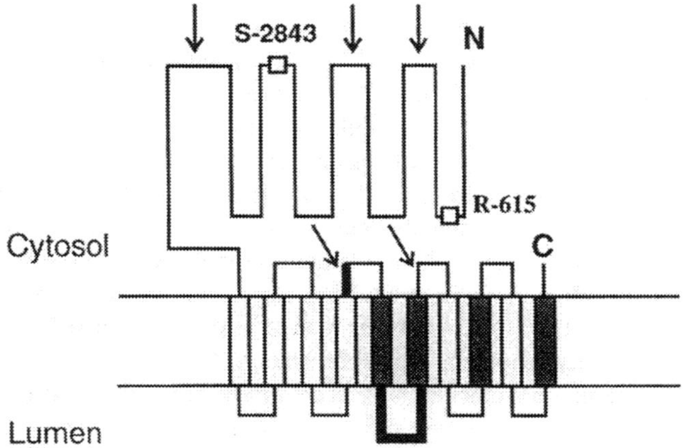

Figure 6. A model for the membrane topology of the RyR of mammalian skeletal muscle. Only one of the four polypeptides making up the intact RyR is shown. The trans-membrane region is predicted to contain either 4 (dark shading; Takeshima et al., 1989) or 10 (Zorzato et al., 1990) α-helices. Also shown are the mutation site (R615C) on the cytoplasmic domain in malignant hyperthermia-susceptible pigs, Ser2843 shown to be phosphorylated by protein kinase, five protease-sensitive sites likely to be on the surface of the receptor (arrows), and two antibody binding sites accessible from the cytoplasmic and lumenal sides of the SR membrane, respectively (thick lines).

cytoplasmic C-terminus and 4 (Takeshima et al., 1989) or 10–12 (Zorzato et al., 1990) transmembrane α-helices in the C-terminal one-fifth of the molecule (Figure 6), these making up the transmembrane pore. There is a marked homology between the putative transmembrane helices of the RyR and those of the inositol-1,4,5-trisphosphate receptor, and it has been suggested that these two channels evolved from a common ancestral cation channel (see Furuichi et al., 1994). Takeshima et al. (1989) noted a sequence homology between helices M2 and M3 of their model for the RyR and helices M2 and M3 of the nicotinic acetylcholine receptor. In the nicotinic acetylcholine receptor, M2 is thought to form the lining of the pore (see Unwin, 1995).

The large N-terminal domain of the RyR is thought to constitute the cytoplasmic foot. An N-terminal signal sequence is not found in any of the receptors, indicating that the N-terminus is cytoplasmic. Antibody-binding studies have shown that both the N- and C-termini of the RyR are cytoplasmic, giving an even number of transmembrane α-helices (Marty et al., 1994b).

C. Regulatory Sites

A number of possible regulatory sites have been identified on the RyR. The nucleotide-binding motif, GXGXXG, is found in all the RyRs, but the secondary structures around these motifs in the RyRs do not match those expected for high-affinity nucleotide binding sites. The skeletal muscle RyR is labeled with azido-ATP or Bz2-ATP, but the location of the binding sites has not been determined (Zarka and Shoshanbarmatz, 1993; Brandt et al., 1992).

Up to nine potential calmodulin-binding sites have been identified in the sequence of the RyR (Coronado et al., 1994). Photoaffinity labeling with azido-[^{125}I]-calmodulin have shown that the RyR is a major calmodulin-binding protein in the junctional SR (Seiler et al., 1984) and, on the basis of experiments with fluorescently labeled calmodulin, it has been suggested that up to six calmodulin molecules could bind per RyR monomer with affinities between 0.1 nM and 0.24 μM, depending on divalent metal ion concentration (Yang et al., 1994). Up to four gold-labeled calmodulin molecules have been reported to bind per RyR tetrameric complex, equivalent to one binding site per monomer (Wagenknecht et al., 1994). Studies of the binding of gold-labeled calmodulin to the RyR using electron microscopy have located a binding site for calmodulin at or near the cleft between domains 4 and 6 in Figure 5 (Wagenknecht et al., 1994). Direct interaction between RyRs and calmodulin has also been demonstrated functionally, since calmodulin decreases the rate of release of Ca^{2+} from heavy SR vesicles (Meissner, 1986a) and decreases the channel open probability for the RyR in planar lipid bilayers (Smith et al., 1989). Studies of binding of calmodulin to trpE fusion proteins containing fragments of the skeletal muscle RyR by Chen and MacLennan (1994) have identified three strong and three weak calmodulin-binding sites. Strong binding domains were located between residues 2063–2091, 3611–3642, and 4303–4328, and weak binding domains between residues 921–1173, 2804–2930, and 2961–3084. Binding to all these domains was only observed in the presence of Ca^{2+} (Chen and MacLennan, 1994). Chen and MacLennan (1994) have pointed out that the strong calmodulin-binding domains are located close to strong Ca^{2+}-binding regions suggesting that interactions between these domains may be important. Menegazzi et al. (1994), using a similar protocol to localize calmodulin-binding sites, identified three calmodulin binding domains between residues 2937–3225, 3610–3629, and 4534–4552. The first two of these domains correspond to two of those detected by Chen and Maclennan (1994), but the third does not.

Of the possible phosphorylation sites, only one has been confirmed *in vitro*. This is Ser2809 in cardiac RyR, phosphorylated by a calmodulin-dependent protein kinase, and the equivalent residue (Ser2843) in skeletal RyR phosphorylated by cAMP-, c-GMP- and calmodulin-dependent protein kinases (Witcher et al., 1991; Suko et al., 1993). However, Hain et al. (1994) have suggested that more than one site on the skeletal muscle RyR can be phosphorylated, although the site(s) of phosphorylation were not determined. In the absence of Mg^{2+}, channel activities were very similar for the phosphorylated or dephosphorylated receptor (Section VI) (Hain et al., 1994). However, phosphorylation of the skeletal muscle RyR by protein kinase A or Ca^{2+}/calmodulin-dependent protein kinase type II was observed to remove the block of the RyR by Mg^{2+} (Hain et al., 1994). This is an important observation since it has been shown that Mg^{2+} inhibits the release of Ca^{2+} from SR vesicles (Meissner et al., 1986) and the free Mg^{2+} concentration is heart and skeletal muscle is ca. 1 mM (Romani and Scarpa, 1992). The RyR is dephosphorylated by acid phosphatase (Hain et al., 1994). Thus, sensitivity to Mg^{2+} block is conferred by dephosphorylation and relief from Mg^{2+} block is conferred by phosphorylation. It has been suggested that phosphorylation/dephosphorylation may be involved in opening and closing the channel during each contraction cycle in E–C coupling (Hain et al., 1994).

The cardiac RyR shows an unusual form of adaptation. Successive increases in the concentration of Ca^{2+} have been shown to repeatedly open RyRs which then close even though the increased concentration of Ca^{2+} is maintained. Adaptation in the absence of Mg^{2+} or ATP is quite slow, occurring on the seconds time-scale. However, in the presence of physiological concentrations of Mg^{2+}, the rate of adaptation is much increased, to the milliseconds time-scale. Phosphorylation of the RyR by protein kinase A increases the rate of adaptation even further. This adaptation allows successive increases in the intracellular concentration of Ca^{2+} to repeatedly, but transiently, activate the RyR (Valdivia et al., 1995). Models for this process of adaptation have been proposed (Cheng et al., 1995; Sachs et al., 1995).

Regions with a sequence similarity to EF-hand type Ca^{2+}-binding sites have been identified in the rabbit skeletal muscle RyR on the N-terminal side of the first predicted transmembrane helix (Takeshima et al., 1989; Zorzato et al., 1990). Experiments using $^{45}Ca^{2+}$ and ruthenium red overlay to identify Ca^{2+}-binding sites on trpE fusion proteins identified three Ca^{2+}-binding peptides in this region, corresponding to amino acid residues 4246–4467, 4382–4417, and 4478–4512 (Chen et al., 1992), and further binding sites have been located between residues 1861–2094 and 3657–3776 (Chen and MacLennan, 1994). An antibody against the 4478–4512 peptide increased the open probability and opening time of the purified RyR incorporated into planar bilayers without altering the channel conductance, suggesting that this site might be involved in Ca^{2+} regulation of the channel (Chen et al., 1992). The observation of an effect of the antibody on the function of the RyR implies that the antibody was able to bind to the receptor and, thus, that this region was surface exposed (Figure 6).

D. Proteolysis

Proteolytic studies with the endoproteases Lys-C and Glu-C identified five protease-sensitive regions (1221–1334, 2725–2929, 3713–3714, 4372–4475, and 4676–4683) which are likely to be exposed on the cytoplasmic surface of the RyR (Figure 6; Marks et al., 1990), and three other protease-sensitive regions have been identified from studies using trypsin (Chen et al., 1993). The C-terminus of the RyR has been shown to be sensitive to proteolysis by carboxypeptidase A in intact vesicles, confirming a location for the C-terminus on the cytoplasmic side of the membrane (Marty et al., 1994b).

The calcium-activated protease, calpain, and its inhibitor, calpastatin, have been found associated with isolated junctional SR (Brandt et al., 1992). Calpain has been reported to cleave the RyR at a single site. The calpain-binding site has been located on the skeletal RyR between residues 1356 and 1367, with the cleavage site between residues 1383 and 1400. Calpain-cleavage of the RyR increases the channel open probability and increases the rate of Ca^{2+} efflux from SR vesicles (Rardon et al., 1990; Shoshan-Barmatz et al., 1994). It has been suggested that calpain-cleavage might occur in muscle under pathological conditions where the intracellular Ca^{2+} concentration is increased (Shoshan-Barmatz et al., 1994).

E. Antibody-binding Studies

An antibody to residues 4581–4640 of the RyR was unable to bind to intact SR vesicles, but was able to bind after permeabilization of the vesicles, suggesting a lumenal location for this epitope (Grunwald et al., 1993). The four-helix model shown in Figure 6 locates the epitope between transmembrane helices 1 and 2. More extensive antibody binding studies favor the four-helix model shown in Figure 6 (Grunwald and Meissner, 1995).

IV. GENES FOR THE RyR

A. *ryr-1*

Three different RyR genes have now been identified in mammals. The skeletal muscle RyR gene (*ryr-1*) encodes a protein containing 5,032 or 5,037 amino acids in human (Zorzato et al., 1990) or rabbit (Takeshima et al., 1989; Zorzato et al., 1990), respectively. Although *ryr-1* is expressed predominantly in skeletal muscle, it is also present in mouse cerebellum (Kuwajima et al., 1992) and may be the gene expressed in sea urchin eggs (McPherson et al., 1992). An alternative start site near the 3′-terminal region of *ryr-1* appears to generate a brain-specific transcript encoding a 75 kD protein suggested to correspond to the pore region of the RyR (Takeshima et al., 1993).

Whereas only a single isoform of the RyR has been identified in mammalian skeletal muscle, two isoforms, α and β, have been identified in a number of avian, amphibian, and piscine skeletal muscles (Olivares et al., 1991; Lai et al., 1992b; Murayama and Ogawa, 1992). It appears that only the α-isoform is found in muscles that are known for their rapid contractions, whereas both the α and β isoforms are found in less specialized muscles (Coronado et al., 1994). When both isoforms are found, they are found to coexist in the same muscle fiber and can be differentiated immunologically or by proteolytic mapping (Coronado et al., 1994). It is currently unclear whether the two isoforms are the result of alternative splicing or result from expression of different genes.

Several pieces of evidence suggest that the two isoforms may play different roles in muscle. In chicken, the α and β isoforms differ in the timing of their initial expression in embryonic muscle (Sutko et al., 1991). Since both isoforms are found in adult avian muscle, it appears that both are required for proper function of adult muscle. In the *Crooked Neck Dwarf* (cn) mutation in chickens, embryonic cn/cn skeletal muscle cells produce normal levels of the β-isoform of the RyR, but fail to make normal α-RyR: cn/cn muscle cells fail to form normal sarcomeres and undergo a degenerative cell death (Airey et al., 1993). It has also been shown that although the two isoforms have similar ion conductance properties, they differ in their gating behavior and in their responses to channel modifiers (Percival et al., 1994). In particular, the β-isoform shows activation at a lower Ca^{2+} concentration than does the α-isoform and only the α-isoform is activated by perchlorate ion (Percival et al., 1994). Perchlorate ion has been shown to bind to the mammalian skeletal muscle RyR at the phosphate-binding site and to cause activation of the receptor (Fruen et al., 1994). These experiments then suggest that the α-isoform of the RyR in chicken skeletal muscle could be activated directly by the DHPRs, as in mammalian skeletal muscle. The greater Ca^{2+} sensitivity of the β-isoform suggests a similarity with the mammalian cardiac muscle, arguing that activation of the β-isoform in chicken skeletal muscle could be by the process of Ca^{2+}-induced Ca^{2+} release (Percival et al., 1994). Thus, both mechanisms of Ca^{2+} release are likely to exist in chicken skeletal muscle, each utilizing one of the two isoforms of the RyR.

B. *ryr-2*

The *ryr-2* gene of rabbit cardiac muscle codes for a protein of 4,968–4,976 amino acids (Nakai et al., 1990; Otsu et al., 1990). The sequence is 66% homologous to that of skeletal muscle, with the same number of predicted transmembrane α-helices and the same predicted cytoplasmic domains, and the isolated RyR from cardiac muscle has been shown to adopt the same tetrameric structure as that from skeletal muscle (Inui et al., 1987a). Nevertheless, the RyRs of skeletal and cardiac muscle are the products of two different genes, the skeletal muscle RyR gene being found at region 19q13.1 on the long arm of human chromosome 19, whereas the cardiac RyR gene has been localized to human chromosome 1 (Mackenzie et al., 1990; Otsu et al., 1990).

Although the *ryr-2* gene is expressed predominantly in heart, it is also found in many regions of the brain, in stomach, and in endothelial cells (Nakai et al., 1990; Kuwajima et al., 1992; Lesh et al., 1993).

C. *ryr-3*

The distribution of RyRs in the brain appears to be particularly complex. Anti-peptide antibodies specific for the skeletal muscle RyR recognize a protein in cerebellum, whereas an antibody specific for the cardiac form recognizes a protein expressed throughout the nervous system (Kuwajima et al., 1992). It has been suggested that the *ryr-2* gene is the major form of the RyR expressed in the brain (McPherson and Campbell, 1993a). However, a third form of the RyR has been identified in brain. This, the *ryr-3* gene, codes for a protein in rabbit brain of 4,872 amino acids, showing ca. 70% homology with the skeletal- and cardiac-muscle forms (Hakamata et al., 1992). Northern blot analysis has shown that the brain RyR is abundantly expressed in corpus striatum, thalamus, hippocampus, and in smooth muscle in aorta, esophagus, taenia coli, urinary bladder, ureter, and uterus (Hakamata et al., 1992). The gene has also been identified in a cDNA library from mink lung epithelial cells (Giannini et al., 1992). The RyR has been localized to the endoplasmic reticulum (Walton et al., 1991) and has been shown to have a distribution throughout the mammalian nervous system that overlaps with, but is distinct from that of the IP_3 receptor (Sharp et al., 1993). Studies of the tissue distribution of mRNA in porcine tissues using the reverse transcription-polymerase chain reaction has identified all three *ryr* isoforms throughout the brain (Ledbetter et al., 1994).

D. *dry*

Part of a gene homologous with the RyR has been cloned from *Drosophila*, and is expressed in somatic muscle and, at a lower level, in neuronal tissue (Hasan and Rosbash, 1992).

V. PURIFICATION OF THE RyR

The use of ryanodine as a specific ligand has allowed the identification and purification of the RyR. The most commonly used detergent for solubilization of the RyR has been 3-[(3-cholamidopropyl)dimethylammonio]-1-propanesulfonate (CHAPS). The detergent-solubilized skeletal receptor has been purified by heparin-agarose chromatography followed by hydroxyapatite chromatography (Inui et al., 1987b), by immunoaffinity chromatography (Imagawa et al., 1987) and, most conveniently, by sucrose gradient centrifugation (Lai et al., 1988, 1989; Lee et al., 1994; Marty et al., 1994b; Shomer et al., 1994). In the latter procedure, the receptor is stabilized in detergent by addition of phospholipid, and purification is performed in the presence of a cocktail of protease inhibitors (Lai et al., 1988, 1989; Timerman et al., 1993). On SDS gels, the purified RyR gave a single band at 350–450 kD, whereas it had an apparent sedimentation coefficient of 30S, indicating a tetrameric structure (Inui et al., 1987b; Lai et al., 1989). The cardiac RyR has also been purified by heparin-agarose chromatography followed by p-aminobenzamidine-agarose and gel-permeation chromatography following solubilization in CHAPS (Inui et al., 1987a), and by density gradient centrifugation (Rardon et al., 1989; Lindsay and Williams, 1991). The brain RyR has been purified from CHAPS solution by a combination of heparin-agarose chromatography, ion-exchange chromatography, and density-gradient centrifugation (McPherson et al., 1991).

VI. CHANNEL PROPERTIES

The Ca^{2+} channel properties of the RyR have been studied by measurement of the rate of Ca^{2+} efflux from loaded SR vesicles and by single-channel recordings following incorporation of the RyR into planar lipid bilayers. As described in Section VIII, [^{3}H]ryanodine binding has also been used to report on conformational changes in the RyR.

For rapid kinetic measurements of Ca^{2+} efflux, heavy SR vesicles are loaded with mM concentrations of $^{45}Ca^{2+}$, either passively by incubation with mM Ca^{2+} for a period of hours, or actively by incubation in the presence of Ca^{2+}, using the Ca^{2+}-ATPase in the SR membrane to load the vesicles. Vesicles are then diluted into the required "release" medium and the rate of efflux of $^{45}Ca^{2+}$ determined (Martonosi, 1984). A complication of such studies is that the Ca^{2+}-ATPase can itself mediate Ca^{2+} release (Gould et al., 1987b) and, under some conditions (e.g., in the presence of sulfydryl reagents) this release can be fast (Gould et al., 1987a).

The rate of efflux of Ca^{2+} shows a bell-shaped dependence on the concentration of myoplasmic Ca^{2+}, with a maximum near 1 μM Ca^{2+} (Figure 7; Meissner et al., 1986). This suggests the presence on the RyR of both high-affinity, activating, and low-affinity, inhibitory sites for Ca^{2+}. Inhibition of Ca^{2+} efflux at high myoplasmic Ca^{2+} concentrations could also follow, at least in part, from a decrease in the driving

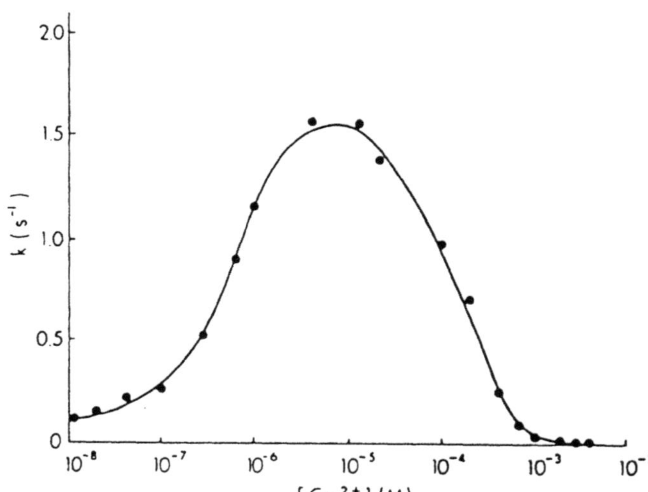

Figure 7. Effect of external Ca^{2+} concentration on the rate of efflux of Ca^{2+} from heavy SR vesicles. SR vesicles were passively loaded with 5 mM $^{45}Ca^{2+}$ at pH 7.0 and diluted into a release medium of 20 mM Pipes, pH 7.0, 0.1 M KCl containing EGTA and Ca^{2+} to give the indicated final concentrations of free Ca^{2+}. Ca^{2+} release followed first order kinetics with the rate constants given. (From Meissner et al., 1986, with permission).

force for Ca^{2+} efflux as lumenal and myoplasmic concentrations of Ca^{2+} become more nearly equal.

As well as Ca^{2+}, cytoplasmic Sr^{2+} activates the channel, whereas Ba^{2+} and Mg^{2+} are inhibitors (Meissner, 1994). Inhibition by mM Mg^{2+} could occur by competition between Mg^{2+} and Ca^{2+} for binding at the high-affinity stimulatory site or by binding of Mg^{2+} at the low-affinity Ca^{2+} inhibitory site (Meissner et al., 1986). It has also been suggested that Mg^{2+} could sterically block the channel by binding to a site near the conduction pathway (Meissner et al., 1986). The inhibitory effects of Mg^{2+} are overcome by phosphorylation (Hain et al., 1994; Section IX) and by ATP (Section VII).

For single-channel recordings, either the purified RyR or SR vesicles are fused with planar lipid bilayers (Coronado et al., 1992). A preformed planar lipid bilayer separating two solutions, referred to as *cis* and *trans* respectively, is exposed on the *cis* side to microgram amounts of SR protein. The RyR incorporates into the bilayer with the myoplasmic side of the receptor protruding into the *cis*-solution. Measurement of the conductance properties of the bilayer then provide information about the conductance properties of the RyR. It has been shown that the conductance properties of the RyR in these systems changes on phosphorylation or dephosphorylation of the receptor in the bilayer membrane (Hain et al., 1994). It is most likely that properties recorded for the RyR in which no precautions are taken to maintain the phosphorylation status of the RyR correspond to properties of the unphosphorylated RyR (Hain et al., 1994).

As shown in Table 1, conductances of the RyR are large and show relatively little discrimination between cations. Divalent cations are more permeable than monovalent ions, but there is little discrimination between Ca^{2+} and Ba^{2+}; permeability to anions is very low (Coronado et al., 1994). For pig skeletal RyR, monovalent cation conductance increases in the order $Li^+ < Na^+ < K^+ < Rb^+ < Cs^+$, with the permeability for Cs^+ being about double that for Li^+ (Shomer et al., 1994). The

Table 1. Conductance Properties of the Purified RyR

	Monovalent Cations		*Divalent Cations*		
Source	*γ (pS)*	*[cis/trans] (mM)*	*γ (pS)*	*[cis/trans] (mM)*	*References*
Rabbit skeletal muscle	496	500 Na^+/500 Na^+	91	125 $Tris^+$/50 Ca^{2+}	Lai et al. (1988)
Dog cardiac muscle	550	500 Na^+/500 Na^+	70	125 $Tris^+$/50 Ca^{2+}	Anderson et al. (1989)
Pig aorta smooth muscle	367	250 K^+/250 K^+	110	250 K^+/100 Ca^{2+}	Hermann-Frank et al. (1991)
Lobster muscle	774	260 K^+/260 K^+	122	260 K^+/50 Ca^{2+}	Seok et al. (1992)
Bovine brain	800	250 K^+/250 K^+	140	250 K^+/50 Ca^{2+}	Lai et al. (1992a)

dependence of Ca^{2+} conductance on the concentration of Ca^{2+} on the luminal side follows simple Michaelis–Menten kinetics with a dissociation constant of 3mM and a maximum conductance of 172pS for the skeletal channel (Smith et al., 1988) and comparable values of 4mM and 80pS, respectively, for the cardiac channel (Lindsay and Williams, 1991). For the skeletal channel, the dissociation constant and maximum conductance for K^+ measured in the absence of divalent metal ions, were 47mM and 1nS, respectively (Smith et al., 1988), giving a ratio of conductances γ_K/γ_{Ca} of 6. This compares with a permeability ratio P_K/P_{Ca}, measured in mixtures of K^+ and Ca^{2+}, of 0.14 (Smith et al., 1988). It has been suggested that this discrepancy indicates a conduction mechanism in which multiple ion-binding sites are arranged in a single-file across the membrane (Smith et al., 1988). For the cardiac channel, a four-barrier three-binding site model has been suggested, with a central well favoring the binding of Ca^{2+} over K^+; the stronger binding of Ca^{2+} in the channel explains the lower Ca^{2+} conductance, but the higher permeability of Ca^{2+} in mixed solutions (Tinker and Williams, 1992).

A selectivity (P_{Ca}/P_K) for Ca^{2+} versus K^+ of ca. 6 for skeletal RyR means that under physiological conditions, with concentrations of Ca^{2+} and K^+ in the lumen of the SR of ca. 1- and 140-mM, respectively, K^+ ions will compete significantly with Ca^{2+} for conductance via the RyR.

Studies of the conductance of the cardiac RyR to tetra-alkyl ammonium salts has provided further information about the nature of the pore. Block of K^+ currents by tetramethyl-, tetraethyl-, or tetrabutyl-ammonium salts were consistent with binding of the blocker at a site in the pore located at 50–90% of the electric field depending on the size of the blocker (Figure 8; Tinker et al., 1992b). The affinity of the tetra-alkyl ammonium salts increased with increasing alkyl group size, suggesting the presence of a single hydrophobic site within the pore to which the blockers could bind. Larger salts such as tetrabutyl- or tetraheptylammonium, resulted in partial occlusion of the channel, suggesting the presence of a larger mouth to the pore to which these ions can bind, but that the ions are too large to enter the pore proper (Tinker et al., 1992a). Tu et al. (1994b) have suggested that the lumenal mouth of the channel contains a large number of negatively charged amino acid residues which could serve to concentrate cations, so that the channel will have a large capture radius for Ca^{2+}.

Less hydrophobic blocker molecules such as $Tris^+$ with a molecular cross-sectional area of ca. 38 $Å^2$ themselves show a small, but distinct conductance, about 20-times lower than the K^+ conductance and 7-times lower than the Ca^{2+} conductance, suggesting a relatively large diameter for the pore (Smith et al., 1988). For the cardiac RyR, it has been suggested that the channel narrows to a radius of ca. 3.4 Å, this narrow region possibly constituting the selectivity filter of the channel (Tinker and Williams, 1993). Measurements of streaming potentials have suggested that this narrow region is relatively short, ca. 9 Å in length, and so constitutes a relatively small part of a long permeation pathway (Figure 6; Tu et al., 1994a). A very similar estimate of 10.4 Å has been made based on studies

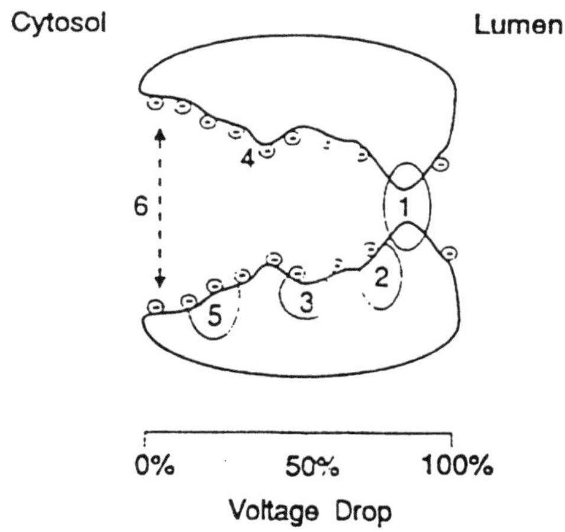

Figure 8. Suggested components of the conduction pathway of the RyR. (**1**) Selectivity filter. (**2**) Hydrophobic cation binding site. (**3**) Hydrophilic cation binding site. (**4**) Negative charge. (**5**) Large organic cation-binding site. (**6**) Shows the capture radius. (From Lindsay et al., 1994, with permission).

of channel blocking by a homologous series of monovalent ($CH_3(CH_2)_nNMe_3^+$) and divalent ($NMe_3^+(CH_2)_nNMe_3^+$)-trimethylammonium derivatives (Tinker and Williams, 1995). It has been suggested that the effect of ryanodine on the RyR is to widen this region of the channel, increasing the permeability to organic cations and decreasing the Ca^{2+} selectivity (Lindsay et al., 1994; Tu et al., 1994a).

A relatively wide channel is consistent with studies of the permeation of uncharged molecules through the channel. Light scatter measurements have shown that the SR membrane is permeable to glucose, xylose, glycine, and glycerol (Kasai et al., 1992). The rate of permeation of glucose was stimulated by μM external Ca^{2+} concentrations, and by mM caffeine or ATP, and inhibited by μM ruthenium red or mM Mg^{2+}: since Ca^{2+}-induced Ca^{2+} release showed the same characteristics, it was concluded that glucose permeated the SR membrane through the RyR pore (Kasai and Kawasaki, 1993; Meissner, 1986b; Kasai et al., 1992).

In experiments with the purified RyR, multiple subconductance states are often observed (Smith et al., 1988; Liu et al., 1989). These could be due to the presence of four separate pores in the homo-tetrameric RyR receptor, or to a single pore in the homo-tetrameric RyR receptor with several discrete conductance states depending on the conformational states of each of the four monomers making up the RyR. The structure of the RyR described above makes this latter possibility the most likely. The FK506-binding protein suppresses these subconductance states (Section X).

The dependence of channel open probability on the Ca^{2+} concentration in the (myoplasmic) *cis*-solution is bell-shaped with a maximum open probability in the

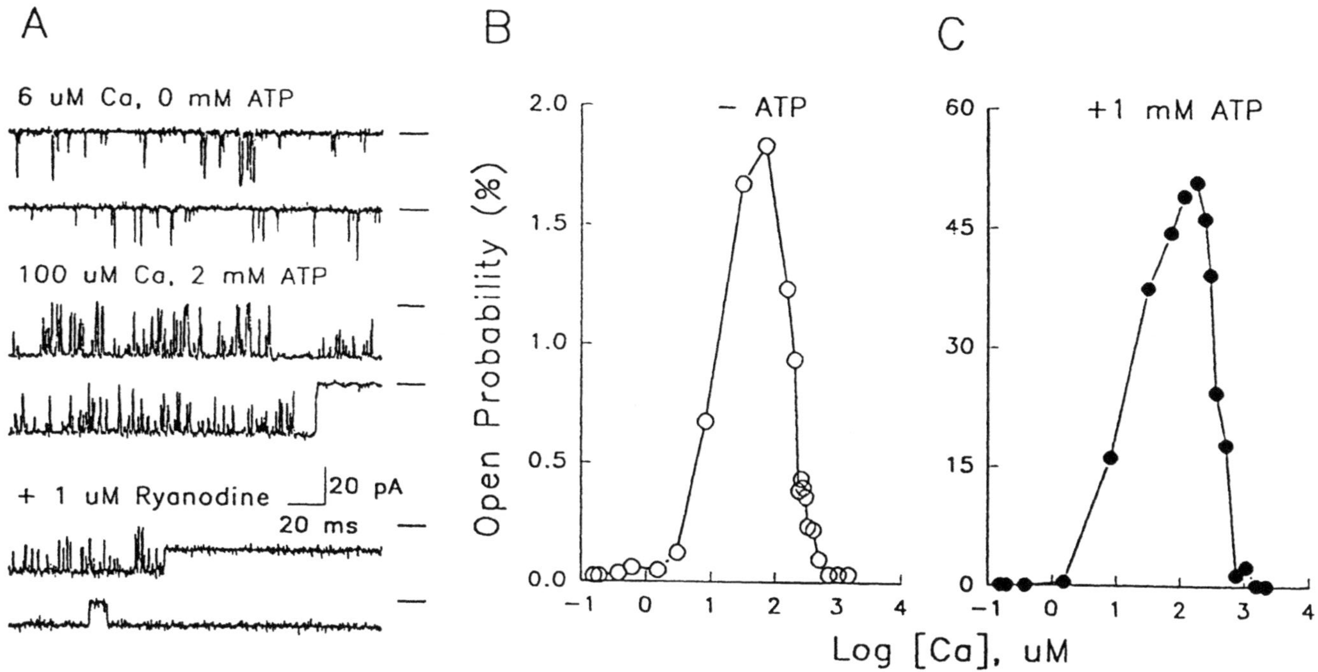

Figure 9. Regulation of the RyR by Ca^{2+} and ATP. (**A**) Single channel currents at −50 mV with Cs^{+} as current carrier (200 mM symmetrical Cs^{+}). (**B,C**) Channel open probabilities as a function of myoplasmic Ca^{2+} concentration, in the absence (**B**) or presence (**C**) of 1 mM ATP. (From Ma and Zhao, 1994, with permission).

Ca^{2+} concentration range 1–10 μM for mammalian skeletal channels when Cs^+ is the current carrier (Figure 9; Fill et al., 1991; Ma and Zhao, 1994). However, when the current carrier was Ca^{2+}, the curve of open probability against Ca^{2+} concentration failed to saturate (Smith et al., 1986). A sigmoidal dependence of open probability on Ca^{2+} concentration saturating at ca. 10 μM Ca^{2+} was observed for the mammalian cardiac RyR with Cs^+ as the current carrier, suggesting the lack of an inhibitory Ca^{2+} site on the cardiac RyR (Chu et al., 1993).

The existence of two open states and two closed states for the skeletal RyR and two open- and three closed-states for the cardiac RyR have been shown by using lifetime analysis (Smith et al., 1986; Ashley and Williams, 1990). The Ca^{2+}-dependence of the open probability for the cardiac channel has been fitted to a branched-gating scheme (Ashley and Williams, 1990).

VII. LIGAND EFFECTS

The RyR, as well as being controlled by depolarization of the T-tubule membrane, is under the control of a variety of endogenous effectors. That is, the RyR has many of the properties of a ligand-gated channel. Ligands affecting the RyR include Ca^{2+}, Mg^{2+}, ATP and related nucleotides, calmodulin, fatty acids and various lipid metabolites, polyamines, and ryanodine. The RyR is also affected by phosphorylation.

The activity of the channel is controlled by myoplasmic Ca^{2+} concentration through activation, at low (nM–μM) Ca^{2+} concentrations, and inactivation, at high (μM–mM) concentrations of Ca^{2+}, as described in Section VI. Ca^{2+} efflux is markedly stimulated by mM ATP and by nonhydrolyzable analogs such as AMP-PCP (Meissner et al., 1986). It is clear, therefore, that effects of ATP do not involve phosphorylation of the RyR, and that there must be an effector site for ATP on the RyR. Single-channel recordings show that ATP (mM) in the myoplasmic solution enhances channel activity without altering the Ca^{2+}-dependent regulation of the channel (Figure 9; Ma and Zhao, 1994). Given the relatively high concentration of Mg^{2+} in the myoplasm, the physiological ligand is likely to be MgATP rather than free ATP. Adenine and adenosine, as well as AMP, ADP, and ATP counter-act the inhibition by Mg^{2+} (Meissner et al., 1986; Moutin and Dupont, 1988).

Ryanodine, at concentrations between 0.01 and 10 μM, stimulates Ca^{2+} release from SR vesicles, whereas at high concentrations, in the range of 10- to 300-μM, it inhibits (Lattanzio et al., 1987; Nelson, 1987; Hasselbach and Migala, 1988; Gilchrist et al., 1992). It has been suggested that these effects follow from binding at functionally independent sites (Humerickhouse et al., 1993). A variety of organic molecules and drugs have also been shown to affect the rate of Ca^{2+} release. Thus anthraquinones, such as doxorubicin, increase the rate of Ca^{2+} release (Abramson et al., 1988) as does caffeine at relatively high (1–100 mM) concentrations (Coronado et al., 1994). The caffeine-binding site on the cardiac RyR has been shown to be distinct from that for ATP (Mcgarry and Williams, 1994).

The polyamines, gentamicin and neomycin, reduce the rate of Ca^{2+} release (Calviello and Chiesi, 1989) as does ruthenium red (Coronado et al., 1994), whereas polylysine and protamine increase it (Cifuentes et al., 1989). Fatty acids and various lipid metabolites have also been shown to affect the RyR. Thus, at high concentrations, sphingosine has been shown to induce release of Ca^{2+}, whereas at lower concentrations, it inhibits caffeine-induced Ca^{2+} release (Dettbarn et al., 1994). The fatty acids, arachidonic acid and stearic acid, stimulate Ca^{2+} release (Dettbarn and Palade, 1993) as do palmitoyl carnitine and palmitoyl coenzyme A (Elhayek et al., 1993; Connelly et al., 1994; Dumonteil et al., 1994). It was found that, in the presence of palmitoyl coenzyme A, the RyR was sensitive to Ca^{2+} in the presence of mM Mg^{2+}, conditions under which the sensitivity to Ca^{2+} is normally lost (Connelly et al., 1994). The possible physiological importance of these observations is still unclear.

The metabolite of NAD^+, cyclic ADP ribose (cADPR) has been shown to increase the open probability of the cardiac RyR in planar lipid bilayers, but has little effect on the skeletal RyR (Galione and White, 1994; Sitsapesan et al., 1994, 1995). Sitsapesan et al. (1994) have shown that NAD^+ and ADP-ribose also stimulate the cardiac RyR, and do so in competition with ATP, presumably binding at the ATP-binding site on the RyR. Since the physiological concentration of cADPR will be much lower than the physiological concentrations of NAD^+ and ATP, it was suggested that cADPR was unlikely to be an endogenous regulator of the RyR (Sitsapesan et al., 1994).

Effects of pH on the conductance properties of the RyR are complex (Ma and Zhao, 1994). Ca^{2+}-efflux rates were found to decrease on decreasing the pH from ca. 7.5 to 6 with little effect on the affinities of either the high-affinity, activating or low-affinity, inactivating sites for Ca^{2+} (Meissner, 1984).

Calmodulin has been reported to decrease the rate of Ca^{2+} efflux from SR vesicles and to decrease single-channel activity in the absence of ATP, suggesting the presence of inhibitory calmodulin-binding sites on the RyR (Meissner, 1986a; Smith et al., 1989). The effect of calmodulin has been shown to be to reduce the mean channel open time, with no effect on single-channel conductance (Smith et al., 1989). The presence of calmodulin-binding sites on the RyR was described in Section IIIC.

VIII. RYANODINE BINDING

The pattern of ryanodine binding to the RyR is rather complex. Both high- and low-affinity binding sites have been detected, in a ratio of 1:3 (Lai et al., 1989; Pessah and Zimanyi, 1991). Dissociation constants determined by Pessah and Zimanyi (1991) for binding of four ryanodine molecules per tetrameric skeletal and cardiac RyR were K_1 = 1–4 nM, K_2 = 30–50 nM, K_3 = 500–800 nM, and K_4 = 2–4 μM. Hill coefficients of less than 1 have been determined for the low-affinity sites, consistent with a model involving sequential binding of 1–3 ryanodine molecules to initially identical sites resulting in a progressive decrease in the affinity of the unbound sites through an allosteric interaction between sites (Lai et al., 1989; Carroll et al., 1991; Pessah and Zimanyi, 1991). Such a model is consistent with a report of a decrease in the rate of binding of [^{3}H]-ryanodine as the concentration of unlabeled ryanodine was increased (Buck et al., 1992) since, if the rate of dissociation remains constant, then a decrease in the rate of binding means a decrease in affinity. However, a reduction in the rate of dissociation of [^{3}H]-ryanodine with increasing concentration of ryanodine has also been reported (Lai et al., 1989; McGrew et al., 1989; Chu et al., 1990a; Wang et al., 1993). The dual effect of ryanodine on Ca^{2+} release described in Section VII has been attributed to binding of ryanodine to the first of the four ryanodine-binding sites with high affinity resulting in opening of the channel to a low-conductance state, with subsequent binding to the low-affinity sites resulting in channel closure (Lai et al., 1989).

On the basis of studies of proteolysis of the ryanodine-bound skeletal receptor, Callaway et al. (1994) have suggested that the ryanodine-binding sites of high- and low-affinity and the pore-forming regions are all located in the C-terminal portion of the receptor following Arg-4475.

A large number of experiments have shown that ryanodine binds preferentially to the open state of the channel, and thus that ryanodine-binding reports on the gating state of the receptor. Ligands that open the channel, such as μM Ca^{2+} and ATP, stimulate ryanodine binding to high-affinity sites, whereas ligands, such as ruthenium red and Mg^{2+}, inhibit binding (Pessah et al., 1986, 1987; Imagawa et al., 1987; Michalak et al., 1988; Holmberg and Williams, 1990).

Binding of ryanodine to the skeletal receptor has a bell-shaped dependence on Ca^{2+} concentration, with high concentrations of Ca^{2+} inhibiting binding (Michalak et al., 1988; Chu et al., 1990a; Zimanyi and Pessah, 1991), as for the dependence of Ca^{2+} channel open probability and Ca^{2+} efflux. For cardiac RyR, binding of ryanodine increases with increasing Ca^{2+} concentration up to ca. 1 mM and shows little inhibition at high Ca^{2+} concentrations (Alderson and Feher, 1987; Michalak et al., 1988; Holmberg and Williams, 1990; Zimanyi and Pessah, 1991; Chu et al., 1993); little inhibition of channel open probability is seen for the cardiac RyR at high Ca^{2+} concentrations (Chu et al., 1993) although Ca^{2+} efflux from cardiac SR was observed to decrease at high concentrations of Ca^{2+} (Chu et al., 1993), probably due to a decrease in driving force for Ca^{2+} efflux as the internal and external concentrations of Ca^{2+} become more nearly equal. Thus, it seems that the cardiac RyR has a high-affinity stimulatory Ca^{2+}-binding site but not a low-affinity inhibitory Ca^{2+}-binding site.

Mg^{2+} has been shown to inhibit ryanodine binding to the skeletal RyR by competition with Ca^{2+} for the Ca^{2+}-activation site (Pessah et al., 1987), but it has relatively little effect on ryanodine binding to the cardiac RyR (Holmberg and Williams, 1990; Michalak et al., 1988). Caffeine increases ryanodine binding to the skeletal RyR through an increase in the rate of binding (Chu et al., 1990a). Adenine nucleotides also increase ryanodine binding to the skeletal RyR receptor (Pessah et al., 1986, 1987; Michalak et al., 1988; Chu et al., 1990a; Zimanyi and Pessah, 1991; Elhayek et al., 1993) with an increase in the binding rate (Chu et al., 1990a), but have relatively little effect on the cardiac RyR (Michalak et al., 1988; Zimanyi and Pessah, 1991). Binding of ryanodine is also reduced in the presence of ruthenium red (Zimanyi and Pessah, 1991; Chu et al., 1990a; Elhayek et al., 1993); it has been suggested that this follows from oxidation of Cys residues in the receptor (Zimanyi et al., 1992).

Phosphate has been shown to bind to the skeletal RyR increasing the open probability of the skeletal RyR and increasing ryanodine binding, but with relatively little effect on the cardiac RyR (Fruen et al., 1994).

IX. PHOSPHORYLATION OF THE RyR

It has been shown that both skeletal- and cardiac-RyR are phosphorylated by calmodulin-dependent kinase and by cAMP-dependent protein kinase (PKA) (Seiler et al., 1984; Chu et al., 1990b; Takasago et al., 1991; Witcher et al., 1991; Wang and Best, 1992; Strand et al., 1993; Suko et al., 1993). Phosphorylation by PKA results in phosphorylation of 1 residue per RyR tetramer whereas calmodulin-dependent kinase phosphorylated 4 residues per RyR tetramer (Takasago et al., 1991). PKA and calmodulin-dependent kinase were reported to phosphorylate different Ser residues in the RyR (Takasago et al., 1991). Phosphorylation of cardiac RyR by calmodulin-dependent kinase has been shown to occur at Ser-2809, and this results in an activation of the cardiac RyR in planar lipid bilayers (Witcher et al., 1991). The same residue is phosphorylated in the brain RyR (Witcher et al., 1992). In skeletal RyR, it has been reported that phosphorylation by calmodulin-dependent kinase and by PKA occurred on the same residue, Ser-2843; calmodulin-dependent kinase was also reported to phosphorylate other unidentified residues, including threonine (Suko et al., 1993). However, other studies found little phosphorylation by either calmodulin-dependent kinase or PKA (Takasago et al., 1991; Witcher et al., 1991; Strand et al., 1993).

Phosphorylation of the RyR has been shown to reduce inhibition of the RyR by Mg^{2+} (Section VI).

X. FK506-BINDING PROTEIN

There is now strong evidence for considering the FK506-binding protein as an integral component of the RyR. The FK506-binding protein (FKBP12) is a 12 kD protein that binds the immuno-suppressant drugs, FK506 and rapamycin. FKBP12 and other members of the immunophilin class, are *cis–trans* peptidyl prolyl isomerases able to catalyze the interconversion of prolyl *cis–trans* isomers of various peptide- and protein-substrates. The first evidence for association of FKBP12 with the RyR came from a study of proteolytic peptides derived from the RyR which identified one peptide which was unrelated to the primary structure of the RyR, but which did correspond to the N-terminus of FKBP12 (Collins, 1991). Subsequent studies showed a tight association of FKBP12 with the RyR. Thus, they copurified on chromatography, were co-immunoprecipitated with antibodies raised either against the RyR or against FKBP12, and both were found to be located at the terminal cisternae of SR (Jayaraman et al., 1992). The stoichiometry of association was found to be one FKBP12 molecule per RyR molecule (i.e., four molecules of FKBP12 per tetramer; Timerman et al., 1993). The effect of FKBP12 on the RyR has been shown to make the channel harder to open, but more stable once open (Brillantes et al., 1994). Thus, in the absence of FKBP12, the RyR was found to exhibit a number of subconductance states, but in the presence of FKBP12, these subconductance states were suppressed. Thus, FKBP12 appears to improve cooperativity among the four RyR subunits, resulting in stable full conductance channels (Brillantes et al., 1994). It was suggested that the subconductance states often seen in experiments with the purified RyR could correspond to partial loss of FKBP12. Excess FKBP12 has been reported to cause complete blockage of the RyR (Chen et al., 1994).

The molecular basis for the effects of FKBP12 are unclear. It has been suggested that the *cis–trans* peptidyl prolyl isomerase activity of FKBP12 is required for its effects, perhaps related to the presence of a proline residue (Pro-4641) near the putative M2 helix in the model of Takeshima et al. (1989). The site of binding of FKBP12 on the RyR has now been identified using cryo-electron microscopy as domain 3, near to domain 9 (Wagenknecht et al., 1996).

XI. MALIGNANT HYPERTHERMIA

Malignant hyperthermia (MH) is a syndrome in which halogenated anesthetics and other stresses cause skeletal muscle contracture and increased muscle metabolism. The mutation responsible for the syndrome in pig has been identified in the *ryr 1* gene and corresponds to the substitution Arg-615 to Cys-615 (Fujii et al., 1991). The homologous mutation in *ryr 1* has been identified in several families of human MH (Gillard et al., 1991) and linkage of human MH to the *ryr 1* gene on chromosome 19 has been demonstrated in a number of other pedigrees (Brum et al., 1987; MacLennan et al., 1990). SR isolated from MH-susceptible pig muscle shows an increased rate of Ca^{2+} efflux and altered kinetics of ryanodine binding (Fill et al., 1990; Mickelson et al., 1990; Louis et al., 1992). Single-channel recordings show that the mutated RyR is less sensitive to the inhibitory effects of high Ca^{2+} (Fill et al., 1990; Shomer et al., 1992), but the conductance of the channel and its ion selectivity are unaltered (Shomer et al., 1994). These observations suggest that Arg-615 is unlikely to be close to the conduction pathway in the channel, for otherwise replacement of the positively charged Arg by a neutral Cys would have been expected to have a significant effect.

In humans, central core disease, an inherited myopathy, has been shown to be closely associated with MH and to be tightly linked to *ryr-1*. Six MH susceptibility/central core disease mutations have been identified close to the pig mutation

(Arg-614 to Cys). Mutations have also been identified at Cys-163 to Arg, Gly-248 to Arg, Met-403 to Ile, Gly-2433 to Arg, and His-2434 to Arg (Quane et al., 1994a, 1994b; Phillips et al., 1994; Moroni et al., 1995).

REFERENCES

Abramson, J. J., Buck, E., Salama, G., Casida, J. E., & Pessah, I. N. (1988). Mechanism of anthraquinone-induced calcium release from skeletal muscle sarcoplasmic reticulum. J. Biol. Chem. 263, 18750–18758.

Airey, J. A., Grinsell, M. M., Jones, L. R., Sutko, J. L., & Witcher, D. R. (1993). Crooked Neck Dwarf (cn) chicken skeletal muscle cells in low density cultures fail to express normal α ryanodine receptor and exhibit a partial mutant phenotype. Dev. Dynam. 197, 189–202.

Alderson, B. H., & Feher, J. J. (1987). The interaction of calcium and ryanodine with cardiac sarcoplasmic reticulum. Biochim. Biophys. Acta 900, 221–229.

Andersen, K., Lai, F. A., Liu, Q. Y., Rousseau, E., Erickson, H. P., & Meissner, G. (1989). Structural and functional characterization of the purified cardiac ryanodine receptor/Ca^{2+} release channel complex. J. Biol. Chem. 264, 1329–1335.

Armstrong, C. M. (1992). Voltage-dependent ion channels and their gating. Physiol. Rev. 72, S5–S14.

Ashley, R. H., & Williams, A. J. (1990). Divalent cation activation and inhibition of single calcium release channels from sheep cardiac sarcoplasmic reticulum. J. Gen. Physiol. 95, 981–1005.

Bean, B. P. (1984). Nitrendipine block of cardiac calcium channels: High-affinity binding to the inactivated state. Proc. Natl. Acad. Sci. USA 81, 6388–6392.

Block, B. A., Imagawa, T., Campbell, K. P., & Franzini-Armstrong, C. (1988). Structural evidence for direct interaction between the molecular components of the transverse tubule/sarcoplasmic reticulum junction in skeletal muscle. J. Cell Biol. 107, 2587–2600.

Brandt, N. R., Caswell, A. H., Brandt, T., Brew, K., & Mellgren, R. L. (1992). Mapping of the calpain proteolysis products of the junctional foot protein of the skeletal muscle triad junction. J. Membrane Biol. 127, 35–47.

Brillantes, A. M. B., Ondrias, K., Scott, A., Kobrinsky, E., Ondriasova, E., Moschella, M. C., Jayaraman, T., Landers, M., Ehrlich, B. E., & Marks, A. R. (1994). Stabilization of calcium release channel (ryanodine receptor) function by FK506-binding protein. Cell 77, 513–523.

Brum, G., Stefani, E., & Rios, E. (1987). Simultaneous measurements of Ca^{2+} currents and intracellular Ca^{2+} concentrations in single skeletal muscle fibres of the frog. Can. J. Physiol. Pharmacol. 65, 681–685.

Brum, G., Rios, E., & Stefani, E. (1988). Effects of extracellular calcium on calcium movements of excitation–contraction coupling in frog skeletal muscle fibers. J. Physiol. 398, 441–473.

Buck, E., Zimanyi, I., Abramson, J. J., & Pessah, I. N. (1992). Ryanodine stabilizes multiple conformational states of the skeletal muscle calcium release channel. J. Biol. Chem. 267, 23560–23567.

Callaway, C., Seryshev, A., Wang, J. P., Slavik, K. J., Needleman, D. H., Cantu, C., Wu, Y. L., Jayaraman, T., Marks, A. R., & Hamilton, S. L. (1994). Localization of the high and low affinity [^{3}H]ryanodine binding sites on the skeletal muscle Ca^{2+} release channel. J. Biol. Chem. 269, 15876–15884.

Calviello, G., & Chiesi, M. (1989). Rapid kinetic analysis of the calcium-release channels of skeletal muscle sarcoplasmic reticulum: The effects of inhibitors. Biochemistry 28, 1301–1306.

Carroll, S., Skarmeta, J. G., Yu, X., Collins, K. D., & Inesi, G. (1991). Interdependence of ryanodine binding, oligomeric receptor interactions, and Ca^{2+} release regulation in junctional sarcoplasmic reticulum. Arch. Biochem. Biophys. 290, 239–247.

Caswell, A. H., Brandt, N. R., Brunschwig, J. P., & Purkerson, S. (1991). Localization and partial characterization of the oligomeric disulfide-linked molecular weight 95,000 protein (triadin) which binds the ryanodine and dihydropyridine receptors in skeletal muscle triadic vesicles. Biochemistry 30, 7507–7513.

Chen, S. R. W., & MacLennan, D. H. (1994). Identification of calmodulin-, Ca^{2+}-, and ruthenium red-binding domains in the Ca^{2+} release channel (ryanodine receptor) of rabbit skeletal muscle sarcoplasmic reticulum. J. Biol. Chem. 269, 22698–22704.

Chen, S. R. W., Zhang, L., & MacLennan, D. H. (1992). Characterization of a Ca^{2+} binding and regulatory site in the Ca^{2+} release channel (ryanodine receptor) of rabbit skeletal muscle sarcoplasmic reticulum. J. Biol. Chem. 267, 23318–23326.

Chen, S. R. W., Airey, J. A., & MacLennan, D. H. (1993). Positioning of major tryptic fragments in the Ca^{2+} release channel (ryanodine receptor) resulting from partial digestion of rabbit skeletal muscle sarcoplasmic reticulum. J. Biol. Chem. 268, 22642–22649.

Chen, S. R. W., Zhang, L., & MacLennan, D. H. (1994). Asymmetrical blockade of the Ca^{2+} release channel (ryanodine receptor) by 12-kDa FK506 binding protein. Proc. Natl. Acad. Sci. USA 91, 11953–11957.

Cheng, H., Fill, M., Valdivia, H., & Lederer, W. J. (1995). Models of Ca^{2+} release channel adaptation. Science 267, 2009–2010.

Chu, A., Diaz Munoz, M., Hawkes, M. J., Brush, K., & Hamilton, S. L. (1990a). Ryanodine as a probe for the functional state of the skeletal muscle sarcoplasmic reticulum calcium release channel. Mol. Pharmacol. 37, 735–741.

Chu, A., Sumbilla, C., Inesi, G., Jay, S. D., & Campbell, K. P. (1990b). Specific association of calmodulin-dependent protein kinase and related substrates with the junctional sarcoplasmic reticulum of skeletal muscle. Biochemistry 29, 5899–5905.

Chu, A., Fill, M., Stefani, E., & Entman, M. L. (1993). Cytoplasmic Ca^{2+} does not inhibit the cardiac muscle sarcoplasmic reticulum ryanodine receptor Ca^{2+} channel, although Ca^{2+}-induced Ca^{2+} inactivation of Ca^{2+} release is observed in native vesicles. J. Membr. Biol. 135, 49–59.

Cifuentes, M. E., Ronjat, M., & Ikemoto, N. (1989). Polylysine induces a rapid Ca^{2+} release from sarcoplasmic reticulum vesicles by mediation of its binding to the foot protein. Arch. Biochem. Biophys. 273, 554–561.

Collins, J. H. (1991). Sequence analysis of the ryanodine receptor—possible association with a 12 kD, FK506-binding immunophilin/protein kinase-C inhibitor. Biochem. Biophys. Res. Commun. 178, 1288–1290.

Connelly, T., Ahern, C., Sukhareva, M., & Coronado, R. (1994). Removal of Mg^{2+} inhibition of cardiac ryanodine receptor by palmitoyl coenzyme A. FEBS Letts. 352, 285–290.

Coronado, R., Kawano, S., Lee, C. J., Valdivia, C., & Valdivia, H. H. (1992). Planar bilayer recording of ryanodine receptors of sarcoplasmic reticulum. Methods Enzymol. 207, 699–707.

Coronado, R., Morrissette, J., Sukhareva, M., & Vaughan, D. M. (1994). Structure and function of ryanodine receptors. Am. J. Physiol. 266, C1485–C1504.

Csernoch, L., Jacquemond, V., & Schneider, M. F. (1993). Microinjection of strong calcium buffers suppresses the peak of calcium release during depolarization in frog skeletal muscle fibres. J. Gen. Physiol. 101, 297–333.

Dettbarn, C., & Palade, P. (1993). Arachidonic acid-induced Ca^{2+} release from isolated sarcoplasmic reticulum. Biochem. Pharmacol. 45, 1301–1309.

Dettbarn, C. A., Betto, R., Salviati, G., Palade, P., Jenkins, G. M., & Sabbadini, R. A. (1994). Modulation of cardiac sarcoplasmic reticulum ryanodine receptor by sphingosine. J. Mol. Cell Cardiol. 26, 229–242.

Dumonteil, E., Herre, H., & Meissner, G. (1994). Effects of palmitoyl carnitine and related metabolites on the avian Ca^{2+}-ATPase and Ca^{2+} release channel. J. Physiol. 479, 29–39.

Elhayek, R., Valdivia, C., Valdivia, H. H., Hogan, K., & Coronado, R. (1993). Activation of the Ca^{2+}-release channel of skeletal muscle sarcoplasmic reticulum by palmitoyl carnitine. Biophys. J. 65, 779–789.

Fill, M., Coronado, R., Mickelson, J. R., Vilven, J., Ma, J. J., Jacobson, B. A., & Louis, C. F. (1990). Abnormal ryanodine receptor channels in malignant hyperthermia. Biophys. J. 57, 471–475.

Fill, M., Stefani, E., & Nelson, T. E. (1991). Abnormal human sarcoplasmic reticulum Ca^{2+} release channels in malignant hyperthermic skeletal muscle. Biophys. J. 59, 1085–1090.

Fleischer, S., & Inui, M. (1989). Biochemistry and biophysics of excitation–contraction coupling. Annu. Rev. Biophys. Bioeng. 18, 333–364.

Franzini-Armstrong, C., & Jorgensen, A. O. (1994). Structure and development of E–C coupling units in skeletal muscle. Annu. Rev. Physiol. 56, 509–534.

Fruen, B. R., Mickelson, J. R., Shomer, N. H., Roghair, T. J., & Louis, C. F. (1994). Regulation of the sarcoplasmic reticulum ryanodine receptor by inorganic phosphate. J. Biol. Chem. 269, 192–198.

Fujii, J., Otsu, K., Zorzato, F., Deleon, S., Khanna, V. K., Weiler, J. E., Obrien, P. J., & MacLennan, D. H. (1991). Identification of a mutation in porcine ryanodine receptor associated with malignant hyperthermia. Science 253, 448–451.

Furuichi, T., Kohda, K., Miyawaki, A., & Mikoshiba, K. (1994). Intracellular channels. Curr. Opinion Neurobiol. 4, 294–303.

Galione, A., & White, A. (1994). Ca^{2+} release induced by cyclic ADP-ribose. Trends Cell Biol. 4, 431–436.

Giannini, G., Clementi, E., Ceci, R., Marziali, G., & Sorrentino, V. (1992). Expression of a ryanodine receptor-Ca^{2+} channel that is regulated by TGF-β. Science 257, 91–94.

Gilchrist, J. S. C., Belcastro, A. N., & Katz, S. (1992). Intraluminal Ca^{2+} dependence of Ca^{2+} and ryanodine- mediated regulation of skeletal muscle sarcoplasmic reticulum Ca^{2+} release. J. Biol. Chem. 267, 20850–20856.

Gillard, E. F., Otsu, K., Fujii, J., Khanna, V. K., Deleon, S., Derdemezi, J., Britt, B. A., Duff, C. L., Worton, R. G., & MacLennan, D. H. (1991). A substitution of cysteine for arginine-614 in the ryanodine receptor is potentially causative of human malignant hyperthermia. Genomics 11, 751–755.

Gould, G. W., Colyer, J., East, J. M., & Lee, A. G. (1987a). Silver ions trigger Ca^{2+} release by interaction with the (Ca^{2+}-Mg^{2+})-ATPase in reconstituted systems. J. Biol. Chem. 262, 7676–7679.

Gould, G. W., McWhirter, J. M., East, J. M., & Lee, A. G. (1987b). A fast passive Ca^{2+} efflux mediated by the (Ca^{2+} + Mg^{2+})-ATPase in reconstituted vesicles. Biochim. Biophys. Acta 904, 45–54.

Grunwald, R., & Meissner, G. (1995). Lumenal sites and C terminus accessibility of the skeletal muscle calcium release channel (ryanodine receptor). J. Biol. Chem. 270 11338–11347.

Grunwald, R., Sealock, R., & Meissner, G. (1993). Membrane topology of the rabbit skeletal muscle ryanodine receptor as determined by site directed antibodies. Biophys. J. 64, A152

Guo, W., & Campbell, K. P. (1995). Association of triadin with the ryanodine receptor and calsequestrin in the lumen of the sarcoplasmic reticulum. J. Biol. Chem. 270, 9027–9030.

Hain, J., Nath, S., Mayrleitner, M., Fleischer, S., & Schindler, H. (1994). Phosphorylation modulates the function of the calcium release channel of sarcoplasmic reticulum from skeletal muscle. Biophys. J. 67, 1823–1833.

Hakamata, Y., Nakai, J., Takeshima, H., & Imoto, K. (1992). Primary structure and distribution of a novel ryanodine receptor/calcium release channel from rabbit brain. FEBS Letts. 312, 229–235.

Hannon, J. D., Lee, N. K. M., Yandong, C., & Blinks, J. R. (1992). Inositol triphosphate ($InsP_3$) causes contraction in skeletal muscle only under artificial conditions: Evidence that Ca^{2+} release can result from depolarization of T-tubules. J. Musc. Res. Cell Motil. 13, 447–456.

Hasan, G., & Rosbash, M. (1992). *Drosophila* homologs of two mammalian intracellular Ca^{2+}-release channels: Identification and expression patterns of the inositol 1,4,5-triphosphate and the ryanodine receptor genes. Dev. Biol. 116, 967–975.

Hasselbach, W., & Migala, A. (1988). Interaction of ryanodine with the calcium releasing system of sarcoplasmic reticulum vesicles. Z. Naturforsch. [C] 43, 140–148.

Herrmann-Frank, A., Darling, E., & Meissner, G. (1991). Functional characterization of the Ca^{2+}-gated Ca^{2+}-release channel of vascular smooth muscle sarcoplasmic reticulum. Pflugers Arch. 418, 353–359.

Holmberg, S. R. M., & Williams, A. J. (1990). The cardiac sarcoplasmic reticulum calcium-release channel: Modulation of ryanodine binding and single-channel activity. Biochim. Biophys. Acta 1022, 187–193.

Humerickhouse, R. A., Besch, H. R., Gerzon, K., Ruest, L., Sutko, J. L., & Emmick, J. T. (1993). Differential activating and deactivating effects of natural ryanodine congeners on the calcium release channel of sarcoplasmic reticulum—evidence for separation of effects at functionally distinct sites. Mol. Pharmacol. 44, 412–421.

Imagawa, T., Smith, J. S., Coronado, R., & Campbell, K. P. (1987). Purified ryanodine receptor from skeletal muscle sarcoplasmic reticulum is the Ca^{2+}-permeable pore of the calcium release channel. J. Biol. Chem. 262, 16636–16643.

Inui, M., Saito, A., & Fleischer, S. (1987a). Isolation of the ryanodine receptor from cardiac sarcoplasmic reticulum and identity with the feet structures. J. Biol. Chem. 262, 15637–15642.

Inui, M., Saito, A., & Fleischer, S. (1987b). Purification of the ryanodine receptor and identity with feet structures of junctional terminal cisternae of sarcoplasmic reticulum from fast skeletal muscle. J. Biol. Chem. 262, 1740–1747.

Jay, S. D., Sharp, A. H., Kahl, S. D., Vedvick, T. S., Harpold, M. M., & Campbell, K. P. (1991). Structural characterization of the dihydropyridine-sensitive calcium channel α2-subunit and the associated δ peptides J. Biol. Chem. 266, 3287–3293.

Jayaraman, T., Brillantes, A. M., Timerman, A. P., Fleischer, S., Erdjumentbromage, H., Tempst, P., & Marks, A. R. (1992). FK506 binding protein associated with the calcium release channel (ryanodine receptor). J. Biol. Chem. 267, 9474–9477.

Jorgensen, A. O., Shen, A. C.-Y., Arnold, W., McPherson, P. S., & Campbell, K. P. (1993). The Ca^{2+} release channel/ryanodine receptor is localized in junctional and corbular sarcoplasmic reticulum in cardiac muscle. J. Cell Biol. 120, 969–980.

Kasai, M., & Kawasaki, T. (1993). Effects of ryanodine on permeability of choline and glucose through calcium channels in sarcoplasmic reticulum vesicles. J. Biochem. (Tokyo) 113, 327–333.

Kasai, M., Kawasaki, T., & Yamamoto, K. (1992). Permeation of neutral molecules through calcium channel in sarcoplasmic reticulum vesicles. J. Biochem. (Tokyo) 112, 197–203.

Knudson, C. M., Stang, K. K., Moomaw, C. R., Slaughter, C., & Campbell, K. P. (1993). Primary structure and topological analysis of a skeletal muscle-specific junctional sarcoplasmic reticulum glycoprotein (triadin). J. Biol. Chem. 268, 12646–12654.

Kuwajima, G., Futatsugi, A., Niinobe, M., Nakanishi, S., & Mikoshiba, K. (1992). Two types of ryanodine receptors in mouse brain: Skeletal muscle type exclusively in Purkinje cells and cardiac muscle type in various neurons. Neuron 9, 133–1142.

Lai, F. A., Erickson, H. P., Rousseau, E., Liu, Q. Y., & Meissner, G. (1988). Purification and reconstitution of the calcium release channel from skeletal muscle. Nature 331, 313–319.

Lai, F. A., Misra, M., Xu, L., Smith, H. A., & Meissner, G. (1989). The ryanodine receptor-Ca^{2+} release channel complex of skeletal muscle sarcoplasmic reticulum. Evidence for a cooperatively coupled, negatively charged homotetramer. J. Biol. Chem. 264, 16776–16785.

Lai, F. A., Dent, M., Wickenden, C., Xu, L., Kumari, G., Misra, M., Lee, H. B., Sar, M., & Meissner, G. (1992a). Expression of a cardiac Ca^{2+} release channel isoform in mammalian brain. Biochem. J. 288, 553–564.

Lai, F. A., Liu, Q. Y., Xu, L., El-Hashem, A., Kramarcy, N. R., Sealock, R., & Meissner, G. (1992b). Amphibian ryanodine receptor isoforms are related to those of mammalian skeletal or cardiac muscle. Am. J. Physiol. 263, C365–C372.

Lattanzio, F. A., Schlatterer, R. G., Nicar, M., Campbell, K. P., & Sutko, J. L. (1987). The effects of ryanodine on passive calcium fluxes across sarcoplasmic reticulum membranes. J. Biol. Chem. 262, 2711–2718.

Ledbetter, M. W., Preiner, J. K., Louis, C. F., & Mickelson, J. R. (1994). Tissue distribution of ryanodine receptor isoforms and alleles determined by reverse transcription polymerase chain reaction. J. Biol. Chem. 269, 31544–31551.

Lee, H. B., Xu, L., & Meissner, G. (1994). Reconstitution of the skeletal muscle ryanodine receptor-Ca^{2+} release channel protein complex into proteoliposomes. J. Biol. Chem. 269, 13305–13312.

Lesh, R. E., Marks, A. R., Somlyo, A. V., Fleischer, S., & Somlyo, A. P. (1993). Anti-ryanodine receptor antibody binding sites in vascular and endocardial endothelium Circ. Res. 72, 481–488.

Lindsay, A. R., & Williams, A. J. (1991). Functional characterization of the ryanodine receptor purified from sheep cardiac muscle sarcoplasmic reticulum. Biochim. Biophys. Acta 1064, 89–102.

Lindsay, A. R. G., Tinker, A., & Williams, A. J. (1994). How does ryanodine modify ion handling in the sheep cardiac sarcoplasmic reticulum Ca^{2+}-release channel? J. Gen. Physiol. 104, 425–447.

Liu, Q. Y., Lai, F. A., Rousseau, E., Jones, R. V., & Meissner, G. (1989). Multiple conductance states of the purified calcium release channel complex from skeletal sarcoplasmic reticulum. Biophys. J. 55, 415–424.

Loesser, K. E., Castellani, L., & Franzini-Armstrong, C. (1992). Disposition of junctional feet in muscles of invertebrates. J. Muscle Res. Cell Motil. 13, 161–173.

Louis, C. F., Zualkernman, K., Roghair, T., & Mickelson, J. R. (1992). The effects of volatile anesthetics on calcium regulation by malignant hyperthermia susceptible sarcoplasmic reticulum. Anesthesiology 77, 114–125.

Lu, X. Y., Xu, L., & Meissner, G. (1994). Activation of the skeletal muscle calcium release channel by a cytoplasmic loop of the dihydropyridine receptor. J. Biol. Chem. 269, 6511–6516.

Ma, J. J., & Zhao, J. Y. (1994). Highly cooperative and hysteretic response of the skeletal muscle ryanodine receptor to changes in proton concentrations. Biophys. J. 67, 626–633.

Mackenzie, A. E., Korneluk, R. G., Zorzato, F., Fujii, J., Phillips, M., Iles, D., Wieringa, B., Leblond, S., Bailly, J., Willard, H. F., Duff, C., Worton, R. G., & MacLennan, D. H. (1990). The human ryanodine receptor gene—its mapping to 19q13.1. Placement in a chromosome-19 linkage group, and exclusion as the gene causing myotonic dystrophy. Amer. J. Human Gen. 46, 1082–1089.

MacLennan, D. H., Duff, C., Zorzato, F., Fujii, J., Phillips, M., Korneluk, R. G., Frodis, W., Britt, B. A., & Worton, R. G. (1990). Ryanodine receptor gene is a candidate for predisposition to malignant hyperthermia. Nature 343, 559–561.

Marks, A. R., Fleischer, S., & Tempst, P. (1990). Surface topography analysis of the ryanodine receptor/junctional channel complex based on proteolysis sensitivity mapping. J. Biol. Chem. 265, 13143–13149.

Martonosi, A. N. (1984). Mechanisms of Ca^{2+} release from sarcoplasmic reticulum of skeletal muscle. Physiol. Rev. 64, 1240–1320.

Marty, I., Robert, M., Villaz, M., Dejongh, K. S., Lai, Y., Catterall, W. A., & Ronjat, M. (1994a). Biochemical evidence for a complex involving dihydropyridine receptor and ryanodine receptor in triad junctions of skeletal muscle. Proc. Natl. Acad. Sci. USA 91, 2270–2274.

Marty, I., Villaz, M., Arlaud, G., Bally, I., & Ronjat, M. (1994b). Transmembrane orientation of the N-terminal and C-terminal ends of the ryanodine receptor in the sarcoplasmic reticulum of rabbit skeletal muscle. Biochem. J. 298, 743–749.

Mcgarry, S. J., & Williams, A. J. (1994). Adenosine discriminates between the caffeine and adenine nucleotide sites on the sheep cardiac sarcoplasmic reticulum calcium-release channel. J. Membrane. Biol. 137, 169–177.

McGrew, S. G., Wolleben, C., Siegl, P., Inui, M., & Fleischer, S. (1989). Positive cooperativity of ryanodine binding to the calcium release channel of sarcoplasmic reticulum from heart and skeletal muscle. Biochemistry 28, 1686–1691.

McPherson, P. S., & Campbell, K. P. (1993a). Characterization of the major brain form of the ryanodine receptor/Ca^{2+} release channel. J. Biol. Chem. 268, 19785–19790.

McPherson, P. S., & Campbell, K. P. (1993b). The ryanodine receptor/Ca^{2+} release channel. J. Biol. Chem. 268, 13765–13768.

McPherson, P. S., Kim, Y. K., Valdivia, H., Knudson, M., Takekura, H., Franzini-Armstrong, C., Coronado, R., & Campbell, K. P. (1991). The brain ryanodine receptor: A caffeine-sensitive calcium release channel. Neuron 7, 17–25.

McPherson, S. M., McPherson, P. S., Mathews, L., Campbell, K. P., & Longo, F. J. (1992). Cortical localization of a calcium release channel in sea urchin eggs. J. Cell Biol. 116, 1111–1121.
Meissner, G. (1984). Adenine nucleotide stimulation of Ca^{2+}-induced Ca^{2+} release in sarcoplasmic reticulum. J. Biol. Chem. 259, 2365–2374.
Meissner, G. (1986a). Evidence of a role for calmodulin in the regulation of calcium release from skeletal muscle sarcoplasmic reticulum. Biochemistry 25, 244–251.
Meissner, G. (1986b). Ryanodine activation and inhibition of the Ca^{2+} release channel of sarcoplasmic reticulum. J. Biol. Chem. 261, 6300–6306.
Meissner, G. (1994). Ryanodine receptor/Ca^{2+} release channels and their regulation by endogenous effectors. Annu. Rev. Physiol. 56, 485–508.
Meissner, G., Darling, E., & Eveleth, J. (1986). Kinetics of rapid Ca^{2+} release by sarcoplasmic reticulum: Effects of Ca^{2+}, Mg^{2+} and adenine nucleotides. Biochemistry 25, 236–244.
Menegazzi, P., Larini, F., Treves, S., Guerrini, R., Quadroni, M., & Zorzato, F. (1994). Identification and characterization of three calmodulin binding sites of the skeletal muscle ryanodine receptor. Biochemistry 33, 9078–9084.
Michalak, M., Dupraz, P., & Shoshan-Barmatz, V. (1988). Ryanodine binding to sarcoplasmic reticulum membranes; comparison between cardiac and skeletal muscle. Biochim. Biophys. Acta 939, 587–594.
Mickelson, J. R., Litterer, L. A., Jacobson, B. A., & Louis, C. F. (1990). Stimulations and inhibition of ^{3}H-ryanodine binding to sarcoplasmic reticulum from malignant hyperthermia susceptible pigs. Arch. Biochem. Biophys. 278, 251–257.
Mikami, A., Imoto, K., Tanabe, T., Niidome, T., Mori, Y., Takeshima, H., Narumiya, S., & Numa, S. (1989). Primary structure and functional expression of the cardiac dihydropyridine-sensitive calcium channel. Nature 340, 230–233.
Miledi, R., Parker, I., & Zhu, P. H. (1984). Extracellular ions and excitation–contraction coupling in frog twitch muscle fibers. J. Physiol. 351, 687–710.
Miller, R. J. (1992). Voltage-sensitive Ca^{2+} channels. J. Biol. Chem. 267, 1403–1406.
Moroni, I., Gonano, E. F., Comi, G. P., Tegazzin, V., Prelle, A., Bordoni, A., Bresolin, N., & Scarlato, G. (1995). Ryanodine receptor gene point mutation and malignant hyperthermia susceptibility. J. Neurol. 242, 127–133.
Moutin, M. J., & Dupont, Y. (1988). Rapid filtration studies of Ca^{2+} induced Ca^{2+} release from skeletal sarcoplasmic reticulum. Role of monovalent ions. J. Biol. Chem. 263, 4228–4235.
Murayama, T., & Ogawa, Y. (1992). Purification and characterization of two ryanodine-binding protein isoforms from sarcoplasmic reticulum of bull-frog skeletal muscle. J. Biochem. (Tokyo) 112, 514–522.
Nakai, J., Imagawa, T., Hakamat, Y., Shigekawa, M., Takeshima, H., & Numa, S. (1990). Primary structure and functional expression from cDNA of the cardiac ryanodine receptor/calcium release channel. FEBS Letts. 271, 169–177.
Nakayama, H., Taki, M., Striessnig, J., Glossmann, H., Catterall, W. A., & Kanaoka, Y. (1991). Identification of 1,4-dihydropyridine binding regions within the α_1 subunit of skeletal muscle Ca^{2+} channels by photoaffinity labeling with diazipine. Proc. Natl. Acad. Sci. USA 88, 9203–9207.
Nelson, T. E. (1987). Ryanodine: antithetical calcium channel effects in skeletal muscle sarcoplasmic reticulum. J. Pharmacol. Exp. Ther. 242, 56–61.
Numa, S., Tanabe, T., Takeshima, H., Mikami, A., Niidome, T., Nishimura, S., Adams, B. A., & Beam, K. G. (1990). Molecular basis of the functional heterogeneity of the muscarinic acetylcholine receptor. Cold Spring Harbor Symp. Quant. Biol. 45, 1–7.
Olivares, E. B., Tanksley, S. J., Airey, J. A., Beck, C. F., Ouyang, Y., Deerinck, T. J., Ellisman, M. H., & Sutko, J. L. (1991). Nonmammalian vertebrate skeletal muscles express two triad junctional foot protein isoforms. Biophys. J. 59, 1153–1163.
Orlova, E. V., Serysheva, I. I., Van Heel, M., Hamilton, S. L., & Chiu, W. (1996). 2 structural configurations of the skeletal-muscle calcium-release channel. Nature Struct. Biol. 3, 547–552.

Otsu, K., Willard, H. F., Khanna, V. K., Zorzato, F., Green, N. M., & MacLennan, D. H. (1990). Molecular cloning of cDNA encoding the Ca^{2+} release channel (ryanodine receptor) of rabbit cardiac muscle sarcoplasmic reticulum. J. Biol. Chem. 265, 13472–13483.

Percival, A. L., Williams, A. J., Kenyon, J. L., Grinsell, M. M., Airey, J. A., & Sutko, J. L. (1994). Chicken skeletal muscle ryanodine receptor isoforms: Ion channel properties. Biophys. J. 67, 1834–1850.

Perez-Reyes, E., Castellano, A., Kim, H. S., Bertrand, P., Baggstrom, E., Lacerda, A. E., Wei, X., & Birnbaumer, L. (1992). Induction of calcium currents by the expression of the α_1 subunit of the dihydropyridine receptor from skeletal muscle. J. Biol. Chem. 267, 1792–1797.

Pessah, I. N., & Zimanyi, I. (1991). Characterization of multiple [^{3}H]ryanodine binding sites on the Ca^{2+} release channel of sarcoplasmic reticulum from skeletal and cardiac muscle: Evidence for a sequential mechanism in ryanodine action. Mol. Pharmacol. 39, 679–689.

Pessah, I. N., Francini, A. O., Scales, D. J., Waterhouse, A. L., & Casida, J. E. (1986). Calcium–ryanodine receptor complex. Solubilization and partial characterization from skeletal muscle junctional sarcoplasmic reticulum vesicles. J. Biol. Chem. 261, 8643–8648.

Pessah, I. N., Stambuk, R. A., & Casida, J. E. (1987). Ca^{2+}-activated ryanodine binding; mechanisms of sensitivity and intensity modulation by Mg^{2+}, caffeine, and adenine nucleotides. Mol. Pharmacol. 31, 232–238.

Phillips, M. S., Khanna, V. K., Deleon, S., Frodis, W., Britt, B. A., & MacLennan, D. H. (1994). The substitution of Arg for Gly(2433) in the human skeletal muscle ryanodine receptor is associated with malignant hyperthermia. Hum. Mol. Genet. 3, 2181–2186.

Quane, K. A., Keating, K. E., Healy, J. M. S., Manning, B. M., Krivosichorber, R., Krivosic, I., Monnier, N., Lunardi, J., & Mccarthy, T. V. (1994a). Mutation screening of the RYR1 gene in malignant hyperthermia: Detection of a novel Tyr to Ser mutation in a pedigree with associated central cores. Genomics 23, 236–239.

Quane, K. A., Keating, K. E., Manning, B. M., Healy, J. M. S., Monsieurs, K., Heffron, J. J. A., Lehane, M., Heytens, L., Krivosichorber, R., Adnet, P., Ellis, F. R., Monnier, N., Lunardi, J., & McCarthy, T. V. (1994b). Detection of a novel common mutation in the ryanodine receptor gene in malignant hyperthermia—implications or diagnosis and heterogeneity studies. Hum. Mol. Genet. 3, 471–476.

Radermacher, M., Rao, V., Grassucci, R., Frank, J., Timerman, A. P., Fleischer, S., & Wagenknecht, T. (1994). Cryo-electron microscopy and three-dimensional reconstruction of the calcium release channel ryanodine receptor from skeletal muscle. J. Cell Biol. 127, 411–423.

Rardon, D. P., Cefali, D. C., Mitchell, R. D., Seiler, S. M., & Jones, L. R. (1989). High molecular weight proteins purified from cardiac junctional sarcoplasmic reticulum vesicles are ryanodine-sensitive calcium channels. Circ. Res. 64, 779–789.

Rardon, D. P., Cefali, D. C., Mitchell, R. D., Seiler, S. M., Hathaway, D. R., & Jones, L. R. (1990). Digestion of cardiac and skeletal muscle junctional sarcoplasmic reticulum vesicles with calpain II. Effects on the Ca^{2+} release channel. Circ. Res. 67, 84–96.

Rios, E., Pizzaro, G., & Stefani, E. (1992). Charge movement and the nature of signal transduction in skeletal. Annu. Rev. Physiol. 54, 109–133.

Rios, E., Karhanek, M., Ma, J., & Gonzalez, A. (1993). An allosteric model of the molecular interaction of excitation–contraction coupling in skeletal muscle. J. Gen. Physiol. 102, 449–482.

Romani, A., & Scarpa, A. (1992). Regulation of cell magnesium. Arch. Biochem. Biophys. 298, 1–12.

Sachs, F., Qin, F., & Palade, P. (1995). Models of Ca^{2+} release channel adaptation. Science 267, 2010–2011.

Schneider, M. F. (1994). Control of calcium release in functioning skeletal muscle fibers. Annu. Rev. Physiol. 56, 463–484.

Schneider, M. F., & Chandler, W. K. (1973). Voltage dependent charge movement in skeletal muscle: A possible step in excitation–contraction coupling. Nature 242, 244–246.

Seiler, S., Wegener, A. D., Whang, D. D., Hathaway, D. R., & Jones, L. R. (1984). High molecular weight proteins in cardiac and skeletal muscle junctional sarcoplasmic reticulum vesicles bind calmodulin, are phosphorylated and are degraded by Ca^{2+}-activated protease. J. Biol. Chem. 259, 8550–8557.

Seok, J. H., Xu, L., Kramarcy, N. R., Sealock, R., & Meissner, G. (1992). The 30S lobster skeletal muscle Ca^{2+} release channel (ryanodine receptor) has functional properties distinct from the mammalian channel proteins. J. Biol. Chem. 267, 15893–15901.

Serysheva, I. I., Orlova, E. V., Chiu, W., Sherman, M. B., Hamilton, S. L., & Van Heel, M. (1995). Electron cryomicroscopy and angular reconstitution used to visualize the skeletal muscle calcium-release channel. Nature Struct. Biol. 2, 18–24.

Sharp, A. H., McPherson, P. S., Dawson, T. M., Aoki, C., Campbell, K. P., & Snyder, S. H. (1993). Differential immunohistochemical localization of inositol 1,4,5- trisphosphate- and ryanodine-sensitive Ca^{2+} release channels in rat brain. J. Neurosci. 13, 3051–3063.

Shomer, N., Louis, C. F., Fill, M., Litterer, L. A., & Mickelson, J. R. (1992). Reconstitution of abnormalities in the malignant hyperthermia susceptible pig ryanodine receptor. Am. J. Physiol. 264, C125–C135.

Shomer, N. H., Mickelson, J. R., & Louis, C. F. (1994). Ion selectivity of porcine skeletal muscle Ca^{2+} release channels is unaffected by the Arg(615) to Cys(615) mutation. Biophys. J. 67, 641–646.

Shoshan-Barmatz, V., Weil, S., Meyer, H., Varsanyi, M., & Heilmeyer, L. M. G. (1994). Endogenous, Ca^{2+}-dependent cysteine-protease cleaves specifically the ryanodine receptor Ca^{2+} release channel in skeletal muscle. J. Membrane Biol. 42, 281–288.

Simon, B. J., & Hill, D. (1992). Charge movement and SR calcium release in frog skeletal muscle can be related by a Hodgkin–Huxley model with four gating particles. Biophys. J. 61, 1109–1116.

Sitsapesan, R., McGarry, S. J., & Williams, A. J. (1994). Cyclic ADP-Ribose competes with ATP for the adenine nucleotide binding site on the cardiac ryanodine receptor Ca^{2+}-release channel. Circ. Res. 75, 596–600.

Sitsapesan, R., McGarry, S. J., & Williams, A. J. (1995). Cyclic ADP-ribose, the ryanodine receptor and Ca^{2+} release. Trends. Pharm. Sci. 16, 386–391.

Smith, J. S., Coronado, R., & Meissner, G. (1986). Single channel measurements of the calcium release channel from skeletal muscle sarcoplasmic reticulum. J. Gen. Physiol. 88, 573–588.

Smith, J. S., Imagawa, T., Ma, J., Fill, M., Campbell, K. P., & Coronado, R. (1988). Purified ryanodine receptor from rabbit skeletal muscle is the calcium-release channel of sarcoplasmic reticulum. J. Gen. Physiol. 92, 1–26.

Smith, J. S., Rousseau, E., & Meissner, G. (1989). Calmodulin modulation of single sarcoplasmic reticulum Ca^{2+}-release channels from cardiac and skeletal muscle. Circ. Res. 64, 352–359.

Sommer, J. R., & Johnson, E. A. (1979). In: Handbook of Physiology Sec. 2 (Berne, R.M., ed.) pp. 113–186. Am. Physiol. Soc., Bethesda.

Stern, M. D., & Lakatta, E. G. (1992). Excitation–contraction coupling in the heart: The state of the question. FASEB J. 6, 3092–3100.

Strand, M. A., Louis, C. F., & Mickelson, J. R. (1993). Phosphorylation of the porcine skeletal and cardiac muscle sarcoplasmic reticulum ryanodine receptor. Biochim. Biophys. Acta 1175, 319–326.

Suko, J., Maurerfogy, I., Plank, B., Bertel, O., Wyskovsky, W., Hohenegger, M., & Hellmann, G. (1993). Phosphorylation of serine-2843 in ryanodine receptor-calcium release channel of skeletal muscle by cAMP-dependent, cGMP-dependent and cam-dependent protein kinase. Biochim. Biophys. Acta 1175, 193–206.

Sutko, J. L., Airey, J. A., Murakami, K., Beck, C. F., Tanksley, S., Takeda, M., Deerinck, T. J., & Ellisman, M. H. (1991). Expression of α-and β-foot protein soforms during embryonic development of avian skeletal muscle. J. Cell Biol. 113, 793–803.

Takasago, T., Imagawa, T., Furukawa, K., Ogurusu, T., & Shigekawa, M. (1991). Regulation of the cardiac ryanodine receptor by protein kinase-dependent phosphorylation. J. Biochem. Tokyo 109, 163–170.

Takeshima, H., Nishimura, S., Matsumoto, T., Ishida, H., Kangawa, K., Minamino, N., Matsuo, H., Ueda, M., Hanaoka, M., Hirose, T., & Numa, S. (1989). Primary structure and expression from complementary DNA of skeletal muscle ryanodine receptor. Nature 339, 439–445.

Takeshima, H., Nishimura, S., Nishi, M., Ikeda, M., & Sugimoto, T. (1993). A brain-specific transcript from the 3′-terminal region of the skeletal muscle ryanodine receptor gene. FEBS Letts. 322, 105–110.

Tanabe, T., Takeshima, H., Mikami, A., Flockerzi, V., Takahashi, H., Kangawa, K., Kojima, M., Matsuo, H., Hirose, T., & Numa, S. (1987). Primary structure of the receptor for calcium channel blockers from skelatal muscle. Nature 328, 313–318.

Tanabe, T., Beam, K. G., Adams, B. A., Niidome, T., & Numa, S. (1990). Regions of the skeletal muscle dihydropyridine receptor critical for excitation–contraction coupling. Nature 346, 567–569.

Timerman, A. P., Ogunbumni, E., Freund, E., Wiederrecht, G., Marks, A. R., & Fleischer, S. (1993). The calcium release channel of sarcoplasmic reticulum is modulated by FK-506-binding protein—dissociation and reconstitution of FKBP-12 to the calcium release channel of skeletal muscle sarcoplasmic reticulum. J. Biol. Chem. 268, 22992–22999.

Timerman, A. P., Jayaraman, T., Wiederrecht, G., Onoue, H., Marks, A. R., & Fleischer, S. (1994). The ryanodine receptor from canine heart sarcoplasmic reticulum is associated with a novel FK-506 binding protein. Biochem. Biophys. Res. Commun. 198, 701–706.

Tinker, A., & Williams, A. J. (1992). Divalent cation conduction in the ryanodine receptor channel of sheep cardiac muscle sarcoplasmic reticulum. J. Gen. Physiol. 100, 479–493.

Tinker, A., & Williams, A. J. (1993). Probing the structure of the conduction pathway of the sheep cardiac sarcoplasmic reticulum calcium-release channel with permeant and impermeant organic cations. J. Gen. Physiol. 102, 1107–1129.

Tinker, A., & Williams, A. J. (1995). Measuring the length of the pore of the sheep cardiac sarcoplasmic reticulum calcium-release channel using related trimethylammonium ions as molecular calipers. Biophys. J. 68, 111–120.

Tinker, A., Lindsay, A. R. G., & Williams, A. J. (1992a). Large tetra-alkyl ammonium cations produce a reduced conductance state in the sheep cardiac sarcoplasmic reticulum Ca^{2+} release channel. Biophys. J. 61, 1122–1132.

Tinker, A., Lindsay, A. R. G., & Williams, A. J. (1992b). Block of the sheep cardiac sarcoplasmic reticulum Ca^{2+}- release channel by tetra-alkyl ammonium cations. J. Membrane Biol. 127, 149–159.

Tu, Q., Velez, P., Brodwick, M., & Fill, M. (1994a). Streaming potentials reveal a short ryanodine-sensitive selectivity filter in cardiac Ca^{2+} release channel. Biophys. J. 67, 2280–2285.

Tu, Q., Velez, P., Cortesgutierrez, M., & Fill, M. (1994b). Surface charge potentiates conduction through the cardiac ryanodine receptor channel. J. Gen. Physiol. 103, 853–867.

Unwin, N. (1995). Acetylcholine receptor channel imaged in the open state. Nature 373, 37–43.

Valdivia, H. H., Kaplan, J. H., Ellisdavies, G. C. R., & Lederer, W. J. (1995). Rapid adaptation of cardiac ryanodine receptors: Modulation by Mg^{2+} and phosphorylation. Science 267, 1997–2000.

Wagenknecht, T., Berkowitz, J., Grassucci, R., Timerman, A. P., & Fleischer, S. (1994). Localization of calmodulin binding sites on the ryanodine receptor from skeletal muscle by electron microscopy. Biophys. J. 67, 2286–2295.

Wagenknecht, T., Grassucci, R., Berkowitz, J., Wiederrecht, G. J., Xin, H. B., & Fleischer, S. (1996). Cryoelectron microscopy resolves FK506-binding protein sites on the skeletal muscle ryanodine receptor. Biophys. J. 70, 1709–1715.

Walton, P. D., Airey, J. A., Sutko, J. L., Beck, C. F., Mignery, G. A., Sidhof, T. C., Deerinck, T. J., & Ellisman, M. H. (1991). Ryanodine and inositol trisphosphate receptors coexist in avian cerebellar Purkinje neurons. J. Cell Biol. 113, 1145–1157.

Wang, J. X., & Best, P. M. (1992). Inactivation of the sarcoplasmic reticulum calcium channel by protein kinase. Nature 359, 739–741.

Wang, J. P., Needleman, D. H., & Hamilton, S. L. (1993). Relationship of low affinity [^{3}H]ryanodine binding sites to high affinity sites on the skeletal muscle Ca^{2+} release channel. J. Biol. Chem. 268, 20974–20982.

Witcher, D. R., Kovacs, R. J., Schulman, H., Cefali, D. C., & Jones, L. R. (1991). Unique phosphorylation site on the cardiac ryanodine receptor regulates calcium channel activity. J. Biol. Chem. 266, 11144–11152.

Witcher, D. R., Strifler, B., & Jones, L. (1992). Cardiac specific phosphorylation site for multifunctional Ca^{2+}/calmodulin-dependent protein kinase is conserved in the brain ryanodine receptor. J. Biol. Chem. 267, 4963–4967.

Yang, H. C., Reedy, M. M., Burke, C. L., & Strasburg, G. M. (1994). Calmodulin interaction with the skeletal muscle sarcoplasmic reticulum calcium channel protein. Biochemistry 33, 518–525.

Yuan, S. H., Arnold, W., & Jorgensen, A. O. (1991). Biogenesis of transverse tubules and triads—immunolocalization of the 1,4-dihydropyridine receptor, Ts28, and the ryanodine receptor in rabbit skeletal muscle developing *in situ*. J. Cell Biol. 112, 289–301.

Zarka, A., & Shoshan-Barmatz, V. (1993). Characterization and photoaffinity labeling of the ATP binding site of the ryanodine receptor from skeletal muscle. Eur. J. Biochem. 213, 147–154.

Zimanyi, I., & Pessah, I. N. (1991). Comparison of <H-3>ryanodine receptors and Ca^{2+} release from rat cardiac and rabbit skeletal muscle sarcoplasmic reticulum. J. Pharmacol. Exp. Ther. 256, 938–946.

Zimanyi, I., Buck, E., Abramson, J. J., Mack, M. M., & Pessah, I. N. (1992). Ryanodine produces persistent inactivation of the Ca^{2+} release channel from skeletal muscle sarcoplasmic reticulum. Mol. Pharmacol. 42, 1049–1057.

Zorzato, F., Fujii, J., Otsu, K., Phillips, M., Green, N. M., Lai, F. A., Meissner, G., & MacLennan, D. H. (1990). Molecular cloning of cDNA encoding human and rabbit forms of the Ca^{2+} release channel (ryanodine receptor) of skeletal muscle arcoplasmic reticulum. J. Biol. Chem. 265, 2244–2256.

THE DYNAMIC NATURE OF GRAMICIDIN

Declan A. Doyle and B. A. Wallace

Biomembranes
Volume 6, pages 327–359.

ISBN: 1-55938-663-0.

I. INTRODUCTION

The plasma membranes of cells consist of a fluid bilayer interspersed with membrane proteins. The hydrophilic lipid head-groups in contact with the aqueous environment shield the lipid hydrocarbon chains which form the hydrophobic bilayer interior. This barrier provides an excellent means for containing cells and creating internal compartments. Cells must, however, for their proper functioning, whether it be cell-to-cell signaling, nerve impulse propagation, ion and solute uptake, or metabolite processing, allow charged particles to cross the membrane bilayer at some stage. Membrane proteins form the passageways which allow signal transduction and solute transfer across the bilayer.

Handling these proteins takes a great deal of care and attention due to their amphipathic nature. Separation is normally achieved using excess detergent to prevent protein aggregation, precipitation, and denaturation. In many cases, the detergent has to be removed and replaced with lipids that form membranes for the protein to function correctly (Cornelius, 1991). Biochemical and biophysical techniques are then applied to these reconstituted membrane proteins in order to elucidate their structure and function. For membrane channels, such studies have provided a wealth of information on the kinetics and specificity of ion transport, but in order to determine the exact mechanism of conductance and ion interaction, high resolution three-dimensional information is required either by X-ray crystallography (e.g., porin; Cowan et al., 1992), electron microscopy (e.g., bacteriorhodopsin; Henderson et al., 1990), or 2D-NMR (as in the case of gramicidin; Arseniev et al., 1986). All of the membrane proteins whose structures have been determined thus far are relatively abundant in their natural membranes (Michel, 1982; Butz et al., 1993), thus providing an easy and plentiful supply of material. For many ion transporting proteins, however, their numbers in membranes are very low (Hille, 1992) as they function very efficiently and so only require relatively few molecules for their desired effect. Hence, to date very little is known structurally about the interaction and transport of ions through this group of proteins. One way to bypass the difficulty of extracting and handling large protein/lipid complexes is to study the behavior of small polypeptides that form active and well-behaved channels in membranes. One such peptide which forms cation-selective channels in lipid bilayers is synthesized by *Bacillus brevis* and is called gramicidin.

II. THE PRIMARY SEQUENCE AND COMPOSITION OF GRAMICIDIN

Gramicidin is a small, hydrophobic polypeptide consisting of fifteen amino acids. An unusual feature of this peptide is the alternating L-,D-configuration of its amino acids, which is important for its function as we shall see later. The sequence of gramicidin A (Sarges and Witkop, 1965) is:

$$\text{HCO–L-Val}^{1}\text{–Gly}^{2}\text{–L-Ala}^{3}\text{–D-Leu}^{4}\text{–L-Ala}^{5}\text{–D-Val}^{6}\text{–L-Val}^{7}\text{–D-Val}^{8}\text{–L-}$$

$$\text{Trp}^{9}\text{–D-Leu}^{10}\text{–L-Trp}^{11}\text{–D-Leu}^{12}\text{–L-Trp}^{13}\text{–D-Leu}^{14}\text{–L-Trp}^{15}\text{–NHCH}_2\text{CH}_2\text{OH.}$$

Both termini are blocked, by a formyl group at the N-terminus and by an ethanolamine group at the C-terminus, hence the peptide is uncharged. Naturally occurring gramicidin, called gramicidin D, is a heterologous mixture of six species due to amino acid substitutions at positions 1 and 11. The three main types, designated gramicidin A (gA), gramicidin B (gB), and gramicidin C (gC), have tryptophan, phenylalanine, and tyrosine, respectively, at position 11. They are present in the approximate ratio of 80:5:15 (Gross and Witkop, 1965; Veatch et al., 1974; Weinstein et al., 1980). For each of these three main types, there are two subtypes depending on the residue present at position 1. A valine is present in 90% of the molecules within the gA form, designated [Val^1]gA, and the remaining 10% having a leucine residues (e.g., [Leu^1]gA).

III. THE VERSATILE NATURE OF THE GRAMICIDIN STRUCTURE

This topic was previously reviewed by Wallace (1990). The unusual L-,D-configuration of the amino acids means that gramicidin is unable to adopt the standard types of protein secondary structures such as α-helices, β-sheets, or turns. In addition its small size allows it to adopt many different conformations in various environments. In the organic solvent methanol, gramicidin was shown to exist as four distinct dimer conformations that differed from one another in helical sense and chain orientation (Veatch et al., 1974). This equilibrium mixture, which can be separated by chromatographic techniques and the individual conformers stabilized in dioxane, was also found to exist in other alcohols. Veatch and his colleagues characterized these by circular dichroism spectroscopy (CD), nuclear magnetic resonance spectroscopy (NMR), and infrared spectroscopy (IR) and proposed models to account for their experimental observations. They suggested that gramicidin could form either parallel (↑↑) or antiparallel (↑↓), left- or right-handed inter-twined double helices with the two monomers joined together in a β-sheet like hydrogen bonding pattern. Their predictions were later confirmed by Bystrov and Arseniev (1988) using 2D-NMR. They found that in ethanol, the short range proton–proton distances matched those calculated from four models: two parallel left-handed double helices that differ in chain overlap (species 1 and 2), one left-handed antiparallel double helix (species 3), and a right-handed antiparallel double helix (species 4). All four species have 5.6 residues per turn with the maximum number of 28 internal hydrogen bonds. In an ordinary, all L-amino acid β-sheet, alternating side-groups lie above and below the plane of the sheet. However the alternating L-,D-configuration of gramicidin places all the side-chains on one side of the sheet and when the two monomers spiral around one another to form a cylinder, all the hydrophobic side chains lie on the exterior of the cylinder and an empty central cavity is formed by the amide and carbonyl-C_α backbone groups. Such a model could allow an ion to cross a lipid bilayer by travelling through the central hydrophilic core while the hydrophobic side-chains stabilize the structure by interactions with the membrane. In this way, gramicidin could form a membrane pore even though it only consists of a dimer of two 15 amino acid monomers.

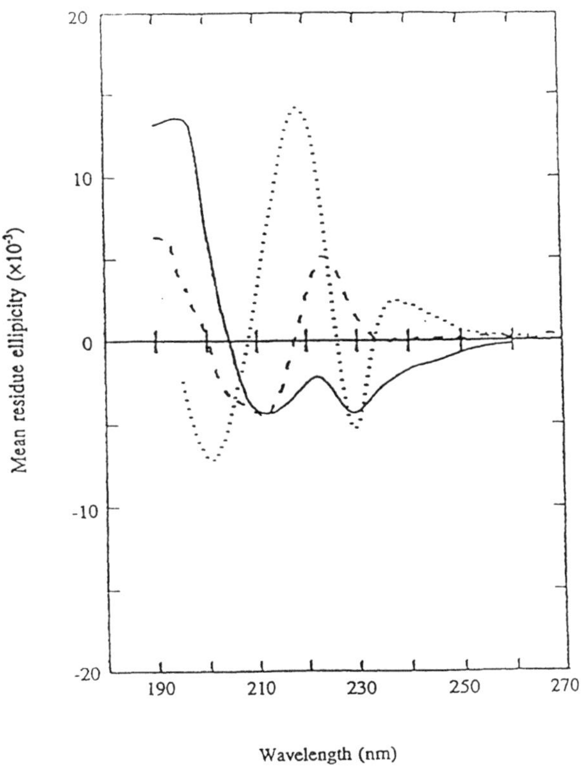

Figure 1. Circular dichroism spectra of gramicidin in lipid and solvent environments: in methanol (———); in trifluoroethanol (– – – –); in dimyristoylphosphatidylcholine vesicles (· · · · ·). (Reproduced from *The Journal of Cell Biology*, 1988, vol. 107, pp. 2587–2600 by copyright permission of The Rockefeller University Press.)

Figure 1 shows the CD spectra of gramicidin in different environments highlighting the versatility of its conformation. The spectrum of gramicidin in methanol, which is produced by the equilibrium mixture of the four species, is shown. This equilibrium can be driven to favor one of the conformers by altering the solvent in which gramicidin is dissolved. At a concentration of 80 mg/ml in an 80:20 mixture of benzene/ethanol, the equilibrium is driven to the left-handed inter-twined antiparallel double helix (species 3) as confirmed by NMR spectroscopy (Zhang et al., 1992).

An alternative model, that of a helical dimer was first proposed by Urry (1972), and, independently, by Ramachandran and Chandrasekaran (1972). This model had to have a central cavity with enough room to accommodate monovalent cations and so a monomer with 6.3 residues per turn, stabilized by twelve parallel intra-hydrogen bonds was proposed. However, the overall size would only be sufficient to span approximately half the thickness of a bilayer. Another monomer on the opposite side of the membrane was necessary to form a conducting unit. This monomer–monomer interaction would be stabilized by six intermolecular-hydrogen bonds and so the two N-terminal formyl groups were positioned in the center of the bilayer with the C-terminal ethanolamine groups on opposite ends of the channel at the membrane/water interface (Figure 2). This arrangement also clusters the tryptophan side-chains near the aqueous interface. Again the hydrogen bonding is in a β-sheet bonding pattern and hence this model has been designated the $\beta^{6.3}$-helix and has proven to be very similar to the dominant conformation found in lipid bilayers containing saturated fatty acids (Weinstein et al., 1980; Prosser and Davies, 1991; Hing and Schaefer, 1993; Ketchem et al., 1993). It has a CD spectrum (Figure 1) which is distinct from that produced by the conformers in methanol (Wallace et al., 1981).

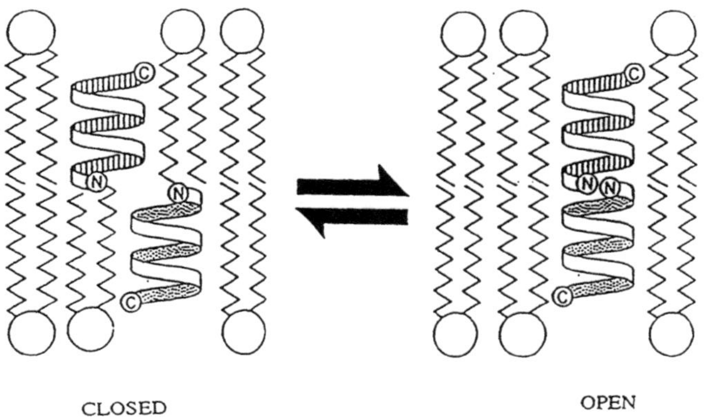

Figure 2. The opening and closing process for the channel form of gramicidin. Two monomers, one from each side of the lipid bilayer, join together at their N-termini and are stabilized by six intermolecular hydrogen bonds forming the "open" state. Breakage of these six hydrogen bonds results in channel closure. (Reproduced from *The Journal of Cell Biology,* 1993, vol. 268, pp. 13765–13768 by The Rockefeller University Press.)

The predominant form that gramicidin adopts in TFE is similar to the $\beta^{6.3}$-monomer, but there are differences as demonstrated by circular dichroism (Figure 1). In TFE, gramicidin appears to exist as three different conformations that rapidly interconvert between one another (Abdul-Manan and Hinton, 1994).

Gramicidin in the polar solvent, dimethylsulfoxide, converts between an ordered $\beta^{6.3}$-monomer and a disordered form (Hawkes et al., 1987). The internal mobility of the ordered state increases progressively towards the N-terminus, but there is still some disagreement as to the structure of the ordered form (Heitz et al., 1986). The polar solvent must be competing for the internal hydrogen bonds formed in the ordered structure, disrupting it and so converting the gramicidin to an unordered structure.

Two different gramicidin structures form in chloroform, depending on the gramicidin concentration used. At low concentrations (0.05 mg/ml) gramicidin adopts a structure similar to the $\beta^{6.3}$-monomer model, but at higher concentrations (>5 mg/ml) gramicidin acquires a hybrid dimer/monomer structure with a left-handed helical sense. In this model, the C-terminus forms a $\beta^{6.3}$-helix while the central section of the dimer is an antiparallel double helix with 5.6 residues per turn (Heitz et al., 1986). This may represent an intermediate form of gramicidin during the conversion from the double helix structure to the two $\beta^{6.3}$-monomers.

In many organic solvents, there also exists an equilibrium between the helical dimers and a monomer form that is in a random coil conformation (Barsukov et al., 1987; Braco et al., 1992). The time required to reach this equilibrium is dependent on the solvent used (e.g., several minutes in methanol to several days in tetrahydrafuran). By using the technique of HPLC with a hydrophobic column, separation based on size exclusion of the double-stranded dimers and monomers provides a quantitative measure of each type at any given time. It was found using a series of alcohols, that the longer the hydrocarbon chain, the slower the monomerization and the greater the proportion of double-stranded dimers at equilibrium. A similar type of equilibrium exists for gramicidin in the membrane environment, but this time it is the $\beta^{6.3}$ form that is favored over the double-helical dimers (Killian et al., 1988; Bano et al., 1991).

The method used for membrane preparation has an influence on the final structure of gramicidin in lipids. The usual vesicle/peptide preparation method involves co-dissolving lipid and peptide in an organic solvent followed by removal of the solvent by evaporation. Vesicles containing the peptide are formed by hydration and sonication. It was found that the final structure in membranes composed of saturated fatty acids was dictated by the organic solvent used (Masotti et al., 1980; Killian et al., 1988; LoGrasso et al., 1988; Bouchard and Auger, 1993). If TFE is used, the sample sonicated, or the temperature raised to 68°C for more than four hours, gramicidin in dimyristoyl phosphatidylcholine (DMPC) bilayers will be converted primarily to the $\beta^{6.3}$-conformation. The kinetics of this conversion were investigated using CD and HPLC and it was shown that the spectrum obtained at any given time during this equilibrium process is a linear combination of two reference

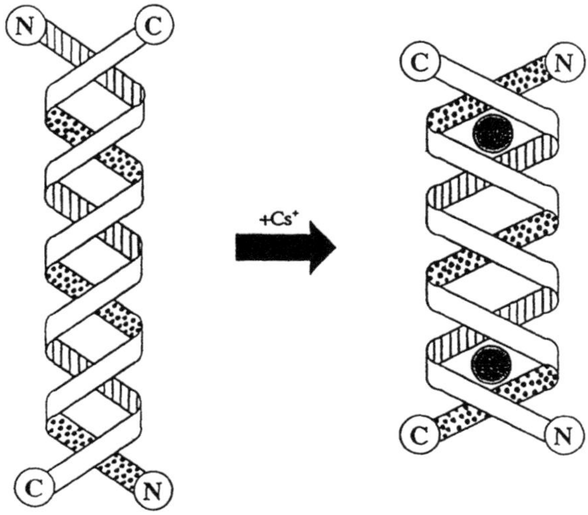

Figure 3. A schematic representation of the conversion of the antiparallel $\beta^{5.6}$ double-helical form to the "pore" form upon addition of cesium cations. These structures may represent the "closed" and "open" states, respectively, of the double helical dimers.

spectra: the double-stranded dimer and the $\beta^{6.3}$-monomer (Takeuchi et al., 1990; Bano et al., 1992).

The addition of ions to gramicidin in alcohols also has a drastic effect on its structure as observed by CD (Wallace, 1984, 1986). The central pore of the double helix must open up, allowing the ion to enter by breaking and reforming all 28 hydrogen bonds during this process (Figure 3). Larger ions, such as cesium, open the central cavity wider, producing a pore with a smaller pitch and a shorter helix-length compared to the ion-free form. The pitch of the helix is related to the magnitude of the peak at 228nm in the respective CD spectra, with the smaller peak height for the ion-free forms due to the larger helix pitches (Wallace, 1986), a result which has since been confirmed by X-ray studies (Langs, 1988; Wallace and Ravikumar, 1988).

For a long time, it was believed that the $\beta^{6.3}$-conformation was the main conducting form in all lipid membranes and that the double-stranded dimer was only a minor component within the membrane. This is true for the case of phospholipid bilayers formed from lipids composed of saturated fatty acid chains, but recently it has been shown that the final stable structure formed after equilibrium in bilayers composed of lipids with unsaturated fatty acids is a double helix even if TFE is used in the liposome preparation and hence the initial structure is that of the $\beta^{6.3}$-monomer form (Sychev et al., 1993). The important prerequisite for this conformation is the presence of two or more double bonds in the hydrocarbon chains of the phospholipids.

In summary, the influence of the surrounding environment determines the final structure that gramicidin will adopt. Peptide concentration, ions, temperature, and the type of fatty acid chains of the lipid have also been shown to play a role in driving gramicidin to one of a number of stable states, several of whose three-dimensional structures have been elucidated, either by X-ray crystallography or NMR spectroscopy. These structures will now be discussed in greater detail.

IV. THE ION-FREE DOUBLE-HELIX STRUCTURE

Gramicidin's size and flexibility produced many problems in attempts to solve its structure using crystallographic techniques (Wallace, 1992). Due to its flexibility, no isomorphous derivatives of the ion-free form were produced and, until recently, it was regarded as too large to solve by direct methods (Langs, 1993). Crystals grown from an ethanol/benzene solution of gramicidin yielded a structure using a modified direct-phasing procedure (Langs, 1988). The double-stranded dimer was found to be a left-handed, antiparallel (↑↓) β-helix with approximately 5.6 residues per turn (Figures 4a and 4b). The $\uparrow\downarrow\beta^{5.6}$-dimer is a thin cylinder, 31Å long and with an uneven central pore diameter with bulges and constrictions along the pore length (Smart et al., 1993) (Figure 4c). It was claimed that there was sufficient space to complex a K^+ ion inside one of the bulges, but no ions or even solvent molecules were present in the central pore and there is no continuous path through the center of sufficient diameter to allow an ion or solvent to traverse the structure. In this structure, the tryptophan side-chains lie mainly parallel to the helix axis. Another ion-free structure crystallized from methanol also formed a left-handed, antiparallel double helix (Langs et al., 1991) with the same helical dimensions as the ethanol structure. The main difference was the altered orientations of the tryptophan side-chains in this solvent. They lie mainly perpendicular to the helix axis allowing π-stacking interactions between neighboring tryptophans in the methanol crystal form and as a result, this helix has a more uniform profile along its length. From these results, it appears that crystal-packing forces can have some influence on the overall structure and the hole diameter. Alternatively, the different solvents used could be promoting differing dimer–dimer interactions and hence producing the different side-chain conformations.

The solution structure of ion-free gramicidin determined from two-dimensional NMR studies in a mixture of benzene/ethanol (80:20) was also a $\uparrow\downarrow\beta^{5.6}$-dimer (Zhang et al., 1992), but its backbone was found to be much more regular, probably reflecting the unconstrained nature of the large tryptophan side-chains in this environment. Although the solvent components were the same as in the crystallization experiment, the ratio of the two solvents were considerably different in the spectroscopic study so the two structures cannot be considered strictly comparable. Thus, the differences seen may arise due to different solvent environments or to different behaviors of the gramicidin in solution as compared to the solid state.

(a)

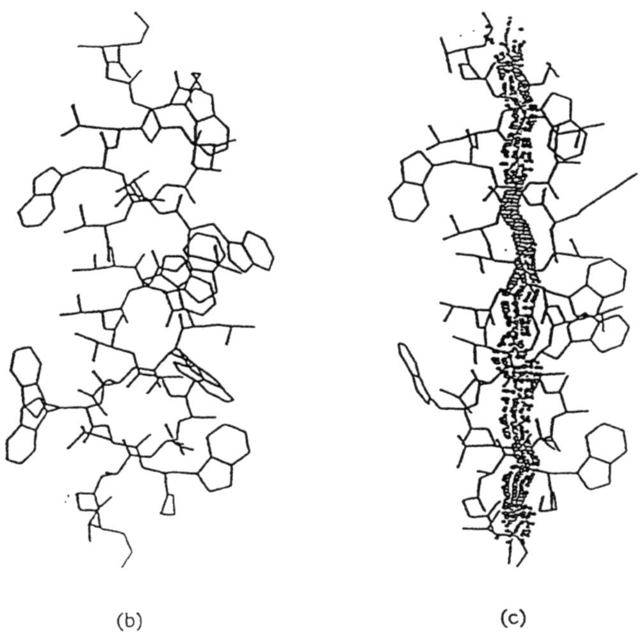

(b) (c)

Figure 4. The crystal structure of the left-handed, antiparallel $\beta^{5.6}$ double-helix looking down (**a**) and along (**b**) the helix axis (Langs, 1988). (**c**) Highlighting the inner pore diameter of the $\beta^{5.6}$-helix using the program HOLE (Smart et al., 1993) shows the non-uniform nature of the hole down the center, which is too constricted to accommodate water molecules or ions in its narrowest regions.

V. THE ION-COMPLEXED DOUBLE-HELIX ("PORE") STRUCTURE

At the same time that the uncomplexed crystal form was solved, a cesium chloride complex of gramicidin (Figure 5) was solved by crystallography using the technique

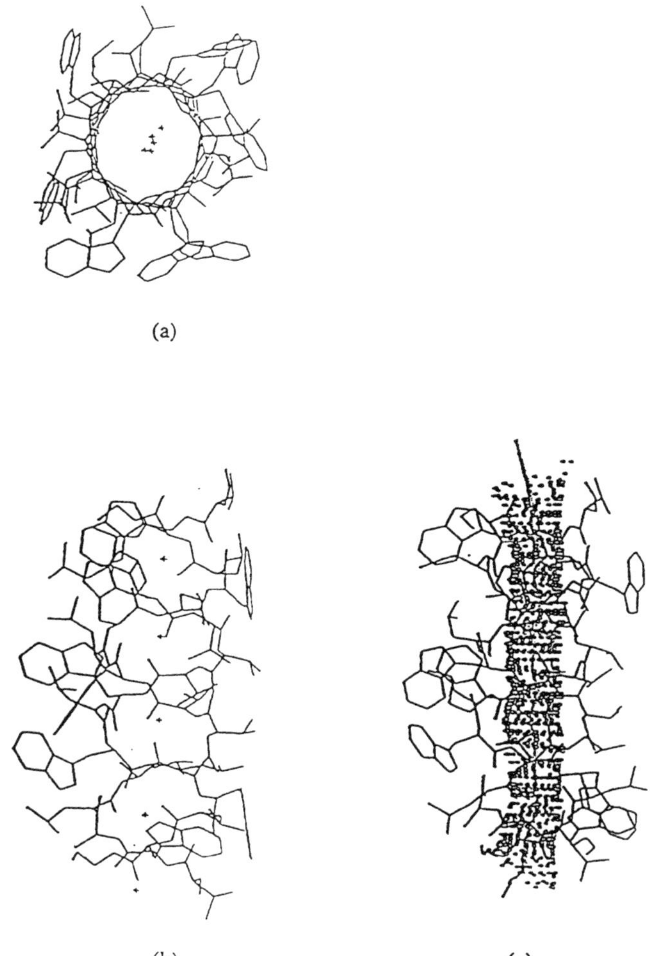

(a)

(b)

(c)

Figure 5. The crystal structure of the double helical gramicidin/CsCl pore looking down (**a**) and along (**b**) the helix axis (Wallace and Ravikumar, 1988). The crosses mark the positions of the ions which are roughly separated from one another by ~4Å along the center of the helix. (**c**) Mapping of the inner pore diameter displays the relatively uniform cylinder shape through which it is possible for ions to traverse the pore (Smart et al., 1993).

of single wavelength anomalous scattering (Wallace and Ravikumar, 1988; Wallace et al., 1990). In this structure, gramicidin is shorter (26Å in length), and has a wider lumen, reflecting a change in helix pitch and hydrogen-bonding pattern upon incorporation of ions inside the lumen (Figure 3). This form is designated the "pore" structure (Wallace, 1986; Wallace and Ravikumar, 1988; Wallace et al., 1990).

Again it is a left-handed antiparallel double helix held together by twenty eight hydrogen bonds. Unwinding the dimer to a linear arrangement shows that there is a three-residue overhang at each end of the sheet. This produces almost flat ends to the pore and with the tryptophan side-chains lying approximately parallel to the helix axis, a relatively symmetrical uniform cylinder is produced (Figure 5, 5a and 5b). It has 6.4 residues per turn and an average internal pore diameter of 4.9Å. Constrictions in the pore diameter occur close to the ion-binding sites and are followed by bulges (Smart et al., 1993; Figure 5c). This suggests that ions may travel through gramicidin pores by a peristaltic wave-like motion with the carbonyl groups tilting first towards the ion then away from it as it passes through the pore.

Two cesium ions and three chloride ions were found inside the pore structure. Each cesium ion is complexed with four backbone carbonyl groups (two from each monomer) and is positioned ~7.2Å from each end of the pore. As gramicidin ordinarily conducts only monovalent cations, the presence of chloride ions inside the pore was unexpected. Sung and Jordan (1987) did calculations which illustrated the main barrier to anion entry was at the mouth of the pore and if this obstacle could be overcome, anions should be quite stable inside the pore. The high salt concentrations used in the crystal growth procedure may have been the driving force that allowed anions to surmount this barrier. Other ion complexes of gramicidin have been grown (Koeppe et al., 1978; Kimball and Wallace, 1984; Wallace, 1990, 1992; Doyle and Wallace, 1994a), several of which have been shown to have cesium ions at different positions along the pore (Doyle and Wallace, 1994b). These other ion/gramicidin complexes offer us, for the first time, successive views of an ion passing from one binding site to another, which will be useful in simulation experiments to study the process of ion translocation through a transmembrane pore.

VI. THE $\beta^{6.3}$-HELICAL DIMER ("CHANNEL") STRUCTURE

Confirmation that in phospholipid bilayers with saturated fatty acid chains, gramicidin primarily forms N-terminal-to-N-terminal helical dimers came from the determination by NMR of the relative accessibility of ^{13}C- and ^{19}F-nuclei to localized paramagnetic probes at the N- and C-termini (Weinstein et al., 1979). Water-soluble probes enhanced the relaxation rates only at the C-termini while a lipid probe produced the same effect only at the N-termini, thus providing the evidence that the N-termini lie in the interior of the membrane and the C-termini are accessible at the membrane surface. In detergent complexes formed in sodium dodecyl sulphate (SDS), 2D-NMR studies detected a right-handed $\beta^{6.3}$-helix (Figures 6a and 6b) that was approximately 26Å long with an inner diameter of 4Å (Figure 6c). This has been designated the "channel" form (Arseniev et al., 1985, Bystrov et al., 1990). Even though the SDS environment appears to mimic the phospholipid bilayer, more evidence was required to confirm the correct structural handedness of this form in phospholipid vesicles. A left-handed helix had originally

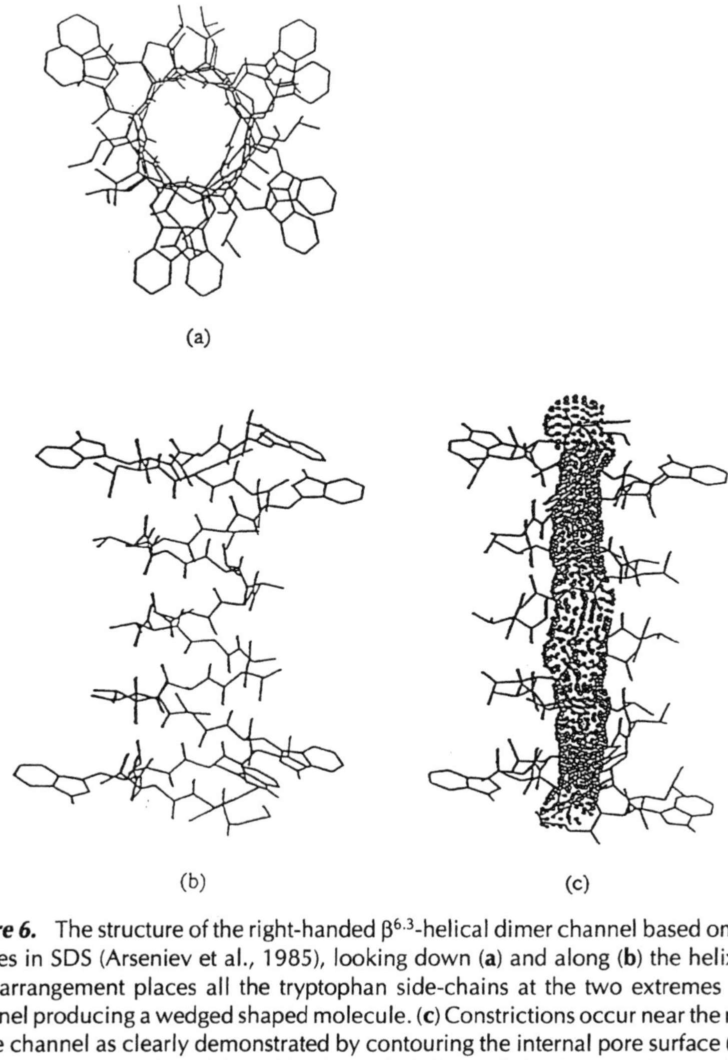

(a)

(b) (c)

Figure 6. The structure of the right-handed $\beta^{6.3}$-helical dimer channel based on NMR studies in SDS (Arseniev et al., 1985), looking down (**a**) and along (**b**) the helix axis. This arrangement places all the tryptophan side-chains at the two extremes of the channel producing a wedged shaped molecule. (**c**) Constrictions occur near the mouth of the channel as clearly demonstrated by contouring the internal pore surface (Smart et al., 1993). (Reproduced from *The Journal of General Physiology,* 1994, vol. 104, pp. 425–447 by copyright permission of The Rockefeller University Press.)

been suggested (Venkatachalam and Urry, 1983) based on CD spectroscopy, conductance measurements, and NMR methods using ion probes, but the single stranded, right-handed $\beta^{6.3}$-helical dimer was verified as the conformation of gramicidin in DMPC multilamellar dispersions (Prosser and Davies, 1991; Hing and Schaefer, 1993; Koeppe et al., 1992; Ketchem et al., 1993) with the two monomers forming the conducting unit being equivalent (Cifu et al., 1992). The tryptophan side-chains are situated near the lipid/solvent interface on both sides of the membrane. These tryptophans may stabilize the $\beta^{6.3}$-structure by electrostatic interactions with the charged lipid head-groups, so special attention has been paid to their relative orientations (Takeuchi et al., 1990; Hu et al., 1993; Koeppe et al., 1994). Other evidence from polarized infrared spectroscopy experiments (Nabedryk et al., 1982) and results from solid-state NMR experiments (Separovic et al., 1993) have indicated the helix axis to be essentially perpendicular to the membrane plane.

The structure (Arseniev et al., 1986) of gramicidin in SDS places the tryptophans closest to the N-terminus deepest in the membrane. This was confirmed for Trp^9 in phospholipid bilayers using Raman spectroscopy (Takeuchi et al., 1990), which also showed that Trp^{15}, the closest to the C-terminus, was always accessible to the surrounding water solvent. For Trp^{11} and Trp^{13}, the thickness of the membrane determined whether or not they were accessible to aqueous solvent. Also, the thicker the membrane, the greater the hydrophobic interaction between the tryptophan residues and the acyl-chains of the phospholipids. CD and fluorescence spectroscopy established that the tryptophan side-chains adopt a number of stable conformational states in lipid membranes (Woolley et al., 1992; Mukherjee and Chattopadhyay, 1994). NMR studies identified Trp^9 as the most mobile (Hu et al., 1993; Koeppe et al., 1994) whereas Trp^{11} movements were found to be restricted. It has been suggested that the indole N–H group of Trp^{11} forms hydrogen bonds with the ester carbonyl of the lipid-head groups, thus limiting its freedom of movement. All of the dipole moments of the indole rings would lie parallel to the channel axis. This preferred orientation of tryptophan residues at a lipid–water interface may be the driving force in the conversion from double helices to monomers as replacement of the tryptophan side-chains by non-polar residues stabilizes the double helical pore form in dioleoylphosphatidylcholine (DOPC) monolayers (Van Mau et al., 1994).

VII. ION INTERACTIONS WITH THE $\beta^{6.3}$-HELICAL DIMER

Unlike in the pore form, ion binding to the $\beta^{6.3}$-helix channel does not alter its overall structure as demonstrated by CD spectroscopy (Wallace et al., 1981) and X-ray diffraction analysis of gramicidin in liquid–crystalline bilayers (Katsaras et al., 1992). Even if the lipid bilayer thickness is increased, the helical pitch of the channel remains constant, but there is a decrease in the number of active dimers present at any one time (Wallace et al., 1981) which correlates with conductance measurements that show active channels with greatly decreased mean channel lifetimes (Hladky and Haydon, 1972). Therefore, ion interaction with the $\beta^{6.3}$-dimer must only involve small alterations in structure such as the tilting of carbonyl groups of the C_α backbone towards the center of the helix channel to complex with the ion. An equilibrium binding study of Tl^+ with gramicidin in DMPC vesicles using ^{205}Tl

NMR demonstrated that when fully occupied, two ions were present inside the channel (Shungu et al., 1986). This technique was also used to competitively displace bound Tl^+ with the group I metal cations to determine their relative strengths of binding to gramicidin (Hinton et al., 1986). The binding constants at 34°C ranged from $32M^{-1}$ to $54M^{-1}$ in the following order:

$$Li^+ < Na^+ < K^+ = Rb^+ = Cs^+.$$

However, there was no simple relationship found between how tightly the cation bound and its respective conductance through the channel.

The location of the Tl^+ ion in the channel was determined by X-ray diffraction of gramicidin in multibilayer membranes, to be either near the bottom of or below the first helix turn, at a distance of 9.6 (± 0.3)Å from the channel midpoint (Olah et al., 1991). In contrast, the divalent cation barium bound 13.0 (± 0.2)Å from the channel midpoint. This places it at the channel entrance where it would block monovalent cation transport through the central cavity. Other evidence that divalent cations bind and block the mouth of the channel was provided by 2D NMR spectroscopy of gramicidin in SDS (Bystrov et al., 1990). The bound divalent cation Mn^{2+} retained its hydration shell, giving it an effective radius of 6Å, which exceeds the inner diameter (~4Å) of the channel. The carbonyl groups of Leu^{10}, Leu^{12}, and Leu^{14} were considered to be the groups that interact with the divalent ions via the water molecules in the ion hydration shell (Golovanov et al., 1991).

Divalent cations are not permanently attached to the channel mouth, but rapidly exchange between the bound (blocked) state and the unbound (open) state. Therefore, the higher the divalent cation concentration around the channel entrance, the larger is the decrease in ion conductance (Golovanov et al., 1991). The interaction energy involved in divalent cation binding is insufficient to dehydrate the ion, but is sufficient to compensate for removal of the water molecules from monovalent cations and so allows them to enter and permeate the channel. The three leucine carbonyl groups which are situated at the ends of the channel have also been shown to play a role in the binding of Na^+ ions (Smith et al., 1990), which corresponds with the location of the monovalent cation-binding site derived from X-ray diffraction studies (Olah et al., 1991).

Gramicidin will transport uncharged atoms as demonstrated by a series of ^{129}Xe-NMR experiments (McKim and Hinton, 1994). These atoms are able to cross a bilayer in the absence of a channel, but transport is increased in the presence of gramicidin. Blocking the channel using divalent cations causes a corresponding drop in ^{129}Xe movement, however, inactivation of the channel to $^{23}Na^+$ transport by ultraviolet photolysis has little effect on ^{129}Xe ability to utilize the channel. Presumably, the inner channel diameter of gramicidin is the only restriction to entry of the uncharged ^{129}Xe atoms, whereas loss of dipole interactions of gramicidin due to photolysis would greatly hinder only cation entry, binding, and movement.

VIII. ION CONDUCTANCE AND ION BLOCK BEHAVIOR

This topic was previously reviewed by Andersen (1989) and Busath (1993). The current profiles for single channel events produced by gramicidin in lipid bilayers are step functions, as shown schematically in Figure 7 (Hladky and Haydon, 1972; Andersen, 1984). Interaction of two $\beta^{6.3}$-monomers to form the dimer creates a continuous passageway across a bilayer (Figure 2). This forms the "open" state of the gramicidin channel through which the ions travel, causing an upward deflection from the baseline on the current trace. The size of the current will depend on the applied potential and the concentration of ions in the bathing solution. Breaking the hydrogen bonds between the monomers closes the channel with the result that the current trace returns to the baseline. At the low gramicidin concentrations used in these experiments, there are considered to be no interactions between monomers on the same side of the bilayer or dimer–dimer interactions. Also the opening of a channel is essentially a chance event as the monomers are randomly distributed in the membrane (He et al., 1993). The two main features that characterize the channels in a particular lipid are the single-channel conductance (the step height of the current trace) and the mean channel lifetime (average channel duration). These values reflect the stresses placed on the monomer–monomer interactions, the

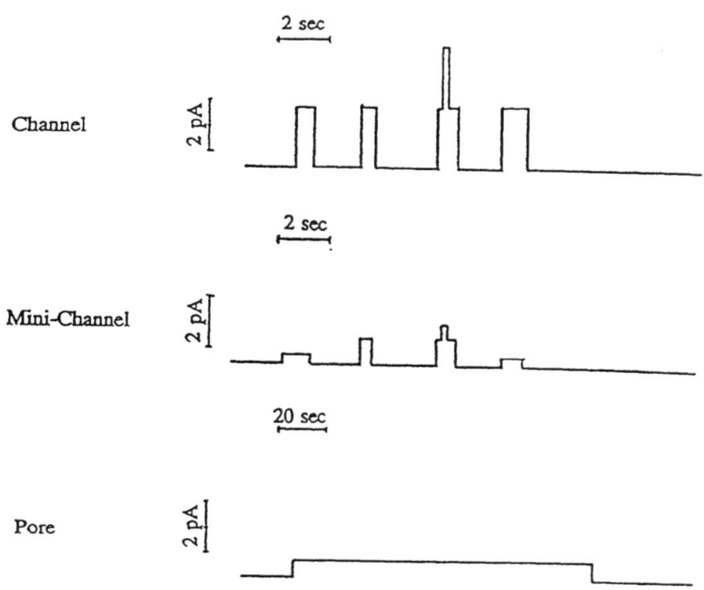

Figure 7. Schematic representations of different types of gramicidin single-channel events. The larger steps are due to the opening of more than one channel over the same period of time. Note the difference in time-scale for the pore form compared to the other types of channels.

influence on gramicidin of its surrounding environment, and if analogs are used, the altered behavior due to the substitutions.

For the majority of channels formed, their conductances fall into a narrow optimum conductance range (Figure 7, top). However, under certain conditions a significant number form "mini" channels that have lower single-channel conductances when compared with the majority of channels although their mean channel lifetimes are similar (Busath et al., 1987; Busath and Szabo, 1988a; Figure 7, middle). These "mini" channels dimerize in the normal fashion and are believed to result from minor conformational changes in the carbonyl groups at the channel entrance that cause a local constriction, enhancing cation binding with the resulting drop in conductance (Busath and Szabo, 1988b). Altering side-chain conformations to a slightly less energetically favorable position for ion transport would result in carbonyl-group misalignment and could be the cause of this behavior (Woolley and Wallace, 1992). As no "maxi" channels with increased conductance have been detected, the $\beta^{6.3}$-dimer is probably in a near optimum conformation for ion translocation and, hence, functions very efficiently.

Normally when the channel is in the open state, small current fluctuations occur producing noise on the current tracing called "shot" noise. Noise greater than "shot" noise also arises especially in the presence of the ions Cs^+, K^+, and Rb^+. It was postulated that this noise was due to fluctuating barriers at the mouth of the channel (Heinmann and Sigworth, 1990). Fluctuations happen preferentially when the channel is empty; when an ion binds, the normal 'high' conductance state is stabilized.

The selectivity of gramicidin from bi-ionic potential experiments is (Myers and Haydon, 1972):

$$H^+ >> NH_4^+ > Cs^+ > Rb^+ > K^+ > Na^+ > Li^+.$$

Ion conductance through gramicidin can be split into two separate mechanisms depending on the type of ion. The rate of translocation of protons is much faster than for the other ions due to the special property of water: if no ions are present inside the $\beta^{6.3}$-dimer then a continuous row of water molecules spans the bilayer through the channel. The proton does not have to physically move around water molecules to travel, but instead is able to "hop" from one water molecule to the next forming hydronium ions in the process (Myer and Haydon, 1972; Pullman, 1987; Deamer and Nichols, 1989; Akeson and Deamer, 1991). Using gramicidin as a model, the mechanism for proton transport in the F_0 subunit of the F_1F_0-ATP synthase could be accounted for by such a "hopping" theory. It was concluded that the rate limiting step for ATP synthesis was proton diffusion to the channel entrance (Akeson and Deamer, 1991; Deamer, 1992). The rate of translocation of other ions in gramicidin is slower, as they must first become partially dehydrated at the channel mouth and then move in single file (Finkelstein and Andersen, 1981), pushing and dragging water molecules along with them.

Enhancement of proton conductance can be achieved by the addition of weak acids, such as formic acid (Decker and Levitt, 1988). A kinetic model was proposed to explain this observation based on the equilibrium-binding of formic acid to the channel mouth. When formic acid is in the neutral state (HCOOH), it is able to bind to the channel mouth. Its proton is released and travels through the channel, while the anion form is repelled and dissociates. Binding of the neutral formic acid to the channel mouth also acts as a blocker and decreases potassium conductance. As for the F_1F_0 ATP synthase, the results supported the finding that the rate of proton conductance was limited by the speed of proton diffusion in the bulk solution to the channel mouth.

The earliest known gramicidin blockers were the divalent cations (Bamberg and Lauger, 1977; Bystrov et al., 1990). They differ in their ability to cause blockage with Ca^{2+} and Ba^{2+} exerting a much stronger block than Zn^{2+} and Mg^{2+}. In the mixed lipid system of DOPC/dioleoylphosphatidylglycerol (DOPG), calcium ions at low concentrations ($<10^{-6}$M) were shown to block Cs^+ conductance through gramicidin (Mittler-Neher and Knoll, 1993). However, increasing the Ca^{2+} ion concentration caused phase-separation with the result that two populations of conducting species formed which also differed in their stabilities. For charged membranes such as those made with phosphatidylserine (PS), part of the decrease in conductance through gramicidin can be attributed to screening of the charge on the lipid head-group by the Ca^{2+} ions. Hence, there is a lower local concentration of monovalent cations at the membrane surface and at the mouth of the channel with the resulting decrease in conductance (Gambale et al., 1987). Ion occupancy stabilizes gramicidin channels in mono-olein/decane bilayers and, as expected, this effect is reversed in the presence of blocking divalent ions (Ring, 1992).

An unusual blocker is the Tl^+ ion (Neher, 1975). Na^+ and K^+ conductances are markedly decreased in its presence, but the Tl^+ ion itself is able to permeate the channel.

Iminium ions cause a transient "flicker" block (Hemsley and Busath, 1991). While not affecting the overall channel structure, it appears that the iminium ions become momentarily lodged within the channel. Conductance returns to normal once the ion frees itself. As the diameters of these ions are larger than the reported inner-channel diameter, the gramicidin backbone must be able to flex, catching and releasing the iminium ion as it passes along. Formamidinium was the only iminium ion that did not produce "flicker" block, but was in fact highly conductive. Empirical energy function calculations using a left-handed $\beta^{6.3}$-dimer, found a residual energy barrier due to hydrogen bond disruption at the C-terminus for all the other iminium ions, but not for formamidinium (Turano et al., 1992). This may explain why formamidinium behaves differently from the other iminium ions.

Small organic cations are also able to travel through gramicidin channels and fall into the following selectivity sequence (Seoh and Busath, 1993a):

$$NH_4^+ > K^+ > \text{hydrazinium} > \text{formamidinium} > Na^+ > \text{methylammonium}.$$

The formamidinium ion was also found to stabilize the conducting dimer, but it was unclear how it produces this effect (Seoh and Busath, 1993b).

Two alternative models have been proposed to account for the mechanism of block (Heinemann and Sigworth, 1988). In both cases, the gramicidin channel was represented by a symmetrical single-file cylinder with two internal ion-binding sites, one at each end of the cylinder. Blocking of the channel occurred by either the blocker binding to two sites outside the channel or competing for the cation-binding sites within the channel, both of which yielded reasonable descriptions for the experimental data.

The local anesthetics dibucaine, butacaine, and tetracaine have a different inhibitory effect on channel activity (Bridal and Busath, 1992). They appear to reduce the number of active monomers within the bilayer, but it is still unknown as to whether this is due to alterations in the gramicidin structure or a change in the lipid bilayer properties.

Most of the conductance studies carried out on gramicidin to date employ conditions that favor the formation of the channel form (low gramicidin/lipid ratio and the use of phospholipids with saturated fatty acid chains), hence very little is know about the conductance properties of the double-helical pore form. The 28 intermolecular hydrogen bonds in the pore structure would ensure that, if open, it would remain so for a much greater period as compared to the channel form (Figure 7, bottom) as has been observed experimentally. The mechanism of ion translocation through the pore should be similar to that of the channel as the carbonyl- and amide-groups of the backbones in both structures would form the hydrophilic environment that would allow ions to progress through them. The double helical $\beta^{5.6}$-dimers may represent the closed state for this form (Figure 3).

In summary, gramicidin shows a variety of conducting states from increased background noise to the formation of "mini" channels. It also has a diverse range of block formation from transient "flicker" block to the more sustained block by the divalent cations and even by a decrease in monomers able to form active channels. Many of these properties appear to be related to the dynamic nature of the gramicidin channel itself, or the ion/polypeptide interactions especially at the mouth of the channel.

IX. THE STUDY OF GRAMICIDIN TRANSPORT THROUGH ALTERATIONS IN ITS PRIMARY SEQUENCE

Solid-phase peptide synthesis has been widely used to create synthetic analogs of gramicidin. Alterations at the C-terminus have been shown to have little effect on the general conductance properties of the channel and, therefore, do not alter the general $\beta^{6.3}$-structure as would be expected for the N-terminal to N-terminal dimer interaction (Morrow et al., 1979; Trudelle et al., 1987; Calmes et al., 1993). Linking two N-terminal formyl groups together produced a stable dimer as demonstrated by single-channel lifetimes of 1–2 secs with the ion selectivity and conductance remaining similar to that of the gA channel (Bamberg and Janko, 1977; Stankovic et al., 1989).

The replacement of the formyl group with an acetyl group at the N-terminus heightens the process of "flicker" block (Seoh and Busath, 1993b). This only occurred when two N-acetyl analogs dimerize, with the resulting disruption at the dimer junction leading to a reduction in the channel lifetime by a factor of 15–50.

Hybrid channels can form between analogs and gramicidin A. These can be distinguished by their different optimum conductance states from the parent compounds. If gA and an analog are added simultaneously to a membrane, then generally three optimum conductance states are observed: one each for the formation of gA dimers and analog dimers and a third due to the hybrid gA/analog dimers (Veatch et al., 1975; Mazet et al., 1984; Cifu et al., 1992; Seoh and Busath, 1993b). For example, heterodimer formation between $[F_6Val^1]$gA and $[Gly^1]$gA created two stable conducting states designated the high (H) conductance and low (L) conductance states (Oiki et al., 1992). The fraction of time spent in the H state was dependent on the voltage applied, which was an unusual finding given that normally gramicidin channels are voltage independent. Another unusual heterodimer formation was between gM^- (replacing the alternating L,D-configuration of gA with D,L-amino acids and also changing all the tryptophans to phenylalanines) and the shortened des-Val^1-$[Ala^2]$gA or des-Val^1-gC (Durkin et al., 1992). Based on the ~1000-fold increase in stability, a double-helix structure was believed to have formed. A CD spectrum of this heterodimer will verify if yet another new structure has been found or just a variation of one of the main conformations so far proposed.

An important question that sequence-substituted analogs have been used to answer is the role that the tryptophans play in ion translocation. It was found that gB and gC had slightly different single-channel conductances than gA even though the type of structure they form is identical, thus indicating that the side-chains do play a role in ion translocation even though they do not directly form the lumen of the channel (Veatch et al., 1975; Sawyer et al., 1990). Two main findings were evident from a number of side-chain substitution experiments. The first is that during the process of ion transport, long-range electrostatic interactions take place between the tryptophan indole groups and the permeating ion (Koeppe et al., 1990; Becker et al., 1991; Fonseca et al., 1992). The second is that the more hydrophobic the side-chain is, even if it is still in possession of a dipole moment similar to tryptophan, the lower is the conductance (Mazet et al., 1984; Daumas et al., 1991; Benamar et al., 1993). It was proposed based on these observations, that the alteration in conductance is directed by the ion-binding process. For the normal gA dimer, there may be a dipole–dipole interaction between the tryptophan side-chain and the C_α-carbonyl groups lining the channel center. This causes a slight destabilization of the ion-binding site as the carbonyl groups are unable to tilt to their maximum degree when complexing with the ion. If the side-chain has no dipole moment or if its increased hydrophobicity draws it further into the bilayer, then the carbonyl groups are free to interact more favorably with the ion thus slowing it down as it travels through the channel.

A relationship was found to exist between the conductance of the single tryptophan-to-phenylalanine substituted analogs and the activation enthalpy for ion transport (Hinton et al., 1993). For these analogs, the conductances decreased in the following order:

$$\text{gA} > \text{Phe}^{15} > \text{Phe}^{13} > \text{Phe}^{11} > \text{Phe}^{9}$$

which was linked to a decrease in activation enthalpy. A measure of the activation enthalpy determines the overall energy barrier for ion transport, hence an exact mechanism for these observations could not be ascertained, but it may be related to the loss of important dipole–dipole interactions as stated above. All these substitutions also altered the average channel lifetime, but no direct link has been found between side-chain structure and channel stability.

X. SIMULATION STUDIES OF ION TRANSPORT

This topic was previously reviewed by Roux and Karplus (1994). Gramicidin is a small polypeptide, but when simulation techniques are utilized to look at its conformation and dynamics in membranes, the large number of interactions possible in the system make it an enormous computational problem. To speed up calculations, many of the gramicidin models used tend to be simplified (Aqvist and Warshal, 1989). Generally the surrounding lipid is ignored or modified, for example by representing the lipid as an area of low dielectric constant (Martinez and Sancho, 1993) or using Lennard–Jones spheres with no charge or polarizability in order to repel water molecules (Roux and Karplus, 1993). A study which represented the surrounding lipid as a series of polarizable dipoles (no water molecules were included and the helix was represented as a polyglycine analog) found that the membrane had a significant effect on the free energy of the ion inside the channel (Aqvist and Warshal, 1989). It was concluded, therefore, that for a complete description of ion/protein interactions it was not wise to ignore the membrane component. Many of the studies to date have used the left-handed helix as the model (Pullman, 1991). However, this structure is significantly different from the right-handed helix, so it may have produced misleading conclusions. Despite these limitations, gramicidin remains one of the most useful model systems for studying the dynamic nature of ion transport in membranes (Jordan, 1987).

All calculations agree on a single-file water structure that is hydrogen bonded to the gA backbone with the water dipoles all aligned in the same direction along the channel axis (Chiu et al., 1989, 1991; Pullman, 1991). Ion-binding sites have been found at the extremities of the channel and the main energy barrier to ion entry is calculated to be at the channel mouth (Sung and Jordan, 1987; Roux and Karplus, 1993). It is interesting to note that constrictions were found near the ends of the channel as shown in Figure 6c when the internal channel diameter was mapped using the program HOLE (Smart et al., 1993). Partial dehydration of the ion occurs as it moves from bulk solvent into the channel over a distance of 5–6Å, hence dissipating the energy involved in dehydration before reaching the channel mouth. Local flexing of the backbone around the permeating ion appears to be an essential factor in the transport process (Roux and Karplus, 1991a, 1991b). Four carbonyl groups tilt away from the helix axis to complex with the ion at any one time (note the similar ion-binding motif to that observed in the crystal structure for the CsCl/gramicidin pore (Wallace and Ravikumar, 1988)).

Through the well-defined and characterized models of gramicidin, it may soon be possible to describe accurately the passage of ions across membranes and the selective nature of channels either with the use of accurate models or representative dielectric models. The rules thus developed can then be applied to more complex biologically important channels to study and predict their behavior.

XI. THE INTERACTION OF GRAMICIDIN WITH LIPIDS

This topic was previously reviewed by Killian (1992).

A. Head-group Effects

The lipid head-group charge has an effect on ion conductance through gramicidin channels. The negatively charged lipid head-group in PS causes a nine-fold increase in Tl^+ conductance through gramicidin as compared to DOPC bilayers. The PS group attracts cations, concentrating them at the lipid-surface region, hence effectively increasing the ion concentration around the channel mouth. This bilayer charge effect can be abolished by the addition of divalent and trivalent cations or lowered by dropping the pH to 3.6 where the PS head-group becomes protonated (Nelson, 1993).

The type of lipid head-group has also been shown to alter the transition between the double-helix form and the $\beta^{6.3}$-monomer. PC bilayers with phosphatidylethanolamine (PE) added formed $\beta^{6.3}$-monomers more readily than PC membranes alone (Cox et al., 1992). This increased transition rate may be related to PE's tendency to form hexagonal phases, Hex_{II}.

B. Shape Effects

The shape of the lipid molecules has been shown to determine the lipid-phase state adopted due to packing considerations. Cylindrically shaped molecules such as PC prefer to form lipid bilayers. For lipids with large head-groups and a single acyl chain, the overall shape is that of a cone and so they tend to form micelles (e.g., lysophosphatidylcholine, LPC). For lipids with a small head group and two acyl chains (e.g., PE), an inverted-cone shape is produced which is then inclined to form a Hex_{II} phase (Cullis and De Kruijff, 1979). All these phases maximize the interactions of like-with-like producing an ordered lipid/water system. Gramicidin in the $\beta^{6.3}$-form possesses a cone-like shape due to the four large tryptophans in the C-terminus which, in the channel, are situated near the lipid/water interface. It has been shown that gramicidin can induce a Hex_{II}-phase change in bilayer-forming lipids. Taking into account the shape concept of molecules, a model was proposed for Hex_{II} induction in the presence of gramicidin that involves the opposite arrangement of lipid and peptide found in saturated bilayers (Killian and De Kruijff, 1988). The gramicidin-induced Hex_{II} state only occurs when the bilayer thickness exceeds the channel length with the corresponding low monomer– monomer stability. The dimer splits, allowing water to fill the gap left between monomers, followed by aggregation of monomers on the same side of the bilayer by lateral diffusion. The next step involves lipid-flip so that the lipid-head groups are situated near the formylated N-termini where they are also in contact with the water that wedged themselves between the monomers when they split. Further peptide aggregation leads to the formation of local pockets of gramicidin-rich and gramicidin-poor regions. This appears to occur only for high gramicidin/lipid ratios with a minimum of 1 gramicidin per 7 lipid molecules.

Conversely, the cone-shaped LPC lipid can be induced to form a lamellar phase in the presence of gramicidin. The ratios of gramicidin/lipid at which this happens varies from 1:4 (Killian et al., 1983) to between 1:8 and 1:10 (Spisni et al., 1983). This is an unusual property of gramicidin as LPC normally behaves like a detergent, solubilizing intrinsic membrane proteins into micelles.

C. Fatty Acid Effects

The structure of the channel form of gramicidin is insensitive to the thickness of bilayers composed of lipids with short-chain fatty acids (Wallace et al., 1981). For longer acyl lipid chain-lengths, the channel is destablized (Wallace et al., 1981) with a resulting decrease in open-channel lifetimes (Hladky and Haydon, 1972). In thicker membranes, a dimpling effect may develop when the two monomers hydrogen bond together at their N-termini to form a conducting dimer. When this occurs, there are three forces acting on the dimer which influence the channel stability (Helfrich and Jakobsson, 1990): (1) compression energy involving changes in membrane thickness, (2) the splay contribution due to lateral interactions between fatty acid chains (this one is affected the most by the overall shape of the molecule), (3) surface tension due to alterations in the area of the membrane–solution interface. To permit dimerization, the bilayer may dimple by a sloping of the lipids in towards the channel rim. This change in gradient is larger for thicker membranes creating an increase in the splay energy during the local thinning of the membrane and, as a result, destabilization of the dimer as shown by CD (Wallace et al., 1981). Detergents reduce this energy resulting in increased dimer stability (Andersen et al., 1992). Acylation of the gramicidin is also believed to lessen the magnitude of the energy of deformation through a similar mechanism. As pointed out earlier, ion occupancy also stabilizes monomer–monomer interaction, but this time it appears that ion interaction eventually leads to a decrease in surface tension which then becomes the major component in the relaxation of membrane deformation (Ring, 1992).

D. Cholesterol Effects

Water molecules were found to associate with gramicidin inside the hydrophobic core of the bilayer at the peptide–lipid interface (Ho and Stubbs, 1992). Recent studies have shown that the addition of cholesterol to the system removed the water molecules in the head-group region as well as those in the hydrophobic core. Cholesterol also has the effect of blocking ion conductance through the gramicidin channel. The "B" ring moiety of cholesterol and not its apolar tail is necessary for this function (Schagina et al., 1992). As cholesterol causes a thickening of the bilayer, there is also a corresponding reduction in channel lifetime. However, some results indicate that gramicidin and cholesterol preferentially interact (Lev, 1990), thus apparently having a role in channel destabilization.

E. Phase Effects

Phase-transitions in lipid bilayers are a consequence of variations in temperature or pressure. At high pressures, a relatively rigid, ordered gel-phase is stabilized which can be converted to the less-dense fluid–liquid crystalline state by decreasing the pressure or by increasing the temperature. The gel-phase is characterized by a straightening of the hydrocarbon chains with a corresponding expansion in membrane thickness. The rotational motion of gramicidin in the fluid liquid-crystalline state of DMPC shows a slight decrease with increasing order. However in the gel-phase, increased packing caused a significant increase in gramicidin rotation in the plane of the bilayer (Scarlata, 1991). No structural changes in gramicidin were seen when the pressure was raised (Teng et al., 1991). This lead to the proposal that due to the increase in bilayer thickness in the gel-phase, hydrogen bonds between the indole groups of the tryptophan side-chains and the lipid head-group oxygens were broken allowing the peptide to rotate more freely.

Gramicidin produces opposite effects on the acyl-chain order in the gel and liquid–crystalline phases of dipalmitoylphosphatidylcholine (DPPC) bilayers (Davies et al., 1990), although there is no effect on the gramicidin structure as the lipid undergoes this phase-transition (Wallace et al., 1981). As expected, gramicidin disrupted the ordered gel-phase, but it caused the acyl chains to become more ordered in the liquid–crystalline phase. Polarized infrared dichroism studies found that gramicidin caused a slight reorientation of the acyl chains in DMPC so that they aligned more closely to the bilayer normal (Nabedryk et al., 1982). Simulation studies of gramicidin in DPPC at temperatures above the gel-to-liquid–crystalline transition produced significantly better fits to experimental ESR data if an increased local ordering of the lipids were included (Ge and Freed, 1993).

F. Summary of Membrane Effects

In lipid systems, gramicidin has been used to study the forces that maintain a lipid bilayer configuration. Its shape influences the phase behavior of lipids and it even has an effect on the ordering of the acyl chains. It has also been shown that the tryptophan side-chains are important in protein/lipid interactions and, as a probe, gramicidin is very effective for studying charge modulation on the membrane surface as well as marking the boundaries between phase-states.

G. Covalently Linked Lipids

One family of gramicidin molecules has fatty acids covalently linked via their ethanolamine groups and is designated the gramicidin "K" family (Koeppe et al., 1985; Williams et al., 1992). These acylated gramicidins appear to form structurally equivalent channels to the normal $\beta^{6.3}$-structure in saturated lipids. This natural modification enhances the lifetime of the channel-open state, but at the same time, the molecules are less effective in channel formation. The typical composition of the naturally bound fatty acids was reported to be 30–35% 12-methyltetradecanoic and hexadecanoic acids and 15–20% of pentadecanoic acid and octadecanoic acid, the larger percentage of fatty acids representing the more abundant fatty acids found naturally in *B. brevis.* Acylated gramicidins using fatty acids more usually associated with eukaryotic systems (lauroyl-12:0, myristoyl-14:0, palmitoyl-16:0, stearoyl-18:0, oleoyl–gramicidin-18:1, chain length:unsaturation) have been prepared in order to study their behavior upon incorporation into lipid bilayers (Vogt et al., 1991, 1992). They appear similar to the natural gramicidin "K" dimers, increasing the average channel duration at a constant level for all the different acyl chain lengths examined.

XII. CONCLUSIONS

First impressions of gramicidin are of a small peptide which has a very flexible and adaptable nature, but which forms several types of well-behaved conducting channels. So, what relevance does gramicidin have to the large complex channels normally associated with ion transport in biological systems (Woolley and Wallace, 1994)?

It is because the conductance properties of gramicidin can be related to its structural features that makes gramicidin an ideal model system. An understanding of the basic events involved in its transport may help to uncover the structural bases of the conductance properties of other intrinsic membrane proteins. Many of gramicidin's conductance characteristics have been observed in large-channel proteins, including open-channel noise in the acetylcholine receptor (Sigworth, 1985) and flicker block in the K^+ rectifier and Ca^{2+} channels (Fukushima, 1982; Lansman et al., 1986). Divalent cation block has been verified in both the K^+

rectifier by Ba^{2+} and in the calcium channel by Cd^{2+}, Mg^{2+}, and Mo^{2+} (Hille, 1992). The water-filled gramicidin channel has been used to help explain proton movement through the F_0F_1 ATP synthase. Small organic cations have also been found to permeate endplate channels (Dwyer et al., 1980) as long as they fit a 6.5Å diameter pore.

Using the endplate as an example of an ion channel, its pore is believed to be formed from five-peptide subunits arranged as a pentagonal complex (Hille, 1992). Unlike in gramicidin, the lumen of the pore is believed to consist of the amino acid side-chains. However, in both cases an ion moving through the channel will experience a series of attractive and repulsive forces as it moves between binding sites and across barriers. For both, the difference in energy between the binding site and the barrier must be relatively small or the ion translocation would fail to take place. The high-resolution three-dimensional structures of gramicidin allow models to be postulated and tested for ion transport through a channel that can ultimately be applied to the more complex biological channels for which we have less detailed structural information.

For an ion channel to be selective, there must be a narrowing of the inner pore where charged groups and/or induced dipole moments are able to interact with the incoming ion to either allow it to pass through or to repel it. This selectivity filter, for the family of K^+ channels, is believed to be formed from a stretch of eight highly conserved amino acids (TXXTXGYG). A number of mutant K^+ channels have been produced, with altered amino acids groups at the two glycine positions, to determine the role of these amino acids during ion transport (Heginbotham et al., 1994). It was found that any mutant channel with changes at these positions was non-selective for K^+ ions. The two glycine residues, which have only a hydrogen as their side-chain, are able to adopt unusual torsion angles. This frees the carbonyl backbone groups to interact with the potassium ion, possibly by similar dipole interactions to those observed in the gramicidin/ion complexes (Wallace and Ravikumar, 1988). The larger substituted side-chains probably prevent the free movement of the carbonyl groups due to steric clashes with their neighboring atoms.

The dynamic nature of gramicidin during ion translocation is probably a feature of other channels, but to a lesser extent. For gramicidin, the groups that principally interact with the ion are the carbonyls of the polypeptide backbone. In other channels, as stated before, they would most probably be side-chains (Lauger, 1987). Comparable fluctuations in side-chains are seen in hemoglobin, where they allow the entry of oxygen into the heme pocket. Also, in other channels electrostatic interactions between side-chains that are not in direct contact with the ion will probably modulate ion conductance through the channel as the tryptophans of gramicidin do during this process.

The tryptophans in the $\beta^{6.3}$-monomer are clustered at the lipid/water interface and appear to interact with the lipid head-groups. This arrangement has also been found in the high resolution structures of other integral membrane proteins solved so far (Deisenhofer et al., 1984; Henderson et al., 1990; Cowan et al., 1992; Daniel et al., 1994). A survey of the putative single transmembrane segments of 115 plasma membrane proteins also appears to suggest that tryptophans are preferentially located at the lipid/water interface (Landolt-Marticorena et al., 1993). Hence the tryptophans may play an important general role in defining this interface and stabilizing the orientation of proteins within a membrane.

In summary, while it may only be a small polypeptide, gramicidin is an important model system for ion transport and is able to simulate many physiological phenomena as seen in more complex biological channels. Most importantly, the high-resolution structural data of gramicidin allows, for the first time, interpretation of ion translocation through a channel on a molecular level.

ACKNOWLEDGMENTS

This work was supported by a grant from the BBSRC (Biomolecular Sciences) Initiative to B.A. Wallace. Declan A. Doyle was supported by an MRC studentship.

REFERENCES

Abdul-Manan, N., & Hinton, J. F. (1994). Conformation states of gramicidin A along the pathway to the formation of channels in model membranes determined by 2D NMR and circular dichroism. Biochemistry 33, 6773–6783.

Akeson, M., & Deamer, D. W. (1991). Proton conductance by the gramicidin water wire. Biophys. J. 60, 101–109.

Andersen, O. S. (1984). Gramicidin channels. Ann. Rev. Physiol. 46, 531–548.

Andersen, O. S. (1989). Kinetics of ion movement mediated by carriers and channels. Methods in Enzymol. 171, 62–112.

Andersen, O. S., Sawyer, D. B., & Koeppe, R. E. II. (1992). Modulation of channel function by the host bilayer. In: Biomembrane Structure and Function—The State of the Art. (Gaber, B. P., & Easwaran, K. D. K., Eds.) pp. 227–244. Adenine Press, Guilderland, N.Y.

Aqvist, J., & Warshel, A. (1989). Energetics of ion permeation through membrane channels. Solvation of Na^+ by gramicidin A. Biophys. J. 56, 171–182.

Arseniev, A. S., Barsukov, I. L., Bystrov, V. F., Lomize, A. L., & Ovchinnikov, Y. A. (1985). ^{1}H-NMR study of gramicidin A transmembrane channel. FEBS Letts. 186, 168–174.

Arseniev, A. S., Lomize, A. L., Barsukov, I. L., & Bystrov, V. F. (1986). Gramicidin A transmembrane ion-channel. Three-dimensional structure reconstruction based on NMR-spectroscopy and energy refinement. Biol. Membr. 3, 1077–1104.

Bamberg, E., & Janko, K. (1977). The action of a carbosuboxide dimerized gramicidin A on lipid bilayer membranes. Biochim. Biophys. Acta 465, 486–499.

Bamberg, E., & Lauger, P. (1977). Blocking of the gramicidin channel by divalent cations. J. Membr. Biol. 35, 351–375.

Bano, M. C., Braco, L., & Abad, C. (1991). Conformation transitions of gramicidin A in phospholipid model membranes. A high performance liquid chromatography accessment. Biochemistry 30, 886–894.

Bano, M. C., Braco, L., & Abad, C. (1992). A semi-empirical approach for the simulation of circular dichroism spectra of gramicidin A in a model membrane. Biophys. J. 63, 70–77.

Barsukov, I. L., Arseniev, A. S., & Bystrov, V. F. (1987). Spatial structures of gramicidin A in organic solvents. ^{1}H-NMR analysis of four species in ethanol. Bioorgan. Khim. 13, 1501–1522.

Becker, M. D., Greathouse, D. V., Koeppe, R. E. II., & Andersen, O. S. (1991). Amino acid sequence modulation of gramicidin channel function: Effects of tryptophan-to-phenylalanine substitutions on the single channel conductance and duration. Biochemistry 30, 8830–8839.

Benamer, D., Daumas, P., Trudelle, Y., Calas, B., Bennes, R., & Heitz, F. (1993). Influence of the nature of the aromatic side-chain on the conductance of the channel of linear gramicidin: Study of a series of 9, 11, 13, 15-Tyr (O-protected) derivatives. Eur. Biophys. J. 22, 145–150.

Bouchard, M., & Auger, M. (1993). Solvent history dependence of gramicidin-lipid interactions: A Raman and infrared spectroscopic study. Biophys. J. 65, 2484–2492.

Braco, L., Bano, M. C., & Abad, C. (1992). Monitoring self-association of a hydrophobic peptide with high performance liquid chromatography. J. Chem. Edu. 69, A113–A116.

Bridal, T. R., & Busath, D. (1992). Inhibition of gramicidin channel activity by local anaesthetics. Biochim. Biophys. Acta 1107, 31–38.

Busath, D., Andersen, O. S., & Koeppe, R. E II. (1987). On the conductance heterogeneity in membrane channels formed by gramicidin A. A cooperative study. Biophys. J. 51, 79–88.

Busath, D., & Szabo, G. (1988a). Low conductance gramicidin channels are head-to-head dimers of $\beta^{6.3}$-helices. Biophys. J. 53, 689–695.

Busath, D., & Szabo, G. (1988b). Permeation characteristics of gramicidin conformers. Biophys J. 53, 697–707.

Busath, D. (1993). The use of physical methods in determining gramicidin channel structure and function. Ann. Rev. Physiol. 55, 473–501.

Butz, S., Benz, R., Wacker, T., Welte, W., Lustig, A., Plapp, R., & Weckesser, J. (1993). Biochemical characterization and crystallization of porin from *Rhodopseudomonas blastica*. Arch. Microbiol. 159, 301–307.

Bystrov, V. F., & Arseniev, A. S. (1988). Diversity of the gramicidin A spacial structure: Two-dimensional ^{1}H NMR study in solution. Tetrahedron 44, 925–940.

Bystrov, V. F., Arseniev, A. S., Barsukov, I. L., Golovanov, A. P., & Maslennikov, I. V. (1990). The structure of the transmembrane channel of gramicidin A: NMR study of its conformation stability and interaction with divalent cations. Gazz. Chim. Ital. 120, 485–491.

Calmes, M., Daunis, J., David, D., & Lazaro, R. (1993). Total synthesis and ionophoric behaviour of gramicidin A analogue. Tetrahedron Letts. 34, 3275–3278.

Chiu, S., Jakobsson, E., Subramaniam, S., & McCammon, J. A. (1991). Time-correlation analysis of a simulated water motion in flexible and rigid gramicidin channels. Biophys. J. 60, 273–285.

Chiu, S., Subramaniam, S., Jakobsson, E., & McCammon, J. A. (1989). Water and polypeptide conformations in the gramicidin channel. A molecular dynamics study. Biophys. J. 56, 253–261.

Cifu, A. S., Koeppe. R. E. II., & Andersen, O. S. (1992). Supramolecular organization of gramicidin channels. The elementary conducting unit is a dimer. Biophys. J. 61, 189–203.

Cornelius, F. (1991). Functional reconstruction of the sodium pump. Kinetics of exchange reactions performed by reconstituted Na^+/K^+ ATPase. Biochim. Biophys. Acta 1071, 19–66.

Cowan, S. W., Schirmer, T., Rummel, G., Steiert, M., Ghosh, R., Pauptit, R. A., Jansonius, J. N., & Rosenbusch, J. P. (1992). Crystal structures explain functional properties of the two *E. coli* porins. Nature 358, 727–733.

Cox, K. J., Ho, C., Lombardi, J. V., & Stubbs, O. D. (1992). Gramicidin conformational studies with mixed-chain unsaturated phospholipid bilayer systems. Biochem. 31, 1112–1118.

Cullis, P. R., & De Kruijff, B. (1979). Lipid polymorphism and the functional role of lipids in biological membranes. Biochim. Biophys. Acta 559, 399–420.

Daniel, P., Loll, P. J., & Garavito, R. M. (1994). The X-ray crystal structure of the membrane protein prostaglandin H_2 synthase-1. Nature 367, 243–249.

Daumas, P., Benamar, D., Heitz, F., Ranjalahy-Rosoloarijao, L., Mouden, R., Lazaro, R., & Pullman, A. (1991). How can aromatic side-chains modulate the conductance of the gramicidin channel? Int. J. Pept. Prot. Res. 38, 218–228.

Davies, M. A., Brauner, J. W., Schuster, H. F., & Mendelsohn, R. (1990). A quantitative infrared determination of acyl chain conformation in gramicidin/dipalmitoylphosphatidylcholine bilayers. Biochem. Biophys. Res. Comm. 168, 85–90.

Deamer, D. W. (1992). Role of water in proton flux mechanisms. In: Biomembranes Structure and Function—The State of the Art (Gaber, B. P., & Easwaran, K. D. K., Eds.) pp. 209–225. Adenine Press, Guilderland, N.Y.

Deamer, D. W., & Nichols, J. W. (1989). Proton flux mechanisms in model and biological membranes. J. Membr. Biol. 107, 91–103.

Decker, E. R., & Levitt, D. G. (1988). Use of weak acids to determine the bulk diffusion limitation of H^+ ion conductance through the gramicidin channel. Biophys. J. 53, 25–32.

Deisenhofer, J., Epp, O., Miki, K., Huber, R., & Michel, H. (1984). X-ray structure analysis of a membrane protein complex: Electron density map at 3Å resolution and a model of the chromophores of the photosynthetic reaction centre from *Rhodopseudomonas viridis*. J. Mol. Biol. 180, 385–398.

Doyle, D. A., & Wallace, B. A. (1994a). The structure of the gramicidin/KSCN complex. Biophys. J. 66, 353.

Doyle, D. A., & Wallace, B. A. (1994b). Caesium-binding sites in the gramicidin pore. Biochem. Soc. Trans. 22, 1043–1045.

Durkin, J. T., Providence, L. L., Koeppe, R. E. II., & Andersen, O. S. (1992). Formation of non-$\beta^{6.3}$-helical gramicidin channels between sequence-substituted gramicidin analogues. Biophys. J. 62, 145–159.

Dwyer, T. M., Adams, D. J., & Hille, B. (1980). The permeability of the endplate channel to organic cations in frog muscle. J. Gen. Physiol. 75, 469–492.

Finkelstein, A., & Andersen, O. S. (1981). The gramicidin A channel: A review of its permeability characteristics with special reference to the single-file aspect of transport. J. Membr. Biol. 59, 155–171.

Fonseca, V., Daumas, P., Ranjalahy-Rosoloarijao, L., Heitz, F., Lazaro, R., Trudelle, Y., & Andersen, O. S. (1992). Gramicidin channels that have no tryptophans. Biochem. 31, 5340–5350.

Fukushima, Y. (1982). Blocking kinetics of the anomalous potassium rectifier of tunicate egg studied by single channel recording. J. Physiol. 331, 311–331.

Gambale, F., Menini, A., & Rauch, G. (1987). Effect of calcium on the gramicidin A single channel in phosphatidylserine membranes. Eur. Biophys. J. 14, 369–374.

Ge, M., & Freed, J. H. (1993). An electron spin resonance study of interactions between gramicidin A and phosphatidylcholine bilayers. Biophys. J. 65, 2106–2123.

Golovanov, A. P., Barsukov, I. L., Arseniev, A. S., Bystrov, V. F., Sukhanov, S. V., & Barsukov, L. I. (1991). The divalent cation-binding sites of gramicidin A transmembrane ion–channel. Biopolymers 31, 425–434.

Gross, E., & Witkop, B. (1965). Gramicidin. IX. Preparation of gramicidin A, B, and C. Biochemistry 4, 2495–2501.

Hawkes, G. E., Lian, L., Randall, E. W., Sales, K. D., & Curson, E. H. (1987). The conformation of gramicidin A in dimethylsulfoxide solution. A full analysis of the one- and two-dimensional ^1H, ^{13}C and ^{15}N nuclear magnetic resonance spectra. Eur. J. Biochem. 166, 437–445.

He, K., Ludtke, S. J., Wu, Y., & Huang, H. W. (1993). X-ray scattering with momentum transfer in the plane of membrane. Application to gramicidin organization. Biophys. J. 64, 157–162.

Heginbotham, L., Lu, Z., Abramson, T., & MacKinnon, R. (1994). Mutations in the K^+ channel signature sequence. Biophys. J. 66, 1061–1067.

Heinmann, S. H., & Sigworth, F. J. (1988). Open channel noise. IV. Estimates of rapid kinetics of formamide block in gramicidin A channel. Biophys. J. 54, 757–764.

Heinmann, S. H., & Sigworth, F. J. (1990). Open channel noise. V. Fluctuating barriers to ion entry in gramicidin A channels. Biophys. J. 57, 499–514.

Heitz, F., Heitz, A., & Trudelle, Y. (1986). Conformations of gramicidin A and its 9, 11, 13, 15-phenylalanyl analog in dimethylsulfoxide and chloroform. Biophys. Chem. 166, 437–445.

Helfrich, P., & Jakobsson, E. (1990). Calculations of deformation energies and conformations in lipid membrane containing gramicidin channels. Biophys. J. 57, 1075–1084.

Hemsley, G., & Busath, D. (1991). Small iminium ions block gramicidin channels in lipid bilayers. Biophys. J. 59, 901–908.

Henderson, R., Baldwin, J. M., Ceska, T. A., Zemlin, F., Beckman, E., & Downing, K. H. (1990). Model for the structure of bacteriorhodopsin based on high-resolution cryo-microscopy. J. Mol. Biol. 213, 899–929.

Hille, B. (1992). Ionic Channels of Excitable Membranes. Sinauer Associates Inc. Publishers, Sunderland, MA.

Hing, A. W., & Schaefer, J. (1993). Two-dimensional rotational-echo double resonance of Val_1–[1-^{13}C]Gly_2–[^{15}N]Ala_3–gA in multilamellar dimyristoylphosphatidylcholine dispersions. Biochemistry 32, 7593–7604.

Hinton, J. F., Easton, P. L., Newkirk, D. A., & Shungu, D. C. (1993). ^{23}Na-NMR study of ion transport across vesicle membranes facilitated by phenylalanine analogs of gramicidin. Biochim. Biophys. Acta 1146, 191–196.

Hinton, J. F., Whaley, W. L., Shungu, D., Koeppe, R. E. II., & Millett, S. (1986). Equilibrium binding constants for the group I metal cations with gramicidin A determined by competitive studies and Tl^+-205 nuclear magnetic resonance spectroscopy. Biophys. J. 50, 539–544.

Hladky, S. B., & Haydon, D. A. (1972). Ion transfer across lipid membranes in the presence of gramicidin A. I. Studies of the unit conductance channel. Biochim. Biophys. Acta 274, 294–312.

Ho, C., & Stubbes, C. D. (1992). Hydration at the membrane protein–lipid interface. Biophys. J. 63, 897–902.

Hu, W., Lee, K. C., & Cross, T. A. (1993). Tryptophans in membrane proteins: Indole ring orientations and functional implications in the gramicidin channel. Biochemistry 32, 7035–7047.

Jordan, P. C. (1987). Microscopic approaches to ion transport through transmembrane channels. The model system gramicidin. J. Phys. Chem. 91, 6582–6591.

Katsaras, J., Prosser, R. S., Stinson, R. H., & Davis, J. H. (1992). Constant helical pitch of the gramicidin channel in phospholipid bilayers. Biophys. J. 61, 827–830.

Ketchem, R. R., Hu, W., & Cross, T. A. (1993). High resolution conformation of gramicidin A in a lipid bilayer by solid-state NMR. Science 261, 1457–1460.

Killian, J. A. (1992). Gramicidin and gramicidin-lipid interactions. Biochim. Biophys. Acta 1113, 391–425.

Killian, J. A., & De Kruijff, B. (1988). Proposed mechanism for H_{II} phase induction by gramicidin in model membranes and its relation to channel function. Biophys. J. 53, 111–117.

Killian, J. A., De Kruijff, B., Van Echteld, C. J. A., Verkleij, A. J., Leunissen-Bijvelt, J., & De Gier, J. (1983). Mixtures of gramicidin and lysophosphatidylcholine form lamellar structures. Biochim. Biophys. Acta 728, 141–144.

Killian, J. A., Prasad, K. U., Hains, D., & Urry, D. W. (1988). The membrane as an environment of minimal interconversion. A circular dichroism study on the solvent dependence of the conformational behaviour of gramicidin in diacylphosphatidylcholine model membranes. Biochemistry 27, 4848–4855.

Kimball, M. R., & Wallace, B. A. (1984). Crystalline ion complexes of gramicidin A. Annals N. Y. Acad. Sci. 435, 551–554.

Koeppe, R. E. II., Hodgson, K. O., & Stryer, L. (1978). Helical channels in crystals of a cesium–gramicidin A complex: An X-ray diffraction study. J. Mol. Biol. 1221, 41–54.

Koeppe, R. E. II., Paczkowski, J. A., & Whaley, W. L. (1985). Gramicidin-K, a new linear channel-forming gramicidin from *Bacillus brevis*. Biochem. 12, 2822–2826.

Koeppe, R. E. II., Providence, L. L., Greathouse, D. V., Heitz, F., Trudelle, Y., Purdie, N., & Andersen, O. S. (1992). On the helix sense of the gramicidin A single channel. Prot. Struct. Funct. Genet. 12, 49–62.

Koeppe, R. E. II., Mazet, J., & Andersen, O. S. (1990). Distinction between dipolar and inductive effects in modulating the conductance of gramicidin. Biochem. 29, 512–520.

Koeppe, R. E. II., Killian, J. A., & Greathouse, D. V. (1994). Orientation of the tryptophan 9 and 11 side chains of the gramicidin channel based on deuterium nuclear magnetic resonance spectroscopy. Biophys. J. 66, 14–24.

Landolt-Marticorena, C., Williams, K. A., Deber, C. M., & Reithmeier, R. A. F. (1993). Non-random distribution of the amino-acids in the transmembrane segments of the human type I single span membrane proteins. J. Mol. Biol. 229, 602–608.

Langs, D. A. (1988). Three dimensional structure at 0.86Å of the uncomplexed form of the transmembrane ion channel peptide gramicidin A. Science 241, 188–191.

Langs, D. A. (1993). *Ab initio* direct methods: Practical advice for getting beyond the first 300 atoms. Acta Cryst. D49, 158–167.

Langs, D. A., Smith, G. D., Courseille, C., Precigoux, G., & Hospital, M. (1991). Monoclinic uncomplexed double-stranded, antiparallel, left-handed $\beta^{5.6}$-helix ($\uparrow\downarrow\beta^{5.6}$) structure of gramicidin A: Alternative patterns of helical association and deformation. Proc. Natl. Acad. Sci. USA 88, 5345–5349.

Lansman, J. B., Hess, P., & Tsien, R. W. (1986). Blockage of current through single calcium channels by Cd^{2+}, Mg^{2+} and Ni^{2+}. Voltage and concentration dependence of calcium entry into the pore. J. Gen. Physiol. 88, 321–347.

Lauger, P. (1987). Dynamics of ion transport systems in membranes. Physiol. Rev. 67, 1296–1331.

Lev, A. A. (1990). Sterol-dependent inactivation of gramicidin-A induced ionic channels in the cell and artificial lipid-bilayer membrane. Biophys. Membr. Transp. 1, 231–250.

LoGrasso, P. V., Moll, F., III. & Cross, T. A. (1988). Solvent history dependence of gramicidin A conformations in hydrated lipid bilayers. Biophys. J. 54, 259–267.

Martinez, G., & Sancho, M. (1993). Electrostatic interactions in gramicidin channels. Three-dielectric model. Eur. Biophys. J. 22, 301–307.

Masotti, L., Spisni, A., & Urry, D. W. (1980). Conformational studies on the gramicidin A transmembrane channel in lipid micelles and liposomes. Cell Biophys. 2, 241–251.

Mazet, J. L., Andersen, O. S., & Koeppe, R. E., II. (1984). Single-channel studies on linear gramicidins with altered amino-acid sequences—a comparison of phenylalanine, tryptophane, and tyrosine substitutions at position-1 and position-11. Biophys. J. 45, 263–276.

McKim, S., & Hinton, J. F. (1994). Evidence for xenon transport through the gramicidin channel: A ^{129}Xe-NMR study. Biochim. Biophys. Acta 1193, 186–198.

Michel, H. (1982). Three-dimensional crystals of a membrane protein complex. J. Mol. Biol. 158, 567–572.

Mittler-Neher, S., & Knoll, W. (1993). Ca^{2+}-induced lateral phase separation in black lipid membranes and its coupling to the ion-translocation by gramicidin. Biochim. Biophys. Acta 1152, 259–269.

Morrow, J. S., Veatch, W. R., & Stryler, L. (1979). Transmembrane channel activity of gramicidin A analogs: Effects of modification and deletion of the amino-terminal residue. J. Mol. Biol. 132, 733–738.

Mukherjee, S., & Chattopadhyay, A. (1994). Motionally restricted tryptophan environments at the peptide–lipid interface of gramicidin channels. Biochemistry 33, 5089–5097.

Myers, V. B., & Haydon, D. A. (1972). Ion transfer across lipid membranes in the presence of gramicidin A. II. The ion selectivity. Biochim. Biophys. Acta 274, 313–322.

Nabedryk, E., Gingold, M. P., & Breton, J. (1982). Orientation of gramicidin A transmembrane channel. Biophys. J. 38, 243–249.

Neher, E. (1975). Ionic selectivity of the gramicidin channel and the thallous ion. Biophys. J. 401, 540–544.

Nelson, A. (1993). Effect of lipid charge and solution composition on the permeability of phospholipid–gramicidin monolayers to Tl. J. Chem. Soc. Far. Trans. 89, 2799–2805.

Oiki, S., Koeppe, R. E. II., & Andersen, O. S. (1992). A dipolar amino acid substitution induces voltage-dependent transitions between two stable conductance states in gramicidin channels. Biophys. J. 62, 28–30.

Olah, G. A., Huang, H. W., & Wu, Y. (1991). Location of ion binding sites in the gramicidin channel by X-ray diffraction. J. Mol. Biol. 218, 847–858.

Prosser, R. S., & Davis, J. H., (1991). ^{2}H nuclear magnetic resonance of the gramicidin A backbone in a phospholipid bilayer. Biochemistry 30, 4687–4696.

Pullman, A. (1987). Energy profiles in the gramicidin A channel. Quart. Rev. Biophys. 20, 173–200.

Pullman, A. (1991). Contribution of theoretical chemistry to the study of ion transport through membranes. Chem. Rev. 91, 793–812.

Ramachandran, G. N., & Chandrasekaran, R. (1972). Conformation of peptide chains containing both L- and D-residues: Part I-helical structures with alternating L- and D-residues with special reference to the LD-ribbon and the LD-helix. Ind. J. Biochem. 9, 1–11.

Ring, A. (1992). Influence of ion occupancy and membrane deformation on gramicidin A channel stability in lipid membranes. Biophys. J. 61, 1305–1315.

Roux, B., & Karplus, M. (1991a). Ion transport in a model gramicidin channel. Biophys. J. 59, 961–981.

Roux, B., & Karplus, M. (1991b). Ion transport in a gramicidin like channel: Dynamics and mobility. J. Phys. Chem. 95, 4856–4868.

Roux, B., & Karplus, M. (1993). Ion transport in the gramicidin channel: Free energy of the solvated right-handed dimer in a model membrane. J. Am. Chem. Soc. 115, 3250–3262.

Roux, B., & Karplus, M. (1994). Molecular dynamic simulations of the gramicidin channel. Ann. Rev. Biophys. Biomol. Struct. 23, 731–761.

Sarges, R., & Witkop, B. (1965). Gramicidin A. V. The structure of valine- and isoleucine-gramicidin A. J. Am. Chem. Soc. 87, 2011–2020.

Sawyer, D. B., Williams, L. P., Whaley, W. L., Koeppe, R. E. II., & Andersen, O. S. (1990). Gramicidin-A, gramicidin-B, and gramicidin-C form structurally equivalent ion channels. Biophys. J. 58, 1207–1212.

Scarlata, S. F. (1991). Effect of increased chain packing on the gramicidin–lipid interactions. Biochemistry 30, 9853–9859.

Sefton, B. M., & Buss, J. E. (1987). The covalent modification of eukaryotic proteins with lipids. J. Cell Biol. 155, 167–183.

Seoh, S., & Busath, D. (1993a). The permeation properties of small organic cations in gramicidin A channels. Biophys. J. 64, 1017–1028.

Seoh, S., & Busath, D. (1993b). Formamidinium-induced dimer stabilization and flicker block behaviour in homo- and heterodimer channels formed by gramicidin A and N-acetyl gramicidin A. Biophys. J. 65, 1817–1827.

Separovic, F., Pax, R., & Cornell, B. (1993). NMR order parameter analysis of a peptide plane aligned in a lyotropic liquid crystal. Mol. Phys. 78, 357–369.

Shungu, D. C., Hinton, J. F., Koeppe, R. E. II., & Millett, F. S. (1986). Investigation of the interaction between thallous ions and gramicidin A in dimyristoylphosphatidylcholine vesicles: A thallium-205 NMR equilibrium study. Biochemistry 25, 6103–6108.

Sigworth, F. J. (1985). Open channel noise. I. Noise in acetylcholine receptor currents suggests conformational fluctuations. Biophys. J. 47, 709–720.

Smart, O. S., Goodfellow, J. M., & Wallace, B. A. (1993). The pore dimensions of gramicidin A. Biophys. J. 65, 2455–2460.

Smith, R., Thomas, D. E., Atkins, A. R., Separovic, F., & Cornell, B. A. (1990). Solid-state ^{13}C-NMR studies of the effects of sodium ions on the gramicidin A ion channel. Biochim. Biophys. Acta 1026, 161–166.

Spisni, A., Pasquali-Ronchetti, I., Casali, E., Linder, L., Cavatorta, P., Masotti, L., & Urry, D. W. (1983). Supramolecular organization of the lysophosphatidylcholine packaged gramicidin A. Biochim. Biophys. Acta 732, 58–68.

Stankovic, C. J., Heinemann, S. H., Delfino, J. M., Sigworth, F. J., & Schreiber, S. L. (1989). Transmembrane channels based on tartaric acid gramicidin-A hybrids. Science 244, 813–817.

Sung, S., & Jordan, P. C. (1987). Why is gramicidin valence selective? Biophys. J. 51, 661–672.

Sychev, S. V., Barsukov, L., & Ivanov, V. Y. (1993). The double $\pi\pi5.6$ helix of gramicidin A predominates in unsaturated lipid membranes. Eur. J. Biochem. 22, 279–288.

Takeuchi, H., Nemoto, Y., & Harada, I. (1990). Environments and conformations of tryptophan side chains of gramicidin A in phospholipid bilayers studied by Raman spectroscopy. Biochemistry 29, 1572–1579.

Teng, Q., Koeppe, R. E. II., & Scarlata, S. F. (1991). Effect of salt and membrane fluidity on fluorophore motions of a gramicidin C derivative. Biochemistry 30, 7984–7990.

Trudelle, Y., Daumas, P., Heitz. F., Etchebest, C., & Pullman, A. (1987). Experimental and theoretical study of gramicidin P, an analog of gramicidin A with a methylamine C-terminal. FEBS Letts. 216, 11–16.

Turano, B., Pear, M., & Busath, D. (1992). Gramicidin channel selectivity. Biophys. J. 63, 152–161.

Urry, D. W. (1972). The gramicidin A transmembrane channel: A proposed $\pi_{(L,D)}$ helix. Proc. Natl. Acad. Sci. USA 68, 672–676.

Van Mau, N., Bonnet, B., Benayad, A., & Heitz, F. (1994). The conformation of linear gramicidin is sequence dependent: A monolayer and infrared study. Eur. Biophys. J. 22, 447–452.

Veatch, W. R., Fossel, E. T., & Blout, E. R. (1974). The composition of gramicidin A. Biochemistry 13, 5249–5256.

Veatch, W. R., Mathies, R., Eisenberg, & Stryer, L. (1975). Simultaneous fluorescence and conductance studies of planar bilayer membranes containing a highly active and fluorescent analog of gramicidin A. J. Mol. Biol. 99, 75–92.

Venkatachalam, C. M., & Urry, D. W. (1983). Theoretical conformational analysis of the gramicidin A transmembrane channel. I. Helix sense and energetics of head-to-head dimerization. J. Comp. Chem. 4, 461–469.

Vogt, T. B. C., Killian, J. A., Demel, R. A., & De Kruijff, B. (1991). Synthesis of acylated gramicidins and the influence of acylation on the interfacial properties and conformational behaviour of gramicidin A. Biochim. Biophys. Acta 1069, 157–164.

Vogt, T. B. C., Killian, J. A., De Kruijff, B., & Andersen, O. S. (1992). Influence of acylation on channel characteristics of gramicidin A. Biochemistry 31, 7320–7324.

Wallace, B. A. (1984). Ion-bound forms of the gramicidin A transmembrane channel. Biophys. J. 45, 114–116.

Wallace, B. A. (1986). Structure of gramicidin A. Biophys. J. 49, 295–307.

Wallace, B. A. (1990). Gramicidin channels and pores. Ann. Rev. Biophys. Biophys. Chem. 19, 127–157.

Wallace, B. A. (1992). Crystallographic studies of a transmembrane ion channel, gramicidin A. Prog. Biophys. Molec. Biol. 57, 59–69.

Wallace, B. A., Hendrickson, W. A., & Ravikumar, K. (1990). The use of single-wavelength anomalous scattering to solve the crystal structure of a gramicidin A/caesium chloride complex. Acta Cryst. B46, 440–446.

Wallace, B. A., & Ravikumar, K. (1988). The gramicidin pore: Crystal structure of a caesium complex. Science 241, 182–187.

Wallace, B. A., Veatch, W. R., & Blout, E. R. (1981). Conformation of gramicidin A in phospholipid vesicles: Circular dichroism studies of effects of ion binding, chemical modification, and lipid structure. Biochemistry 20, 5754–5760.

Weinstein, S., Wallace. B. A., Blout. E. R., Morrow, J. S., & Veatch, W. (1979). Conformation of gramicidin A channel in phospholipid vesicles: A ^{13}C and ^{19}F nuclear magnetic resonance study. Proc. Natl. Acad. Sci. USA 76, 4230–4234.

Weinstein, S., Wallace. B. A., Morrow, J. S., & Veatch, W. (1980). Conformation of gramicidin A transmembrane channel: a ^{13}C nuclear magnetic resonance study of ^{13}C-enriched gramicidin in phosphatidylcholine vesicles. J. Mol. Biol. 143, 1–19.

Williams, L. P., Narcessian, E. J., Andersen, O. S., Walker, G. R., Taylor, M. J., Lazenby, J. P., Hinton. J. F., & Koeppe, R. E. II. (1992). Molecular and channel forming characteristics of gramicidin K's: A family of naturally occurring acylated gramicidins. Biochemistry 31, 7311–7319.

Woolley, G. A., Dunn, A., & Wallace, B. A. (1992). Gramicidin–lipid interactions induce specific tryptophan side chain conformations. Biochem. Soc. Trans. 20, 864–867.

Woolley, G. A., & Wallace, B. A. (1992). Model ion channels: gramicidin and alamethicin. J. Membr. Biol. 129, 109–136.

Woolley, G. A., & Wallace, B. A. (1994). Membrane protein structure: lessons from gramicidin. In: Membrane Protein Structure: Experimental Approaches (White, S. H., Ed.) pp. 314–334. Oxford University Press, New York.

Zhang, Z., Pascal, S. M., & Cross, T. A. (1992). A conformational rearrangement in gramicidin A: From a double-stranded left-handed helix to a single-stranded right-handed helix. Biochemistry 31, 8822–8828.

THE MIP FAMILY OF INTEGRAL MEMBRANE CHANNEL PROTEINS

Milton H. Saier, Jr., Aiala Reizer, and
Jonathan Reizer

Biomembranes
Volume 6, pages 361–375.

ISBN: 1-55938-663-0.

I. INTRODUCTION

In earlier reports we and others have described a family of integral membrane channel proteins from bacteria, animals, yeast, and plants that function, for example, in (1) metazoan development and neurogenesis, (2) communication between host plant cells and symbiotic bacteria, (3) organellar transport during plant seed germination and in mature plants, and (4) nutrient transport in single-celled bacteria (Shiels et al., 1988; Baker and Saier, 1990; Pao et al., 1991). We named this protein family the MIP family, after its first sequenced and best characterized member, the major intrinsic protein or main intrinsic polypeptide (MIP) of the bovine lens fiber cell membrane. The MIP family currently has 18 sequenced members.

Members of this family have been identified in both Gram-negative and Gram-positive bacteria as well as in yeast, plants, and animals (see Table 1). Five of these proteins (MIP, ChIP, TIP, NOD, and GLP) have been reported to have transmembrane-channel activities for various substrates (ions, water, dicarboxylates, and straight-chain carbon compounds, respectively; Heller et al., 1980; Nikaido and Rosenberg, 1985; Ehring et al., 1990, 1993; Ouyang et al., 1991; Preston et al., 1992; van Hoek and Verkman, 1992; Zeidel et al., 1992; Maurel et al., 1993, 1994; and see Ehring et al., 1993 for a concise summary of functional studies with MIP family proteins).

All MIP family members are integral membrane proteins presumed to have essentially the same topology (Doolittle, 1986). Each subunit of the MIP, ChIP, and NOD polypeptide chains appears to span the membrane as α-helices six times (Takemoto and Takehana, 1986; Horwitz and Bok, 1987; Takemoto et al., 1987, 1988; Zampighi et al., 1989; Verma, 1992; Preston et al., 1994), and evidence suggests that the native proteins exist in the membrane as tetrameres (Aerts et al.,

Table 1. Major Types of MIP Family Proteins

Abbreviations	*Name*	*Function*	*Organism*
MIP	Major intrinsic protein	Lens biogenesis ? (Ion transport?)	Animal
BIB	Big brain	Brain development	Animal
ChIP	Channel intrinsic protein	Osmoregulation (Water transport)	Animal
NOD	Nodulation protein	Communication (Dicarboxylate transport?)	Plant
TIP	Tonoplast intrinsic protein	α: seed germination w: Water stress resistance (Water transport)	Plant Plant
FPS	Fermentable sugar Defect suppressor (fdp-1)	?	Yeast
GLP	Glycerol facilitator	Carbohydrate supply (Glycerol transport)	Bacteria

1990; Smith and Agre, 1991; Verma, 1992). By contrast, the glycerol facilitator of *E. coli* may exist as a homodimer (Voegele et al., 1992). Biologically-active two-dimensional crystals of ChIP have recently been prepared, leading to the hope that three-dimensional structural features will soon be available (Walz et al., 1994).

Because the biochemical functions of the eukaryotic proteins are not well established, we have recently extended our earlier studies to provide: (1) a multiple alignment of the eighteen sequenced MIP family proteins with identification and discussion of fully or largely conserved residues, (2) statistical analyses of corresponding binary comparisons that establish homology for all members of this protein family, (3) a phylogenetic tree defining the relatedness of the currently known members of the family, (4) a signature sequence allowing rapid identification of new members of this family, (5) a pathway of MIP family protein evolution, and (6) a summary of analyses leading to an expanded understanding of structure–function relationships (Reizer et al., 1993). In this last respect, we showed that the first and second halves of these proteins, segments 1 and 2, respectively, which probably arose by tandem duplication of a common genetic element in prokaryotes prior to the advent of eukaryotes on earth, have diverged from each other in ways which suggest that they have acquired distinct functions. Thus, in closely related proteins, segments 2 are more similar to each other than are segments 1, but in distantly related proteins, segments 1 are more similar to each other than are segments 2. These observations suggest that segments 1 are more important for the generalized or common structure/function of these proteins (i.e., biogenesis or channel formation), whereas segments 2 are more important for the specialized or dissimilar functions of the proteins (i.e., channel specificity or regulation).

Based on published reports and the analyses described by Reizer et al. (1993), we propose that all members of the MIP family serve a common function in transmembrane transport of small molecules, although each class of these proteins may exhibit distinctive specificities and characteristics. Some of these proteins are regulated by protein kinase-mediated phosphorylation, whereas others are post-translationally modified by palmitoylation. Some of these proteins have distinctive N- and/or C-terminal extensions that are not homologous to other proteins in the database. BIB and FPS, for example, have extensive hydrophilic N- and C-terminal regions of this type (Rao et al., 1990; van Aelst et al., 1991). The degrees of conservation of the proposed sites of phosphorylation between members of the family are also discussed. Computer methods used in our studies have been described (Reizer et al., 1993).

II. PHYLOGENETIC TREE OF THE MIP FAMILY PROTEIN MEMBERS

Figure 1 shows a phylogenetic tree of 16 currently sequenced members of the MIP family. The major classes of these proteins are described in Table 1, while individual proteins are characterized in Reizer et al. (1993). The members include: (1) MIP

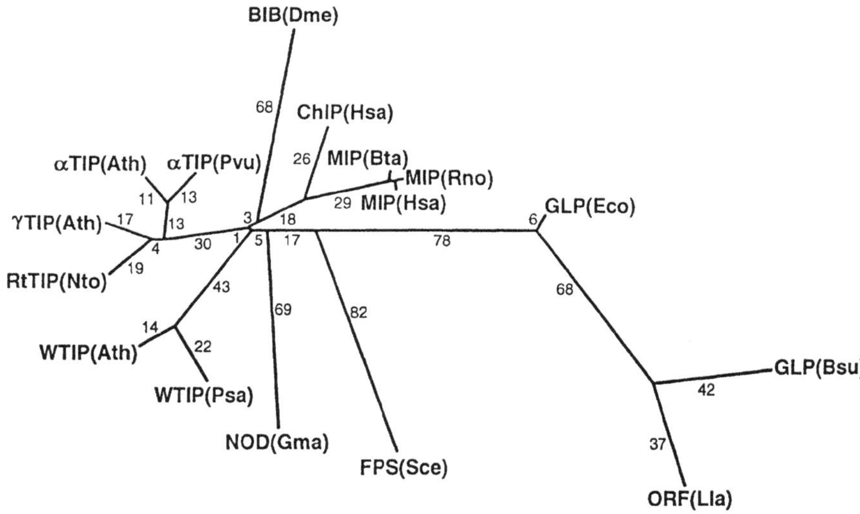

Figure 1. Phylogenetic tree of 16 members of the MIP family of presumed channel-forming proteins. Relative evolutionary distances are given adjacent to the branches unless less than one arbitrary unit. The program of Doolittle and Feng (1990) was used to calculate the relative branch lengths. The abbreviations used and references of published protein sequences are described in Reizer et al. (1993).

(major intrinsic protein of the lens fiber cell of the cow, human, and rat that appears to function in lens biogenesis or maintenance; partial sequence of the chicken protein is also available); (2) BIB (big brain, a neurogenic protein in *Drosophila melanogaster*; partial sequence of the human protein is also available) (Adams et al., 1992); (3) ChIP (channel intrinsic protein) of the human erythrocyte that can serve as a mercury-sensitive water channel; also found in kidney tubules (Preston et al., 1992, 1993; van Hoek and Verkman, 1992); (4) NOD (nodulin-26, a legume-encoded protein that allows communication between plant cells and bacteria in the nitrogen fixing bacteroid); (5) TIP (tonoplast intrinsic protein of various plants and plant tissues; the different TIP isoforms are designated Wsi or W (water stress induced), Rt (root), α (seed), and γ (all plant vegetative tissues, but not seeds; Höfte et al., 1992; Ludevid et al., 1992); (6) FPS (the fdp1 suppressor of *Saccharomyces cerevisiae* (FPS1) that plays a role in sugar fermentation and possibly in carbon catabolite repression); (7) GLP (the glycerol facilitator (GlpF) from three evolutionarily divergent bacteria); and (8) ORF1 (an open reading frame in *Lactococcus lactis* with no known physiological function; possibly a glycerol facilitator).

As shown in Figure 1, plant members of the MIP family cluster together. Thus, six different TIP proteins have been sequenced, and they fall into two major subclusters. The clustering pattern generally correlates with their cellular locations and induction patterns. The two water stress-induced proteins (W-TIP) cluster tightly together, while the remaining TIP proteins (α- and γ-TIP) form a second loose cluster. NOD26 comprises a branch close to the W-TIP branch (Figure 1). Animal proteins, MIP, ChIP, and BIB, cluster together. ChIP is the closest relative of MIP, while BIB is more distant, but still on the same major branch. All remaining proteins, from yeast and bacteria, are relatively distant from each other. The yeast FPS1 suppressor protein from *Saccharomyces cerevisiae* branches from the trunk of the tree after NOD, and it is followed by the three fully sequenced bacterial proteins.

III. HYDROPATHY ANALYSES OF REPRESENTATIVE MEMBERS OF THE MIP FAMILY

Several regions in the MIP family proteins are highly conserved. The dual appearance of the highly conserved NPA sequence in the N- and C-terminal halves of the proteins reflect a tandem intragenic duplication (Pao et al., 1991). The topological consequences of this duplication were considered briefly in an earlier report (Saier and Reizer, 1991) and are illustrated in Figure 2.

We previously published the hydropathy plots of eight dissimilar, but representative members of the MIP family, one from each of the major clusters of this protein family (Reizer et al., 1993; see Figure 1). The eight hydropathy plots showed striking similarities, but also interesting differences. First, BIB showed an extensive hydrophilic C-terminal domain with tandemly repeated glutamine residues. This domain is not homologous to any protein in the current databases. Short N-terminal hydrophilic regions are also observed for the plant and yeast proteins, but not for MIP, ChIP, or the bacterial proteins. Second, there is substantial variation in the hydropathy profiles of the different proteins. In the animal proteins, the first of these repeats (segments or repeats 1) are more hydrophobic than the second of these repeats (segments or repeats 2). The first two putative transmembrane-helical spanners are close together, while the third is more distant. In some cases, the intervening region between spanners 2 and 3 is quite hydrophobic (MIP, BIB, ChIP, and GLP) in segments 1, but not in segments 2. Third, the region between the repeated segments 1 and 2 is somewhat variable in length, being very short for αTIP, longer for the yeast and bacterial proteins, and intermediate and of fairly constant length for the remaining proteins (Figure 1).

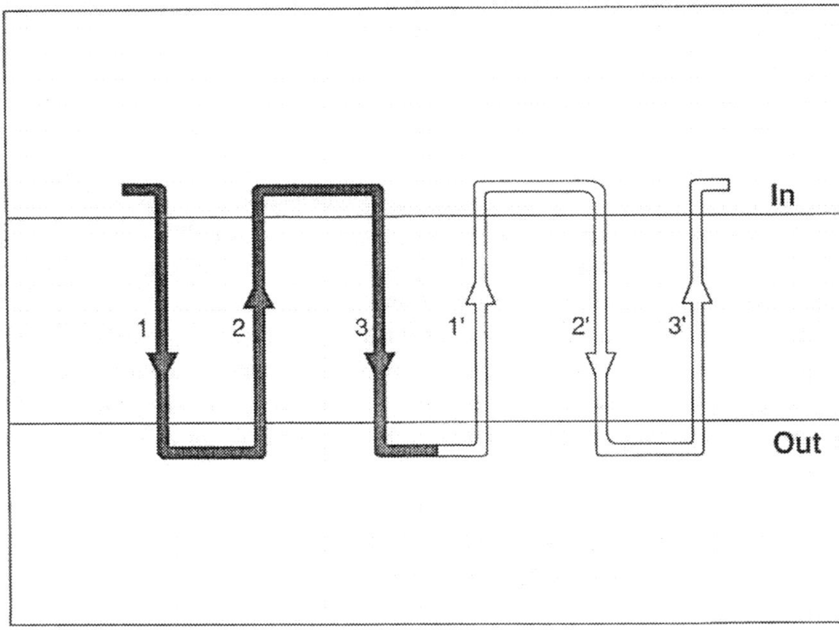

Figure 2. Topology (transmembrane orientation) of MIP and other members of the MIP family in the membrane showing the consequence of the tandem intragenic duplication event which is believed to have given rise to the primordial gene of the MIP family. Segments 1 (transmembrane spanners 1–3) are homologous to segments 2 (transmembrane spanners 4–6); yet homologous transmembrane helices are believed to be of opposite orientation in the membrane (Saier and Reizer, 1991; Pao et al., 1991).

IV. CORRELATION OF THE AVERAGE HYDROPATHY PLOT WITH THE AVERAGE SEQUENCE SIMILARITY PLOT

The eight hydropathy plots described in the previous section were averaged following multiple alignment of the sequences (Figure 3, upper panel). The three putative transmembrane α-helical spanners are clearly revealed for both segments 1 (N-terminal halves) and segments 2 (C-terminal halves). Furthermore, the average hydropathy plot clearly shows that the regions between spanners 2 and 3 in segments 1 are substantially more hydrophobic than the regions between spanners 5 and 6 in segments 2. Otherwise the two halves show remarkable similarity, not only in degree of hydrophobicity, but also in the average spacing between peaks.

In Figure 3 (lower panel), the average similarity scores for the same eight proteins are plotted as a function of position. Comparison of the data shown in Figures 3a and 3b clearly demonstrates that the six hydrophobic spanners consistently show

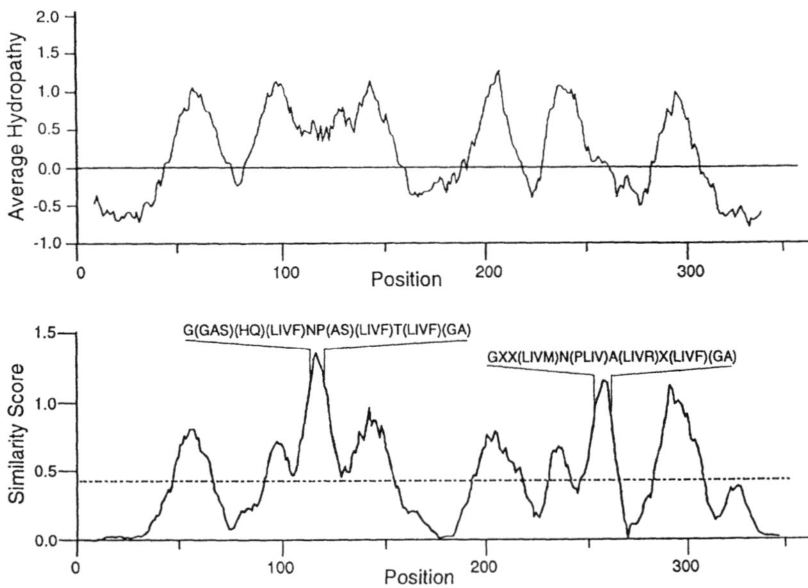

Figure 3. Average hydropathy plot (**A**, upper panel) and average similarity plot (**B**, lower panel) for eight representative proteins of the MIP family. The average hydropathy and average similarity plots were calculated as described in Reizer et al. (1993). The average similarity across the entire alignment in (**B**) is shown as a dotted line. The plot reveals the coincidence of hydropathy and similarity plots except for the peaks of similarity between spanners 2 and 3 and spanners 5 and 6. The degenerate sequences of these two hydrophilic regions were derived from all protein members of the MIP family and are provided above the similarity plot. In these sequences, ambiguous residues at a particular position are given in parentheses, whereas an X indicates that any residue can occur at the specified position.

peaks of sequence similarity while the loop regions between them usually show lower degrees of similarity. However, the hydrophilic regions between spanners 2 and 3 in segments 1 and spanners 5 and 6 in segments 2 show maximal degrees of sequence similarity. The first of these two regions is more conserved than the second, but both exhibit similar sequences which are characteristic of the currently known MIP family proteins. The sequences shown above these two peaks of sequence similarity in Figure 3b reveal the degree to which they resemble each other.

V. MULTIPLE ALIGNMENT OF SEQUENCED MIP FAMILY PROTEINS

The multiple alignment of 18 sequenced MIP family proteins was published by Reizer et al. (1993). Following a variable N-terminal region is a conserved, hydrophobic stretch of about 18 residues containing a single, fully conserved glutamyl residue. These 18 residues correspond to putative transmembrane spanner 1, and the conserved glutamyl residue is in the N-terminal portion of this spanner, near the cytoplasmic side of the membrane. A largely conserved threonyl residue occurs four residues after the fully conserved glutamyl residue. The loop region between spanners 1 and 2 shows many gaps, but the entire region from the beginning of putative transmembrane spanner 2 to the end of spanner 3 is well conserved. The consensus sequence reveals two five-residue sequences which are largely conserved: ISGAH and NPAVT, separated by a single variable hydrophobic residue (positions 95–105 in the multiple alignment). The NPAVT sequence is particularly well conserved. Within putative transmembrane spanner 3 are fully conserved glutaminyl and glycyl residues.

Following the central loop region where numerous gaps occur, putative transmembrane spanner 4 with its fully conserved glutamyl residue is found. Like the homologous spanner 1 in segment 1, a largely conserved threonyl residue occurs 4 residues after the conserved glutamyl residue. At the beginning of the loop region between spanners 4 and 5 is a nearly fully conserved aspartyl residue not found between spanners 1 and 2 in segment 1. However, greater variability is observed in the regions before and after spanners 4 and 5 as compared to the corresponding regions in segments 1. The NPAVT sequence of segment 1 is largely conserved as NPARS (Figure 3).

VI. SITES OF PHOSPHORYLATION OF MIP FAMILY PROTEINS

A site of phosphorylation of MIPBta is serine residue 243 near the C-terminus of the protein (Lampe and Johnson, 1989; Johnson et al., 1986; Lampe et al., 1986, 1988). This residue is not conserved in other MIP family members (not shown). The phosphorylation of this residue occurs by a cyclic AMP-dependent (A-type) kinase, and its phosphorylation by protein kinase A converts a voltage-independent Na^+ channel into a voltage-dependent channel (Ehring et al., 1991). MIPBta can also be phosphorylated by a Ca^{2+}-calmodulin-type (C-type) kinase, but the position of this latter phosphorylation is not known (Lampe et al., 1986; Lampe and Johnson, 1989). α-TIP from the kidney bean (TIPPvu; Johnson and Chrispeels, 1992) and NOD from soybeans (NODGma; Weaver et al., 1991) are also known to be phosphorylated. The position of phosphorylation in NOD by a Ca^{2+}-dependent calmodulin-independent kinase (Weaver et al., 1991; Weaver and Roberts, 1992; Miao et al., 1992) is a C-terminal seryl residue (S262). α-TIP from kidney bean is phosphorylated on an N-terminal serine (S7; Johnson and Chrispeels, 1992). These phosphorylated residues are always present in the N- and C-terminal portions of the MIP family proteins that are poorly conserved. In the case of NOD, phosphorylation correlated with stimulation of malate uptake across the peribacteroid membrane of the soybean nodule (Ouyang et al., 1991). Site-specific mutagenesis analyses of these phosphorylation sites in only one member of the MIP family, α-TIP of kidney bean, have been reported, and in this case, the S7A mutation abolished phosphorylation of the protein as expected (M. Chrispeels, personal communication). These results suggest that the poorly conserved N- and C-termini of MIP family proteins can serve in regulation rather than catalysis.

VII. SIGNATURE SEQUENCE FOR THE MIP FAMILY

Employing the complete multiple alignment of 18 MIP family proteins, the following signature sequence was derived: (HQ) (LIVMF) N P (AST) (LIVMF) T (LIVMF) (GA) (see Figure 3 and Reizer et al., 1993). Ambiguous residues at a specific position are given in parentheses. This signature sequence should assist in the identification of new protein members of this family (Bairoch, 1992).

VIII. MULTIPLE ALIGNMENT OF SEGMENTS 1 WITH SEGMENTS 2

When segments 1 and 2 were multiply aligned, only five residues were found to be conserved (Reizer et al., 1993). These residues are the glutamyl residue at the beginning of spanners 1 and 4, the NPA sequence between spanners 2 and 3 (segments 1) as well as spanners 5 and 6 (segments 2), and the glycyl residue found just after spanners 3 and 6. However, the fourth residues following the conserved glutamyl residues at the beginnings of spanners 1 in segments 1 and spanners 4 in segments 2 are threonyl residues in all but four of the proteins. In three of these proteins, a seryl residue replaces the conserved threonyl residue. Only in the second half of FPS(Sce) is a nonhydroxyl amino acid present at this position. Since segments 1 are oriented in the membrane from inside to out, and segments 4 are of the opposite orientation, these hydrophilic residues could partially comprise the hydrophilic lining of a transmembrane channel with the symmetrical transmembrane sequence: ETTE. No other conserved transmembrane hydrophilic residues are good candidates for potential hydrophilic channel liners.

A phylogenetic tree was generated for the two segments of MIP family proteins (Reizer et al., 1993). The results argued for the parallel evolution of segments 1 and 2 from a common ancestral sequence at rates which were at least comparable. The configuration of the tree implies that the tandem intragenic duplication that gave rise to the primordial gene encoding the 6-spanner precursor of the MIP family occurred before any of the duplication and divergence events that led to the different members of the family.

IX. STATISTICAL ANALYSES OF THE BINARY COMPARISONS FOR SEGMENTS 1 AND SEGMENTS 2

A matrix of binary comparison scores for representative segments 1 against each other, for corresponding segments 2 against each other, and for segments 1 against segments 2 was generated (Reizer et al., 1993). When segments 1 or 2 were compared with segments 1 or 2, respectively, clear homology was demonstrated. By contrast, when segments 1 were compared with segments 2, scores were much lower. The results substantiated the conclusion that divergence of segments 1 from segments 2 following the intragenic duplication event which gave rise to the intact protein of six-transmembrane spanners occurred substantially before the divergence events that gave rise to the different members of the MIP family.

Distance scores for all segments 1 against each other, for all segments 2 against each other, and for all segments 1 against all segments 2 were calculated and tabulated in decreasing order of the distance scores for the intact proteins being compared and analyzed. As shown in Figure 4, when the intact protein distance scores were large (i.e., for distantly related proteins), the corresponding values for the segment 1 versus segment 1 comparisons were of lesser magnitude than those for the segment 2 versus segment 2 scores. By contrast, when the intact protein distance scores were small (i.e., for closely related proteins), the distance scores for segments 1 versus segments 1 were generally larger than the segment 2 versus segment 2 scores. The curves presented in Figure 4 reveal that segments 1 are more similar to each other than are the segments 2 when they are derived from distantly related proteins, but that segments 2 are more similar to each other than are segments 1 when they are derived from closely related proteins.

Also presented in Figure 4 are the results of a similar averaging process when segments 1 are compared with segments 2. It can be seen that segments 1 are more similar to segments 2 for similar proteins than for more divergent proteins. This observation suggests that while segments 1 and 2 play different functional roles, they nevertheless function in some common capacity. This proposed common function must account for the evolutionary pressure which caused similar proteins to exhibit segment 1 sequences which more closely resemble segment 2 sequences than for the evolutionarily divergent proteins of the MIP family.

These observations lead to the conclusion that segments 1 must be more important for the generalized or common structure/function of the entire family of MIP proteins while segments 2 are more important for the specific or dissimilar functions of these proteins, shared only by closely related proteins. A common function primarily involving segments 1 might be proper membrane insertion, subunit interactions or channel formation, while a dissimilar function, determined more by the segments 2, might be solute specificity or channel regulation. Both segments might function cooperatively, for example, to create the channel, but with a greater dependency on segment 1 than on segment 2. The results of site-specific mutagene-

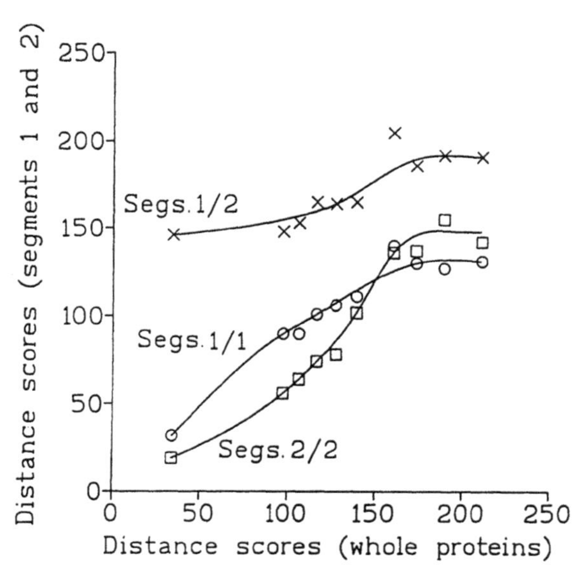

Figure 4. Correlation of the averaged distance scores between intact protein members of the MIP family (plotted on the X-axis) with the corresponding segment 1:segment 1 comparisons (Segs. 1/1), the corresponding segment 2:segment 2 comparisons (Segs. 2/2), and the corresponding segment 1:segment 2 comparisons (Segs. 1/2), plotted on the Y axis. See Reizer et al. (1993) for details.

sis studies are likely to provide answers as to the nature of the functional differentiation that these repeat sequences have undergone during evolutionary history.

X. CONCLUSIONS

The two halves of all MIP family proteins were apparently derived from a common ancestral gene, half as big as the current genes, and this ancestral gene tandemly duplicated internally to give rise to the primordial gene of the MIP family proteins. We presume that this event occurred in prokaryotes, well before divergence of the species. MIP family proteins are roughly related to each other as are the organisms in which they are found (Figure 1). Since only one such gene has been found in any one bacterium, but several have been found in single eukaryotes, we propose that a single MIP family gene was transmitted vertically from the prokaryotic ancestor to each of the eukaryotic kingdoms (animals, plants, and fungi), and that these genes then duplicated and diverged within eukaryotes to yield current MIP protein subfamilies.

Initial statistical analyses revealed that segments 2 exhibit a higher degree of average similarity (i.e., they have lower average distance scores) than do the segments 1, but that the segments 2 also exhibit more variability. The N-terminal halves apparently evolved for a function largely common to all MIP family proteins, while the second halves differentiated to play a subfamily specific role. Our statistical analyses appear to provide the first evidence for regional differentiation of permease proteins derived from repeated internal structural elements. However, the two segments have apparently retained at least one common function which depends on some degree of sequence similarity between the two segments.

We have recently conducted comparative functional analyses employing two of the MIP family proteins, that is, γ-TIPAth and GLPEco, following expression in *Xenopus* oocytes (Maurel et al., 1993, 1994). The results of [^{14}C]glycerol-uptake measurements have established that GLPEco, but not γ-TIPAth, can transport this straight-chain carbon compound. The results show, however, that of these two proteins, only γ-TIPAth readily transports water. Detailed analyses of [^{14}C]glycerol-uptake via GLPEco has confirmed virtually all of the functional characteristics reported by Heller et al. (1980). Thus, GLPEco is a simple channel protein capable of facilitating the nonspecific passage of straight-chain compounds across the membrane. We anticipate that all protein members of the MIP family will prove to exhibit simple channel characteristics, but for differing substrates.

The analyses reported previously (Pao et al., 1991; Reizer et al., 1993) have also led to the proposal that MIP family proteins evolved by duplication of a triplex, that is, a three-spanner precursor (Figure 2), about 2.5 billion years ago, shortly before the advent of eukaryotes on earth. By contrast, the mitochondrial carrier family (MCF) apparently evolved more recently by triplication of a duplex, that is, a two-spanner precursor, probably about 1.5 billion years ago, *after* the advent of mitochondria within eukaryotes (Kuan and Saier, 1993). All members of the latter family occur in eukaryotic organelles (most are within mitochondria), and none have yet been found in prokaryotes. Finally, it has been suggested that the major facilitator superfamily (MFS; Marger and Saier, 1993) evolved much earlier, possibly by duplication of a six-spanner precursor (Henderson and Maiden, 1990; Henderson, 1991; Griffith et al., 1992). This last-mentioned duplication event probably occurred greater than 3.5 billion years ago, long before the advent of eukaryotes on earth. It was probably only in this last family, the MFS, that the duplication and divergence events, giving rise to currently recognized protein members of the MFS, occurred long before the prokaryotic–eukaryotic divergence (Marger and Saier, 1993). Thus, although these three transport-protein families (the MIP family, the MCF, and the MFS) evolved independently of each other, by distinct routes, and at different times in evolutionary history, they all generated fundamental structural units of six-transmembrane spanners. This observation disproves the proposal of several investigators suggesting that transport proteins exhibiting the characteristic six-spanner unit (Saier and Reizer, 1991; Nikaido and Saier, 1992) might share a common evolutionary origin (Krupinski et al., 1989; Maloney, 1990; Saier, 1990; Jan and Jan, 1992). Why this six-transmembrane structural element is so often used for transport function has yet to be ascertained.

NOTE ADDED IN PROOF

The work reported in this chapter was completed as of 1994. Since then the MIP family has increased in size to nearly 100 sequenced proteins (Spring, 1996). In a recent topical review (J.H. Park and M.H. Saier, Jr. (1996), Phylogenetic characterization of the MIP family of transmembrane channel proteins, *Journal of Membrane Biology*) we have multiply aligned more than 50 representative, divergent, fully-sequenced members, used the resultant multiple alignment to derive current MIP family-specific signature sequences, and constructed a phylogenetic tree. The tree revealed novel features relevant to the evolutionary history of this protein family. These features plus an evaluation of functional studies lead to the postulates (1) that all current MIP family proteins derived from two or possibly three divergent bacterial paralogues, one a glycerol facilitator, the second an aquaporin, the third, possibly a cation channel, and (2) that most or all current members of the family have retained these or closely related biochemical functions.

REFERENCES

Adams, M. D., Dubnick, M., Kerlavage, A. R., Moreno, R., Kelley, J. M., Utterback, T. R., Nagle, J. W., Fields, C., & Venter, J. C. (1992). Sequence identification of 2,375 human brain genes. Nature 355, 632–634.

Aerts, T., Xia, J.-Z., Slegers, H., de Block, J., & Clauwaert, J. (1990). Hydrodynamic characterization of the major intrinsic protein from the bovine lens fiber membranes. J. Biol. Chem. 265, 8675–8680.

Bairoch, A. (1992). Prosite—a dictionary of sites and patterns in proteins. Nucl. Acids Res. 20(Suppl.), S2013–S2018.

Baker, M. E., & Saier, M. H., Jr. (1990). A common ancestor for bovine lens fiber major intrinsic protein, soybean nodulin-26 protein, and *E. coli* glycerol facilitator. Cell 60, 185–186.

Doolittle, R. F. (1986). Of Urfs and Orfs. A Primer on How to Analyze Derived Amino Acid Sequences. University Science Books, Mill Valley, CA.

Doolittle, R. F., & Feng, D.-F. (1990). Nearest neighbor procedure for relating progressively aligned amino acid sequences. Meth. Enzymol. 183, 659–669.

Ehring, G. R., Zampighi, G., Horwitz, J., Bok, D., & Hall, J. E. (1990). Properties of channels reconstituted from the major intrinsic protein of lens fiber membranes. J. Gen. Physiol. 96, 631–664.

Ehring, G. R., Lagos, N., Zampighi, G. A., & Hall, J. E. (1991). Phosphorylation modulates the voltage dependence of channels reconstituted from the major intrinsic protein of lens fiber membranes. J. Membr. Biol. 126, 75–88.

Ehring, G. R., Zampighi, G. A., & Hall, J. E. (1993). Does MIP play a role in cell–cell communication? In Gap Junctions (Hall, J. E., Zampighi, G. A., and Davis, R. M., Eds.) pp. 143–152. Amsterdam: Elsevier Publishers.

Griffith, J. K., Baker, M. E., Rouch, D. A., Page, M. G. P., Skurray, R. A., Paulsen, I. T., Chater, K. F., Baldwin, S. A., & Henderson P. J. F. (1992). Membrane transport proteins: Implications of sequence comparisons. Curr. Opin. Cell Biol. 4, 684–695.

Heller, K. B., Lin, E. C. C., & Wilson, T. H. (1980). Substrate specificity and transport properties of the glycerol facilitator of *Escherichia coli*. J. Bacteriol. 144, 274–278.

Henderson, P. J. F. (1991). Sugar transport proteins. Curr. Opin. Struct. Biol. 1, 590–601.

Henderson, P. J. F., & Maiden, M. C. J. (1990). Homologous sugar transport proteins in *Escherichia coli* and their relatives in both prokaryotes and eukaryotes. Phil. Trans. R. Soc. Lond. B 326, 391–410.

Höfte, H., Hubbard, L., Reizer, J., Ludevid, D., Herman, E. M., & Chrispeels, M. J. (1992). Vegetative and seed-specific forms of tonoplast intrinsic protein in the vacuolar membrane of *Arabidopsis thaliana*. Plant Physiol. 99, 561–570.

Horwitz, J., & Bok, D. (1987). Conformational properties of the main intrinsic polypeptide (MIP26) isolated from lens plasma membranes. Biochemistry 26, 8092–8098.

Jan, L. Y., & Jan, Y. N. (1992). Tracing the roots of ion channels. Cell 69, 715–718.

Johnson, K. D., & Chrispeels, M. J. (1992). A tonoplast-bound protein kinase phosphorylates tonoplast intrinsic protein. Plant Physiol. 100, 1787–1795.

Johnson, K. R., Lampe, P. D., Hur, K. C., Louis, C. F., & Johnson, R. G. (1986). A lens intracellular junction protein, MP26, is a phosphoprotein. J. Cell. Biol. 102, 1334–1343.

Krupinski, J., Coussen, F., Bakalyar, H. A., Tang, W.-J., Feinstein, P. G., Orth, K., Slaughter, C., Reed, R. R., & Gilman, A. G. (1989). Adenylyl cyclase amino acid sequence: Possible channel- or transporter-like structure. Science 244, 1558–1564.

Kuan, J., & Saier, M. H., Jr. (1993). The mitochondrial carrier family of transport proteins: Structural, functional and evolutionary relationships. Crit. Revs. Biochem. Mol. Biol. 28, 209–233.

Lampe, P. D., & Johnson, R. G. (1989). Phosphorylation of MP26, a lens junction protein, is enhanced by activators of protein kinase C. J. Membr. Biol. 107, 145–155.

Lampe, P. D., Bazzi, M. D., Nelsestuen, G. L., & Johnson, R. G. (1986). Phosphorylation of lens intrinsic membrane proteins by protein kinase C. Eur. J. Biochem. 156, 351–357.

Lampe, P. D., Eccleston, E. D., Howard, J. B., & Johnson, R. G. (1988). The amino acid sequence of *in vivo* phosphorylation site of lens MP26. J. Cell. Biol. 107, 558.

Ludevid, D., Höfte, H., Himelblau, E., & Chrispeels, M. J. (1992). The expression pattern of the tonoplast intrinsic protein γ-TIP in *Arabidopsis thaliana* is correlated with cell enlargement. Plant Physiol. 100, 1633–1639.

Maloney, P. C. (1990). A consensus structure for membrane transport. Res. Microbiol. 141, 374–383.

Marger, M. D., & Saier, M. H., Jr. (1993). A major superfamily of transmembrane facilitators that catalyse uniport, symport and antiport. TIBS 18, 13–20.

Maurel, C., Reizer, J., Schroeder, J. I., & Chrispeels, M. J. (1993). The vacuolar membrane protein γ-TIP creates water specific channels in *Xenopus* oocytes. EMBO J. 12, 2241–2247.

Maurel, C., Reizer, J., Schroeder, J. I., Chrispeels, M. J., & Saier, M. H., Jr. (1994). Functional characterization of the *Escherichia coli* glycerol facilitator, GlpF, in *Xenopus* ooxytes. J. Biol. Chem. 269 (in press).

Miao, G.-H., Hong, Z., & Verma, D. P. S. (1992). Topology and phosphorylation of soybean nodulin-26, an intrinsic protein of the peribacteroid membrane. J. Cell Biol. 118, 481–490.

Nikaido, H., & Rosenberg, E. Y. (1985). Functional reconstitution of lens gap junction proteins into proteoliposomes. J. Membr. Biol. 85, 87–92.

Nikaido, H., & Saier, M. H., Jr. (1992). Transport proteins in bacteria: Common themes in their design. Science 258, 936–942.

Ouyang, L., Whelan, J., Weaver, C. D., Roberts, D. M., & Day, D. A. (1991). Protein phosphorylation stimulates the rate of malate uptake across the peribacteroid membrane of soybean nodules. FEBS Letts. 293, 188–190.

Pao, G. M., Wu, L.-F., Johnson, K. D., Höfte, H., Chrispeels, M. J., Sweet, G., Sandal, N. N., & Saier, M. H., Jr. (1991). Evolution of the MIP family of integral membrane transport proteins. Mol. Microbiol. 5, 33–37.

Preston, G. M., Carroll, T. P., Guggino, W. B., & Agre, P. (1992). Appearance of water channels in *Xenopus* oocytes expressing red cell CHIP28 protein. Science 256, 385–387.

Preston, G. M., Jung, J. S., Guggino, W. B., & Agre, P. (1993). The mercury-sensitive residue at cysteine 189 in the CHIP28 water channel. J. Biol. Chem. 268, 17–20.

Preston, G. M., Jung, J. S., Guggino, W. B., & Agre, P. (1994). Membrane topology of aquaporin CHIP. Analysis of functional epitope-scanning mutants by vectorial proteolysis. J. Biol. Chem. 269, 1668–1673.

Rao, Y., Jan, L. Y., & Jan, Y. N. (1990). Similarity of the product of the *Drosophila* neurogenic gene *big brain* to transmembrane channel proteins. Nature 345, 163–167.

Reizer, J., Reizer, A., & Saier, M. H., Jr. (1993). The MIP family of integral membrane channel proteins: Sequence comparisons, evolutionary relationships, reconstructed pathway of evolution, and proposed functional differentiation of the two repeated halves of the proteins. Crit. Rev. Biochem. Mol. Biol. 28, 235–257.

Saier, M. H., Jr. (1990). Evolution of permease diversity and energy-coupling mechanisms: An introduction. Res. Microbiol. 141, 281–286.

Saier, M. H., Jr., & Reizer, J. (1991). Families and superfamilies of transport proteins common to prokaryotes and eukaryotes. Curr. Opin. Struct. Biol. 1, 362–368.

Shiels, A., Kent, N. A., McHale, M., & Bangham, J. A. (1988). Homology of MIP26 to Nod26. Nucl. Acids Res. 16, 9348.

Smith, B. L., & Agre, P. (1991). Erythrocyte M28,000 transmembrane protein exists as a multi-subunit oligomer similar to a channel protein. J. Biol. Chem. 266, 6407–6415.

Takemoto, L., & Takehana, M. (1986). Major intrinsic polypeptide (MIP26K) from human lens membrane: Characterization of low-molecular-weight forms in the aging human lens. Exp. Eye Res. 43, 661–667.

Takemoto, L., Kodama, T., & Takemoto, D. (1987). Antisera to synthetic peptides of MIP26K as probes of changes in opaque vs. transparent regions within the same human cataractous lens. Exp. Eye Res. 45, 179–183.

Takemoto, L., Kuck, J., & Kuck, K. (1988). Changes in the major intrinsic polypeptide (MIP26K) during opacification of the Emory mouse lens. Exp. Eye Res. 47, 329–336.

Van Aelst, L., Hohmann, S., Zimmermann, F. K., Jans, A. W. H., & Thevelein, J. M. (1991). A yeast homologue of the bovine lens fibre MIP gene family complements the growth defect of a *Saccharomyces cerevisiae* mutant on fermentable sugars but not its defect in glucose-induced RAS-mediated cAMP signalling. EMBO J. 10, 2095–2104.

van Hoek, A. N., & Verkman, A. S. (1992). Functional reconstitution of the isolated erythrocyte water channel CHIP28. J. Biol. Chem. 267, 18267–18269.

Verma, D. P. S. (1992). Soybean nodulin-26: A channel protein conserved from bacteria to mammals. In: Transport and Receptor Proteins of Plant Membranes (Cooke, D. T., and Clarkson, D. T., Eds.), pp. 113–117. Plenum Press, New York.

Voegele, R., Sweet, G., & Boos, W. (1992). Isolation and characterization of the glycerol transport protein, GlpF, from *Escherichia coli*. BioEngineering February, p. 57.

Walz, T., Smith, B. L., Zeidel, M. L., Engel, A., & Agre, P. (1994). Biologically active two-dimensional crystals of aquaporin CHIP. J. Biol. Chem. 269, 1583–1586.

Weaver, C. D., & Roberts, D. M. (1992). Determination of the site of phosphorylation of nodulin 26 by the calcium-dependent protein kinase from soybean nodules. Biochemistry 31, 8954–8959.

Weaver, C. D., Crombie, B., Stacey, G., & Roberts, D. M. (1991). Calcium-dependent phosphorylation of symbiosome membrane proteins from nitrogen-fixing soybean nodules. Plant Physiol. 95, 222–227.

Zampighi, G. A., Hall, J. E., Ehring, G. R., & Simon, S. A. (1989). The structural organization and protein composition of lens fiber junctions. J. Cell Biol. 108, 2255–2276.

Zeidel, M. L., Ambudkar, S. V., Smith, B. L., & Agre, P. (1992). Reconstitution of functional water channels in liposomes containing purified red cell CHIP28 protein. Biochemistry 31, 7436–7440.

ION CHANNELS OF MITOCHONDRIAL MEMBRANES

Carmen A. Mannella and Kathleen W. Kinnally

Biomembranes
Volume 6, pages 377–410.

ISBN: 1-55938-663-0.

I. INTRODUCTION

A. Mitochondrial Membrane Transport and Oxidative Phosphorylation

The mitochondrion is arguably the most heavily studied cellular organelle, befitting its central role in eukaryotic energy metabolism. It has an outer and an inner membrane which are normally closely apposed along the periphery of the organelle, with intermembrane contacts occurring at frequent intervals. The inner membrane encloses a protein-rich matrix that contains (among other things) enzymes of the citric acid cycle and the organelle's own genetic and protein synthetic machinery. The inner membrane has numerous invaginations called cristae which can assume a variety of shapes depending on osmotic and metabolic conditions. The *coupling event* of mitochondrial energy metabolism is the establishment across the inner membrane of ion gradients, a process driven by substrate oxidation and which, in turn, drives ADP phosphorylation and other energy-requiring reactions. Thus, oxidative phosphorylation is intimately connected to the transmembrane flux of ions (especially protons) and metabolites (such as carboxylic acids and nucleotide cofactors) between the mitochondrion and cytosol, as well as between the different internal compartments of the mitochondrion.

There are several classes of membrane transport proteins involved in the intricate and incompletely understood events of mitochondrial energy metabolism. These include the respiratory chain complexes that contain the proton pumps needed to generate the chemiosmotic gradients across the inner mitochondrial membrane. There is also a family of carrier proteins in this membrane that are involved in the transport of specific metabolites between the matrix and the intermembrane/intracristal spaces. The topic of the present review is a third class of membrane transporters, comprised of channels or pores, entities that define water-filled transmembrane pathways of varying size and selectivity. In current usage, the term "channel" is often associated with such interesting characteristics as ion-specificity and gating, while "pores" are often thought of as large and static holes. No such distinction is made here, since several large-conductance mitochondrial pores exhibit both selectivity and gating.

B. Channels of Mitochondrial Membranes

One of the first voltage-gated channels discovered with bilayer technology in the mid-1970s was a mitochondrial pore-former that was named VDAC for its voltage-dependent closure and anion selectivity (Schein et al., 1976). The 30 KD VDAC protein was found to be a major component of the outer mitochondrial membrane, occurring at densities of 10^3 to 10^4 per square μm of the native membrane's surface (Freitag et al., 1982; DePinto et al., 1987). The relative abundance of the VDAC protein has aided in its isolation and eventually in the identification and cloning of its gene. The high density of VDAC in the outer mitochondrial membrane also is a factor in the relative ease with which two-dimensional crystals of the channel can be grown. This combination of factors has made VDAC a useful and convenient model for structure–function studies of large-conductance protein pores (see following; also Vodyanoy and Bezrukov, 1992; Zambrowicz and Colombini, 1993).

The existence of many copies of a large-conductance pore like VDAC in the outer mitochondrial membrane is not surprising since this membrane has long been known to be very permeable to small solutes. Thus, it is commonly presumed that VDAC's main physiological role is to serve as the general permeability pathway for ions and metabolites across the outer membrane (Colombini, 1979). More recent experiments using a variety of electrophysiological techniques have produced evidence for several other classes of mitochondrial channel activities occurring in both the outer and inner membranes of this organelle. Unlike VDAC, these channels have not yet been identified at the molecular level, nor is their physiological function obvious. Given that, VDAC is only weakly selective and present in great abundance in the outer mitochondrial membrane, why are additional channels present in this membrane? Likewise, if maintenance of ion gradients across the inner mitochondrial membrane is essential to oxidative phosphorylation, why does this membrane contain channels which (if opened) would uncouple the process?

This review is intended to summarize current knowledge about the channel activities present in mitochondrial outer and inner membranes. It describes the progress being made towards a molecular characterization of the VDAC channel and provides at least tentative answers to the above questions about the other channel activities that are associated with this organelle.

II. OUTER MITOCHONDRIAL MEMBRANE CHANNELS

A. The Voltage-dependent Anion-selective Channel (VDAC)

The VDAC channel is formed by a protein of about 280 amino acids encoded by nuclear DNA, synthesized on cytoplasmic ribosomes and incorporated into the outer mitochondrial membrane via a receptor-mediated mechanism (Mihara and Sato, 1987; Kleene et al., 1987; Pfaller and Neupert, 1987; Takahashi et al., 1993). Several different VDAC genes have been cloned and the cDNA sequenced, including those of bakers yeast (Mihara and Sato, 1987; Forte et al., 1987), *Neurospora* (Kleene et al., 1987), and slime mold (Troll et al., 1992). In addition, the amino acid sequence has been directly determined for a VDAC protein isolated from human lymphocytes (Kayser et al. , 1989). It was subsequently found that there are two closely related human genes for VDAC that are differentially expressed in various tissues (Blachly-Dyson et al.,1993). Interspecies similarity among VDAC sequences is low (e.g., 43% overall amino acid identity between *Neurospora* and yeast, 29% between *Neurospora* and man), despite the remarkable similarity in electrophysiological characteristics of VDAC from different organisms (DePinto et al., 1987). This suggests that VDAC may be one of a growing list of proteins that have highly conserved structure in the absence of strict sequence preservation.

Examination of the amino-acid sequences of different species of VDAC reveals numerous stretches of alternating polar and nonpolar residues. This amphipathic pattern led H.R. Guy to suggest that the polypeptide might fold as a cylindrical β-sheet (i.e., β-barrel) in the membrane bilayer, with nonpolar residues forming a hydrophobic exterior and polar residues lining a hydrophilic channel (Forte et al., 1987). The existence of β-barrel pores has since been confirmed by electron and x-ray crystallography in the case of bacterial porins, which are large-conductance channels found in the outer envelope of Gram-negative bacteria (Weiss et al., 1990; Jap et al., 1991). While there is no sequence homology between VDAC and bacterial porins (Forte et al., 1987; cf. Mannelle et al., 1996), evidence from circular dichroism indicates that VDAC's secondary structure is predominantly β-sheet, making a β-barrel structure very likely (Shao et al., 1996).

Electrophysiological Behavior in Planar Bilayers

When detergent-solubilized outer mitochondrial membranes or purified VDAC are added to one side of a planar phospholipid bilayer under voltage–clamp conditions, incremental increases in current are observed corresponding to insertion of channels with unit conductances of approximately 650 pS (where S = Siemen = 1/Ohm) in physiological salt. As the transmembrane potential is increased in either direction, transitions are observed corresponding to partial closure of the channels. There appear to be several different partly closed substates of the channel, with the most commonly observed conductance decrease equal to approximately 300 pS (Colombini, 1979; Freitag et al.,1982). In bilayers, VDAC displays symmetrical voltage gating, that is, amplitude of V_0, the voltage at which half the channels partially close (or for which a given channel will be open half the time), is typically in the range 20–40 mV for potentials applied in either direction across the membrane. In their fully open states, most VDAC pores are slightly more permeable towards anions than cations. Typically, the ratio P_{Cl^-}/P_{K^+} (where P_X is the permeability for ion X) is 1.5 to 2 (based on measurements of reversal potentials), although it is reported to be zero for some species of VDAC (Doring and Colombini, 1985; DePinto et al., 1987). The ion selectivity reverses when the channel occupies its partly closed substates, for which P_{K^+}/P_{Cl^-} has been measured at 8 (Colombini, 1989).

As noted previously, the basic electrical parameters (unit conductances, V_0, selectivity) of VDAC incorporated into phospholipid bilayers are remarkably consistent for detergent-extracts of outer membranes from numerous sources assayed in different laboratories. This is also true of a fourth parameter, n, which is related to voltage sensitivity and which is derived from the steepness of the slopes of current vs. voltage (I/V) curves (Ehrenstein et al., 1970; Schein et al., 1976). In terms of a conventional model of voltage gating, n is the equivalent number of charges that move completely across the membrane electric field during the gating process. This parameter varies for different species of VDAC, usually falling within the range 2–4 (see summaries in Colombini, 1989 and DePinto et al., 1987). The interspecies variation in n appears systematic, as pointed out by Sorgato and Moran (1993), such that n increases with decreasing V_0, and vice versa. The net effect is to keep the energy needed to partly close the different species of VDAC channels (related to the product nV_0) more or less constant.

The kinetics of the transitions between the fully open and partly closed conductance states of single VDAC channels also have been characterized in bilayer systems (Ludwig et al., 1986). Closing is much slower than opening, with half-times on the order of seconds and milliseconds, respectively. Also, the speed of channel closure increases with the magnitude of the potential difference across the membrane.

Patch–clamp Studies

Unlike the situation described above for bilayer-reconstituted VDAC ($VDAC_B$), there is less consensus about the behavior of VDAC when incorporated into liposomes ($VDAC_L$) and assayed by patch–clamp techniques. In several labs, liposomes containing mitochondrial outer membranes or isolated VDAC have been

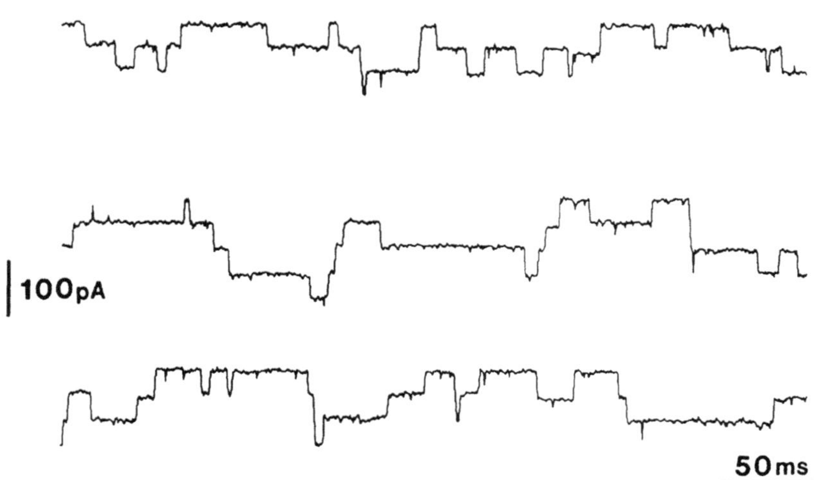

Figure 1. Current recording from a membrane patch excised from a liposome fused (according to Criado and Keller, 1987) with outer membranes isolated from *Neurospora crassa* mitochondria (by the procedure of Mannella, 1982). Under voltage–clamp conditions (in this case, +90 mV, measured relative to the bath), VDAC-size transitions (averaging 450 pS in 0.15 M KCl) are readily observed. (Source: DeBritz, Kinnally, Mannella, unpublished data.)

found to display conductances with a mean size comparable to that seen with planar bilayer systems. This is true for liposomes that have been reconstituted with VDAC both in the presence (Wunder and Colombini, 1991; Shao et al., forthcoming) and absence of detergent (Figure 1). However, at least one group has obtained discordant results. When applying patch–clamp techniques to detergent-free liposomes containing rat-brain outer mitochondrial membranes, Moran and Sorgato (1992) observed a variety of conductances in the range 100- to 500-pS, but none typical of the fully-open-to-fully-closed transition of VDAC in bilayers (650 pS).

Even in those cases where the observed mean conductance values of $VDAC_L$ and $VDAC_B$ are similar, there is a significant difference in the gating behavior of the two kinds of channel activities. Unlike $VDAC_B$, liposome-incorporated VDAC frequently exhibits markedly asymmetric voltage gating. This is demonstrated in Figure 2, which shows the current–voltage (I/V) curve for a liposome patch containing a single channel reconstituted from a fraction of purified VDAC protein. While the channel closes for both positive and negative voltages, the V_0 values on the two sides of the curve are very different (–25 mV vs. +5 mV). Furthermore, the

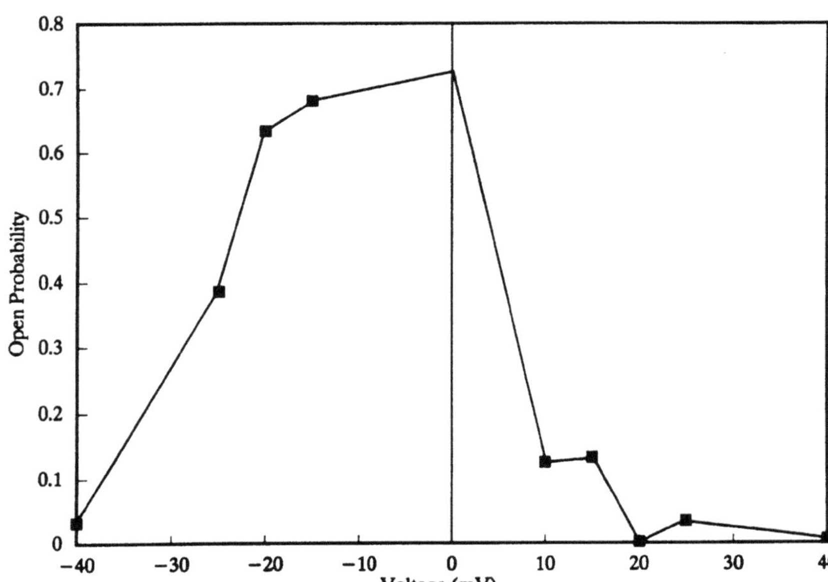

Figure 2. Voltage dependence of a single VDAC channel. Current was recorded under voltage–clamp conditions (Kinnally et al., 1993) from a patch excised from a liposome reconstituted with purified *N. crassa* VDAC (Shao et al., 1996) and bathed in symmetrical 0.15 M KCl 5 mM HEPES pH 7.4. Shown is a plot of the probability that the channel is open at various voltages, calculated as the percent of total time spent in the fully open (608 pS) state.

voltage-dependence of single VDAC channels in outer-membrane-fused liposomes is even more asymmetric—frequently these channels close for voltages applied in only one direction (Kinnally, unpublished observations). These findings are consistent with the general notion that VDAC must have (at least) two different gating mechanisms responding to voltages of opposite polarity. Why the difference between these two gating mechanisms is more exaggerated in liposome than in planar bilayer systems remains to be explained.

The Functional State of Native VDAC

At present, the inability to detect the 650 pS transitions characteristic of $VDAC_B$ in some patch–clamp studies of outer-membrane-fused liposomes has not been satisfactorily explained. Also, the 650 pS transition has not been observed in native (not liposome-fused) mitochondrial outer membranes, although the same group observes this transition in planar bilayers reconstituted with VDAC in the absence of detergent (Mirzabekov et al., 1993), ruling out detergent as a factor influencing the occurrence of this transition. The suggestion has been made (Mirzabekov et al., 1993) that, perhaps, VDAC in the native outer membrane does not occupy the same fully open conductance state observed consistently in planar bilayers and less consistently in liposome-reconstituted systems.

There are several other possible explanations for these apparent discrepancies. One is that the 650 pS transition of VDAC is sensitive to one or more as yet unidentified experimental parameters in the patch–clamp technique. Another possibility is that outer membrane regions containing several fully open VDAC are so highly conducting that the formation of high-resistance (10 GΩ) seals is impossible. If, in native outer membranes and outer-membrane-fused liposomes, the fully open channels occur in clusters, the low-resistance patches formed at such regions would be unsuitable for single-channel studies and so would be rejected. In the first reported patch–clamp studies of mitochondria, Tedeschi et al. (1987) measured the currents across low-resistance (0.1–1 GΩ) patches on liver mitochondria and isolated fungal outer mitochondrial membranes. While the observed currents were too large to allow detection of single channels, the conditions did permit measurement of I/V curves, some of which were found to resemble the symmetric bell-shaped I/V curves characteristic of VDAC in bilayers. More commonly, the I/V curves were atypical of VDAC, showing a voltage-induced conductance *increase* in one voltage direction, which was attributed to a novel, non-VDAC outer membrane channel (see following). Interestingly, the conductance of these outer-membrane patches at near-zero voltage was approximately that expected for the known surface density of VDAC protein if most or all of the channels were in a conducting state.

As the previous discussion suggests, the issue of defining the functional state of native VDAC is largely unresolved. Two points should be made in this regard: First, given the large density of VDAC proteins in the outer mitochondrial membrane, only a small fraction need to be open at any given time for the membrane to be very permeable to metabolites. For example, based on the permeability of VDAC to a dicarboxylic acid (fumarate; Colombini, 1979), it can be estimated that less than 1% of the VDAC channels need to be fully open in order to sustain maximum steady-state respiration rates when substrate pools are in the millimolar range (Mannella, unpublished calculation). Second, it is important to avoid the conceit that a particular *in vitro* technique (electrophysiological or other) necessarily yields the correct answer about the *in vivo* behavior of VDAC or any other channel. Different techniques provide complementary information about channels under given sets of conditions, none of which exactly mimics that of the native membrane *in situ*. This caveat extends even to the patch–clamp technique, which assays a piece of membrane that has been severely distended (disrupting its normal shape and its associations with other cellular or organellar components) and sealed against a glass micropipette by a process that is poorly understood. In the case of VDAC, the information gained from different *in vitro* techniques is providing important insights into what this channel is capable of doing, but does not yet allow us to say with certainty how the channel actually behaves *in situ*.

Effectors of Voltage-dependent Gating

Information about the factors and conditions that alter VDAC's voltage dependence have provided valuable insight into the underlying mechanism of gating, as well as its possible physiological significance. Several classes of effectors of VDAC's voltage-dependent closure have been described.

1. *pH*: Voltage gating of VDAC is markedly pH dependent. Alkaline pH reversibly inhibits voltage-induced closure of the VDAC channel (Bowen et al., 1985). The midpoint of inhibition was found to be near the pK_a of lysine (10.6), suggesting the involvement of this positively charged amino acid in VDAC's response to transmembrane potentials. Lysines are also implicated by the loss of voltage-dependence caused by succinylation (Doring and Colombini, 1985), which reverses the charge on exposed lysines. Neither elevated pH nor succinylation cause large-scale changes in the structure of crystalline VDAC, (Mannella and Frank, 1986; Guo and Mannella, 1992) suggesting that their effects on gating are not due to gross conformational changes in the protein. At the other extreme of pH, exposure of VDAC to acid conditions (pH 3–4) has been reported to enhance voltage-dependent closure, as does amidation of the protein which reverses the charge on carboxyl groups of acidic residues (Mirzabekov and Ermishkin, 1989). Thus, it is likely that acidic as well as basic amino acids play a role in the voltage sensing mechanism of this channel.

2. *Polyanions*: A copolymer of methacrylate, styrene, and maleate (molecular weight = 10,000), commonly referred to as Konig's polyanion after its inventor, has been found to alter VDAC's voltage sensitivity at nanomolar concentrations

(Colombini et al., 1987). This polyanion's effects are somewhat complex and depend on the polarity of the applied potential with respect to the side on which the polyanion is added. Konig's polyanion reduces V_0 from (approximately) −30 mV to −5 mV when added to the negative-potential side, while increasing V_0 for positive potentials to 80–100 mV (Benz et al., 1990). Another polyanion, dextran sulfate (molecular weight = 8000), exerts a similar effect at higher (micromolar) levels on VDAC at negative potentials, increasing n by up to 14-fold, but maintaining nV_0 constant (Mangan and Colombini, 1987). However, unlike Konig's polyanion, dextran sulfate does not stabilize the open state when added to the positive side of the bilayer. There is, as yet, no accepted explanation for the effects of these two polyanions on VDAC. Since both anionic polymers are too large to diffuse across the pore, they might set up a local Donnan field that adds to the applied potentials, thereby inducing channel closure at lower applied negative potentials (Colombini, 1987). However, this simple mechanism would predict that both Konig's polyanion and dextran sulfate should stabilize VDAC's open state at positive potentials when, in fact, only the former polymer does so. Apparently, the mechanism is more complex and may involve specific interactions of one or both polymers with the channel protein or with membrane lipid. (Konig's polyanion is considerably lipophilic since one-half of its residues are comprised of styrene.)

3. *Aluminum*: Like the effects of polyanions, those of aluminum and other group IIIA metals on VDAC's voltage-gating characteristics are complex. Dill et al. (1987) were the first to show that aluminum (10–100 μM), when added symmetrically on both sides of a bilayer containing VDAC, inhibits voltage-dependent closure for potentials of either polarity. However, when effects of one-sided additions were explored with indium, Zhang and Colombini (1990) found that the metal inhibits closure when added to the negative-potential side of the bilayer, but that it increases the extent of closure (reducing V_0 and speeding up the kinetics of closing) when added to the positive side. More recently, Mirzabekov et al. (1993) monitored the effects of aluminum on VDAC in planar bilayers that were reconstituted with rat outer mitochondrial membranes using a detergent-free protocol (involving osmotic-gradient-induced fusion). Micromolar aluminum was found to irreversibly increase the voltage sensitivity of VDAC when added to the positive-potential side of the bilayer, but not when addition was to the negative side. Titration of the effects on VDAC of different group IIIA metals indicates that the active species are the uncharged metal trihydroxides and not the metal cations which are rare at neutral pH. As with polyanions, interaction of the metal hydroxides with VDAC has been suggested to involve lysines on the voltage sensor (with binding shifting equilibrium from $R\text{-}NH_3^+$ to $R\text{-}NH_2$ at neutral pH; Zhang and Colombini, 1989), although interactions with membrane lipids cannot be ruled out (Mirzabekov et al., 1993).

4. *NADH*: Zizi et al. (1994) showed that NADH has a specific effect on VDAC from several species, increasing n from 2–3 to 4–6 without altering V_0 at concentrations of 100 μM. A candidate subsequence for the dinucleotide-binding site has

been found near the C-terminus of the polypeptide in a region previously shown to have significant sequence similarity with the adenine nucleotide transporter (Mannella, 1990; Mannella and Auger, 1986). Regulation of outer membrane permeability by interaction of the VDAC channel with NADH might have important physiological implications. For example, as Zizi et al. (1994) point out, NADH is a by-product of glycolysis and glucose is known to suppress mitochondrial respiration in several tissues (the Crabtree effect).

5. *Mitochondrial "modulator protein"*: A soluble protein fraction released from lysed mitochondria has been found to affect VDAC's voltage dependence in a manner similar to polyanions (Holden and Colombini, 1988). V_0 is decreased about one-half while n is increased 2–3 fold, the net effect being a slight increase in nV_0 (Liu and Colombini, 1992). This modulator activity appears to be associated with a protein of molecular weight 54,000 and pI around 5 (Liu et al., 1994).

The existence of endogenous modulators of VDAC's permeability (NADH and the modulator protein) lends further credence to the physiological significance of the channel's ability to switch to lower conductance states. In fact, there is evidence that diffusion of ATP across the outer membrane is greatly restricted by preincubation of intact mitochondria with the polyanion, presumably due to the switching of VDAC to its less conducting, cation-selective substrate (Benz et al., 1988; Liu and Colombini, 1992).

Structure of VDAC

Most of our present knowledge of the structure of VDAC has come from electron microscopic studies of 2-D crystals of the channel (Mannella, 1982, 1989, 1990; Mannella and Guo, 1992). VDAC forms extended periodic arrays when lipid is depleted from the mitochondrial outer membrane by the action of phospholipase A_2 (Mannella, 1984, 1986). Similar (though perhaps less extensive) ordering of VDAC may occur in the outer membrane *in situ*, since mitochondria contain a calcium-activated phospholipase A_2 (Waite and Sisson, 1971). The usually observed crystal unit in phospholipase-treated outer mitochondrial membranes (Figure 3) is a parallelogram with sides of 13.3 nm and 11.5 nm, separated by an angle of 109°. Inside the parallelogram, six VDAC channels are arranged in a group with two-fold rotational symmetry. (Thus, the arrays are said to exhibit p2 symmetry.) The channels within each hexameric group exhibit center-to-center spacings that vary (depending on position) from 4.3 nm to 5.3 nm. Crystal density and space-filling considerations indicate that each VDAC channel is likely formed by a single 31 kD polypeptide (Mannella, 1987), which was later confirmed by STEM mass measurements on freeze–dried membrane crystals (Thomas et al., 1991; also, H. Chen and C. Mannella, unpublished observations) and by experiments with engineered yeast strains expressing both wild-type and mutant VDAC genes (Peng et al., 1992a).

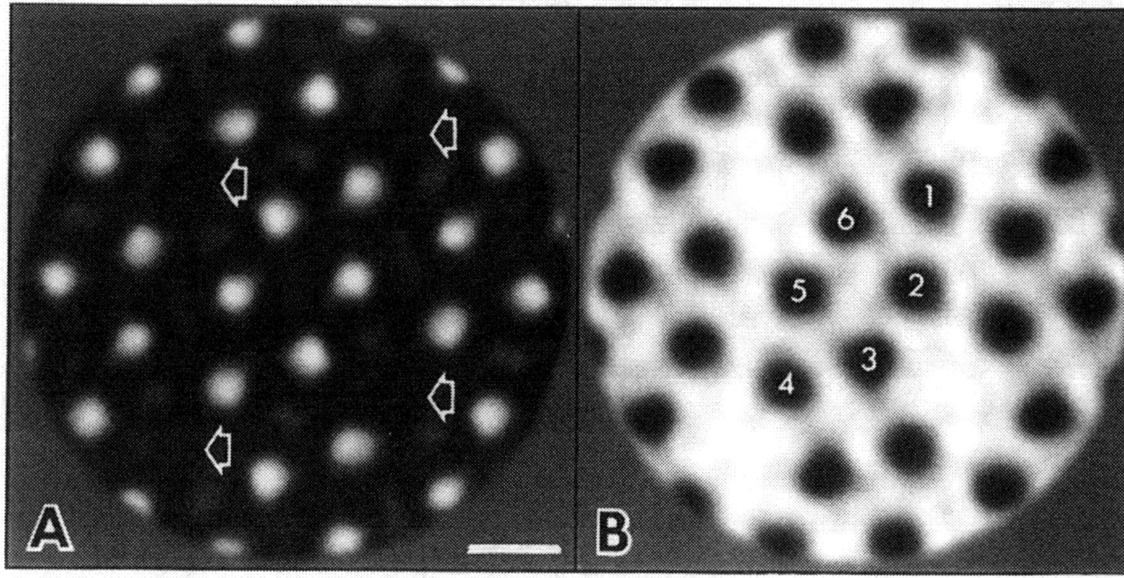

Figure 3. Computer averages of low-dose electron microscopic images of crystalline *N. crassa* VDAC embedded in (**A**) vitreous ice and (**B**) aurothioglucose. The six channels in the central oblique unit cell are numbered in (**B**) The hydrophilic lumens are white in (**A**) because they are filled with water (low density) and black in (**B**) because they contain gold atoms (attached to glucose). The projected density of the protein wall of each lumen is clearly visible in (**A**), as are the lateral arms at the four corners of the unit cell (arrows). (Reproduced with permission from Mannella et al., 1992.)

Three-dimensional reconstruction of VDAC membrane crystals embedded in negative stain indicates that each pore has a lumen that traverses the outer mitochondrial membrane with an inner diameter of about 3 nm (Mannella et al., 1984). There is no indication of merger of adjacent lumens on one side of the membrane as seen with some bacterial porins (Engel et al., 1985). The close packing of the VDAC channels and their wide inner diameter suggest that the wall forming the lumen is thin (see following). This was corroborated by cryo- electron microscopy of unstained, frozen-hydrated VDAC arrays (Mannella et al., 1989; Guo and Mannella, 1993) which allows direct visualization of the protein, lipid, and water domains in the crystals (Figure 3a). Several important aspects of VDAC's structure have been provided by cryo-imaging of VDAC arrays. For example, the mean diameter of the wall forming the lumen can be accurately determined to be 3.8 +/– 0.1 nm. If the lumen is a roughly circular β-barrel like bacterial porin's (see preceding), this is the diameter of the β-barrel measured at the C_α backbone. This lumen diameter is somewhat wider than that of bacterial porin, which is composed of a β-barrel with 16 transmembrane strands (Weiss et al., 1990; Jap et al., 1991). The uneven density in the thin dark shell of protein around each pore in the cryo-image of Figure 3a suggests that the lumen wall is not uniform in height, which has since been confirmed by three-dimensional reconstruction of the VDAC array embedded in aurothioglucose (Guo et al., 1995). Also, there are dense "arms" of protein that extend away from the channel hexamers into the corner of the crystal cell. These arms are labeled by antibodies directed against the first twenty residues of the *Neurospora* VDAC polypeptide (Stanley et al., 1995), a subsequence predicted to fold as an amphipathic α-helix (Kleene at al., 1987). Thus, the lateral arms might be composed (at least in part) of N-terminal α-helices that do not form part of the lumen, but instead interact with the phospholipid bilayer.

Assuming that VDAC's lumen is formed by a β-barrel with a mean C_α diameter of 3.8 nm and given that amino acid residues typically extend about 0.5 nm from the backbone of a β-sheet, VDAC's mean outer diameter can be predicted to be:

$$\text{O.D.} = 3.8 \text{ nm} + (2 \times 0.5) \text{ nm} = 4.8 \text{ nm} \tag{1}$$

which is consistent with the packing of the channels in the two-dimensional arrays. Likewise, VDAC's inner diameter is predicted to be:

$$\text{I.D.} = 3.8 \text{ nm} - (2 \times 0.5) \text{ nm} = 2.8 \text{ nm}. \tag{2}$$

This lumen size is consistent with the known permeability properties of VDAC, for example, it is very permeable to many metabolites (molecular weight < 1000) but impermeable towards holocytochrome c (shortest linear dimension = 3 nm). This inner diameter falls between estimates based on VDAC's permeability towards flexible polymers (4.0 nm; Colombini, 1980) and on the ion conductance of VDAC's open state (2.0 nm; DePinto et al., 1987).

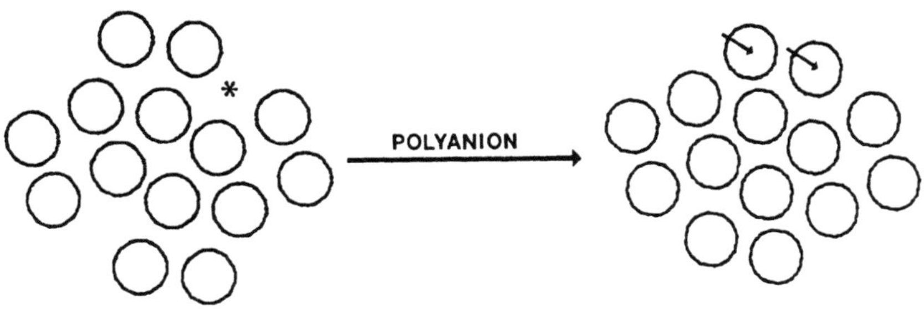

Figure 4. The oblique-to-contracted lateral phase transition of *N. crassa* VDAC crystals. The central unit cell containing six channels is shown, along with the eight nearest-neighbor channels. (Note, these arrays are rotated 90° counter-clockwise with respect to Figure 3.) One of the corner regions in the oblique array (left) occupied by protein arms (see Figure 3a) is marked by an asterisk. Small arrows in the contracted array (right) indicate the distance and direction in which the channels in adjacent rows move during lattice contraction. (Reproduced with permission from Mannella and Guo, 1992 and the American Crystallographic Association.)

The periodic 2-D arrays formed by VDAC are polymorphic, that is, they can exhibit more than one crystal habit. The parallelogram array described above is the most common crystal type observed in fresh preparations. These so-called *oblique* arrays can contract in such a way that the lattice angle decreases from 109° to 99° (Figure 4). While the intra-hexamer packing and the mean projected lumen diameter of the channels in the contracted array are the same as those of the oblique array, the lateral protein arms are no longer visible (Guo and Mannella, 1993). In fact, the space that they occupy in the corner of the oblique array is dramatically reduced after contraction (Figure 4). Interestingly, incubation of the VDAC crystals with Konig's polyanion induces rapid lattice contraction, suggesting that one effect of this modulator is to cause the arm (N-terminal helix) to detach from the membrane surface. It has been suggested (Guo and Mannella, 1993) that this event destabilizes the open channel and so might be the first step in channel closure.

Genetically-engineered VDAC

In a series of elegant experiments combining the techniques of electrophysiology with those of molecular genetics, Forte, Colombini, and coworkers have constructed and functionally characterized plasmid-encoded mutant VDACs expressed in yeast strains in which the nuclear VDAC gene has been deleted (Blachly-Dyson et al., 1989, 1990). In this work, positions along the VDAC sequence have been classified as falling into one of two categories, depending on whether or not the reversal potential (ion selectivity) of the channel is detectably altered when the charge at that position is changed. These results provide a valuable functional map of the channel sequence which, when combined with other information, can be used to model the folding of the VDAC polypeptide. One such model has been proposed (Blachly-Dyson et al., 1990) based on this functional map, hydropathy analysis (to identify likely transmembrane β-strands), and two basic assumptions: (1) that the lumen is a right circular cylinder formed by a β-barrel, and (2) that the residues that affect ion selectivity are on transmembrane β-strands while those that do not affect selectivity are outside the barrel. The resulting model for VDAC consists of 12 β-strands and a transmembrane α-helix formed by the N-terminal sequence.

While this is an important working model for VDAC and serves as a basis for a proposed gating mechanism (see the following section), some of its aspects are controversial. For example, in order for the proposed 13 transmembrane segments (12 β-strands + 1 α-helix) to form a pore of the appropriate diameter for VDAC (see above), all the segments must be tilted approximately 60° with respect to the cylinder axis, which would define a wall too short (under 3 nm) to span a bilayer (see discussion in Mannella et al., 1992). By contrast, 12 of the 16 β-strands in bacterial porin are tilted only 30°. Also, results from several labs (DePinto et al., 1991; Stanley et al., 1996) suggest that the N-terminal segment of the VDAC protein in the mitochondrial outer membrane is accessible to antibodies, which would be unexpected if this segment were embedded inside the membrane at a 60° tilt. However, there is evidence that the N-terminus may be flexible and move between the lumen and the bilayer (Guo and Mannella, 1993; also see following).

It may be that the folding scheme of Blachly-Dyson et al. (1990) underestimates the number of strands in VDAC's β-barrel by incorrectly assigning residues in "non-ideal" regions of the pore to be outside the lumen. For example, the pore profile in high-resolution projection images of crystalline VDAC suggests that the lumen wall is not circular (as assumed for the above model), but polygonal. Charged residues in the corners of the lumen would have less affect on ion selectivity than those on segments of the wall closer to the center of the lumen. It is also possible that some residues might be masked by loops of the polypeptide that fold back inside the lumen, as occurs with bacterial porin. Attempts are underway to reconcile the functional map of the VDAC polypeptide with a folding scheme that fits the molecular envelope of a single VDAC channel determined by three-dimensional electron microscopy (Guo et al., 1995).

Mechanism of Voltage Gating

By measuring the osmotic pressure needed to induce VDAC to close, Zimmerberg and Parsegian (1986) could show that this functional change involves a large decrease in lumen volume. This result suggests that closure involves a large-scale conformational change in the protein and not, for example, a small, localized constriction or obstruction. This conclusion has been supported by the results of Peng et al. (1992b) which showed that numerous charged residues that affect reversal potential in the open state have no effect in the partially closed state, suggesting a large-scale rearrangement of protein domains. This group has proposed a gating mechanism involving removal of protein from the wall of the lumen, thereby reducing its physical diameter. The putative transmembrane strands involved in this process have been identified by determining which charged residues assigned to the lumen of the open state (as described previously) affect V_0 when genetically substituted (Thomas et al., 1993). The results suggest that the gating mechanism involves protein domains at the N- and C-termini of the VDAC polypeptide. However, alternative gating mechanisms can be proposed which are also consistent with the general nature of these observations. In particular, movement into the lumen of segments or loops of protein in response to voltage would both partially obstruct the pore (i.e., lower its conductance) and mask residues on the wall of the lumen (i.e., reduce their effects on ion selectivity). The first step in this alternative mechanism might be the insertion of the N-terminal region into the lumen following its detachment from the lipid bilayer (see Mannella, 1990). Evidence is being sought to distinguish between the two types of gating models from several kinds of experiments, for example, accessibility of sequence-specific antibodies to their epitopes in different functional states of the VDAC channel (Stanley et al., 1995).

VDAC as a Receptor: Extramitochondrial Expression

Rose and Warms (1967) found that hexokinase was associated with mitochondria in certain mammalian tissues and that the bound enzyme did not display product inhibition characteristic of the soluble enzyme. This binding was found to be elevated in some tumor tissues where it was linked to elevated levels of aerobic glycolysis (Bustamante et al., 1981). Not long after VDAC's discovery, the channel was determined to be the *hexokinase binding protein* of the outer mitochondrial membrane (Fiek et al., 1982; Linden et al., 1982). There is evidence that hexokinase bound to VDAC may have preferred access to intramitochondrially generated ATP (Arora and Pedersen, 1988; BeltranddelRio and Wilson, 1992; cf. Kabir and Nelson, 1991). The suggestion has been made that there is a microcompartmentation of adenine nucleotides within the mitochondrial intermembrane space that is controlled by VDAC and hexokinase at the outer membrane, kinases (like creatine kinase) in the intermembrane space, and the adenine nucleotide transporter (ANT) at the inner membrane (see review by Brdiczka, 1991).

Experiments of a different nature provide additional evidence for close communication between the two mitochondrial membranes. Ligand-binding experiments (McEnery, 1992; McEnery et al., 1992) indicate that VDAC and ANT are two components of a benzodiazepine receptor complex in animal mitochondria (the so-called *peripheral benzodiazepine receptor*), with a third component being an 18 kD outer-membrane polypeptide that has been cloned and sequenced (e.g., Sprengel et al., 1989). While the precise physiological function of this drug receptor complex is unclear at present, several interesting leads are being followed. For example, there is considerable evidence that the receptor is involved in the regulation of steroid metabolism, with the 18 kD component possibly functioning as a cholesterol carrier (Krueger and Papadopolous, 1992). There is also evidence that benzodiazepine ligands have a direct effect on oxidative phosphorylation, i.e., they cause loss of respiratory control with IC_{50} values that are similar to their receptor-binding constants (Hirsch et al., 1989a, 1989b). The same drugs also have been shown to modulate the activity of inner membrane ion channels (see following).

Another putative VDAC-containing benzodiazepine-receptor complex has been reported to occur in the *plasma membrane* of mammalian cells (Bureau et al., 1992), following reports by Thinnes et al. (1989) that the VDAC protein is located in the plasmalemma of several human cell types. The latter conclusion was derived largely from light- and electron-microscopic immunolabeling experiments, although Thinnes (1992) points out that several types of plasma membrane channel activities detected by patch–clamping bear functional resemblances to VDAC. While the possibility cannot be excluded that the co-isolation of so-called *human porin* with a brain GABA receptor (Bureau et al., 1992) is an artifact, the intriguing possibility remains that VDAC is expressed in the plasma membrane of some mammalian cells, perhaps complexed with receptors or other proteins that regulate its conductance.

B. Non-VDAC Outer Membrane Channels

Evidence for one or more large-conductance cation-selective channels in the outer mitochondrial membrane has come from two sources: bilayer studies (Dihanich et al., 1989) and the so-called tip–dip technique, in which a bilayer is formed across the opening of a patch-pipette and allowed to fuse with mitochondrial proteoliposomes (Chich et al., 1991). Both channel activities display conductance steps of about 300 pS, with the channel detected by the tip–dip technique also showing 100 pS and 540 pS transitions, so it is possible that the two activities are manifestations of the same channel. Interestingly, these channel activities bear a resemblance (in terms of step conductance size and ion selectivity) to VDAC's predominant partly closed state. However, since both activities occur in mitochondria of yeast strains in which the VDAC gene has been disrupted, it is clear that they do not represent a functionally altered form of VDAC. The cation-selective channel activity detected by the tip–dip method is activated by transmembrane potentials (Chich et al., 1991), suggesting that it might be responsible for the voltage-induced outer membrane currents detected in patch–clamp studies by Tedeschi et al., (1987). A possible function for this cation-selective channel has been suggested by the finding that it undergoes transient blockage by *signal peptides*, that is, the cationic N-terminal presequences of mitochondrial precursor proteins involved in targeting of the imported proteins to the organelle (Henry et al., 1989). Thus, this peptide sensitive channel (PSC) may be the pore through which proteins are transported across the outer mitochondrial membrane.

Finally, there have been reports of outer membrane channel activities with step conductances in the 10–50 pS range (e.g., Moran and Sorgato, 1992). These channel activities have not been characterized in detail.

III. INNER MITOCHONDRIAL MEMBRANE CHANNELS

A. Detection of Channels in the Energy Transducing Membrane

As noted above, the existence of channels or pores in the inner mitochondrial membrane was considered improbable because they would be expected to short-circuit this membrane's energy-transducing ion gradients. In fact, thermogenin, the uncoupling protein of brown adipose tissue functions in just this way, generating a membrane leak that dissipates the mitochondrial ion gradients as heat for the process of nonshivering thermogenesis (Klaus et al., 1991). Aside from this specialized transport protein (which has been isolated and cloned), several other mitochondrial inner membrane channels have been postulated to exist in order to explain permeability and swelling data for mitochondria in suspension. Two classes of channels, in particular, have been functionally characterized by these techniques, a calcium-induced *permeability transition pore* (PTP; see review by Gunter and Pfeiffer, 1990) and an *inner membrane anion channel* (IMAC, see review by Beavis, 1992). The most likely physiological role for the anion channel appears to be volume homeostasis, that is, regulation of mitochondrial volume by allowing exchange of particular ionic species (e.g., Bernardi et al., 1992; Garlid and Beavis, 1996). In contrast, PTP is associated with pathological changes in mitochondrial inner membrane integrity (Gunter and Pfeiffer, 1990) and does not have an obvious normal physiological function.

The earliest electrophysiological demonstration of a putative inner mitochondrial membrane channel was that of a K^+ channel in a bilayer-reconstitution study by Mironova et al. (1981). Direct demonstration of the existence of ion channels in the *native* inner mitochondrial membrane did not happen until 1987 with the application of patch–clamp techniques to mitoplasts, which are mitochondria that have had their outer membranes removed (Sorgato et al., 1987; Kinnally et al., 1989). However, it should be noted that the techniques used to make mitoplasts (hypo-osmotic swelling, French press, digitonin treatment) generally leave behind pieces of the outer membrane attached at so-called contact points between the two mitochondrial membranes (e.g., Miller and Hackenbrock, 1975).

Patch–clamping of mitoplasts indicates that the mitochondrial inner membrane has a high resistance (in the 1–10 GΩ range) and is usually electrically silent (Sorgato et al., 1987; Kinnally et al., 1989, 1992, 1996). However, it was found that a variety of conditions (divalent ions, pH, transmembrane voltage) can activate at least five distinct classes of channel activities, which are summarized in Table 1. These channel activities display a wide range of ion selectivities and voltage dependence. In the main, each activity is influenced by at least one physiological

Table 1. Summary of Mitoplast Channel Activities

Channel Activity	*Size[a] (pS)*	*Voltage Dependence*	*Selectivity*	*Effectors*
MCC	10 to >1000	Yes	Slight cation to none	Ca^{2+}, Mg^{2+}, ADP, pH, voltage
mCS	≈100	Yes	Slight anion	Ca^{2+}, voltage
K^+	9	No	K^+	ATP
ACA	15	No	Slight cation	pH, Mg^{2+}
AAA	45	No	Slight anion	pH, Mg^{2+}

Note: [a]All measurements made in symmetrical 0.15 M KCl except K^+ channel (salt gradient described in Figure 9).

effector, for example, ATP, Ca^{2+}, or pH. Although great strides have been made in recent years in characterizing the activities of these channels, none of them have been identified with a particular mitochondrial protein.

B. The 100 pS Voltage-gated Channel (mCS)

The first activity detected in *single channel* recordings from the native inner mitochondrial membrane is a voltage-gated, 100 pS conductance (Sorgato et al, 1987) now called mCS (for *mitochondrial centum picosiemen* activity). While discovered with mouse liver mitoplasts, this activity has since been recorded in several labs using mitoplasts from a variety of tissues including rat and mouse liver, heart and kidney, as well as rat brown adipose (Sorgato et al., 1989; Kinnally et al., 1991, 1993; Costa et al., 1991; Szabo and Zoratti, 1991; Campo, et al., 1992). However, mCS has not as yet been detected in yeast mitoplasts (Lohret and Kinnally, 1995). In some laboratories, mCS activity is present upon patching (Sorgato et al., 1987), while in others, the channel is quiescent and is only activated after exposure of the cytoplasmic side of the inner membrane to Ca^{2+}-chelators (Kinnally et al., 1991).

The conductance of mCS ranges from about 90–140 pS with most reported values around 110 pS. Substates on the order of 0.25, 0.5, 0.7, and 1.3 times the primary conductance value also have been reported (Kinnally et al., 1993; Klitsch and Siemen, 1991). mCS is anion selective ($P_{Cl^-}/P_{K^+} = 4.5$) in its main 110 pS conductance state and in at least one (50 pS) substate (Kinnally and Tedeschi, 1994). This channel is strongly voltage-dependent, closing with matrix negative potentials as shown in Figure 5. Kinetic analysis of this bursting activity indicates that mCS has multiple open- and closed-states as well as intermediate subconductance states (Klitsch and Siemen, 1991; Campo et al., 1992).

Interestingly, Kinnally et al. (1991) have found that, once mCS activity is activated by removing Ca^{2+} from the cytoplasmic face of the inner membrane, this

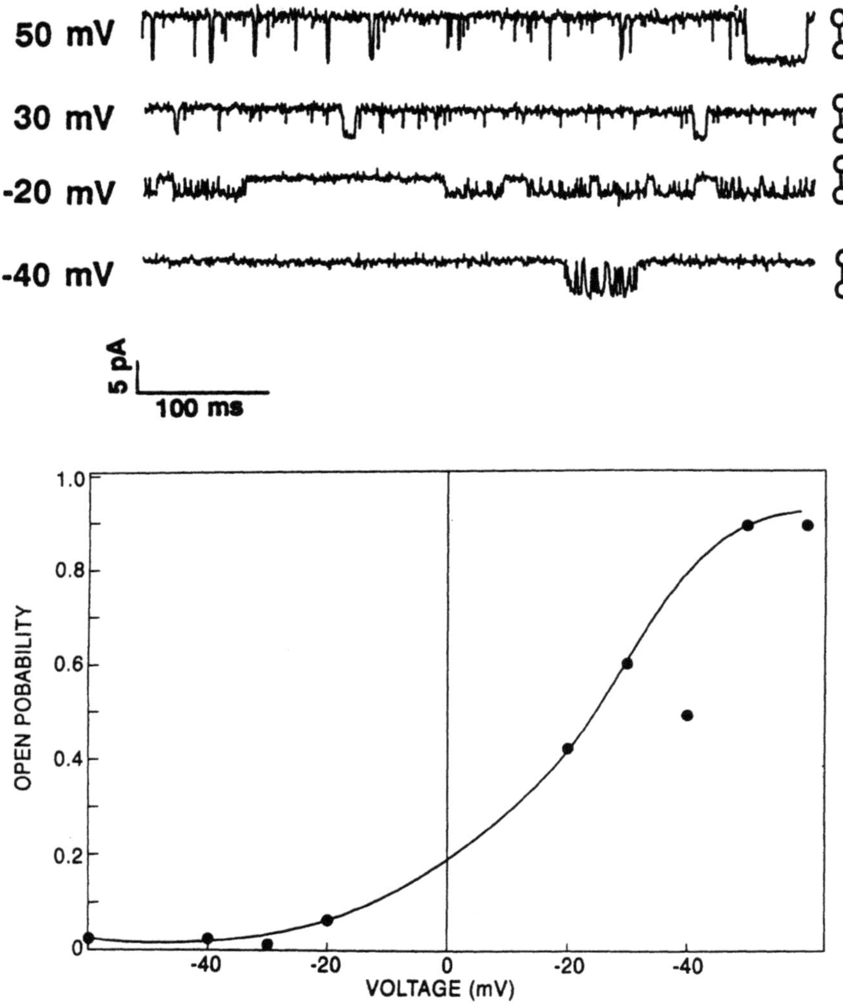

Figure 5. mCS activity and voltage dependence. (**A**) Current traces were recorded from a patch attached to a mouse liver mitoplast in symmetrical 150 mM KCl. (Reproduced with permission from Tedeschi and Kinnally, 1994 and Academic Press.) (**B**) Voltage dependence of mCS activity recorded from a patch excised from a rat-heart mitoplast in symmetrical 150 mM KCl, 5 mM HEPES, 0.75 mM $CaCl_2$, 1 mM EGTA, pH 7.4. Open probability was calculated as in Figure 2.

activity is insensitive to Ca^{2+}. mCS is also insensitive to changes in matrix pH from 6–9 (Sorgato et al., 1987) and is not affected by millimolar Mg^{2+} or ATP and micromolar ADP if applied on the matrix side of excised patches (Inoue et al., 1991). Klitsch and Seimen (1991) have reported that mCS is inhibited by submillimolar levels of not only di- and tri-nucleotide phosphates, but also of GMP when added to the outside of patched mitoplasts, providing evidence that this channel activity is not related to the uncoupling protein, thermogenin, which is insensitive to GMP.

Besides Ca^{2+}, a variety of metals and organic compounds have been tested for their effects on mCS activity. For example, Sorgato et al. (1989) have reported that several effectors of other types of channels, Zn^{2+}, Gd^{3+}, and DIDS, have no effect on mCS. However, Antonenko et al. (1991) found that this channel activity was inhibited by the amphiphilic cationic drugs amiodarone and propranolol. Campo et al. (1992) have described an inhibition of mCS by antimycin A binding at a site distinct from that involved with respiratory inhibition, as indicated by differences in IC_{50}. Kinnally et al. (1993) reported that ligands of the mitochondrial benzodiazepine receptor (mBzR), such as RO5-4864, PK11195, and protoporphyrin IX (PPIX), are high-affinity effectors, but that clonazepam, a central benzodiazepine receptor ligand, is without effect.

Early efforts to partially purify the mCS channel indicate that it co-isolates with the F_0 region of the ATP synthase (Sorgato et al., 1989). It is *not*, however, thought to be a component of the ATP synthase since its activity is sensitive to neither oligomycin nor DCCD. The mCS channel is closed by potentials with the polarity expected to be generated across the inner membrane during oxidative phosphorylation, so it is unlikely to be responsible for uncoupling or thermogenesis (Klitsch and Siemen, 1991). Some postulated roles for this channel include volume regulation (Klitsch and Siemen, 1991) and protein import (e.g., Sorgato and Moran, 1993).

C. The Multiple Conductance Channel (MCC)

The *multiple conductance channel* activity of mitoplasts, MCC, (also called MMC for *mitochondrial megachannel* by Petronilli et al., 1989) is characterized by a variety of transition sizes (10–1000 pS) and a peak conductance of 1000–1500 pS. The largest (nS) transitions are typically observed in about 25% of the patches from mammalian and yeast mitoplasts (Zorov et al., 1992; see Figure 6). While these multiple transition levels may correspond to more than one discrete protein channel, they have been grouped as a single class because they (1) frequently occur in a single patch, (2) can be induced together by elevated Ca^{2+} levels or voltages greater than ±60 mV, and (3) have the same pharmacology (Kinnally et al., 1992). The conductance levels between 300- and 1300-pS are probably substates of one channel since multiple small openings can close simultaneously with a single large

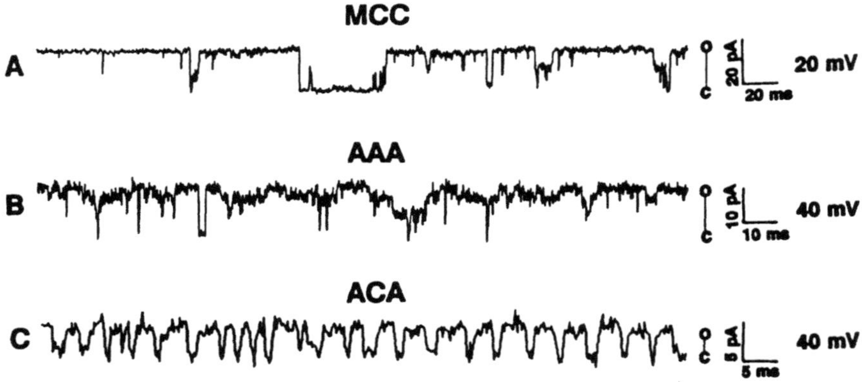

Figure 6. Sample current traces of mitoplast channel activities. (**A**) MCC activity was recorded from a rat-heart mitoplast patch as in Figure 5b. (**B**) AAA was recorded from a patch excised from a mouse-liver mitoplast in symmetrical 1 M KCl, 5 mM HEPES, 10 μM $CaCl_2$ with the pipette medium at pH 6.8 and bath medium at pH 8.2. (**C**) ACA was recorded under the same conditions as in (**B**). (Reproduced with permission from Tedeschi and Kinnally, 1994 and Academic Press.)

transition (Petronilli et al., 1989). MCC shows little or no ion selectivity except in some of its lower-conductance substates in which it is slightly cation selective.

Once activated (by Ca^{2+} or large positive potentials) MCC is voltage- dependent, opening with negative matrix potentials and closing to lower conductance substates for small positive potentials as shown in Figure 7. Increasing the transmembrane voltage above 50 mV usually reopens the channel. The highest-conductance states often have extremely long open times, on the order of seconds to minutes. Recently, however, Szabó and Zoratti (1993) report closure of the large-conductance mitoplast channel they call MMC at both negative- and positive-potentials, a voltage dependence reminiscent of VDAC. However, the recent finding that an MCC activity occurs in mitoplasts of yeast strains lacking the VDAC gene suggests that VDAC is not required for MCC activity (Lohret and Kinnally, 1995a). Nonetheless, there are several indications of a functional interaction between the outer membrane and mitoplast channels, including the inhibition by mBzR ligands of MCC (as well as mCS; see preceding), and a difference in the behavior at large negative or positive potentials of the MCC activity of wild-type and VDAC-less yeast strains. Thus, while MCC is not itself VDAC, it appears to interact with VDAC, probably at intermembrane contact sites.

A variety of physiologically significant effectors for MCC activity have been defined. As already noted, one of the most striking characteristics of MCC is its activation by Ca^{2+}. As shown in Table 2, mitoplast patches are, for the most part, electrically silent if Ca^{2+} levels are kept below micromolar and the pH is close to neutral during mitoplast preparation. Exposure of mitoplasts to higher Ca^{2+} levels

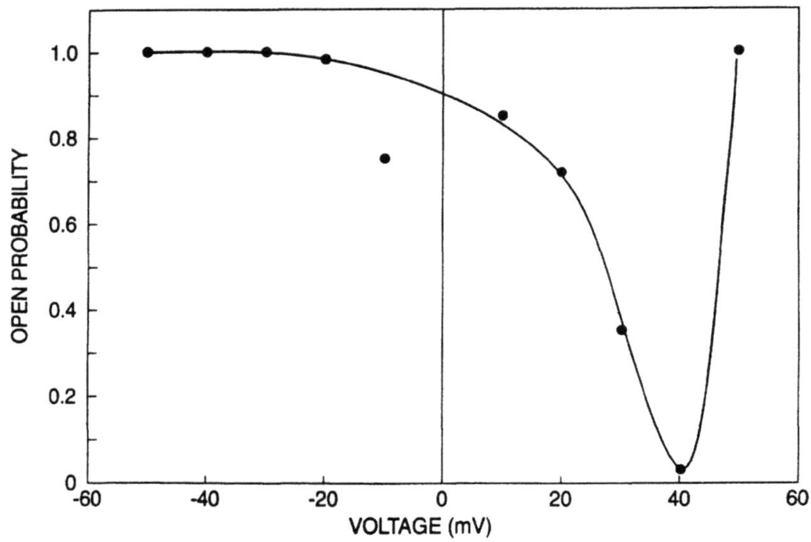

Figure 7. Voltage dependence of MCC. Current was recorded from a patch excised from a rat-heart mitoplast in symmetrical 150 mM KCl, 5 mM HEPES, 0.75 mM $CaCl_2$, 1 mM EGTA, pH 7.4. Open probability was calculated as in Figure 2b.

activates MCC, and subsequent treatment with the chelator EGTA will, under some conditions, reverse this activation (Kinnally et al., 1991; Szabó and Zoratti, 1992). This inner membrane channel activity is also regulated by such physiological effectors as Mg^{2+}, acidic pH, and ADP (Szabó and Zoratti, 1992), suggesting that it is normally under tight control.

Several pharmacological agents have been shown to affect MCC activity. The high-affinity ligands of mBzR (e.g., RO5-4864 and PPIX) affect MCC activity in

Table 2. Calcium Activation of MCC in Mouse-liver mitoplasts[a]

Free [Ca^{2+}][b] *(M)*	*Number of Patches*	*Percent Inactive*	*Percent with MCC Activity*
10^{-5}	50	4	96
10^{-7}	8	100	0
$<10^{-10}$	79	85	15
$10^{-5} \rightarrow 10^{-9}$[c]	59	7	76[d]

Notes: [a]Currents recorded from patches excised from mouse-liver mitoplasts that were voltage-clamped in the range ± 50 mV in symmetrical 150 mM KCl containing the indicated concentration of free Ca^{2+}.

[b]Free [Ca^{2+}] present in the mitochondrial isolation media and patching media.

[c]Mitochondria isolated in media containing 10^{-5} free [Ca^{2+}] and subsequently washed in buffer containing 10^{-9} free [Ca^{2+}].

[d]The remaining 17% of these patches had mCS activity only.

a biphasic manner, inhibiting at nanomolar levels and activating at higher concentrations (Kinnally et al., 1993). However, the structurally similar compounds clonazepam and RO15-1788 have no effects on MCC in the same range of concentrations. Several other MCC effectors including antimycin A, cyclosporin A, and the uncoupler CCCP, are also mBzR ligands (Hirsch et al., 1988b). In addition, the list of compounds affecting MCC activity include the amphiphilic cations amiodarone, propranolol, and quinine, as well as dibucaine (see Kinnally et al., 1992 for summary).

There is little doubt that MCC is an inner mitochondrial membrane channel activity since it is recorded from patches which also contain mCS activity (see Figure 8). Kinnally et al. (1992) have proposed that MCC is associated with mitochondrial contact sites at which the inner- and outer-membranes closely adhere. In this model, inner- and outer-membrane components would act in series (like gap-junction connexons) and allow communication directly between the matrix and cytoplasm. Support for such a model is provided by the reports of MCC-like transitions in reconstituted contact site preparations (Moran et al., 1990) as well as by the fact that levels of Ca^{2+} which activate MCC also increase the density of mitochondrial contact sites (Bakker et al., 1993). While an intermembrane protein complex has been proposed for mBzR (McEnery, 1992), it should be noted that MCC is present in knock-out strains of the adenine nucleotide translocator and of VDAC (Lohret and Kinnally, 1995a, 1996).

A large-scale calcium-induced permeability increase in the mitochondrial inner membrane has been studied for many years and has been linked to such pathological events as ischemia–reperfusion injury (Gunter and Pfeiffer, 1990). There are several interesting correlations between MCC activity observed in patch–clamp experiments and the so-called *permeability transition pore* (PTP) deduced from suspension studies. For example, both MCC and PTP are activated by Ca^{2+} and inhibited by Mg^{2+}, ADP, H^+, amiodarone, and cyclosporin A, with most effects occurring at comparable levels for the two channels/pores. Thus, it has been suggested that MCC and PTP are the same entity (e.g., Szabo and Zoratti, 1992; Zoratti and Szabo, 1995). In terms of a normal physiological role, involvement of MCC/PTP in Ca^{2+} homeostasis should not yet be ruled out. Also, MCC has several functional similarities to PSC, the peptide-sensitive outer-membrane cation channel (i.e., sizes of conductance transitions, blockade by targeting peptides; Kinally et al., 1996), implicating a possible role for MCC in protein import (which likely occurs at the sites of intermembrane contact; Hartl et al., 1989; Brdiczka, 1991).

D. ATP-Sensitive K^+ Channels

Since mitochondria are the primary producers of ATP in the cell, it is not surprising that they have ATP-sensitive channels in their inner membrane. Inoue et al. (1991) originally described the inhibition of a K^+ channel in fused mitoplasts by

Figure 8. MCC and mCS activities coexist in the mitoplast membrane. Current trace revealing transitions corresponding to MCC (approx. 1300 pS) and mCS (110 pS) was recorded from an excised mouse liver mitoplast in symmetrical 150 mM KCl, 5 mM HEPES, 0.95 mM $CaCl_2$, 1 mM EGTA, pH 7.4.

ATP. This activity has a conductance of about 10 pS in 0.1 M salt (Figure 9) and is voltage independent. That this channel is localized in the inner mitochondrial membrane is supported by detection of its activity in patches also containing mCS activity. The channel activity is inhibited by known effectors of the plasma membrane ATP-sensitive K^+ channels, 4-aminopyridine, and glibenclamide plus ATP.

Voltage-dependent cation-selective channel activities with step conductances in the range 24–175 pS have been reported to be associated with mitochondrial membrane protein fractions prepared by ethanol extraction (Mironova et al., 1981), quinine affinity chromatography (Costa et al., 1991), and ion exchange chromatography (Paucek et al., 1992). However, the specific protein(s) responsible for the glibenclamide-sensitive, ATP-dependent K^+ transport of mitochondria has not yet been unambiguously identified. It has been suggested by Inoue et al. (1991) and Mironova et al. (1981) that the ATP-sensitive K^+ channel might form multimeric complexes capable of coordinated gating, which would explain the range of step conductances observed in reconstituted systems. A possible role for the ATP-dependent K^+ channel is the regulation or fine-tuning of the mitochondrial membrane potential by changing K^+ permeability in response to the availability of ATP.

E. Alkaline pH-Sensitive Channels

Two pH-sensitive mitoplast channel activities have been described, both displaying greater open probabilities at alkaline pH and both being activated by depletion of Mg^{2+} (Antonenko et al., 1994). One of these channel activities is cation selective (ACA, alkaline-induced cation activity), while the other is slightly anion selective (AAA, alkaline-induced anion activity) (Kinnally and Tedeschi, 1994). The two pH-sensitive channel activities were originally inferred from whole patch currents which were later resolved into discrete transitions by high ionic strength (0.5–2 M salt) and high voltage (around ± 100 mV) as shown in Figure 6 (Antonenko et al., 1994).

The cation channel, ACA, has a conductance of about 15 pS in 0.15 M KCl and is relatively voltage insensitive. Its voltage dependence and unit conductance are similar to those of the ATP-sensitive K^+ channel. However, unlike the latter channel, ACA is relatively nonselective for cations and is not affected by 4-aminopyridine and glibenclamide plus ATP. Instead, like mCS and MCC, ACA is inhibited by amiodarone and propranolol. Mg^{2+}-depletion by EDTA is used to induce ACA activity which itself is insensitive to Ca^{2+}. ACA's cation selectivity and Mg^{2+} sensitivity suggest that it may correspond to one of the cation uniporters whose presence has been inferred from solution studies and which are implicated in volume homeostasis (e.g., Bernardi et al., 1992).

The anionic channel, AAA, has an open-state conductance of about 45 pS and, like ACA, is relatively voltage insensitive. AAA has two conductance substates of one-third and two-thirds the fully open state and is only slightly selective for different anions. Based on similarities in activation requirements, ACA initially was

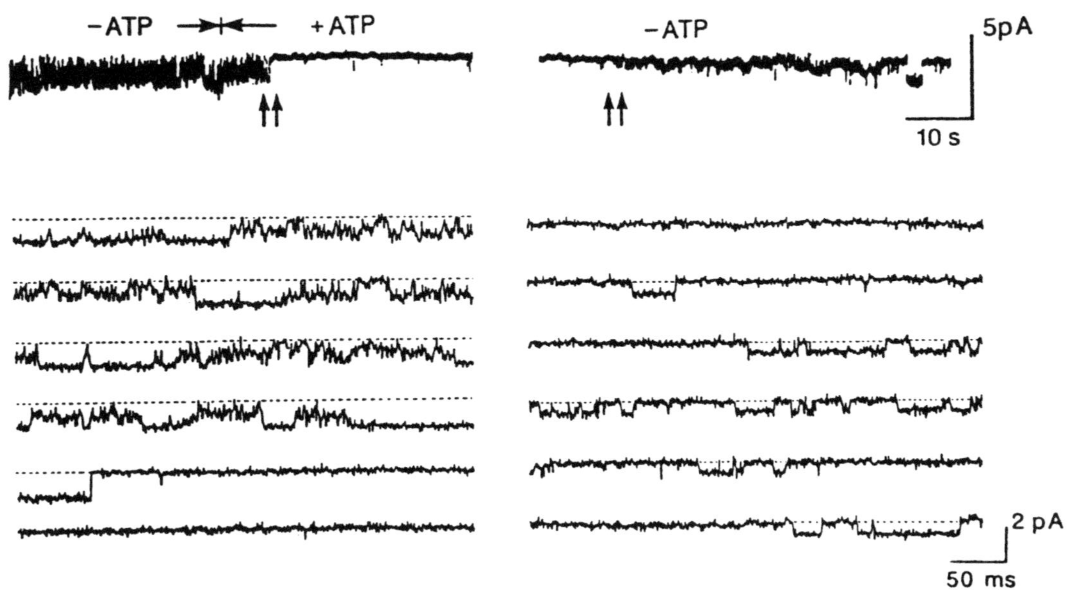

Figure 9. Current traces showing reversible suppression by ATP of a K^+ channel in an excised patch from fused rat-liver mitoplasts. The lower traces show an expanded scale of the region between the arrows. Pipette buffer contained 100 mM KCl; external bath contained 33 mM KCl, 67 mM NaCl. (Reproduced with permission from Inoue et al., 1991 and *Nature* © 1991 Macmillan Magazines Limited.)

thought to correspond to IMAC, the inner membrane anion channel inferred from mitochondrial suspension studies. However, there are discrepancies in several functional characteristics of AAA and IMAC, for example, estimated pore size and degree of anion selectivity. AAA may correspond to a low-conductance anion channel activity reconstituted in bilayers by Hayman et al. (1993).

F. A Note on the Pharmacology of Mitoplast Channels

Drugs which have proven invaluable in sorting out the numerous channel activities of plasma membranes have not been as useful when applied to mitochondria. With a few notable exceptions (such as the specific inhibition by cyclosporin A of MCC and PTP) most of the pharmacologic agents tested affect multiple channel activities—or none at all. Amiodarone and propranolol influence mCS, MCC, ACA, and AAA with similar efficacy, while the latter three activities are similarly affected by quinine (for a summary see Kinnally et al., 1992, 1996). MCC and mCS are both affected by RO5, PPIX, antimycin A, and CCCP, with the high-affinity benzodiazepine ligands influencing both channels at concentrations resembling their affinities for the mitochondrial receptor. Hence, it may be that either homologous binding sites occur on different proteins or the same proteins are involved in producing these channel activities. Alternatively, it may be that one or more drug binding proteins exist in the inner membrane (analogous to the 18 kD outer-membrane protein) which interact with the various channel-forming proteins in this membrane (Kinnally and Tedeschi, 1994).

IV. CONCLUSION

Numerous channel activities have been detected by application of bilayer and patch–clamp methodologies to mitochondrial membranes. In the outer membrane, the main channel activity is that of the porin, VDAC, a voltage-gated large-conductance (650 pS) channel, structure/function studies of which are ongoing. It is likely that this channel represents the primary (albeit not the only) permeability pathway through the outer membrane for metabolites and ions. The modulation of the VDAC channel by endogenous effectors (NADH and a soluble protein) suggests a regulatory role for outer membrane permeability in mitochondrial metabolism. A second large-conductance outer-membrane channel activity, PSC, has been detected and may be involved in protein import. Five inner-membrane channel activities have been described based on patch–clamp studies of mitoplasts. The first discovered was mCS, a voltage-dependent 100 pS anion channel activity that co-isolates with (but does not appear to be part of) the ATP synthase. The multi-conductance channel activity, MCC, has a maximum conductance twice that of VDAC and appears to correspond to the calcium- and cyclosporin-sensitive permeability transition pore (PTP) deduced from swelling studies. There are several indications that MCC and VDAC interact, perhaps at a drug-receptor complex located at intermembrane contact sites. Three low-conductance (<100 pS) channel activities have been detected in the mitochondrial inner membrane: a glibenclamide- and ATP-sensitive K^+ channel, and two pH-sensitive channel activities with opposite ion selectivity (ACA and AAA). The inner membrane ion channels appear to be highly regulated (e.g., by divalent ions, pH, ATP, etc.) and are not expected to be open continuously under normal conditions (which would, of course, short circuit oxidative phosphorylation). The role of these inner membrane channel activities in mitochondrial physiology is not clear at present. One or more of the channels (in particular, MCC) may be involved in mitochondrial protein import. However, it is likely that some or all play a role in regulation of mitochondrial volume by allowing exchange of specific ions.

REFERENCES

Antonenko, Y. N., Kinnally, K. W., Perini, S., & Tedeschi, H. (1991). Selective effects of inhibitors on mitochondrial inner membrane channels. FEBS Letts. 285, 89–93.

Antonenko, Yu., Smith, D., Kinnally, K. W., & Tedeschi, H. (1994). Single channel activity induced in mitoplasts by alkaline pH. Biochim. Biophys. Acta 1194, 247–254.

Arora, K. K., & Pedersen, P. L. (1988). Functional significance of mitochondrial bound hexokinase in tumor cell metabolism. J. Biol. Chem. 263, 17422–17428.

Bakker, A., De Bie, I., Bernaert, T., Ravingerova, A., Ziegelhoffer, H., Van Belle, H., & Jacob, W. (1993). Increased extracellular calcium concentrations in the rat isolated heart: Effects on mitochondrial contact site formation. Eur. J. Morphol. 31, 46–50.

Beavis, A. D. (1992). Properties of the inner membrane anion channel in intact mitochondria. J. Bioenerg. Biomembr. 24, 77–90.

BeltrandelRio, H., & Wilson, J. E. (1992). Coordinated control of cerebral glycolytic and oxidative metabolism, mediated by mitochondrially bound hexokinase dependent on intramitochondrially generated ATP. Arch. Biochem. Biophys. 296, 667–677.

Benz, R., Wojtczak, L., Bosch, W., & Brdiczka, D. (1988). Inhibition of adenine nucleotide transport through the mitochondrial porin by a synthetic polyanion. FEBS Letts. 231, 75–80.

Benz, R., Kottke, M., & Brdiczka, D. (1990). The cationically selective state of the mitochondrial outer membrane pore: A study with intact mitochondria and reconstituted mitochondrial porin. Biochim. Biophys. Acta 1022, 311–318.

Bernardi, P., Zoratti, M., & Azzone, G. F. (1992). Mitochondrial volume homeostasis: Regulation of cation transport systems. In: NATO ASI Series, Vol. H 64, Mechanics of Swelling (Karalis, T. K., Ed.), pp. 357–377. Springer-Verlag, Berlin Heidelberg.

Blachly-Dyson, E., Peng, S. Z., Colombini, M., & Forte, M. (1989). Probing the structure of the mitochondrial channel, VDAC, by site-directed mutagenesis: A progress report. J. Bioenerg. Biomembr. 21, 471–483.

Blachly-Dyson, E., Peng, S. Z., Colombini, M., & Forte, M. (1990). Selectivity changes in site-directed mutants of the VDAC ion channel: Structural implications. Science 247, 1233–1236.

Blachly-Dyson, E., Zambronicz, E. B., Yu, W. H., Adams, V., McCabe, E. R. B., Adelman, J., Colombini, M., & Forte, M. (1993). Cloning and functional expression in yeast of two human isoforms of the outer mitochondrial membrane channel, the voltage-dependent anion channel. J. Biol. Chem. 268, 1835–1841.

Bowen, K. A., Tam, K., & Colombini, M. (1985). Evidence for titratable gating charges controlling the voltage dependence of the outer mitochondrial channel, VDAC. J. Membr. Biol. 86, 51–59.

Brdiczka, D. (1991). Contact sites between mitochondrial envelope proteins. Structure and function in energy- and protein-transfer. Biochim. Biophys. Acta 1071, 291–312.

Bureau, M. H., Khrestchatisky, M., Heeren, M., Zambrowicz, E. B., Kim, H., Grisar, T. M., Colombini, M., Tobin, A. J., & Olsen, R. W. (1992). Isolation and cloning of a voltage-dependent anion channel-like Mr 36,000 polypeptide from mammalian brain. J. Biol. Chem. 267, 8679–8684.

Bustamante, E., Morris, H. P., & Pedersen, P. L. (1981). Energy metabolism of tumor cells. Requirement for a form of hexokinase with a propensity for mitochondrial binding. J. Biol. Chem. 256, 8699–8704.

Campo, M. L., Kinnally, K. W., Perini, S., & Tedeschi, H. (1992). The effect of antimycin A on mouse liver inner mitochondrial membrane channel activity. J. Biol. Chem. 267, 8123–8127.

Campo, M. L., Kinnally, K. W., Perini, S., & Tedeschi, H. (1994). Respiratory uncouplers affect mitoplast channel activity. Biophys. J. 66, A22.

Chich, J.-F., Goldshmidt, D., Thieffry, M., & Henry, J.-P. (1991). A peptide-sensitive channel of large conductance is localized on mitochondrial outer membrane. Eur. J. Biochem. 196, 29–35.

Colombini, M. (1979). A candidate for the permeability pathway of the outer mitochondrial membrane. Nature 279, 643–645.

Colombini, M. (1980). Pore size and properties of channels from mitochondria isolated from *Neurospora crassa*. J. Membr. Biol. 53, 79–84.

Colombini, M. (1987). Regulation of the mitochondrial outer membrane channel, VDAC. J. Bioenerg. Biomembr. 19, 309–320.

Colombini, M. (1989). Voltage gating in the mitochondrial channel, VDAC. J. Membr. Biol. 111, 103–111.

Colombini, M., Yeung, C. L., Tung, J., & König, T. (1987). The mitochondrial outer membrane channel, VDAC, is regulated by a synthetic polyanion. Biochim. Biophys. Acta 905, 279–286.

Costa, G., Kinnally, K. W., & Diwan, J. J. (1991). Patch clamp analysis of a partially purified ion channel from rat liver mitochondria. Biochem. Biophys. Res. Comm. 175, 305–310.

Criado, M., & Keller, B. U. (1987). A membrane fusion strategy for single-channel recordings of membranes usually non-accessible to patch–clamp pipette electrodes. FEBS Letts. 224, 172–176.

De Pinto, V., Ludwig, O., Krause, J., Benz, R., & Palmieri, F. (1987). Porin pores of mitochondrial outer membranes from high and low eukaryotic cells: Biochemical and biophysical characterization. Biochim. Biophys. Acta 894, 109–119.

De Pinto, V., Prezioso, G., Thinnes, F., Link, T. A., & Palmieri, F. (1991). Peptide-specific antibodies and proteases as probes of the transmembrane topology of the bovine heart mitochondrial porin. Biochemistry 30, 10191–10200.

Dihanich, M., Schmid, A., Oppliger, W., & Benz, R. (1989). Identification of a new pore in the mitochondrial outer membrane of a porin-deficient yeast mutant. Eur. J. Biochem. 181, 703–708.

Dill, E. T., Holden, M. J., & Colombini, M. (1987). Voltage gating in VDAC is markedly inhibited by micromolar quantities of aluminum. J. Membr. Biol. 99, 187–196.

Doring, C., & Colombini, M. (1985). The mitochondrial voltage-dependent channel, VDAC, is modified asymmetrically by succinic anhydride. J. Membr. Biol. 83, 87–94.

Ehrenstein, G., Lecar, H., & Nossal, R. (1970). The nature of the negative resistance in bimolecular lipid membranes containing excitability-inducing material. J. Gen. Physiol. 55, 119–133.

Engel, A., Massalski, A., Schindler, H., Dorset, D. L., & Rosenbusch, J. P. (1985). Porin channel triplets merge into single outlets in *Escherichia coli* outer membranes. Nature 317, 643–645.

Fiek, C., Benz, R., Roos, N., & Brdiczka, D. (1982). Evidence for identity between the hexokinase-binding protein and the mitochondrial porin in the outer membrane of rat liver mitochondria. Biochim. Biophys. Acta 688, 429–440.

Forte, M., Guy, H. R., & Mannella, C. A. (1987). Molecular genetics of the VDAC ion channel: Structural model and sequence analysis. J. Bioenerg. Biomembr. 19, 341–350.

Freitag, H., Neupert, W., & Benz, R. (1982). Purification and characterization of a pore protein of the outer mitochondrial membrane from *Neurospora crassa*. Eur. J. Biochem. 123, 629–639.

Garlid, K., & Beavis, A. D. (1986). Evidence for the existence of an inner membrane anion channel in mitochondria. Biochim. Biophys. Acta 853, 187–204.

Gunter, T. E., & Pfeiffer, D. R. (1990). Mechanisnms by which mitochondria transport calcium. Am. J. Physiol. Cell Physiol. 27, C755–C786.

Guo, X. W., & Mannella, C. A. (1992). Classification of projection images of crystalline arrays of the mitochondrial channel, VDAC, embedded in aurothioglucose. Biophys. J. 63, 418–427.

Guo, X. W., & Mannella, C. A. (1993). Conformational change in the mitochondrial channel, VDAC, detected by electron cryo-microscopy. Biophys. J. 64, 545–549.

Guo, X. W., Smith, P. R., Cognon, B., D'Arcangelis, D., Dolginova, E., & Mannella, C. A. (1995). Molecular design of the voltage-dependent, anion-selective channel in the mitochondrial outer membrane. J. Struct. Biol. 114, 41–59.

Hartl, F. U., Pfanner, N., Nicholson, D. W., & Neupert, W. (1989). Mitochondrial protein import. Biochim. Biophys. Acta 988, 1–45.

Hayman, K. A., Spurway, T. D., & Ashley, R. H. (1993). Single anion channels reconstituted from cardiac mitoplasts. J. Membr. Biol. 136, 181–190.

Henry, J.-P., Chich, J.-F., Goldschmidt, D., & Thieffry, M. (1989). Blockade of a mitochondrial cationic channel by an addressing peptide: An electrophysiological study. J. Membr. Biol. 112, 139–147.

Hirsch, J. D., Beyer, C. F., Malkowitz, L., Beer, B., & Blume, A. J. (1988a). Mitochondrial benzodiazepine receptors mediate inhibition of mitochondrial respiratory control. Molec. Pharmacol. 34, 157–163.

Hirsch, J. D., Beyer, C. F., Malkowitz, L., Loullis, C. C., & Blume, A. J. (1988b). Characterization of ligand binding to mitochondrial benzodiazepine receptors. Molec. Pharmacol. 34, 164–172.

Holden, M. J., & Colombini, M. (1988). The mitochondrial outer membrane channel, VDAC, is modulated by a soluble protein. FEBS Letts. 241, 105–109.

Inoue, I., Nagase, H., Kishi, K., & Higuti, T. (1991). ATP-sensitive K^+ channel in the mitochondrial inner membrane. Nature 352, 244–247.

Jap, B. K., Walian, P., Gehring, K., & Earnest, T. (1991). Three-dimensional map of PhoE porin at 3.5Å based on electron crystallography. Biophys. J. 59, 15a.

Kabir, F., & Nelson, B. D. (1991). Hexokinase bound to rat brain mitochondria uses externally added ATP more efficiently than internally generated ATP. Biochim. Biophys. Acta 1057, 147–150.

Kayser, H., Kratzin, H. D., Thinnes, F. P., Gotz, H., Schmidt, W. E., Eckart, K., & Hilschmann, N. (1989). To the knowledge of human porins. II. Characterization and primary structure of a 31-kDa porin from human B-lymphocytes (Porin 31HL). Biol. Chem. Hoppe-Seyler 370, 1265–1278.

Kinnally, K. W., & Tedeschi, H. (1994), Mitochondrial channels: An integrated view. In: Molecular Biology of Mitochondrial Transport Systems (Forte, M., & Colombini, M., Eds.). Springer Verlag, Berlin, New York, pp. 169–198.

Kinnally, K. W., Campo, M. L., & Tedeschi, H. (1989). Mitochondrial channel activity studied by patch-clamping mitoplasts. J. Bioenerg. Biomembr. 21, 497–506.

Kinnally, K. W., Zorov, D. B., Antonenko, Yu., & Perini, S. (1991). Calcium modulation of inner mitochondrial membrane channel activity. Biochem. Biophys. Res. Comm. 176, 1183–1188.

Kinnally, K. W., Antonenko, Y., & Zorov, D. B. (1992). Modulation of inner mitochondrial membrane channel activity. J. Bioenerg. Biomembr. 24, 99–110.

Kinnally, K. W., Zorov, D. B., Antonenko, Yu., Snyder, S., McEnery, M. W., & Tedeschi, H. (1993). Mitochondrial benzodiazepine receptor linked to inner membrane ion channels by nM actions of ligands. Proc. Natl. Acad. Sci. USA 90, 1374–1378.

Kinnally, K. W., Lohret, T. A., Campo, M. L., & Mannella, C. A. (1996). Perspectives on the mitochondrial multiple conductance channel. J. Bioenerg. Biomemb. 28, 113–121.

Klaus, S., Casteilla, L., Bouillaud, F., & Ricquier, D. (1991). The uncoupling protein UCP: A membranous mitochondrial ion carrier exclusively expressed in brown adipose tissue. Int. J. Biochem. 23, 791–801.

Kleene, R., Pfanner, N., Pfaller, R., Link, T., Sebald, W., Neupert, W., & Tropschug, M. (1987). Mitochondrial porin of *Neurospora crassa*: cDNA cloning, *in vitro* expression, and import mitochondria. EMBO J. 6, 2627–2633.

Klitsch, T., & Siemen, D. (1991). Inner mitochondrial membrane anion channel is present in brown adipocytes but is not identical with the uncoupling protein. J. Membr. Biol. 122, 69–75.

Krueger, K. E., & Papadopoulos, V. (1992). Mitochondrial benzodiazepine receptors and the regulation of steroid biosynthesis. Ann. Rev. Pharmacol. Toxicol. 32, 211–237.

Linden, M., Gellerfors, P., & Nelson, B. D. (1982). Pore protein and the hexokinase-binding protein from the outer membrane of rat liver mitochondria are identical. FEBS Letts. 141, 189–192.

Liu, M.-Y., & Colombini, M. (1992). A soluble mitochondrial protein increases the voltage dependence of the mitochondrial channel, VDAC. J. Bioenerg. Biomembr. 24, 41–46.

Liu, M. Y., Togrimson, A., & Colombini, M. (1994). Characterization and partial purification of the VDAC-channel-modulating protein from calf liver mitochondria. Biochim. Biophys. Acta 1185, 203–212.

Lohret, T. A., & Kinnally, K. W. (1995a). Multiple conductance channel activity of wild-type and VDAC-less yeast mitochondria. Biophys. J. 68, 2299–2309.

Lohret, T. A., & Kinnally, K. W. (1995b). Targeting peptides transiently block a mitochondrial channel. J. Biol. Chem. 270, 15950–15953.

Lohret, T. A., Murphy, R. C., Drgoñ, T., & Kinnally, K. W. (1996). Evidence that the mitochondrial multiple conductance channel, MCC, is not related to the adenine nucleotide translocator. J. Biol. Chem. 271, 4846–4849.

Ludwig, O., De Pinto, V., Palmieri, F., & Benz, R. (1986). Pore formation by the mitochondrial porin of rat brain in lipid bilayer membranes. Biochim. Biophys. Acta 860, 268–276.

Mangan, P. S., & Colombini, M. (1987). Ultrasteep voltage dependence in a membrane channel. Proc. Natl. Acad. Sci. USA 84, 4896–4900.

Mannella, C. A. (1982). Structure of the outer mitochondrial membrane: Ordered arrays of porelike subunits in outer-membrane fractions from *Neurospora crassa* mitochondria. J. Cell Biol. 94, 680–687.

Mannella, C. A. (1984). Phospholipase-induced crystallization of channels in mitochondrial outer membranes. Science 224, 165–166.

Mannella, C. A. (1986). Mitochondrial outer membrane channel (VDAC, porin) two-dimensional crystals from *Neurospora.* In: Methods in Enzymology (Fleischer, S., & Fleischer, B., Eds.), pp. 595–610. Academic Press, Inc, London.

Mannella, C. A. (1987). Electron microscopy and image analysis of the mitochondrial outer membrane channel, VDAC. J. Bioenerg. Biomembr. 19, 329–340.

Mannella, C. A. (1989). Structure of the mitochondrial outer membrane channel derived from electron microscopy of 2D crystals. J. Bioenerg. Biomembr. 21, 427–437.

Mannella, C. A. (1990). Structural analysis of mitochondrial pores. Experientia 46, 137–145.

Mannella, C. A., & Auger, I. (1986). Sequence homologies between the mitochondrial outer-membrane pore-protein and inner-membrane transport proteins. Biophys. J. 49, 272a.

Mannella, C. A., & Frank, J. (1986). Effect of succinylation on images of negatively stained arrays of mitochondrial outer membrane channels. J. Ultrastruct. Molec. Struct. Res. 96, 31–40.

Mannella, C. A., & Guo, X. W. (1992). Model for a gated mitochondrial channel based on electron microscopy of 2D crystal polymorphs. Trans. ACA 28, 176–182.

Mannella, C. A., Guo, X. W., & Cognon, B. (1989). Diameter of the mitochondrial outer membrane channel: Evidence from electron microscopy of frozen-hydrated membrane crystals. FEBS Letts. 253, 231–234.

Mannella, C. A., Radermacher, M., & Frank, J. (1984). Three-dimensional structure of mitochondrial outer-membrane channels from fungus and liver. In: Proceedings of the 42nd Annual Meeting of the Electron Microscopy Society of America (Bailey, G. W., Ed.), pp. 644–645. San Francisco Press, Inc., San Francisco.

Mannella, C. A., Forte, M., & Colombini, M. (1992a). Toward the molecular structure of the mitochondrial channel, VDAC. J. Bioenerg. Biomembr. 24, 7–19.

Mannella, C. A., Neuwald, A. F., & Lawrence, C. E. (1996). Detection of likely transmembrane beta-strand regions in sequences of mitochondrial pore proteins using the Gibbs sampler. J. Bioenerg. Biomembr. 28, 161–167.

McEnery, M. W. (1992). The mitochondrial benzodiazapene receptor: Evidence for association with the voltage dependent anion channel (VDAC). J. Bioenerg. Biomembr. 24, 63–69.

McEnery, M. W., Snowman, A. M., Trifiletti, R. R., & Snyder, S. H. (1992). Isolation of the mitochondrial benzodiazepine receptor: Association with the voltage-dependent anion channel and the adenine nucleotide carrier. Proc. Natl. Acad. Sci. USA 89, 3170–3174.

Mihara, K., & Sato, R. (1985). Molecular cloning and sequencing of cDNA for yeast porin, an outer mitochondrial membrane protein: A search for targeting signal in the primary structure. EMBO J. 4, 769–774.

Miller, K. J., & Hackenbrock, C. R. (1975). The distribution of anionic sties on the surfaces of mitochondrial membranes. J. Cell Biol. 65, 615–630.

Mironova, G. D., Fedotcheva, N. I., Makarov, P. R., Pronevich, L. A., & Mironov, G. P. (1981). A protein of the bovine heart mitochondria inducing channel potassium conductivity in bilayer lipid membranes. Biofizika 26, 451–457.

Mirzabekov, T. A., & Ermishkin, L. N. (1989). The gate of mitochondrial porin channel is controlled by a number of negative and positive charges. FEBS Letts. 249, 375–378.

Mirzabekov, T., Ballarin, C., Nicolini, M., Zatta, P., & Sorgato, M. C. (1993). Reconstitution of the native mitochondrial outer membrane in planar bilayers. Comparison with the outer membrane in a patch pipette and effect of aluminum compounds. J. Membr. Biol. 133, 129–143.

Moran, O., & Sorgato, M. C. (1992). High conductance pathways in mitochondrial membranes. J. Bioenerg. Biomembr. 24, 91–98.

Moran, O., Sandri, G., Panfili, G., Stuhmer, W., & Sorgato, M. C. (1990). Electrophysiological characterization of contact sites in brain mitochondria. J. Biol. Chem. 265, 908–913.

Paucek, P., Mironova, G., Mahdi, F., Beavis, A. D., Woldegiorgis, G., & Garlid, K. D. (1992). Reconstitution and partial purification of the glibenclamide-sensitive, ATP-dependent K^+ channel from rat liver and beef heart mitochondria. J. Biol. Chem. 267, 26062–26069.

Peng, S., Blachly-Dyson, E., Colombini, M., & Forte, M. (1992a). Determination of the number of polypeptide subunits in a functional VDAC channel from *Saccharomyces cerevisiae*. J. Bioenerg. Biomembr. 24, 27–31.

Peng, S., Blachly-Dyson, E., Forte, M., & Colombini, M. (1992b). Large-scale rearrangement of protein domains is associated with voltage gating of the VDAC channel. Biophys. J. 62, 123–135.

Petronilli, V., Szabo, I., & Zoratti, M. (1989). The inner mitochondrial membrane contains ion-conducting channels similar to those found in bacteria. FEBS Letts. 259, 137–143.

Pfaller, R., & Neupert, W. (1987). High-affinity binding sites involved in the import of porin into mitochondria. EMBO J. 6, 2635–2642.

Rose, I. A., & Warms, J. V. B. (1967). Mitochondrial hexokinase. Release, rebinding, and location. J. Biol. Chem. 242, 1635–1645.

Schein, S. J., Colombini, M., & Finkelstein, A. (1976). Reconstitution in planar lipid bilayers of a voltage-dependent anion-selective channel obtained from *Paramecium* mitochondria. J. Membr. Biol. 30, 99–120.

Shao, L., Kinnally, K. W., & Mannella, C. A. (Forthcoming). Circular dichroism studies of the mitochondrial channel, VDAC, from Neurospora crassa. Biophys. J. 71.

Sorgato, M. C., & Moran, O. (1993). Channels in mitochondrial membranes: Knowns, unknowns, and prospects for the future. Crit. Rev. Biochem. Molec. Biol. 18, 127–171.

Sorgato, M. C., Keller, B. U., & Stuhmer, W. (1987). Patch clamping of the inner mitochondrial membrane reveals a voltage dependent channel. Nature 330, 498–500.

Sorgato, M. C., Moran, O., De Pinto, V., Keller, B. U., & Stuhmer, W. (1989). Further investigations on the high-conductance ion channel of the inner membrane of mitochondria. J. Bioenerg. Biomembr. 21, 485–496.

Sprengel, R., Werner, P., Seeburg, P. H., Mukhin, A. G., Santi, M. R., Grayson, D. R., Guidotti, A., & Krueger, K. E. (1989). Molecular cloning and expression of cDNA encoding a peripheral-type benzodiazepine receptor. J. Biol. Chem. 264, 20415–20421.

Stanley, S., Dias, J. A., D'Arcangelis,D., & Mannella, C. A. (1995). Peptide-specific antibodies as probes of the topography of the voltage-gated channel in the mitochondrial outer membrane of Neurospora crassa. J. Biol. Chem. 270, 16694–16700.

Szabo, I., & Zoratti, M. (1991). The giant channel of the inner mitochondrial membrane is inhibited by cyclosporine A. J. Biol. Chem. 266, 3376–3379.

Szabo, I., & Zoratti, M. (1992). The mitochondrial megachannel is the permeability transition pore. J. Bioenerg. Biomembr. 24, 111–117.

Szabo, I., & Zoratti, M. (1993). The mitochondrial permeability pore may comprise VDAC molecules. I. The electrophysiological properties of VDAC are compatible with those of the mitochondrial megachannel. FEBS Letts. 330, 201–205.

Takahashi, K., Ono, H., Yoshida, T., & Tuboi, S. (1993). Integration of porin into the outer mitochondrial membrane through the common import machinery in contact sites. Biochem. Molec. Biol. Internatl. 29, 429–434.

Tedeschi, H., & Kinnally, K. W. (1994), Mitochondrial channels. In: *Membrane Channels—Molecular and Cellular Physiology* (Peracchia, C., Ed.). Academic Press, New York, London, pp. 529–548.

Tedeschi, H., Mannella, C. A., & Bowman, C. L. (1987). Patch clamping the outer mitochondrial membrane. J. Membr. Biol. 97, 21–29.

Thinnes, F. P. (1992). Evidence for extra-mitochondrial localization of the VDAC/porin channel in eucaryotic cells. J. Bioenerg. Biomembr. 24, 71–75.

Thinnes, F. P., Gotz, H., Kayser, H., Benz, R., Schmidt, W. E., Kratzin, H. D., & Hilschmann, N. (1989). To the knowledge of human porins. I. Purification of a porin from human B-lymphocytes (Porin31HL) and the topochemical proof of its expression on the plasmalemma of the parental cell line (German). Biol. Chem. Hoppe-Seyler 370, 1253–1264.

Thomas, L., Kocsis, E., Colombini, M., Erbe, E., Trus, B. L., & Steven, A. C. (1991). Surface topography and molecular stoichiometry of the mitochondrial channel, VDAC, in crystalline arrays. J. Struct. Biol. 106, 161–171.

Thomas, L., Blachly-Dyson, E., Colombini, M., & Forte, M. (1993). Mapping of residues forming the voltage sensor of the voltage-dependent anion-selective channel. Proc. Natl. Acad. Sci. USA 90, 5446–5449.

Troll, H., Malchow, D., Muller-Taubenberger, A., Humbel, B., Lottspeich, F., Ecke, M., Gerisch, G., Scmid, A., & Benz, R. (1992). Purification, functional characterization, and cDNA sequencing of mitochondrial porin from *Dictyostelium discoideum*. J. Biol. Chem. 267, 21072–21079.

Vodyanoy, I., & Bezrukov, S. M. (1992). Sizing of an ion pore by access resistance measurements. Biophys. J. 62, 10–11.

Waite, M., & Sisson, P. (1971). Partial purification and characterization of the phospholipase A_2 from rat liver mitochondria. Biochemistry 10, 2377–2383.

Weiss, M. S., Wacker, T., Weckesser, J., Welte, W., & Schulz, G. E. (1990). The three-dimensional structure of porin from *Rhodobacter capsulatus* at 3 Å resolution. FEBS Letts. 267, 268–272.

Wunder, U. R., & Colombini, M. (1991). Patch clamping VDAC in liposomes containing whole mitochondrial membranes. J. Membr. Biol. 123, 83–91.

Zambrowicz, E. B., & Colombini, M. (1993). Zero-current potentials in a large membrane channel: A simple theory accounts for complex behavior. Biophys. J. 65, 1093–1100.

Zhang, D.-W., & Colombini, M. (1989). Inhibition by aluminum hydroxide of the voltage-dependent closure of the mitochondrial channel, VDAC. Biochim. Biophys. Acta 991, 68–78.

Zhang, D.-W., & Colombini, M. (1990). Group III-A-metal hydroxides indirectly neutralize the voltage sensor of the voltage-dependent mitochandrial channel, VDAC, by interacting with a dynamic binding site. Biochim. Biophys. Acta 1025, 127–134.

Zimmerberg, J., & Parsegian, V. A. (1986). Polymer inaccessible volume changes during opening and closing of a voltage-dependent ionic channel. Nature (London) 323, 36–39.

Zizi, M., Forte, M., Blachly-Dyson, E., & Colombini, M. (1994). NADH regulates the gating of VDAC, the mitochondrial outer membrane channel. J. Biol. Chem. 269, 1614–1616.

Zoratti, M., & Szabó, I. (1995). The mitochondrial permeability transition. Biochem. Biophys. Acta 1241, 139–176.

Zorov, D., Kinnally, K. W., & Tedeschi, H. (1992). Voltage activation of heart inner mitochondrial membrane channels. J. Bioenerg. Biomembr. 24, 119–124.

INDEX

www.ingramcontent.com/pod-product-compliance
Ingram Content Group UK Ltd.
Pitfield, Milton Keynes, MK11 3LW, UK
UKHW020229290926
15470UKWH00002B/435

9 781559 386630